T0348742

VOLUME ONE HUNDRED AND TWENTY ONE

PROGRESS IN
MOLECULAR BIOLOGY AND TRANSLATIONAL SCIENCE

Glucose Homeostatis and the
Pathogenesis of Diabetes Mellitus

VOLUME ONE HUNDRED AND TWENTY ONE

PROGRESS IN
MOLECULAR BIOLOGY AND TRANSLATIONAL SCIENCE

Glucose Homeostatis and the Pathogenesis of Diabetes Mellitus

Edited by

YA-XIONG TAO

Department of Anatomy, Physiology and Pharmacology, College of Veterinary Medicine, Auburn University, Auburn, Alabama, USA

AMSTERDAM • BOSTON • HEIDELBERG • LONDON
NEW YORK • OXFORD • PARIS • SAN DIEGO
SAN FRANCISCO • SINGAPORE • SYDNEY • TOKYO
Academic Press is an imprint of Elsevier

Academic Press is an imprint of Elsevier
The Boulevard, Langford Lane, Kidlington, Oxford, OX5 1GB, UK
32 Jamestown Road, London NW1 7BY, UK
Radarweg 29, PO Box 211, 1000 AE Amsterdam, The Netherlands
225 Wyman Street, Waltham, MA 02451, USA
525 B Street, Suite 1800, San Diego, CA 92101-4495, USA

First edition 2014

Library of Congress Cataloging-in-Publication Data
A catalog record for this book is available from the Library of Congress

British Library Cataloguing in Publication Data
A catalogue record for this book is available from the British Library

ISBN: 978-0-12-800101-1
ISSN: 1877-1173

For information on all Academic Press publications
visit our website at store.elsevier.com

Printed and bound by CPI Group (UK) Ltd, Croydon, CR0 4YY

Working together
to grow libraries in
developing countries

www.elsevier.com • www.bookaid.org

CONTENTS

CONTRIBUTORS

Rajesh Amin
Department of Pharmacal Sciences, Harrison School of Pharmacy, Auburn University, Auburn, Alabama, USA

Subhrajit Bhattacharya
Department of Pharmacal Sciences, Harrison School of Pharmacy, Auburn University, Auburn, Alabama, USA

Jenna Bloemer
Department of Pharmacal Sciences, Harrison School of Pharmacy, Auburn University, Auburn, Alabama, USA

Yongchang Chang
Barrow Neurological Institute, St. Joseph's Hospital and Medical Center, Phoenix, Arizona, USA

Guoxun Chen
Department of Nutrition, University of Tennessee at Knoxville, Knoxville, Tennessee, USA

Oleg G. Chepurny
Department of Medicine, State University of New York (SUNY), Upstate Medical University, Syracuse, New York, USA

Cory C. Cortez
Laboratory of Nutrient Sensing and Adipocyte Signaling, Pennington Biomedical Research Center, Baton Rouge, Louisiana, USA

Meng-Hong Dai
Department of Anatomy, Physiology and Pharmacology, College of Veterinary Medicine, Auburn University, Auburn, Alabama, USA, and Department of Basic Veterinary Medicine, College of Veterinary Medicine, Huazhong Agricultural University, Wuhan, Hubei, China

Yishu Ding
Department of Nutrition Sciences, University of Alabama at Birmingham, Birmingham, Alabama, USA

Yuchang Fu
Department of Nutrition Sciences, School of Health Professions, University of Alabama at Birmingham, Birmingham, Alabama, USA

Thomas W. Gettys
Laboratory of Nutrient Sensing and Adipocyte Signaling, Pennington Biomedical Research Center, Baton Rouge, Louisiana, USA

Hong-Ping Guan
Department of Diabetes, Merck Research Laboratories, Kenilworth, New Jersey, USA

Margarethe Hoenig
College of Veterinary Medicine, University of Illinois, Urbana, Illinois, USA

George G. Holz
Department of Medicine, and Department of Pharmacology, State University of New York
(SUNY), Upstate Medical University, Syracuse, New York, USA

Hui Huang
Department of Anatomy, Physiology and Pharmacology, College of Veterinary Medicine,
Auburn University, Auburn, Alabama, USA

Yao Huang
Department of Obstetrics and Gynecology, St. Joseph's Hospital and Medical Center,
Phoenix, Arizona, USA

Xu-Fang Liang
College of Fisheries, Key Lab of Freshwater Animal Breeding, Ministry of Agriculture,
Huazhong Agricultural University, Wuhan, Hubei, China

Mahmoud Mansour
Department of Anatomy, Physiology and Pharmacology, College of Veterinary Medicine,
Auburn University, Auburn, Alabama, USA

Xiu-Lei Mo
Department of Anatomy, Physiology and Pharmacology, College of Veterinary Medicine,
Auburn University, Auburn, Alabama, USA

Prashant Nadkarni
Department of Medicine, and Joslin Diabetes Center, State University of New York
(SUNY), Upstate Medical University, Syracuse, New York, USA

Manda L. Orgeron
Laboratory of Nutrient Sensing and Adipocyte Signaling, Pennington Biomedical Research
Center, Baton Rouge, Louisiana, USA

Lyudmila I. Rachek
Department of Cell Biology and Neuroscience, College of Medicine, University of South
Alabama, Mobile, Alabama, USA

Kirsten P. Stone
Laboratory of Nutrient Sensing and Adipocyte Signaling, Pennington Biomedical Research
Center, Baton Rouge, Louisiana, USA

Vishnu Suppiramaniam
Department of Pharmacal Sciences, Harrison School of Pharmacy, Auburn University,
Auburn, Alabama, USA

Ya-Xiong Tao
Department of Anatomy, Physiology and Pharmacology, College of Veterinary Medicine,
Auburn University, Auburn, Alabama, USA

Nancy T. Van
Laboratory of Nutrient Sensing and Adipocyte Signaling, Pennington Biomedical Research
Center, Baton Rouge, Louisiana, USA

Desiree Wanders
Laboratory of Nutrient Sensing and Adipocyte Signaling, Pennington Biomedical Research
Center, Baton Rouge, Louisiana, USA

Jinzeng Yang
Department of Human Nutrition, Food and Animal Sciences, University of Hawaii at Manoa, Honolulu, Hawaii, USA

Kevin D. Yang
Department of Nutrition Sciences, University of Alabama at Birmingham, Birmingham, Alabama, USA

Qinglin Yang
Department of Nutrition Sciences, University of Alabama at Birmingham, Birmingham, Alabama, USA

Zhao Yang
Department of Anatomy, Physiology and Pharmacology, College of Veterinary Medicine, Auburn University, Auburn, Alabama, USA

PREFACE

Diabetes mellitus is an ancient disease. The earliest description of polyuria in diabetic patients dates back to 1550 B.C. by Imhotep. Greek physician Arateus of Cappadocia first coined the term "diabetes," meaning "siphon," the melting of flesh and bones into urine. It was also known for a long time that the urine from diabetic patients is sweet, with the earliest description in about 400 B.C. in India ("mellitus" in diabetes mellitus means "sweet" or "honey"). Apollinaire Bouchardat first noticed the existence of two types of diabetes, the more severe diabetes in younger patients (*diabète maigre*, diabetes of the thin), currently called type 1 diabetes, and diabetes in obese adults (*diabète gras*, diabetes of the fat), currently called type 2 diabetes. The discovery of insulin and its availability to diabetics represent one of the true milestones in medical history that provided "greatest benefit to mankind" (for a thorough description of diabetes, please consult the chapter "The History of Diabetes" by Donald M. Barnett and Leo P. Krall in the 14th Edition of *Joslin's Diabetes Mellitus*, edited C. Ronald Kahn and others, and published by Lippincott Williams & Wilkins in 2005).

Type 1 diabetes accounts for less than 10% whereas type 2, frequently associated with obesity, accounts for more than 90% of diabetes cases. With the current obesity epidemic, type 2 diabetes is also becoming an epidemic. Together with its comorbidities, especially cardiovascular disease, diabetes represents a tremendous health, economic, and social burden. Intensive studies are being conducted to gain a better understanding of the regulation of glucose homeostasis, the pathogenesis of diabetes, and identifying novel therapeutic targets.

This volume provides a selected glimpse of some of these studies. Starting with a general introduction of G protein-coupled receptors as regulators of glucose homeostasis and potential therapeutics of diabetes, three chapters summarized the studies on three G protein-coupled receptors, glucagon-like peptide 1 receptor, GPR40, and GPR119. Other chapters describe the involvement of skeletal muscle and liver in glucose homeostasis, and the regulation of these physiological processes by nuclear receptors, fatty acids, adiponectin receptors, and growth factor receptors. One chapter addresses how dietary methionine restriction can be used to study metabolic syndrome. Hoenig's chapter summarizes the current status of research on diabetes mellitus in companion animals. The last two chapters address insulin resistance in the brain, sometimes called type 3 diabetes, and cardiovascular dysfunction.

I take this opportunity to thank all authors in this volume for taking time out of their hectic schedules to write their outstanding contributions. I apologize for the occasional nagging e-mails reminding them of the impending due dates. I also thank the Chief Editor of the *Progress in Molecular Biology and Translational Science* series, Dr. P. Michael Conn, for always being supportive and ready to help. It has been wonderful to work with the colleagues at Elsevier, including Mary Ann Zimmerman, Sarah Lay, and Helene Kabes. Their help in moving the project along is sincerely appreciated.

Last but not the least, I thank my wife Zhen-Fang for taking care of our three daughters when I was working late in the office, and Nancy, Rachel, and Lily for their unconditional love.

YA-XIONG TAO
Auburn, Alabama

CHAPTER ONE

G Protein-Coupled Receptors as Regulators of Glucose Homeostasis and Therapeutic Targets for Diabetes Mellitus

Ya-Xiong Tao*, Xu-Fang Liang†

*Department of Anatomy, Physiology and Pharmacology, College of Veterinary Medicine, Auburn University, Auburn, Alabama, USA
†College of Fisheries, Key Lab of Freshwater Animal Breeding, Ministry of Agriculture, Huazhong Agricultural University, Wuhan, Hubei, China

Contents

Abstract

As critical regulators of almost all physiological processes in the body, G protein-coupled receptors (GPCRs) are important regulators of glucose homeostasis. Some of the newer drugs for treating diabetes mellitus or in development are also targeting GPCRs. This chapter provides a summary of some of the GPCRs that have been shown to be involved in regulating glucose homeostasis.

ABBREVIATIONS

ACh acetylcholine
AgRP Agouti-related peptide
cAMP cyclic AMP
CCK cholecystokinin
DAG diacylglycerol
DM diabetes mellitus

Progress in Molecular Biology and Translational Science, Volume 121
ISSN 1877-1173
http://dx.doi.org/10.1016/B978-0-12-800101-1.00001-6

1

DPP-IV dipeptidyl peptidase IV
ERK1/2 extracellular signal-regulated kinase 1/2
FKN fractalkine
GCGR glucagon receptor
GIP glucose-dependent insulinotropic polypeptide
GIPR GIP receptor
GLP-1 glucagon-like peptide-1
GLP-1R glucagon-like peptide-1 receptor
GLUT-4 glucose transporter-4
GPCR G protein-coupled receptor
GPRC6A G protein-coupled receptor family C group 6 subtype A
GSIS glucose-stimulated insulin secretion
icv intracerebroventricular
IP3 inositol-1,4,5-phosphate
JNK c-Jun N-terminal kinase
K_{ATP} channel ATP-sensitive K^+ channel
KO knockout
mAChR muscarinic acetylcholine receptor
MC3R melanocortin-3 receptor
MC4R melanocortin-4 receptor
MCH melanin-concentrating hormone
MSH melanocyte-stimulating hormone
NPY neuropeptide Y
OXM oxyntomodulin
PC prohormone convertase
PKA protein kinase A
PKB protein kinase B
PKC protein kinase C
PI3K phosphoinositol-3-kinase
PLC phospholipase C
POMC proopiomelanocortin
T1DM type 1 diabetes mellitus
T2DM type 2 diabetes mellitus
VIP vasoactive intestinal peptide

1. INTRODUCTION

Diabetes mellitus (DM) has become an epidemic worldwide. According to estimates from International Diabetes Federation, in 2012, more than 371 million people had DM globally, and it is estimated to affect 552 million in 2030. The prevalence of DM is increasing in every country and region. In 2012, DM caused 4.8 million deaths and 471 billion US dollars were spent on health care for DM. Worldwide, diabetes is currently the third most common disease and fourth leading cause of death, and by 2030, it is projected to be 1 of the 5 leading causes of death in high-income countries and 1 of the 10 leading causes of death worldwide.[1] DM is associated with significantly increased risks for cardiovascular disease, as well as other morbidities including kidney failure, blindness, nontraumatic limb amputation, and mortality. Therefore, the epidemic of DM represents tremendous health and economic burdens.

DM is a disorder of glucose homeostasis. In healthy animals and humans, coordinate actions of several peripheral organs, including the pancreas, liver, adipose tissue, and skeletal muscle, together with central nervous system, tightly control blood glucose levels at a relatively small range of 70–110 mg per deciliter. Blood glucose levels after meal ingestion, fasting, or strenuous exercise are maintained within this range. When one or several organs fail to contribute normally to maintain glucose homeostasis, glucose intolerance, prediabetes, and diabetes will ensue. Impaired insulin secretion from pancreatic β-cells and insulin resistance in peripheral tissues including the muscle, liver, and adipose tissue cause DM. Several chapters in this volume discuss insulin resistance. This chapter is focused on the regulation of glucose homeostasis by G protein–coupled receptors (GPCRs) including the regulation of insulin and incretin secretion by these receptors.

2. REGULATION OF GLUCOSE HOMEOSTASIS BY GPCRs IN THE CENTRAL NERVOUS SYSTEM

Hormonal (such as insulin, leptin, and glucagon-like peptide 1 (GLP-1)) and nutritional (such as glucose, fatty acids, and amino acids) signals from peripheral metabolic tissues, including the pancreas, gastrointestinal tract, adipose tissue, muscle, and liver, are sent to the central nervous system for generating behavioral and metabolic responses to maintain energy homeostasis. In a parallel fashion, using the same circuits, the central nervous system also integrates these signals and generates responses that coordinate pancreatic insulin secretion, hepatic glucose production, and peripheral glucose uptake to maintain glucose homeostasis.[2–4]

The best-studied hormones that are involved in the central regulation of energy homeostasis are the adiposity signals leptin and insulin. They do not act through GPCRs and therefore will not be discussed in detail here. However, there are a number of GPCRs that are involved in sensing the peripheral signals and generating the responses to maintain energy homeostasis.[5] These GPCRs include the melanocortin-4 receptor (MC4R), ghrelin receptor, neuropeptide Y (NPY) receptors, orexin receptors, galanin receptors, corticotropin-releasing hormone receptor, melanin–concentrating hormone receptor, neurotensin receptor, and cannabinoid CB1 receptor, among others.[5] Many of these receptors are also directly involved in regulating glucose homeostasis, independent of its effect on energy homeostasis and body weight.

One example of a GPCR that regulates both energy and glucose homeostasis is MC4R. The MC4R mediates leptin's action in the arcuate nucleus in regulating food intake and energy expenditure. When leptin level is increased (with body weight gain), leptin receptor activation results in increased production of proopiomelanocortin (POMC) that can be cleaved to generate α-melanocyte-stimulating hormone activating the MC4R, leading to decreased food intake and increased energy expenditure and therefore weight loss, maintaining a stable body weight.[6] The MC4R also directly regulates glucose homeostasis.[7] Direct and acute effects of MC4R activation on glucose homeostasis and insulin sensitivity, independent of its effect on body weight, have been clearly demonstrated in rodents. Longitudinal studies in Mc4r knockout (KO) mice showed that hyperinsulinism already exists before the onset of hyperphagia and body weight divergence.[8] Central administration of MC3/4R agonists inhibits basal insulin secretion and enhances insulin sensitivity by increasing glucose disposal.[8–11] Pancreatic MC4R might also regulate insulin secretion.[12] In vitro, MC4R activation inhibits c-Jun N-terminal kinase (JNK) activity and serine 307 phosphorylation at insulin receptor substrate-1.[13] In vivo, MC4R activation enhances insulin-stimulated protein kinase B (PKB) phosphorylation in the rat hypothalamus. These results suggest that the MC4R, through changing JNK activity, interacts with insulin signaling.[13] Mc4r-null mice have hyperinsulinemia and hyperglycemia.[14] Restoring MC4R signaling only in cholinergic neurons in Mc4r-null mice attenuates hyperglycemia and hyperinsulinemia, with enhanced hepatic insulin action and insulin-mediated suppression of gluconeogenesis.[15]

The central melanocortin receptors are also important for mediating leptin's suppression of circulating insulin levels.[16] Central but not β–cell leptin receptors are responsible for mediating leptin's effect (for a comprehensive review, see Ref. 2). Similarly, MC4R (but not MC3R) also mediates serotonin 2C receptor agonist-induced improvements in glucose homeostasis.[17] Roux-en-Y gastric bypass surgery improves glucose homeostasis independent of weight change,[18] and it was recently demonstrated that MC4Rs in autonomic neurons mediate these beneficial effects in both mice and humans.[19]

GLP-1 is a well-established incretin critical for maintaining glucose homeostasis.[20] Most of the GLP-1 actions were believed to be in pancreatic islets. However, central actions of GLP-1 have been demonstrated for both energy and glucose homeostasis. Relevant to the topic here, intra-cerebroventricular (icv) administration of GLP-1R agonist or antagonist to mice in different metabolic states demonstrated that central GLP-1 signaling regulates peripheral insulin secretion and glucose disposal to increase hepatic glycogen storage.[21] Another study showed that GLP-1R is expressed in the majority of POMC neurons (but not in NPY neurons) in the arcuate nucleus; icv administration of low doses of GLP-1R antagonist causes relative hyperglycemia after an intraperitoneal glucose load (the same dose given peripherally has no effect); and icv administration of GLP-1 enhances glucose-stimulated insulin secretion (GSIS) and decreases hepatic glucose production.[22] These results suggest that the arcuate GLP-1R regulates both insulin secretion and glucose production.[22]

Cholecystokinin (CCK) is a satiety hormone produced by I cells in the duodenum in response to nutrient intake that decreases meal size and increases intermeal interval.[23] Lam and colleagues[24] suggested that CCK activates local receptors that then stimulates vagal afferent feedback to the hindbrain, which then sends signals to the liver to decrease glucose production, improving hepatic insulin sensitivity. In carnivorous rainbow trout, a so-called "glucose intolerant" fish species,[25] intraperitoneal or icv administration of CCK results in significant hyperglycemia, not hypoglycemia as observed in mammals. Part of the signal is transmitted from the gastrointestinal tract through the vagal and splanchnic afferents to the brain, suggesting the existence of an ancestral gut–brain axis.[26] CCK-induced hyperglycemia in hypothalamus and hindbrain and direct action in hindbrain also contribute to the hyperglycemia response. Consistent with a central site of CCK action, high-affinity binding sites in the forebrain and hindbrain were observed in fish.[27] There are also extensive literature on the central control of food intake by CCK in fishes (e.g., see Ref. 28).

The central regulation of glucose homeostasis, similar to central regulation of energy homeostasis, is also dysfunctional in diet-induced obesity.[4,29] The substantial overlap between the CNS circuits that regulate energy balance and those that regulate glucose levels (with only a few exceptions) suggests that their dysregulation could link obesity and diabetes. New therapeutics might target these CNS circuits that can treat both obesity and T2DM.[4,30]

3. REGULATION OF GLUCOSE HOMEOSTASIS BY GPCRs FOR PEPTIDES FROM ENDOCRINE PANCREAS

Glucose homeostasis is controlled by the balance between the actions of insulin and glucagon. Insulin, secreted by β-cells in the islets of Langerhans, decreases blood glucose level, whereas glucagon, secreted by α-cells present at the periphery of the islets of Langerhans, increases blood glucose levels. Insulin acts through a tyrosine kinase receptor, whereas glucagon acts through the glucagon receptor (GCGR), a member of family B GPCRs. Glucagon, a peptide of 29 amino acids with a molecular weight of 3485, increases blood glucose by promoting liver glycogenolysis, the breakdown of glycogen, and liver gluconeogenesis, the synthesis of new glucose molecules. It also decreases glycogenesis and glycolysis.[31] GCGR couples to the stimulatory heterotrimeric G protein G_s; therefore, receptor activation results in increased intracellular cyclic AMP (cAMP) levels.[32] The cAMP then activates the protein kinase A (PKA), also known as cAMP-dependent protein kinase, by binding to the regulatory subunits and releasing the catalytic subunits. Active PKA then activates phosphorylase *b* kinase to phosphorylase *a* kinase; phosphorylase *a* kinase activates glycogen phosphorylase *a*, which promotes glycogen degradation into glucose-1-phosphate that is then dephosphorylated by glucose-6-phosphatase to glucose in the lumen of endoplasmic reticulum. Glucose is transported from the endoplasmic reticulum into the cytoplasm by glucose transporters located in the endoplasmic reticulum membrane. This system represents a well-described example of signal amplification, elegantly elucidated by Earl W. Sutherland, Jr, highlighting the potent effect of glucagon on increasing blood glucose levels. Activation of GCGR also increases intracellular calcium levels,[33,34] although the physiological relevance of calcium signaling is not clear. It has been suggested to be involved in the stimulation of glutamate metabolism.[35]

Although far less attention has been paid to the study of glucagon compared to the study of insulin, a significant body of literature highlights that glucagon is a major player in the pathogenesis of diabetes. Unger and Orci[36] proposed the bihormonal-abnormality hypothesis: both lack of insulin or insulin action and excess of glucagon contribute to type 2 DM (T2DM), with the major consequence of absolute or relative insulin deficiency being glucose underutilization and absolute or relative glucagon excess being the principal contributor to the increased hepatic glucose production. Data in both animals and humans have provided support for this hypothesis. For example, sustained hyperglucagonemia continues to increase hepatic glucose production.[37] Intact glucagon secretion is necessary for normal glucose counterregulation.[38] In DM, glucagon is required for hyperglycemia when insulin availability is limited.[39]

There are three strategies targeting glucagon for T2DM treatment: suppressing glucagon-secreting α-cells, blocking GCGR,[40] and inhibiting adenylate cyclase.[41] Several hormones including somatostatin, GLP-1, and amylin modulate glucagon secretion. Somatostatin, a 14-amino acid polypeptide secreted by the δ-cells of the islets of Langerhans, has paracrine effect decreasing both insulin and glucagon secretion.[42] When glucagon secretion is suppressed by somatostatin in type 1 DM (T1DM) subjects, blood glucose decreases quickly, with excessive glucagon secretion estimated to account for about 25% of the fasting plasma glucose levels in these patients.[43]

Since the first report of GCGR antagonists decreasing blood glucose levels in diabetic animals three decades ago,[44] both peptide and small-molecule GCGR antagonists have been described in the literature.[45–50] The first peptide antagonist synthesized by Hruby, (Des-His[1]) (N^ε-phenylthiocarbamoyl Lys[12])-glucagon was shown to be able to decrease blood glucose levels *in vivo* in streptozotocin-induced diabetic rats when infused intravenously.[44] Recently, mirror-image DNA aptamers (Spiegelmer®) that bind and inhibit glucagon were discovered.[51] A mixed DNA/RNA Spiegelmer binds glucagon with a Kd of 3 nM with no cross-reactivity with GLP-1, GLP-2, glucose-dependent insulinotropic polypeptide (GIP), and vasoactive intestinal peptide (VIP). *In vitro*, it inhibits glucagon-stimulated cAMP production in cells transfected with human GCGR with an IC_{50} of 3.4 nm. *In vivo*, a single injection of this DNA/RNA Spiegelmer ameliorates glucose excursions in intraperitoneal glucose tolerance tests in both T1DM and T2DM mouse models, highlighting its therapeutic potential to treat hyperglycemia in both types of DM.[51]

Metformin is the most frequently prescribed T2DM medicine as a first-line therapeutics with an outstanding safety record. It has antihyperglycemic action primarily by decreasing hepatic glucose production. Its mechanism of action has been extensively studied but still not completely understood. Recently, Birnbaum and colleagues[41] showed that this biguanides antagonize glucagon action by increasing levels of intracellular AMP and related nucleotides, leading to decreased activation of adenylyl cyclase by glucagon and subsequent decreased activation of PKA and phosphorylation of downstream PKA protein targets of glucagon action. Biguanides do not directly modulate GCGR or G protein signaling.

4. REGULATION OF INSULIN SECRETION BY GPCRs

Glucose is considered the only physiological initiator of insulin secretion from pancreatic β-cells in adult mammals.[52] During GSIS, glucose is metabolized to produce ATP increasing the ATP–ADP ratio that results in inhibition of ATP-sensitive K^+ channel (K_{ATP} channel) activity, depolarizing β-cell. Voltage-gated calcium channels are then activated, increasing intracellular Ca^{2+} levels, resulting in insulin secretion. Multiple hormones, acting through GPCRs, modulate GSIS. By themselves, they are not able to depolarize the β-cells enough to open the voltage-dependent Ca^{2+} channels to cause the influx of Ca^{2+} and insulin release. However, when there is a threshold concentration of glucose, the small effect the potentiators have on membrane potential can increase Ca^{2+} influx by activating additional Ca^{2+} channels.[52]

There are numerous GPCRs that are expressed in pancreatic islets.[53–55] Most of these receptors are present in insulin-producing β-cells. In these cells, as well as in other islet cells, GPCRs mediate sympathetic and parasympathetic actions and the effects of incretins, such as GLP-1 and GIP. However, the physiological and therapeutic relevance of the many GPCRs expressed in pancreatic islets is still unknown.

At the molecular level, ligand binding at GPCRs leads to conformational changes at the receptors to cause activation of heterotrimeric guanine nucleotide-binding proteins (G proteins). There are three major family of G proteins activated in β-cells. The activation of the stimulatory G protein, G_s, activates adenylyl cyclase to raise intracellular cAMP level, whereas the activation of the inhibitory G protein, G_i, inhibits adenylyl cyclase, decreasing intracellular cAMP level. Activation of G_q increases phospholipase C (PLC) activity, producing two major second messengers, inositol-1,4,5-phosphate (IP3) and diacylglycerol (DAG). IP3 can activate IP3 receptor in the endoplasmic reticulum, causing Ca^{2+} release from the endoplasmic reticulum to the cytosol. DAG can activate protein kinase C (PKC) that also results in insulin release. In β-cells, GPCRs coupled to G_s and $G_{q/11}$ usually stimulate insulin secretion, whereas GPCRs coupled to $G_{i/o}$ inhibit insulin secretion.[54]

Earlier studies identified a number of GPCRs in β-cells that inhibit insulin secretion.[56] Recently, agonists at several GPCRs are found to enhance

GSIS, including GLP-1R, GIP receptor (GIPR), VIP receptor, and pituitary adenylate cyclase-activating polypeptide receptor. All these receptors belong to Family B GPCRs. Small-molecule ligands for these receptors have proven extremely difficult to identify.

Family B GPCRs bind to the hormones in two steps. The C-terminal portion of the peptide hormones first contacts the extracellular domain; this facilitates the binding of the N-terminal portion of the hormones with the transmembrane domains that lead to conformational change in the transmembrane domains activating the receptors. Recently, the crystal structures of the transmembrane portion of the two family B GPCRs were published, those of human corticotropin-releasing hormone receptor 1 and human GCGR.[57,58] In addition to the seven transmembrane helices, the crystal structure of the GCGR also provides new insights into glucagon binding to the receptor. For example, compared to family A GPCRs, the binding pocket is large, perhaps explaining the difficulty in identifying small-molecule ligands for the receptor.[59] Transmembrane domain 1 extends three α-helical turns above the plane of the membrane; this "stalk" coordinates the extracellular domain with the membrane to capture and bind C-terminal portion of glucagon, facilitating the insertion of the N-terminus unto the seven transmembrane domain core.[58]

GLP-1 and GIP are incretins. The incretin concept was first proposed more than a century ago by Moore and colleagues,[60] suggesting that duodenal mucous membrane might secrete hormones regulating the pancreatic endocrine secretion, analogous to the regulation of pancreatic exocrine secretion by secretin. Incretin was rediscovered in 1960s after Yalow and Berson[61] developed radioimmunoassay to measure blood insulin levels. When insulin secretion was compared with glucose given either orally or intravenously, it was found that oral glucose load leads to a much greater insulin response than intravenous glucose injection.[62,63] Since then, two incretins have been discovered: GIP (also called gastric inhibitory polypeptide because it was first found to inhibit gastric acid secretion)[64] and GLP-1. As messengers in the enteroinsular axis, both incretins are secreted after glucose or nutrient ingestion and amplify GSIS at physiological concentrations. The roles of GLP-1 in maintaining glucose homeostasis are reviewed in detail in a chapter in this volume[20] and will not repeated here. Here, we present a brief description of GIP physiology.

GIP is secreted by endocrine K cells in duodenum and jejunum. The major circulating form of GIP is a 42-amino acid peptide [GIP(1–42)]; there is also a shorter form lacking the C-terminal 12 amino acids

[GIP(1–30)]. Both forms of GIP have similar insulinotropic effects at β-cells.[65] Like GLP-1, GIP is also quickly inactivated by the ubiquitous serine protease dipeptidyl peptidase IV (DPP-IV) that cleaves the N-terminal dipeptide. The half-life of GIP is about 7 min, slightly longer than that of GLP-1. Like the GLP-1R, the GIPR is primarily coupled to Gs; therefore, receptor activation leads to increased intracellular cAMP level and subsequent activation of PKA, resulting in increased Ca^{2+} influx through voltage-gated Ca^{2+} channels and insulin release.[66] Activation of phosphoinositol-3-kinase (PI3K) has also been shown to be involved in mediating GIP's insulinotropic effect, since wortmannin, a PI3K inhibitor, inhibits GIP-stimulated but not glucose- or forskolin-stimulated insulin secretion.[67] GIP also stimulates cell growth. Multiple signaling pathways are involved in mediating GIP's mitogenic and antiapoptotic effects at β-cells, including cAMP/PKA, PI3K/PKB, and extracellular signal-regulated kinase 1/2 (ERK1/2).[68,69]

In addition to its effect on β-cells, GIP also acts on adipose tissue to increase glucose uptake by promoting glucose transporter-4 (GLUT-4) translocation to the plasma membrane through activating PKB, fatty acid synthesis, and fatty acid incorporation into lipids with insulin-like lipogenic effects.[70,71] In T2DM, there is decreased GIPR expression and GIP-stimulated insulin release.[72] In human studies, GIP cannot stimulate insulin secretion significantly or normalize glucose level in diabetic patients[73]; together with its effect on lipid physiology, these results raised questions regarding its potential as a new T2DM therapeutic target,[74] despite the success of its cousin GLP-1 enjoys.

The effects of VIP and pituitary adenylate cyclase-activating peptide in modulating GSIS and the therapeutic potential in treating diabetes have been recently reviewed and will not be discussed here.[75]

Although receptors for the incretins are highly expressed in islets and their physiology well understood, there are many other additional GPCRs highly expressed in the islets. Glucagon, acetylcholine (ACh), β-adrenergic agonists, and several gastrointestinal hormones (including gastrin, CCK, and secretin) increase insulin secretion, whereas somatostatin (produced by δ-cells in the islets of Langerhans) and α-adrenergic agonists decreases insulin secretion. All of these ligands exert their actions by interacting with GPCRs. There are several fatty acid receptors that are highly expressed in islets. Two, GPR40 and GPR119, are reviewed in this volume.[76,77] Here, we list a few additional hormones that have been shown to be important regulators of insulin secretion.

Both arginine vasopressin and oxytocin increase GSIS and V_{1b} vasopressin receptor and oxytocin receptor mediate their actions, respectively.[78–82] $G_{q/11}$-mediated increases in IP3 are involved in mediating the potentiation of GSIS by both arginine vasopressin and oxytocin. In mouse islets, oxytocin also increases glucagon and somatostatin release.[81] Recent *in vivo* experiments showed that oxytocin and oxytocin analogs are effective in both humans and rodents at correcting glucose dyshomeostasis by reversing insulin resistance and glucose intolerance and enhancing insulin secretion before any significant change in body weight, highlighting the therapeutic potential of targeting oxytocin receptor for obesity and DM.[83]

Bradykinin stimulates insulin secretion in perfused rat pancreas dose-dependently.[84] Bradykinin also stimulates insulin secretion in clonal β-cell lines.[85,86] The potentiation of GSIS by bradykinin is blocked by B_2 bradykinin receptor antagonist, suggesting that B_2 receptor likely mediates bradykinin action.[84] *In vivo*, it was suggested that kinins contribute to the improvement of insulin sensitivity during treatment with angiotensin-converting enzyme inhibitor.[87]

ACh, released by intrapancreatic nerve endings during the preabsorptive and absorptive phases of feeding, has been shown to have glucose-dependent insulinotropic effect in different species.[88,89] ACh exerts its effects through five subtypes of muscarinic acetylcholine receptors (mAChRs), M_1 through M_5. Of the five subtypes, although both M_1 and M_3 subtypes are expressed in β-cells, the M_3 mAChR has been shown to be responsible for mediating the increased GSIS from pancreatic β-cells by ACh in both pharmacological and *in vivo* transgenic animal studies.[90–93] M_3 mAChR is coupled to G_q; therefore, IP3 and DAG signaling is involved in mediating ACh-induced insulin secretion; however, arrestin-mediated signaling may also be involved in this process.[94]

In vitro insulin release experiments were done using islets prepared from mice lacking either M_1 or M_3 mAChR. Oxotremorine-M, a non-subtype-selective muscarinic agonist, was used. There is no difference in insulin release from islets prepared in M_1 mAChR KO mice. However, oxotremorine-M-induced insulin release is abolished in homozygous M_3 mAChR KO mice and is attenuated in heterozygous mice, demonstrating a gene dosage-dependent effect on GSIS.[92] These mice also have decreased food intake and increased energy expenditure, leading to decreased fat mass,[95,96] It is suggested to be downstream of the hypothalamic leptin–melanocortin system that is known to be critical for regulating energy homeostasis. They are protected from diet-induced obesity and associated metabolic deficits. Therefore, in these mice, improved glucose tolerance and increased insulin sensitivity (with decreased insulin level) are observed, masking any effect of direct action of M_3 mAChR in the β-cells.

Wess and colleagues generated a series of new mutant mouse models of M_3 mAChR, including β-cell-specific KO, transgenic overexpression of the wild-type or constitutively active mutant receptor in β-cell, and β-cell-specific transgenic mice that express a designer receptor that is activated by an otherwise pharmacologically inert drug (clozapine-N-oxide) but not activated by the endogenous agonist.[97–101] Studies with these mouse models demonstrate that M_3 mAChR in the β-cells plays a central role in maintaining normal glucose homeostasis. Chronic activation of the M_3 mAChR results in significant improvement in glucose homeostasis (including challenge with streptozotocin injection, a T1DM model, or with high-fat diet, a T2DM model) and does not lead to significant desensitization. These results suggest that agonists activating the receptor might be used to treat T2DM, although issues regarding potential side effects need to be addressed since mAChR is also expressed in a number of other peripheral tissues.[101]

Recently, it was shown that chemokine receptor CX3CR1 expressed in β-cells is necessary for normal insulin secretory response to glucose, arginine, and GLP-1 challenge.[102] Both the ligand fractalkine (FKN) and the receptor are expressed in human and rodent islets. Mice lacking *Cx3cr1* have marked defect in glucose- and GLP-1-stimulated insulin secretion, whereas administration of FKN increases insulin secretion and glucose tolerance.[102] High-fat feeding and obesity as well as aging decrease *FKN* (but not *CX3CR1*) expression. The authors suggest that decreased signaling in the FKN/CX3CR1 system may contribute to the β-cell dysfunction in type 2 diabetics and agonist for the CX3CR1 might be a therapeutic for T2DM.[102] Human genetic studies identified two SNPs in *CX3CR1*, V249I and T280M, associated with increased incidence of obesity and T2DM.[103,104] These SNPs have lower FKN binding affinity and are therefore partially defective in function.

The GPCRs listed earlier represent just a sampling of GPCRs expressed in islets that can potentially regulate secretion of pancreatic hormones, including insulin, glucagon, and somatostatin. Recently, comprehensive expression profiling analyses have identified GPCRs expressed in rodent and human islets.[55,105,106] These studies show a high percentage of non-odorant GPCRs to be expressed in pancreatic islet. For example, in human islets, of the 384 nonodorant GPCRs screen, 293 receptors are expressed at different levels. The functions and physiological relevance of these receptors in regulating pancreatic hormone secretion, and pancreatic cell proliferation and apoptosis, as well as the therapeutic values and mediation of side effects of current and future therapeutics, largely remain to be explored. Although these receptors are expressed in pancreatic β-cells, all of them are also expressed, albeit to a lesser extent, in other tissues. Their functions in non-β-cells will have to be better understood if these GPCRs are to be targeted for drug development. For example, if receptor activation leads to increased secretion of both insulin and glucagon, its therapeutic potential would be limited due to the counteraction of insulin and glucagon on blood glucose level.

5. REGULATION OF INCRETIN SECRETION BY GPCRs

The incretins are critically involved in regulating glucose homeostasis. The secretion of these hormones is also regulated by a number of GPCRs. Several previously orphan but recently deorphanized receptors for free fatty acids (GPR40, GPR41, GPR43, GPR84, and GPR120) or fatty acid amides (GPR119) are found to be potent regulators of incretin secretion. Several recent reviews on these receptors including chapters in this volume have appeared. Readers interested in these fatty acid receptors are directed to these articles.[76,77,107–113]

Several family C GPCRs, including calcium-sensing receptor, the GPRC6A (G protein-coupled receptor family C group 6 subtype A), and the T1R1/T1R3 (heterodimeric umami taste receptor), mediate broad-spectrum L-amino acid sensing in the body.[114–117] These receptors are all intimately involved in regulating glucose homeostasis (recently reviewed by Cobb and associates)[117] The following is a brief summary on one of these receptors, GPRC6A.

GPRC6A was deorphanized in 2005 as a promiscuous receptor for L-α-amino acid, especially for basic amino acids.[118] It also senses cations including calcium, magnesium, strontium, aluminum, and gadolinium and responds to the bone hormone osteocalcin.[119] Recently, it was shown that GPRC6A, through activation of G_q-PLC pathway, mediates amino acid-induced GLP-1 secretion.[120] Two GPRC6A KO studies were published with discrepant results. One study showed that the KO mice have a mild metabolic syndrome phenotype, with hyperglycemia, insulin resistance, and glucose intolerance,[121] whereas another study found no difference between the *Gprc6a* KO mice and the wild-type mice in blood insulin and glucose levels, insulin sensitivity, and glucose tolerance.[122] The reasons for the different results are not clear, although differences in gene targeting strategy and mouse background differences may be contributing factors.[117] Further studies are also needed to clarify how GPRC6A affects glucose metabolism, through regulating GLP-1 secretion or directly on insulin secretion.[117]

6. REGULATION OF GLUCOSE HOMEOSTASIS BY ORPHAN GPCRs

A number of GPCRs recently deorphanized, including fatty acid receptors described earlier, are important regulators of glucose homeostasis. There are still ~100 non-chemosensory GPCRs remaining as orphan receptors with the endogenous ligands still unknown.[123] Some of these GPCRs also regulate glucose homeostasis, representing potential therapeutic targets for diabetes treatment. The following are just two examples.

GPR27, an intronless gene encoding a receptor of 379 amino acids, was first cloned in 1998.[124] A recent siRNA screening strategy identified GPR27 as a potent stimulator of insulin secretion.[125] More detailed studies need to be done, for example, in genetically modified animals.

On the other hand, obestatin, originally thought to act through GPR39,[126] is a peptide derived from preproghrelin with some functions on regulating energy balance and gastric emptying.[127] It has an unknown receptor and GPR39 goes back to be an orphan receptor. *GRP39* expression is restricted to endocrine and metabolic organs including the pancreas, the liver, the gastrointestinal tract, and white adipose tissue.[126] In the pancreas, GPR39 is expressed in β-cells of the pancreatic islets as well as in the duct cells of the exocrine pancreas.[128,129] Several KO studies have reported the phenotypes of *Gpr39*-null mice. *Gpr39*-null mice have impaired GSIS both *in vivo* and *in vitro*,[129–131] suggesting that GPR39 is important in regulating GSIS. GPR39 has significant basal activity.[132] Overexpression of GPR39 specifically in β-cells (leading to increased activity) fully protects the mice from the gradual hyperglycemia after streptozotocin treatment.[133] These results together suggest that although its role in energy homeostasis is contradictory, GPR39 may play a more important role in glucose homeostasis and represents a potential therapeutic target for treatment of both type 1 and type 2 DM.[134]

7. SUMMARY

GPCRs, as the largest family of membrane proteins and versatile signaling molecules mediating diverse extracellular signals, regulate almost every physiological function in the body. Glucose homeostasis is no exception. Numerous GPCRs regulate different aspects of glucose homeostasis, including insulin and incretin secretion, glucose uptake, and fat metabolism. Historically, GPCRs have proven to be very successful therapeutic targets. The GPCRs regulating glucose homeostasis, not of all of them listed here, represent important therapeutic targets. Recent successes with GLP-1R and GPR40 support this enthusiasm.

ACKNOWLEDGMENTS

Funding for our work on obesity and diabetes was provided by American Diabetes Association Grant 1-12-BS212, National Institutes of Health Grant R15-DK077213, Auburn University Intramural Grant Program, and Interdisciplinary Grant and Animal Health and Diseases Research Program of College of Veterinary Medicine at Auburn University (to Y.-X.T.).

REFERENCES

1. Tabak AG, Herder C, Rathmann W, Brunner EJ, Kivimaki M. Prediabetes: a high-risk state for diabetes development. *Lancet*. 2012;379:2279–2290.
2. Morton GJ, Schwartz MW. Leptin and the central nervous system control of glucose metabolism. *Physiol Rev*. 2011;91:389–411.
3. Chambers AP, Sandoval DA, Seeley RJ. Integration of satiety signals by the central nervous system. *Curr Biol*. 2013;23:R379–R388.
4. Grayson BE, Seeley RJ, Sandoval DA. Wired on sugar: the role of the CNS in the regulation of glucose homeostasis. *Nat Rev Neurosci*. 2013;14:24–37.
5. Tao YX, Yuan ZH, Xie J. G protein-coupled receptors as regulators of energy homeostasis. *Prog Mol Biol Transl Sci*. 2013;114:1–43.
6. Tao YX. Mutations in melanocortin-4 receptor and human obesity. *Prog Mol Biol Transl Sci*. 2009;88:173–204.
7. Tao YX. The melanocortin-4 receptor: physiology, pharmacology, and pathophysiology. *Endocr Rev*. 2010;31:506–543.
8. Fan W, Dinulescu DM, Butler AA, Zhou J, Marks DL, Cone RD. The central melanocortin system can directly regulate serum insulin levels. *Endocrinology*. 2000;141:3072–3079.
9. Obici S, Feng Z, Tan J, Liu L, Karkanias G, Rossetti L. Central melanocortin receptors regulate insulin action. *J Clin Invest*. 2001;108:1079–1085.
10. Heijboer AC, van den Hoek AM, Pijl H, et al. Intracerebroventricular administration of melanotan II increases insulin sensitivity of glucose disposal in mice. *Diabetologia*. 2005;48:1621–1626.
11. Banno R, Arima H, Hayashi M, et al. Central administration of melanocortin agonist increased insulin sensitivity in diet-induced obese rats. *FEBS Lett*. 2007;581:1131–1136.

12. Mansour M, White D, Wernette C, et al. Pancreatic neuronal melanocortin-4 receptor modulates serum insulin levels independent of leptin receptor. *Endocrine*. 2010;37:220–230.
13. Chai B, Li JY, Zhang W, Wang H, Mulholland MW. Melanocortin-4 receptor activation inhibits c-Jun N-terminal kinase activity and promotes insulin signaling. *Peptides*. 2009;30:1098–1104.
14. Huszar D, Lynch CA, Fairchild-Huntress V, et al. Targeted disruption of the melanocortin-4 receptor results in obesity in mice. *Cell*. 1997;88:131–141.
15. Rossi J, Balthasar N, Olson D, et al. Melanocortin-4 receptors expressed by cholinergic neurons regulate energy balance and glucose homeostasis. *Cell Metab*. 2011;13:195–204.
16. Muzumdar R, Ma X, Yang X, et al. Physiologic effect of leptin on insulin secretion is mediated mainly through central mechanisms. *FASEB J*. 2003;17:1130–1132.
17. Zhou L, Sutton GM, Rochford JJ, et al. Serotonin 2C receptor agonists improve type 2 diabetes via melanocortin-4 receptor signaling pathways. *Cell Metab*. 2007;6:398–405.
18. Stefater MA, Wilson-Perez HE, Chambers AP, Sandoval DA, Seeley RJ. All bariatric surgeries are not created equal: insights from mechanistic comparisons. *Endocr Rev*. 2012;33:595–622.
19. Zechner JF, Mirshahi UL, Satapati S, et al. Weight-independent effects of roux-en-Y gastric bypass on glucose homeostasis via melanocortin-4 receptors in mice and humans. *Gastroenterology*. 2013;144(580–590):e7.
20. Nadkarni P, Chepurny OG, Holz GG. Regulation of glucose homeostasis by GLP-1. *Prog Mol Biol Transl Sci*. 2014;121:23–65.
21. Knauf C, Cani PD, Perrin C, et al. Brain glucagon-like peptide-1 increases insulin secretion and muscle insulin resistance to favor hepatic glycogen storage. *J Clin Invest*. 2005;115:3554–3563.
22. Sandoval DA, Bagnol D, Woods SC, D'Alessio DA, Seeley RJ. Arcuate glucagon-like peptide 1 receptors regulate glucose homeostasis but not food intake. *Diabetes*. 2008;57:2046–2054.
23. Sayegh A. The role of cholecystokinin receptors in the short-term control of food intake. *Prog Mol Biol Transl Sci*. 2013;114:277–316.
24. Cheung GW, Kokorovic A, Lam CK, Chari M, Lam TK. Intestinal cholecystokinin controls glucose production through a neuronal network. *Cell Metab*. 2009;10:99–109.
25. Moon TW. Glucose intolerance in teleost fish: fact or fiction? *Comp Biochem Physiol B*. 2001;129:243–249.
26. Polakof S, Miguez JM, Soengas JL. Cholecystokinin impact on rainbow trout glucose homeostasis: possible involvement of central glucosensors. *Regul Pept*. 2011; 172:23–29.
27. Himick BA, Vigna SR, Peter RE. Characterization of cholecystokinin binding sites in goldfish brain and pituitary. *Am J Physiol*. 1996;271:R137–R143.
28. Kang KS, Yahashi S, Azuma M, Matsuda K. The anorexigenic effect of cholecystokinin octapeptide in a goldfish model is mediated by the vagal afferent and subsequently through the melanocortin- and corticotropin-releasing hormone-signaling pathways. *Peptides*. 2010;31:2130–2134.
29. Sandoval D, Cota D, Seeley RJ. The integrative role of CNS fuel-sensing mechanisms in energy balance and glucose regulation. *Annu Rev Physiol*. 2008;70:513–535.
30. Myers Jr MG, Olson DP. Central nervous system control of metabolism. *Nature*. 2012;491:357–363.
31. Jiang G, Zhang BB. Glucagon and regulation of glucose metabolism. *Am J Physiol Endocrinol Metab*. 2003;284:E671–E678.
32. Mayo KE, Miller LJ, Bataille D, et al. International Union of Pharmacology. XXXV. The glucagon receptor family. *Pharmacol Rev*. 2003;55:167–194.

33. Charest R, Blackmore PF, Berthon B, Exton JH. Changes in free cytosolic Ca^{2+} in hepatocytes following α_1-adrenergic stimulation. Studies on Quin-2-loaded hepatocytes. *J Biol Chem.* 1983;258:8769–8773.

34. Jelinek LJ, Lok S, Rosenberg GB, et al. Expression cloning and signaling properties of the rat glucagon receptor. *Science.* 1993;259:1614–1646.

35. Sistare FD, Picking RA, Haynes Jr RC. Sensitivity of the response of cytosolic calcium in Quin-2-loaded rat hepatocytes to glucagon, adenine nucleosides, and adenine nucleotides. *J Biol Chem.* 1985;260:12744–12747.

36. Unger RH, Orci L. The essential role of glucagon in the pathogenesis of diabetes mellitus. *Lancet.* 1975;1:14–16.

37. Rizza RA, Gerich JE. Persistent effect of sustained hyperglucagonemia on glucose production in man. *J Clin Endocrinol Metab.* 1979;48:352–355.

38. Rizza RA, Cryer PE, Gerich JE. Role of glucagon, catecholamines, and growth hormone in human glucose counterregulation. Effects of somatostatin and combined alpha- and beta-adrenergic blockade on plasma glucose recovery and glucose flux rates after insulin-induced hypoglycemia. *J Clin Invest.* 1979;64:62–71.

39. Shah P, Basu A, Basu R, Rizza R. Impact of lack of suppression of glucagon on glucose tolerance in humans. *Am J Physiol.* 1999;277:E283–E290.

40. Dunning BE, Gerich JE. The role of alpha-cell dysregulation in fasting and postprandial hyperglycemia in type 2 diabetes and therapeutic implications. *Endocr Rev.* 2007;28:253–283.

41. Miller RA, Chu Q, Xie J, Foretz M, Viollet B, Birnbaum MJ. Biguanides suppress hepatic glucagon signalling by decreasing production of cyclic AMP. *Nature.* 2013;494:256–260.

42. Sakurai H, Dobbs R, Unger RH. Somatostatin-induced changes in insulin and glucagon secretion in normal and diabetic dogs. *J Clin Invest.* 1974;54:1395–1402.

43. Gerich JE, Lorenzi M, Schneider V, et al. Effects of somatostatin on plasma glucose and glucagon levels in human diabetes mellitus. Pathophysiologic and therapeutic implications. *N Engl J Med.* 1974;291:544–547.

44. Johnson DG, Goebel CU, Hruby VJ, Bregman MD, Trivedi D. Hyperglycemia of diabetic rats decreased by a glucagon receptor antagonist. *Science.* 1982;215:1115–1116.

45. Bregman MD, Hruby VJ. Synthesis and isolation of a glucagon antagonist. *FEBS Lett.* 1979;101:191–194.

46. Madsen P, Knudsen LB, Wiberg FC, Carr RD. Discovery and structure–activity relationship of the first non-peptide competitive human glucagon receptor antagonists. *J Med Chem.* 1998;41:5150–5157.

47. Mu J, Jiang G, Brady E, et al. Chronic treatment with a glucagon receptor antagonist lowers glucose and moderately raises circulating glucagon and glucagon-like peptide 1 without severe alpha cell hypertrophy in diet-induced obese mice. *Diabetologia.* 2011;54:2381–2391.

48. Xiong Y, Guo J, Candelore MR, et al. Discovery of a novel glucagon receptor antagonist N-[(4-{(1S)-1-[3-(3, 5-dichlorophenyl)-5-(6-methoxynaphthalen-2-yl)-1H-pyrazol-1-yl]ethyl}phenyl)carbo nyl]-beta-alanine (MK-0893) for the treatment of type II diabetes. *J Med Chem.* 2012;55:6137–6148.

49. Guzman-Perez A, Pfefferkorn JA, Lee EC, et al. The design and synthesis of a potent glucagon receptor antagonist with favorable physicochemical and pharmacokinetic properties as a candidate for the treatment of type 2 diabetes mellitus. *Bioorg Med Chem Lett.* 2013;23:3051–3058.

50. Irwin N, Franklin ZJ, O'Harte FP. desHis^{1}Glu^9-glucagon-[mPEG] and desHis^{1}Glu^9(Lys^{30}PAL)-glucagon: Long-acting peptide-based PEGylated and acylated glucagon receptor antagonists with potential antidiabetic activity. *Eur J Pharmacol.* 2013;709:43–51.

51. Vater A, Sell S, Kaczmarek P, et al. A mixed mirror-image DNA/RNA aptamer inhibits glucagon and acutely improves glucose tolerance in models of Type 1 and Type 2 diabetes. *J Biol Chem.* 2013;288:21136–21147.
52. Henquin JC. Cell biology of insulin secretion. In: Kahn CR, Weir GC, King GL, Jacobson AM, Moses AC, Smith RJ, eds. *Joslin's Diabetes Mellitus 14th edit.* Philadelphia: Lippincott Williams & Wilkins; 2005:83–107.
53. Winzell MS, Ahren B. G-protein-coupled receptors and islet function-implications for treatment of type 2 diabetes. *Pharmacol Ther.* 2007;116:437–448.
54. Ahren B. Islet G protein-coupled receptors as potential targets for treatment of type 2 diabetes. *Nat Rev Drug Discov.* 2009;8:369–385.
55. Amisten S, Salehi A, Rorsman P, Jones PM, Persaud SJ. An atlas and functional analysis of G-protein coupled receptors in human islets of Langerhans. *Pharmacol Ther.* 2013;139:359–391.
56. Robertson RP, Seaquist ER, Walseth TF. G proteins and modulation of insulin secretion. *Diabetes.* 1991;40:1–6.
57. Hollenstein K, Kean J, Bortolato A, et al. Structure of class B GPCR corticotropin-releasing factor receptor 1. *Nature.* 2013;499:438–443.
58. Siu FY, He M, de Graaf C, et al. Structure of the human glucagon class B G-protein-coupled receptor. *Nature.* 2013;499:444–449.
59. Sexton PM, Wootten D. Structural biology: meet the B family. *Nature.* 2013;499:417–418.
60. Moore B, Edie ES, Abram JH. On the treatment of diabetus mellitus by acid extract of duodenal mucous membrane. *Biochem J.* 1906;1:28–38.
61. Yalow RS, Berson SA. Immunoassay of endogenous plasma insulin in man. *J Clin Invest.* 1960;39:1157–1175.
62. Elrick H, Stimmler L, Hlad Jr CJ, Arai Y. Plasma insulin response to oral and intravenous glucose administration. *J Clin Endocrinol Metab.* 1964;24:1076–1082.
63. McIntyre N, Holdsworth CD, Turner DS. New interpretation of oral glucose tolerance. *Lancet.* 1964;2:20–21.
64. Brown JC, Mutt V, Pederson RA. Further purification of a polypeptide demonstrating enterogastrone activity. *J Physiol.* 1970;209:57–64.
65. Morrow GW, Kieffer TJ, McIntosh CH, et al. The insulinotropic region of gastric inhibitory polypeptide; fragment analysis suggests the bioactive site lies between residues 19 and 30. *Can J Physiol Pharmacol.* 1996;74:65–72.
66. Lu M, Wheeler MB, Leng XH, Boyd 3rd AE. The role of the free cytosolic calcium level in β-cell signal transduction by gastric inhibitory polypeptide and glucagon-like peptide I(7-37). *Endocrinology.* 1993;132:94–100.
67. Straub SG, Sharp GW. Glucose-dependent insulinotropic polypeptide stimulates insulin secretion via increased cyclic AMP and $[Ca^{2+}]_i$ and a wortmannin-sensitive signalling pathway. *Biochem Biophys Res Commun.* 1996;224:369–374.
68. Trumper A, Trumper K, Trusheim H, Arnold R, Goke B, Horsch D. Glucose-dependent insulinotropic polypeptide is a growth factor for β (INS-1) cells by pleiotropic signaling. *Mol Endocrinol.* 2001;15:1559–1570.
69. Trumper A, Trumper K, Horsch D. Mechanisms of mitogenic and anti-apoptotic signaling by glucose-dependent insulinotropic polypeptide in β(INS-1)-cells. *J Endocrinol.* 2002;174:233–246.
70. Getty-Kaushik L, Song DH, Boylan MO, Corkey BE, Wolfe MM. Glucose-dependent insulinotropic polypeptide modulates adipocyte lipolysis and reesterification. *Obesity (Silver Spring).* 2006;14:1124–1131.
71. Song DH, Getty-Kaushik L, Tseng E, Simon J, Corkey BE, Wolfe MM. Glucose-dependent insulinotropic polypeptide enhances adipocyte development and glucose uptake in part through Akt activation. *Gastroenterology.* 2007;133:1796–1805.

72. Holst JJ, Gromada J, Nauck MA. The pathogenesis of NIDDM involves a defective expression of the GIP receptor. *Diabetologia*. 1997;40:984–986.
73. Vilsboll T, Krarup T, Madsbad S, Holst JJ. Defective amplification of the late phase insulin response to glucose by GIP in obese Type II diabetic patients. *Diabetologia*. 2002;45:1111–1119.
74. Meier JJ, Nauck MA. GIP as a potential therapeutic agent? *Horm Metab Res*. 2004;36:859–866.
75. Sanlioglu AD, Karacay B, Balci MK, Griffith TS, Sanlioglu S. Therapeutic potential of VIP vs PACAP in diabetes. *J Mol Endocrinol*. 2012;49:R157–R167.
76. Huang H, Dai MH, Tao YX. Physiology and therapeutics of the free fatty acid receptor GPR40. *Prog Mol Biol Transl Sci*. 2014;121:67–94.
77. Mo XL, Yang Z, Tao YX. Targeting GPR119 for the potential treatment of type 2 diabetes mellitus and obesity. *Prog Mol Biol Transl Sci*. 2014;121:95–131.
78. Kaneto A, Kosaka K, Nakao K. Effects of the neurohypophysial hormones on insulin secretion. *Endocrinology*. 1967;81:783–790.
79. Monaco ME, Levy BL, Richardson SB. Synergism between vasopressin and phorbol esters in stimulation of insulin secretion and phosphatidylcholine metabolism in RIN insulinoma cells. *Biochem Biophys Res Commun*. 1988;151:717–724.
80. Gao ZY, Drews G, Nenquin M, Plant TD, Henquin JC. Mechanisms of the stimulation of insulin release by arginine-vasopressin in normal mouse islets. *J Biol Chem*. 1990;265:15724–15730.
81. Gao ZY, Drews G, Henquin JC. Mechanisms of the stimulation of insulin release by oxytocin in normal mouse islets. *Biochem J*. 1991;276(Pt 1):169–174.
82. Lee B, Yang C, Chen TH, al-Azawi N, Hsu WH. Effect of AVP and oxytocin on insulin release: involvement of V1b receptors. *Am J Physiol*. 1995;269:E1095–E1100.
83. Zhang H, Wu C, Chen Q, et al. Treatment of obesity and diabetes using oxytocin or analogs in patients and mouse models. *PLoS One*. 2013;8:e61477.
84. Yang C, Hsu WH. Stimulatory effect of bradykinin on insulin release from the perfused rat pancreas. *Am J Physiol*. 1995;268:E1027–E1030.
85. Saito Y, Kato M, Kubohara Y, Kobayashi I, Tatemoto K. Bradykinin increases intracellular free Ca^{2+} concentration and promotes insulin secretion in the clonal β-cell line, HIT-T15. *Biochem Biophys Res Commun*. 1996;221:577–580.
86. Yang C, Lee B, Chen TH, Hsu WH. Mechanisms of bradykinin-induced insulin secretion in clonal *beta* cell line RINm5F. *J Pharmacol Exp Ther*. 1997;282:1247–1252.
87. Tomiyama H, Kushiro T, Abeta H, et al. Kinins contribute to the improvement of insulin sensitivity during treatment with angiotensin converting enzyme inhibitor. *Hypertension*. 1994;23:450–455.
88. Miller RE. Pancreatic neuroendocrinology: peripheral neural mechanisms in the regulation of the Islets of Langerhans. *Endocr Rev*. 1981;2:471–494.
89. Gilon P, Henquin JC. Mechanisms and physiological significance of the cholinergic control of pancreatic β-cell function. *Endocr Rev*. 2001;22:565–604.
90. Henquin JC, Nenquin M. The muscarinic receptor subtype in mouse pancreatic B-cells. *FEBS Lett*. 1988;236:89–92.
91. Boschero AC, Szpak-Glasman M, Carneiro EM, et al. Oxotremorine-m potentiation of glucose-induced insulin release from rat islets involves M3 muscarinic receptors. *Am J Physiol*. 1995;268:E336–E342.
92. Duttaroy A, Zimliki CL, Gautam D, Cui Y, Mears D, Wess J. Muscarinic stimulation of pancreatic insulin and glucagon release is abolished in m3 muscarinic acetylcholine receptor-deficient mice. *Diabetes*. 2004;53:1714–1720.
93. Gautam D, Han SJ, Duttaroy A, et al. Role of the M_3 muscarinic acetylcholine receptor in β-cell function and glucose homeostasis. *Diabetes Obes Metab*. 2007; 9(Suppl 2):158–169.

94. Nakajima K, Wess J. Design and functional characterization of a novel, arrestin-biased designer G protein-coupled receptor. *Mol Pharmacol*. 2012;82:575–582.

95. Yamada M, Miyakawa T, Duttaroy A, et al. Mice lacking the M3 muscarinic acetylcholine receptor are hypophagic and lean. *Nature*. 2001;410:207–212.

96. Gautam D, Gavrilova O, Jeon J, et al. Beneficial metabolic effects of M3 muscarinic acetylcholine receptor deficiency. *Cell Metab*. 2006;4:363–375.

97. Gautam D, Han SJ, Hamdan FF, et al. A critical role for beta cell M3 muscarinic acetylcholine receptors in regulating insulin release and blood glucose homeostasis *in vivo*. *Cell Metab*. 2006;3:449–461.

98. Guettier JM, Gautam D, Scarselli M, et al. A chemical-genetic approach to study G protein regulation of β cell function *in vivo*. *Proc Natl Acad Sci U S A*. 2009;106:19197–19202.

99. Gautam D, Ruiz de Azua I, Li JH, et al. Beneficial metabolic effects caused by persistent activation of β-cell M_3 muscarinic acetylcholine receptors in transgenic mice. *Endocrinology*. 2010;151:5185–5194.

100. Jain S, Ruiz de Azua I, Lu H, White MF, Guettier JM, Wess J. Chronic activation of a designer G_q-coupled receptor improves β cell function. *J Clin Invest*. 2013;123:1750–1762.

101. Nakajima K, Jain S, Ruiz de Azua I, McMillin SM, Rossi M, Wess J. Minireview: Novel aspects of M3 muscarinic receptor signaling in pancreatic β-cells. *Mol Endocrinol*. 2013;27:1208–1216.

102. Lee YS, Morinaga H, Kim JJ, et al. The fractalkine/CX3CR1 system regulates β cell function and insulin secretion. *Cell*. 2013;153:413–425.

103. Sirois-Gagnon D, Chamberland A, Perron S, Brisson D, Gaudet D, Laprise C. Association of common polymorphisms in the fractalkine receptor (CX3CR1) with obesity. *Obesity (Silver Spring)*. 2011;19:222–227.

104. Shah R, Hinkle CC, Ferguson JF, et al. Fractalkine is a novel human adipochemokine associated with type 2 diabetes. *Diabetes*. 2011;60:1512–1528.

105. Regard JB, Sato IT, Coughlin SR. Anatomical profiling of G protein-coupled receptor expression. *Cell*. 2008;135:561–571.

106. Regard JB, Kataoka H, Cano DA, et al. Probing cell type-specific functions of Gi *in vivo* identifies GPCR regulators of insulin secretion. *J Clin Invest*. 2007;117:4034–4043.

107. Stoddart LA, Smith NJ, Milligan G. International Union of Pharmacology. LXXI. Free fatty acid receptors FFA1, -2, and -3: pharmacology and pathophysiological functions. *Pharmacol Rev*. 2008;60:405–417.

108. Talukdar S, Olefsky JM, Osborn O. Targeting GPR120 and other fatty acid-sensing GPCRs ameliorates insulin resistance and inflammatory diseases. *Trends Pharmacol Sci*. 2011;32:543–550.

109. Blad CC, Tang C, Offermanns S. G protein-coupled receptors for energy metabolites as new therapeutic targets. *Nat Rev Drug Discov*. 2012;11:603–619.

110. Ulven T. Short-chain free fatty acid receptors FFA2/GPR43 and FFA3/GPR41 as new potential therapeutic targets. *Front Endocrinol*. 2012;3:111.

111. Ryan KK, Seeley RJ. Physiology. Food as a hormone. *Science*. 2013;339:918–919.

112. Mo XL, Wei HK, Peng J, Tao YX. Free fatty acid receptor GPR120 and pathogenesis of obesity and type 2 diabetes mellitus. *Prog Mol Biol Transl Sci*. 2013;114:251–276.

113. Hudson BD, Ulven T, Milligan G. The therapeutic potential of allosteric ligands for free fatty acid sensitive GPCRs. *Curr Top Med Chem*. 2013;13:14–25.

114. Conigrave AD, Hampson DR. Broad-spectrum L-amino acid sensing by class 3 G-protein-coupled receptors. *Trends Endocrinol Metab*. 2006;17:398–407.

115. Wellendorph P, Brauner-Osborne H. Molecular basis for amino acid sensing by family C G-protein-coupled receptors. *Br J Pharmacol*. 2009;156:869–884.

116. Conigrave AD, Hampson DR. Broad-spectrum amino acid-sensing class C G-protein coupled receptors: molecular mechanisms, physiological significance and options for drug development. *Pharmacol Ther*. 2010;127:252–260.

117. Wauson EM, Lorente-Rodriguez A, Cobb MH. Minireview: nutrient sensing by G protein-coupled receptors. *Mol Endocrinol*. 2013;27:1188–1197.

118. Wellendorph P, Hansen KB, Balsgaard A, Greenwood JR, Egebjerg J, Brauner-Osborne H. Deorphanization of GPRC6A: a promiscuous L-alpha-amino acid receptor with preference for basic amino acids. *Mol Pharmacol*. 2005;67:589–597.

119. Pi M, Faber P, Ekema G, et al. Identification of a novel extracellular cation-sensing G-protein-coupled receptor. *J Biol Chem*. 2005;280:40201–40209.

120. Oya M, Kitaguchi T, Pais R, Reimann F, Gribble F, Tsuboi T. The G protein-coupled receptor family C group 6 subtype A (GPRC6A) receptor is involved in amino acid-induced glucagon-like peptide-1 secretion from GLUTag cells. *J Biol Chem*. 2013;288:4513–4521.

121. Pi M, Chen L, Huang MZ, et al. GPRC6A null mice exhibit osteopenia, feminization and metabolic syndrome. *PLoS One*. 2008;3:e3858.

122. Smajilovic S, Clemmensen C, Johansen LD, et al. The L-alpha-amino acid receptor GPRC6A is expressed in the islets of Langerhans but is not involved in L-arginine-induced insulin release. *Amino Acids*. 2013;44:383–390.

123. Chung S, Funakoshi T, Civelli O. Orphan GPCR research. *Br J Pharmacol*. 2008; 153(Suppl 1):S339–S346.

124. O'Dowd BF, Nguyen T, Marchese A, et al. Discovery of three novel G-protein-coupled receptor genes. *Genomics*. 1998;47:310–313.

125. Ku GM, Pappalardo Z, Luo CC, German MS, McManus MT. An siRNA screen in pancreatic beta cells reveals a role for Gpr27 in insulin production. *PLoS Genet*. 2012;8:e1002449.

126. Zhang JV, Ren PG, Avsian-Kretchmer O, et al. Obestatin, a peptide encoded by the ghrelin gene, opposes ghrelin's effects on food intake. *Science*. 2005;310:996–999.

127. Zhang JV, Li L, Huang Q, Ren PG. Obestatin receptor in energy homeostasis and obesity pathogenesis. *Prog Mol Biol Transl Sci*. 2013;114:89–107.

128. Moechars D, Depoortere I, Moreaux B, et al. Altered gastrointestinal and metabolic function in the GPR39-obestatin receptor-knockout mouse. *Gastroenterology*. 2006;131:1131–1141.

129. Holst B, Egerod KL, Jin C, et al. G protein-coupled receptor 39 deficiency is associated with pancreatic islet dysfunction. *Endocrinology*. 2009;150:2577–2585.

130. Tremblay F, Richard AM, Will S, et al. Disruption of G protein-coupled receptor 39 impairs insulin secretion *in vivo*. *Endocrinology*. 2009;150:2586–2595.

131. Verhulst PJ, Lintermans A, Janssen S, et al. GPR39, a receptor of the ghrelin receptor family, plays a role in the regulation of glucose homeostasis in a mouse model of early onset diet-induced obesity. *J Neuroendocrinol*. 2011;23:490–500.

132. Holst B, Holliday ND, Bach A, Elling CE, Cox HM, Schwartz TW. Common structural basis for constitutive activity of the ghrelin receptor family. *J Biol Chem*. 2004;279:53806–53817.

133. Egerod KL, Jin C, Petersen PS, et al. β-Cell specific overexpression of GPR39 protects against streptozotocin-induced hyperglycemia. *Int J Endocrinol*. 2011;2011:401258.

134. Depoortere I. GI functions of GPR39: novel biology. *Curr Opin Pharmacol*. 2012;12:647–652.

CHAPTER TWO

Regulation of Glucose Homeostasis by GLP-1

Prashant Nadkarni[*,†], **Oleg G. Chepurny**[*], **George G. Holz**[*,‡,1]

[*]Department of Medicine, State University of New York (SUNY), Upstate Medical University, Syracuse, New York, USA
[†]Joslin Diabetes Center, State University of New York (SUNY), Upstate Medical University, Syracuse, New York, USA
[‡]Department of Pharmacology, State University of New York (SUNY), Upstate Medical University, Syracuse, New York, USA
[1]Corresponding author: e-mail address: holzg@upstate.edu

Contents

Abstract

Glucagon-like peptide-1(7–36)amide (GLP-1) is a secreted peptide that acts as a key determinant of blood glucose homeostasis by virtue of its abilities to slow gastric emptying, to enhance pancreatic insulin secretion, and to suppress pancreatic glucagon secretion. GLP-1 is secreted from L cells of the gastrointestinal mucosa in response to a meal, and the blood glucose-lowering action of GLP-1 is terminated due to its enzymatic degradation by dipeptidyl-peptidase-IV (DPP-IV). Released GLP-1 activates enteric and autonomic reflexes while also circulating as an incretin hormone to control endocrine pancreas function. The GLP-1 receptor (GLP-1R) is a G protein-coupled receptor that is activated directly or indirectly by blood glucose-lowering agents currently in use for the treatment of type 2 diabetes mellitus (T2DM). These therapeutic

Progress in Molecular Biology and Translational Science, Volume 121
ISSN 1877-1173
http://dx.doi.org/10.1016/B978-0-12-800101-1.00002-8

23

agents include GLP-1R agonists (exenatide, liraglutide, lixisenatide, albiglutide, dulaglutide, and langlenatide) and DPP-IV inhibitors (sitagliptin, vildagliptin, saxagliptin, linagliptin, and alogliptin). Investigational agents for use in the treatment of T2DM include GPR119 and GPR40 receptor agonists that stimulate the release of GLP-1 from L cells. Summarized here is the role of GLP-1 to control blood glucose homeostasis, with special emphasis on the advantages and limitations of GLP-1-based therapeutics.

LIST OF ABBREVIATIONS

ACh acetylcholine
ANF atrial natriuretic factor
CICR Ca^{2+}-induced Ca^{2+} release
CNS central nervous system
CBP CREB-binding protein
CRE cAMP response element
CREB cAMP response element-binding protein
CRTC CREB-regulated transcriptional coregulator
DPP-IV dipeptidyl-peptidase-IV
Epac cAMP-regulated guanine nucleotide exchange factor
ER extended release
GLP-1R GLP-1 receptor
GPCR G protein-coupled receptor
GSIS glucose-stimulated insulin secretion
HFD high-fat diet
IRS-2 insulin receptor substrate 2
K_{ATP} ATP-sensitive K^+ channel
KO knockout
MEN2 multiple endocrine neoplasia syndrome type 2
NDM neonatal diabetes mellitus
NPY neuropeptide tyrosine
PACAP pituitary adenylyl cyclase-activating peptide
PG proglucagon
PIP_2 phosphatidylinositol 4,5-bisphosphate
PKA protein kinase A
PTH parathyroid hormone
PYY peptide tyrosine tyrosine
RCT randomized control trial
RYGB Roux-en-Y gastric bypass
SNARE soluble *N*-ethylmaleimide-sensitive factor attachment protein receptor
SUR1 sulfonylurea receptor type 1
T2DM type 2 diabetes mellitus
TMAC transmembrane adenylyl cyclase
TxNIP thioredoxin-interacting protein
VDCC voltage-dependent Ca^{2+} channel
WT wild type

1. INTRODUCTION

Systemic blood glucose homeostasis in humans is under the control of glucagon-like peptide-1(7–36)amide (GLP-1), a peptide secreted from intestinal enteroendocrine L cells in response to a meal.[1] L cells are located within the gastrointestinal mucosa and they act as nutrient sensors to release GLP-1 in response to luminal sugars, amino acids, and fatty acids.[2] Released GLP-1 acts locally within the intestinal wall to activate enteroenteric reflexes important to the control of gastric motility, thereby slowing gastric emptying.[3] Simultaneously, released GLP-1 activates vagal sensory nerve terminals that innervate the intestinal wall, and in this manner, GLP-1 initiates vagal–vagal autonomic reflexes that control endocrine pancreas function.[4] Circulating GLP-1 also acts as a hormone at the islets of Langerhans in the endocrine pancreas to stimulate the release of insulin, while suppressing the release of glucagon.[5] During the postprandial phase of blood glucose control, these immediate and multiple actions of GLP-1 act in concert to lower levels of blood glucose.

Clinical studies demonstrate that the blood glucose-lowering action of GLP-1 is itself glucose-dependent.[6–8] More specifically, GLP-1 reduces levels of blood glucose only when concentrations of blood glucose are elevated above fasting levels, as is the case after a meal. As the postprandial blood glucose levels fall in response to GLP-1, the blood glucose-lowering action of GLP-1 is self-terminating. This remarkable glucose-dependent property of GLP-1 action results in a situation in which intravenously administered GLP-1 fails to reduce levels of blood glucose below fasting levels.[6–8] Since administered GLP-1 does not produce hypoglycemia, these clinical findings have led to the use of GLP-1 receptor (GLP-1R) agonists as a new class of blood glucose-lowering agents for use in the treatment of type 2 diabetes mellitus (T2DM).[9,10]

2. GLP-1 BIOSYNTHESIS, SECRETION, AND DEGRADATION

Proglucagon gene expression in the intestinal L cells generates proglucagon (PG) that is processed by prohormone convertases (PC1/3) to liberate the GLP-1(1–37) peptide precursor.[11,12] Endopeptidase-catalyzed cleavage of GLP-1(1–37) generates two peptides with insulin secretagogue properties. These are GLP-1(7–37) that is processed by amidating enzyme to generate GLP-1(7–36)amide.[13–15] Although glucagon gene expression also generates PG in islet α-cells, it was thought that α-cells fail to synthesize GLP-1 due to the fact that these endocrine cells contain a prohormone convertase (PC2) that preferentially processes PG to glucagon.[16] However, it is now apparent that endocrine cell "plasticity" exists within the islets such that α-cells synthesize GLP-1 under stressful or pathophysiological conditions including T2DM.[17] Thus, it seems likely that GLP-1 can also act as an intraislet paracrine hormone but in a context-dependent manner.

GLP-1 is packaged in secretory granules and it is released from intestinal L cells by exocytosis in response to an elevation of cytosolic Ca^{2+} and cAMP.[18] In this regard, it is important to note that L cells are electrically excitable and that glucose transporter-mediated uptake of glucose by L cells is Na^+-dependent and electrogenic. Thus, L cells respond to orally administered glucose by generating action potentials that trigger depolarization-induced Ca^{2+} influx through voltage-dependent Ca^{2+} channels (VDCCs).[19] Ca^{2+} mobilized from intracellular Ca^{2+} stores is also a stimulus for GLP-1 secretion, and this Ca^{2+} mobilization is initiated by the binding of fatty acids to a receptor designated as GPR40 located on L cells.[20] GLP-1 secretion is also stimulated by fatty acid amides (oleoylethanolamide) and monoacylglycerols (2-oleoyl glycerol) that activate GPR119, a receptor that is positively coupled to cAMP production in L cells.[20]

GLP-1 released from L cells acts locally within the hepato-portal circulation to activate the GLP-1R located on vagal sensory neurons. These neurons constitute the hepato-portal glucose sensor that communicates with brainstem neurons in order to regulate whole-body metabolism.[4] Highest concentrations of released GLP-1 are found immediately within the hepato-portal circulation since GLP-1 is quickly metabolized to GLP-1 (9–36)amide by dipeptidyl-peptidase-IV (DPP-IV).[21] DPP-IV exists in two forms—a 766-amino-acid transmembrane protein and a smaller soluble form found in the plasma.[22] Both forms of DPP-IV have enzymatic activity, and the preferred substrates are peptides such as GLP-1 that contain penultimate N-terminus alanine or proline residues. The half-life for intravenously administered GLP-1 is less than 5 min owing to its rapid degradation by DPP-IV.[23] However, picomolar concentrations of GLP-1 activate the GLP-1R,[24] and concentrations of circulating GLP-1 are sufficiently high to allow it to activate the GLP-1R on islet β-cells.

Exenatide is a DPP-IV-resistant peptide that shares structural homology with GLP-1.[25] It is the synthetic form of exendin-4, a peptide isolated from the Gila monster lizard *Heloderma*.[26] Exenatide is a GLP-1R agonist in humans, and it has a half-life of ca. 20 min when it is administered by the intravenous route.[9] Patients with T2DM receive exenatide by subcutaneous injection, and circulating exenatide produces a blood glucose-lowering effect since it directly activates the GLP-1R.[9,10] Orally administrable DPP-IV inhibitors such as sitagliptin and vildagliptin are also useful for the treatment of T2DM.[27,28] By slowing degradation of GLP-1, these DPP-IV inhibitors enhance the action of endogenous GLP-1 to activate vagal sensory neurons that compose the hepato-portal glucose sensor.[29] This intestinal action of DPP-IV inhibitors is of significance since it is the predominant means by which low concentrations of DPP-IV inhibitors exert a blood glucose-lowering effect.[29]

3. INSULINOTROPIC AND GROWTH FACTOR PROPERTIES OF GLP-1

Soon after the cloning of the anglerfish *PG* gene by the Habener laboratory in 1982,[11] the sequence of a hamster *PG* cDNA was reported by Bell and coworkers.[12] Bioinformatics analysis of the hamster *PG* cDNA revealed that it encoded GLP-1(1–37) derived from exon 4 of the *PG* gene.[12] Subsequently, it was demonstrated by the laboratories of Creutzfeldt,[30] Holst,[31] Habener,[13] Weir,[32] and Bloom[15] that pancreatic insulin secretion could be stimulated in a glucose-dependent manner by derivates of GLP-1(1–37) that included GLP-1(1–36), GLP-1(7–36)amide, and GLP-1(7–37). Cloning of the 463-amino-acid rat pancreatic islet GLP-1R by Thorens in 1992 revealed that nanomolar high-affinity agonist binding to the recombinant GLP-1R was measurable using a truncated metabolite of GLP-1(1–37) that corresponded to GLP-1(7–36)amide.[33]

It is now recognized that GLP-1(7–36)amide is the predominant bioactive GLP-1 present in human serum. It has an MW of 3298 Da and is composed of 30-amino-acid residues with the sequence HAEGTFTSDVSSYLEGQAAKEFIAWLVKGR-NH$_2$.[34] This truncated and fully bioactive GLP-1 binds to the human GLP-1R[35,36] expressed on islet β-cells in order to potentiate glucose-stimulated insulin secretion *in vivo* (GSIS).[6–8] Importantly, the insulin secretagogue action of GLP-1 is accompanied by its ability to stimulate insulin gene transcription, insulin mRNA translation, and proinsulin biosynthesis in β-cells.[37–40] Potentially just as important, GLP-1 also acts as a β-cell growth factor so that in mice, it stimulates β-cell proliferation while also exerting a prosurvival (antiapoptosis) action to protect against β-cell death.[41–45] Thus, there is great interest to determine whether such preclinical findings concerning GLP-1 are applicable to T2DM patients treated with GLP-1R agonists.

In addition to improving β-cell insulin secretion and insulin biosynthesis, a modern treatment for T2DM might also be capable of increasing β-cell "mass," either by stimulating β-cell replication or by slowing β-cell death. Furthermore, for certain forms of T2DM, it might be useful to identify GLP-1R agonists that selectively "bias" GLP-1R signal transduction in order to achieve a desired therapeutic outcome. In this regard, attention has recently focused on whether it might be possible to synthesize GLP-1R agonists that allosterically induce a GLP-1R conformation that enhances receptor coupling to select downstream effectors such as GTP-binding proteins, mitogen-activated protein kinases, c-src kinase, and β-arrestin.[46] In this context, it is valuable to summarize what is currently known concerning signal transduction pathways activated by the GLP-1R.

The GLP-1R is a G protein-coupled receptor (GPCR) that is a member of the secretin receptor-like family of seven transmembrane-spanning domain proteins. These group B receptors include GPCRs that selectively bind secretin, glucagon, glucose-dependent insulinotropic peptide (GIP), vasoactive intestinal peptide (VIP), and pituitary adenylyl cyclase-activating peptide (PACAP).[47] Heterotrimeric G_S GTP-binding proteins are activated in response to agonist binding to the GLP-1R, and they couple GLP-1R agonist occupancy to the stimulation of transmembrane adenylyl cyclases (TMACs). In β-cells, TMACs catalyze conversion of ATP to cytosolic cAMP, a second messenger that activates either protein kinase A (PKA) or the cAMP-regulated guanine nucleotide exchange factor designated as Epac2.[48] PKA is a serine/threonine protein kinase that phosphorylates key substrate proteins of the β-cell stimulus-secretion coupling and gene regulatory networks. In contrast, Epac2 acts via Rap1 GTPase to activate a novel phospholipase C-epsilon (PLCε) that specifically hydrolyzes phosphatidylinositol 4,5-bisphosphate (PIP_2).[49]

In β-cells, there exists a PKA-mediated action of GLP-1R agonists to phosphorylate Snapin.[50] Snapin is a protein that associates with SNAP-25, a component of the soluble N-ethylmaleimide-sensitive factor attachment protein receptor (SNARE) complex that couples an increase of cytosolic Ca^{2+} concentration to insulin secretory granule exocytosis. By phosphorylating Snapin, GLP-1R agonists potentiate GSIS, thus lowing levels of blood glucose.[50] An additional PKA-mediated action of GLP-1R agonists controls β-cell gene expression by phosphorylating CREB, a cAMP response element-*b*inding protein.[51] Activated CREB binds cAMP response elements (CREs) located in 5′ gene promoters, and it couples PKA activation to the stimulation of gene transcription. This CREB-mediated action of PKA is facilitated by CREB coactivators such as p300, CREB-binding protein (CBP), and CRTC (CREB-regulated transcriptional coregulator).[52] Numerous CREB-regulated genes are under the control of GLP-1R agonists in β-cells, as demonstrated for CREB-dependent stimulation of insulin gene expression and insulin receptor substrate-2 (IRS-2) gene expression.[51–53]

Particularly interesting is the role of Epac2 in the control of pancreatic insulin secretion. Studies of mice with a knockout (KO) of Epac2 gene expression demonstrate that Epac2 mediates the action of GLP-1R agonist exendin-4 to potentiate the first phase kinetic component of GSIS.[54] Since first phase GSIS is defective in patients with T2DM,[55] and since exendin-4 restores first phase GSIS under conditions of T2DM,[56] it appears that Epac2 activation might play an especially important role when considering how GLP-1-based therapeutic agents restore normoglycemia in patients with T2DM. Interestingly, Epac proteins may also play a role in the central nervous system (CNS) control of glucose homeostasis and energy expenditure since their activation appears to reduce leptin sensitivity in neural networks controlling feeding behavior.[57]

When considering how GLP-1R agonists act as β-cell trophic factors to increase β-cell mass, studies of β-cell lines or neonatal mouse β-cells indicate that it is PKA that mediates transcriptional induction of cyclin D1 expression by GLP-1 in order to stimulate β-cell proliferation.[58] Interestingly, a proliferative action of GLP-1 also results from PKA-mediated phosphorylation of β-catenin, thereby indicating that the β-cell cAMP–PKA signaling branch exhibits signal transduction cross talk with a noncanonical Wnt signaling pathway that uses the transcription factor TCF7L2 to control gene expression.[59] An additional surprising finding is that a truncated GLP-1 designated as GLP-1(28–36)amide stimulates cAMP production in β-cells, thereby activating the β-catenin/TCF7L2 signaling pathway.[60] Furthermore, GLP-1(28–36)amide protects against β-cell glucotoxicity by improving mitochondrial function.[61] GLP-1(28–36)amide is a cell-penetrating peptide that does not exert its effects by binding to the GLP-1R, but instead acts intracellularly.[61] Thus, it is not clear how GLP-1(28–36)amide stimulates cAMP production.

PKA-mediated induction of IRS-2 expression also promotes β-cell growth in response to GLP-1,[53,62] and PKA mediates the action of

GLP-1 to promote translocation of transcription factor PDX-1 to the nucleus, thereby enhancing the differentiated state of β-cells.[63] In contrast, Epac2 participates in the protection of β-cells from cytotoxicity induced by reactive oxygen species (ROS).[64,65] Redox control in β-cells is under the control of thioredoxin (TxN), and TxNIPs are thioredoxin–interacting proteins that downregulate the ROS buffering capacity of thioredoxin.[66] Thus, it is significant that GLP-1 acts via Epac2 to suppress TxNIP expression and to enhance ROS buffering in β-cells.[64] To what extent Epac2 also mediates actions of GLP-1R agonists to control β-cell mass and/or survival remains an active field of investigation.

4. MOLECULAR BASIS OF GLP-1 RECEPTOR ACTIVATION

Like other group B GPCRs, the GLP-1R possesses a long extracellularly oriented N-terminus of ca. 150 amino acids in which three pairs of disulfide bonds create a secondary structure important to ligand binding. This N-terminal extracellular domain is connected to a core domain of the receptor consisting of seven transmembrane α-helices interconnected by three extracellular loops (ECL 1–3) and three intracellular loops (ICL 1–3). Based on findings originally obtained in studies of the group B GPCR for parathyroid hormone (PTH),[67] a "two–domain model" for GLP-1R activation exists in which the α-helical C-terminus of GLP-1 (7–36)amide interacts with the receptor's N-terminal domain, while the N-terminus of GLP-1(7–36)amide interacts with ECL-1 and ECL-2 of the GLP-1R.[68–72] The affinity and selectivity of ligand binding is determined by interactions of the GLP-1(7–36)amide C-terminus with the receptor's N-terminal domain, whereas coupling of the GLP-1R to intracellular signaling pathways is strongly influenced by the receptor core domain and its intracellular loops. In this regard, ICL-3 is of major importance to GLP-1R-stimulated adenylyl cyclase activity.[73]

There exists an alternative model of GLP-1R activation in which it is proposed that binding of GLP-1(7–36)amide to the receptor results in a structural rearrangement of the receptor's N-terminal domain so that a pentapeptide NRTFD signature sequence within the N-terminus of the receptor acts as an "endogenous agonist" at ECL-2 or ECL-3 of the receptor.[74] In this model, the signature sequence acts as a "tethered ligand" to promote GLP-1R activation.[74] Adding to this complexity, the signaling properties of the GLP-1R are also dictated by its ability to form homodimers in which receptor dimerization occurs at the interface of transmembrane helix four of each receptor protomer.[75]

Since small molecule GLP-1R agonists are highly desired for the treatment of T2DM, considerable effort has been exerted in an attempt to identify the precise mechanisms of ligand binding to the GLP-1R. Nuclear magnetic resonance (NMR) analysis using isolated N-terminal extracellular domains of group B GPCRs reveals that these ligand-binding domains contain a core structure composed of two antiparallel β-sheets stabilized by three disulfide bonds. X-ray crystallographic analysis also reveals that these β-sheets are linked to an N-terminal α-helix in order to form a "fold" that is highly conserved among the group B GPCRs.[69] Binding of GLP-1(7–36) amide within this fold leads to a structural rearrangement of GLP-1(7–36) amide so that it transitions from its disordered solution structure to an induced α-helical conformation with hydrophobic residues buried within the fold.[69]

Since binding of GLP-1(7–36)amide to the GLP-1R is accompanied by a structural rearrangement of the peptide in order for it to stimulate receptor signaling, it is understandable that the rationale design of small molecule GLP-1R agonists is complex and is not guided simply by the disordered solution structure of GLP-1(7–36)amide. However, the flexibility of GLP-1(7–36)amide to adopt multiple conformations might be of significance in view of current efforts to design small molecule allosteric modulators of the GLP-1R.[76–78] These modulators bind regions of the receptor distinct from GLP-1(7–36)amide, and they might cause GLP-1(7–36)amide to adopt a conformation that "biases" its signal transduction properties so that it activates select downstream pathways. Given that potentially dangerous islet cell hyperplasia is reported to occur in some T2DM patients treated with GLP-1R agonists,[79] it might be possible to design allosteric modulators of the GLP-1R that preferentially stimulate insulin secretion rather than islet growth.

5. CONTROL OF PANCREATIC β-CELL INSULIN SECRETION BY GLP-1

In studies of isolated islets, a stepwise increase of the glucose concentration from 2.8 to 16.7 mM leads to an initial first phase kinetic component of insulin secretion, followed by a delayed second phase, and the amplitudes of both phases of GSIS are potentiated by GLP-1.[80–83] These insulin secretagogue actions of GLP-1 are also measurable *in vivo* under conditions of a glucose clamp in which a GLP-1R agonist is infused intravenously while raising the blood glucose concentration in a stepwise manner.[56] In patients with prediabetes, there is a characteristic loss of first phase GSIS that can be restored quickly under conditions of administered GLP-1. As T2DM progresses, there is an additional loss of second phase GSIS, and it too can be restored under conditions of acute GLP-1 administration. Such findings indicate that in T2DM, GLP-1 has the capacity to quickly restore GSIS independently of any long-term action to increase islet insulin content. Presumably, such acute actions of GLP-1 reflect, at least in part, its physiological role as an incretin hormone in which it activates the GLP-1R located on islet β-cells.

When considering the acute insulin secretagogue action of GLP-1 *in vivo*, it is also thought that oral administration of glucose leads to activation of vagal–vagal reflexes that allow GLP-1 to control insulin exocytosis indirectly. Thus, GLP-1 released from L cells activates vagal sensory neurons that project to the brainstem in order to initiate efferent vagal reflexes via the parasympathetic branch of the autonomic nervous system. Parasympathetic neurons release acetylcholine (ACh) in the islets in order to activate muscarinic cholinergic receptors that stimulate Ca^{2+} mobilization in β-cells, and these neurons also release PACAP to stimulate cAMP production in β-cells. The net effect is an indirect and neurally mediated action of GLP-1 to potentiate GSIS.[84]

As illustrated in Fig. 2.1, the direct action of GLP-1 to activate the GLP-1R on β-cells leads to dual activation of the PKA and Epac2 branches of the cAMP signaling mechanism. In this manner, GLP-1 facilitates glucose-dependent closure of ATP-sensitive K^+ channels (K_{ATP}).[85] The net effect is β-cell depolarization with consequent activation of VDDCs in order to allow Ca^{2+} influx that stimulates Ca^{2+}-dependent insulin secretion. Simultaneously, GLP-1 enhances a mechanism of Ca^{2+}-induced Ca^{2+} release (CICR) in which Ca^{2+} influx triggers the release of Ca^{2+} from intracellular Ca^{2+} stores.[86–90] This mobilized Ca^{2+} then acts as an additional stimulus for Ca^{2+}-dependent insulin secretion. In patients with T2DM, β-cell glucose metabolism is dysfunctional so that glucose is not fully capable of closing K_{ATP} channels in order to stimulate Ca^{2+} influx.[84] Under these pathophysiological conditions, glucose alone fails to generate the critically important cytosolic Ca^{2+} signal that initiates insulin exocytosis. By facilitating glucose-dependent K_{ATP} channel closure and by enhancing CICR, GLP-1 restores the Ca^{2+} signal, thereby allowing GSIS to occur.

When considering how GLP-1 potentiates GSIS from β-cells of healthy individuals, a different scenario exists. Under these nonpathological conditions, coupling of glucose metabolism to K_{ATP} channel closure is not

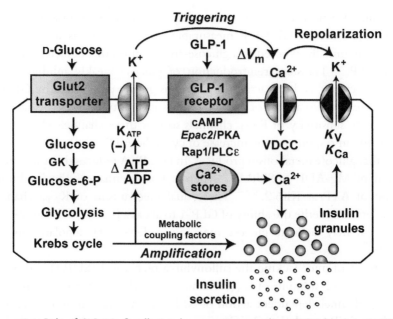

Figure 2.1 Role of GLP-1 in β-cell stimulus-secretion coupling. GLP-1 binds to its GPCR in order to stimulate cAMP production and to potentiate GSIS. One cAMP-dependent action of GLP-1 is mediated by PKA that phosphorylates secretory granule-associated proteins (e.g., Snapin) in order to facilitate Ca^{2+}-dependent exocytosis of insulin. The PKA-independent action of GLP-1 is mediated by the cAMP-regulated guanine nucleotide exchange factor Epac2. Binding of cAMP to Epac2 results in sequential activation of Rap1 GTPase and PLCε, thereby promoting PIP_2 hydrolysis and intracellular Ca^{2+} mobilization. GLP-1 also exerts PKA and Epac2-mediated actions to enhance glucose-dependent K_{ATP} channel closure, thereby promoting Ca^{2+} influx through VDCCs. The primary role of GLP-1 relevant to insulin secretion is to act as a β-cell glucose sensitizer in order to enhance insulin exocytosis mediated by the triggering and amplification pathways of GSIS. Abbreviations: GK, glucokinase; ΔV_m, depolarization; K_v, voltage-dependent K^+ channel, K_{Ca}, calcium-activated K^+ channel. (See color plate.)

disturbed, so that glucose is fully capable of generating the cytosolic Ca^{2+} signal that stimulates insulin exocytosis. Importantly, single cell studies demonstrate that this Ca^{2+} signal is a more efficient stimulus for insulin exocytosis under conditions in which β-cells are treated with GLP-1.[91] Such a facilitation of exocytosis by GLP-1 is explained by its PKA- and Epac2-mediated actions that occur at "late" steps of β-cell stimulus–secretion coupling and that promote Ca^{2+}-dependent fusion of secretory granules with the plasma membrane (Fig. 2.1).

Since a K_{ATP} channel-dependent action of GLP-1 is likely to explain insulin secretagogue properties of GLP-1 in patients with T2DM, it is useful to summarize what is known concerning this effect. Restoration of K_{ATP} channel closure by GLP-1 is measurable under conditions in which rat β-cells are initially exposed to a glucose-free solution that depletes intracellular ATP.[85] Transient reintroduction of glucose weakly inhibits K_{ATP} channel activity, and this action of glucose is greatly potentiated by GLP-1. Such a restorative action of GLP-1 reflects its ability to alter the adenine nucleotide sensitivity of K_{ATP} channels so that these channels close more efficiently in response to the increase of cytosolic ATP/ADP concentration ratio that glucose metabolism produces. In fact, PKA reduces the stimulatory action of Mg-ADP at SUR1,[92] whereas Epac2 enhances the inhibitory action of ATP at Kir6.2.[93,94] These dual mechanisms of K_{ATP} channel modulation underlie the ability of GLP-1 to act as a β-cell glucose sensitizer so that it may facilitate glucose metabolism-dependent depolarization of β-cells.[91]

Studies of mice lacking the sulfonylurea receptor-1 (SUR1) and pore-forming Kir6.2 subunits of K_{ATP} channels provide additional evidence for a K_{ATP} channel-dependent action of GLP-1 to stimulate insulin secretion. In these SUR1 and Kir6.2 KO mice, potentiation of GSIS by GLP-1 is absent[95,96] or reduced.[97] Furthermore, in mice harboring a tyrosine to stop codon (Y12STOP) mutation in the gene coding for Kir6.2, K_{ATP} channel expression and GLP-1-stimulated insulin secretion are absent.[98] Important findings are also provided by a study of patients with neonatal diabetes mellitus (NDM) owing to gain-of-function mutations (C435R; R1380H) in the gene coding for SUR1.[99] These mutations lead to overactive K_{ATP} channels and a consequent reduction of GSIS. Remarkably, administration of a GLP-1R agonist restores insulin secretion in these patients.

6. ALTERED GLP-1 ACTION IN A RODENT MODEL OF INSULIN RESISTANCE

It is interesting to note that expression of Epac2 within β-cells is of critical importance to β-cell compensation that occurs in mice fed with a high-fat diet (HFD).[54] The HFD induces insulin resistance, and under these conditions, GSIS is enhanced in order to compensate for insulin resistance.[100] When comparing wild-type (WT) and Epac2 KO mice fed with the HFD, it is possible to demonstrate that compensatory GSIS is lost in Epac2 KO mice.[54] This unexpected role of Epac2 to *enable* GSIS is measurable in isolated islets and it does not require treatment of islets with GLP-1R agonists. Furthermore, this compensation under conditions of the HFD results from alterations of β-cell Ca^{2+} handling such that there is enhanced glucose-dependent Ca^{2+} influx and Ca^{2+} mobilization.[54] Since Epac2 mediates stimulatory effects of GLP-1 on Ca^{2+} influx and mobilization,[49,91,101–103] it appears that "plasticity" exists in the β-cell cAMP signaling network such that the HFD leads to an unexpected coupling of glucose metabolism to Epac2 activation and insulin secretion.[54,104]

Since GLP-1R expression is elevated in islets of mice fed with the HFD,[105] it is evident that GLP-1 also participates in the functional adaptation of islets to diet-induced insulin resistance. In this regard, it is interesting to note that under conditions of the HFD, *in vivo* administration of a GLP-1R agonist leads to an additional compensatory increase of GSIS that is "durable" in that it is measurable in isolated islets in the complete absence of *in vitro* GLP-1R stimulation.[106] This finding indicates that a GLP-1R agonist has the capacity to upregulate the expression and functionality of key components of the β-cell stimulus-secretion coupling mechanism, most likely including glucose-sensing, oxidative glucose metabolism, ion channel regulation, and Ca^{2+}-dependent exocytosis.

When considering how the HFD also induces islet hyperplasia with a compensatory increase of β-cell mass, it could be that increased GLP-1R expression on β-cells plays a role. Thus, increased GLP-1R expression might favor increased β-cell sensitivity to circulating GLP-1, thereby allowing GLP-1 to efficiently signal through cAMP, PKA, and Epac2 to upregulate β-cell proliferation and survival. Although this is an attractive hypothesis, recent studies indicate that in mice fed with a normal diet, enhanced PKA activity per se does not increase β-cell mass.[50,107] Furthermore, a KO of Epac2 expression does not lead to decrease of β-cell mass in mice fed with a normal diet.[54] However, it could be that in mice fed with an HFD, a role for PKA and Epac2 in the induction of islet hyperplasia might be revealed.

Since cAMP-independent actions of GLP-1 exist, such actions might also play a role in promoting adaptive responses of β-cells under conditions of the HFD. Thus, it is of interest to summarize what is known concerning cAMP-independent actions of GLP-1 that allow it to act as a β-cell trophic factor. Studies performed primarily with β-cell lines or mouse β-cells demonstrate cAMP-independent actions of GLP-1R agonists to counteract endoplasmic reticulum stress[108] and to signal via the GLP-1R through β-arrestin[109,110] and epidermal growth factor (EGF) receptor transactivation[111] in order to downregulate the activities of proapoptotic protein

BAD,[112] the SirT1 deacetylase,[113] and transcription factor FoxO1.[114] GLP-1 also upregulates the activities of c-Src kinase,[115] phosphatidylinositol 3-kinase (PI-3-kinase),[116] protein kinase B (PKB),[117] protein kinase c-zeta (PKC-ζ),[118] and extracellular signal-regulated protein kinases (ERK1/2).[119] Conceivably, these growth factor-like signaling pathways might be selectively activated by allosteric GLP-1R agonists that have "biased" signal transduction properties and that bind the GLP-1R in order to promote β-cell compensation under conditions of the HFD.

7. GLUCOREGULATORY PROPERTIES OF GLP-1 MEDIATED BY THE NERVOUS SYSTEM

A high-profile area of research concerns GLP-1 action in the nervous system, and it is recently appreciated that such actions of GLP-1 are of importance to glucoregulation. Vagal–vagal reflexes are activated by injection of a GLP-1R agonist into the hepatic portal vein, and they are measurable as increased electrical activity in vagal sensory afferent neurons and also in vagal efferent neurons.[120] By measuring insulin secretion induced by intraportal administration of glucose, it is also possible to demonstrate that GSIS is potentiated by coadministration of glucose with GLP-1.[121] This action of GLP-1 to potentiate GSIS is reduced by a ganglionic blocker,[121] as expected if vagal–vagal reflexes activate neurons within pancreatic ganglia in order to stimulate insulin secretion. Vagal sensory neurons express GLP-1 receptors,[122–124] and direct application of GLP-1 to vagal afferent neuron cell bodies within the nodose ganglion leads to action potential generation.[125]

Vagal sensory neurons activated by GLP-1 project to the brainstem and hypothalamus, and under conditions in which the GLP-1R agonist exendin-4 is administered intraperitoneally, a surgical subdiaphragmatic vagotomy blunts activation of neurons located within the hypothalamic and paraventricular nuclei.[126] Such findings obtained with rats indicate that peripherally administered exendin-4 acts via the vagus nerve to stimulate neural activity within the brain and that this effect of exendin-4 complements its more direct action to cross the blood–brain barrier in order to activate CNS GLP-1 receptors.[127–129] A vagus nerve-mediated action of GLP-1R agonists also occurs in humans since in vagotomized patients treated for pyloroplasty, there is a reduced ability of intravenously infused GLP-1 to suppress appetite, to slow gastric emptying, to stimulate insulin secretion, and to suppress glucagon secretion.[130]

GLP-1 receptors are widely expressed within the brain where they are activated by neuronally released GLP-1. Thus, GLP-1 is a neuropeptide, and neuroanatomical studies demonstrate that it is contained within neuronal cell bodies located in the medullary caudal nucleus tractus solitarius (NTS), the raphe obscurus, and the intermediate reticular nucleus.[131,132] Axons of these neurons project to regions of the brain that are involved in the control of appetite, metabolism, water intake, stress, and cardiovascular functions.[133–136] These regions include the dorsal vagal nucleus, dorsomedial and paraventricular hypothalamic nuclei, ventrolateral periaqueductal gray, and thalamic paraventricular nucleus.[131,132] GLP-1-containing neurons project to the brainstem where they synapse on cholinergic vagal motor neurons, some of which project to the pancreas.[137,138] Collectively, these findings indicate three mechanisms by which GLP-1 controls vagal efferent activity: (1) vagal–vagal reflexes in which GLP-1 initially activates the GLP-1R located on vagal sensory nerve terminals, (2) an action of circulating GLP-1 that requires its action at, or transit across, the blood–brain barrier in order to activate brainstem neural circuits, and (3) direct or indirect synaptic relays in which GLP-1 released within the brain activates vagal motor neurons.

When considering the physiological significance of such neural influences of GLP-1, it is important to note that neural control of insulin secretion is not an absolute requirement in order to measure an insulin secretagogue action of a GLP-1R agonist. This fact is demonstrated in studies of glucoregulation using Pdx1-hGLP1R:Glp1r −/− mice administered with the DPP-IV-resistant GLP-1R agonist exendin-4.[139] These engineered mice do not express the mouse GLP-1R in any tissue, whereas they express recombinant human GLP-1 receptors only in the pancreas. In such mice, exendin-4 exerts its normal action to potentiate GSIS and to improve glucose tolerance in the absence of neural GLP-1R activation.[139]

Despite the fact that GLP-1R agonist action is preserved in Pdx1-hGLP1R:Glp1r −/− mice, there is reason to believe that the nervous system does in fact mediate important glucoregulatory actions of GLP-1.[140–145] For example, under conditions in which pancreatic insulin secretion is induced by intragastric infusion of mice with glucose, an intracerebroventricular (i.c.v.) injection of GLP-1R antagonist exendin(9–39) results in less insulin being secreted.[142,146] Furthermore, glucose uptake and glycogen synthesis within skeletal muscle are enhanced by intragastric infusion of glucose, and this effect of glucose is blocked by exendin(9–39) delivered by the i.c.v. route.[142,146] Such findings indicate that during the initial prandial state of intestinal glucose absorption, GLP-1 "primes" whole-body metabolism by acting within the brain to facilitate pancreatic insulin secretion while also enhancing skeletal muscle glucose disposal.

Interestingly, the neurally mediated action of GLP-1 that is important to glucoregulation may be different under conditions that mimic the postprandial state when levels of blood glucose are rising. In studies of mice using infusion clamp techniques that elevate levels of blood glucose and insulin, it is reported that i.c.v. administration of GLP-1R agonist exendin-4 reduces blood flow and glucose uptake within skeletal muscle.[142,146] Simultaneously, insulin secretion is stimulated in order to enhance insulin-dependent hepatic glucose uptake.[142,146] Thus, in contrast to the initial prandial state of glucose absorption described earlier, GLP-1 acts in the postprandial state to shift glucose disposal from muscle to liver. Resultant hepatic glycogen synthesis allows for sufficient glycogen mobilization and hepatic glucose production during the subsequent fasting state. What remains to be demonstrated is that such neurally mediated effects of GLP-1 occur in healthy humans and/or patients with T2DM.

It is also recognized that GLP-1 receptors located on neurons within the arcuate nucleus (Arc) are activated in order for GLP-1 to stimulate insulin secretion while also suppressing hepatic glucose production.[147–149] One mechanism that may explain how GLP-1 alters neural function in the Arc is provided by the finding that transmission in these neural circuits is modulated by neuropeptides, nutrients, and hormones that control K_{ATP} channels.[149–151] In this regard, GLP-1 may inhibit K_{ATP} channel activity within the Arc in order to regulate blood glucose homeostasis. It will be interesting to assess whether this action of GLP-1 is selective for glucose-responsive neurons in the Arc and whether the inhibition of K_{ATP} channel activity results from Epac2 activation, as described for β-cells. Furthermore, since leptin activates K_{ATP} channels in the Arc,[150,151] it could be that GLP-1 and leptin act as counterregulatory hormones to control Arc circuits important to glucoregulation.

Finally, it is interesting to note that GLP-1R agonists are under evaluation for use in the treatment of neurological disorders.[152] In an *in vitro* model of Alzheimer's disease, GLP-1 protects hippocampal neurons from cytotoxicity induced by amyloid-beta peptide.[153] Also surprising is the report that such neuroprotection is conferred by GLP-1(9–36)amide, which is the metabolite generated by DPP-IV-catalyzed degradation of GLP-1 (7–36)amide.[154] This finding suggests the existence of a nonconventional GLP-1R, although its identity remains unknown. Just as interesting, there is a potential usefulness of GLP-1R agonists to treat Parkinson's disease.[155,156] Collectively, the available evidence suggests that these neuroprotective actions of GLP-1 might be secondary to its ability to alter glucose homeostasis in the brain. For example, under conditions of hyperglycemia, peripherally administered GLP-1 increases the phosphorylation velocity (V_{max}) of neuronal hexokinase while also increasing blood–brain glucose transport capacity (T_{max}).[157,158]

8. GLP-1 RECEPTOR AGONISTS

Drug development strategies have led to the identification of GLP-1R agonists that are either peptide-based or small molecule-based. For peptide-based GLP-1R agonists, a further subdivision exists in order to classify "incretin mimetics" or "GLP-1 analogs," as summarized in Table 2.1. Exenatide, also known as Byetta, is the prototypical incretin mimetic and it is the synthetic form of exendin-4 (Ex-4). In contrast, the prototypical GLP-1 analog is liraglutide, also known as Victoza. Liraglutide is structurally equivalent to GLP-1(7–37) except that lysine residue 26 is acylated by its conjugation to a hexadecanoyl (C16) side chain, whereas residue 34 contains arginine rather than the lysine residue found within native GLP-1. Exenatide and liraglutide are both high-affinity agonists at the GLP-1R, yet they are relatively resistant to hydrolysis by DPP-IV. For example, after intravenous administration, the half-life of circulating GLP-1 is only 1.5–5.0 min, whereas the half-lives of exenatide and liraglutide are 26 min and 8 h, respectively. Thus, exenatide and liraglutide exert prolonged blood glucose-lowering actions when they are administered by subcutaneous injection to patients with T2DM.

Mechanistically, the hexadecanoyl side chain of liraglutide allows this peptide to bind to plasma albumin via hydrophobic interactions, thereby minimizing hydrolysis by DPP-IV. In the GLP-1 analog albiglutide, two molecules of GLP-1(7–36) are fused in tandem, and the tandem is covalently conjugated to recombinant human albumin in order to achieve DPP-IV resistance. Simultaneously, a glycine substitution is introduced at residue 8 in order to improve DPP-IV resistance. In dulaglutide, a different approach is taken in which GLP-1(7–36) is fused to human immunoglobulin heavy constant γ4 chain (IgGγ4-Fc) to create a monomer that then dimerizes with itself in order to generate the DPP-IV-resistant GLP-1R agonist.

Table 2.1 Pharmacological properties of GLP-1R agonists currently in use or under study for the treatment of T2DM

GLP-1R agonist	Parental peptide	Modifications	Half-life	Route of administration	HbA1c reduction	Weight reduction	Refs.
Exenatide BID	Ex-4	None	2.4 h	SC 5 or 10 µg BID	0.7–0.9%	2.8–3.1 kg	159
Lixisenatide	Ex-4	Proresidue deleted from C-terminus, six Lys residues added to C-terminus	3 h	SC 20 µg QD	0.8–0.9%	1.8–3.0 kg	160
Exenatide LAR	Ex-4	Injectable microspheres of biopolymer with entrapped exenatide	5–6 d	SC 2 mg QW	1.3–1.9%	3.6 kg	161,162
Liraglutide	GLP-1	Palmitic acid conjugated to Lys-26, Lys-34/Arg substitution	11–13 h	SC 1.2 mg QD	1.1–1.8%	2.0–3.0 kg	163,164
Semaglutide	GLP-1	Palmitic acid conjugated to Lys-26, Gly-8/aminoisobutyric acid, and Lys-34/Arg substitutions	6–7 d	SC 0.1–1.6 mg QW	1.7%	4.8 kg	165
Albiglutide	GLP-1	Two molecules of GLP-1 fused as a tandem and conjugated to albumin; Ala-2/Gly substitution	6–8 d	SC 50 mg QW	0.8%	0.6 kg	166–168
CJC-1134-PC	Ex-4	Peptide coupled to albumin by a linker	8 d	SC 2 mg QW	1.4%		169
Dulaglutide	GLP-1	Two molecules of GLP-1 covalently linked to a IgG4-Fc heavy chain; Ala-8/Gly,Gly-26/Glu, Arg-36/Gly substitutions	4 d	SC 1.5 mg QW	1.5%	NS	170,171
Langlenatide	Ex-4	Peptide fused to Fc region	6 d	SC 1–4 mg QW SC 8–16 mg QMT	NDA	NDA	172
VRS-859	Ex-4	Peptide fused to Xten protein	3 d	SC 200 mg QMT	NDA	NDA	173

Ex-4, exendin-4; BID, twice daily dosing; QD, once daily dosing; QW, once weekly dosing; QMT, once monthly dosing; SC, subcutaneous administration; HbA1c, hemoglobin A1c.

Attempts to identify small molecule GLP-1R agonists are complicated by the complex ligand–receptor interactions that are characteristic of group B GPCRs.[46,174] Despite this complication, new "ago-allosteric" modulators of the GLP-1R are described. These small molecules not only act as GLP-1R agonists (ago control) but also modify the ability of GLP-1 itself to activate the GLP-1R (allosteric control). Synthetic ago-allosteric modulators currently under preclinical investigation include substituted quinoxaline[76,175–177] and cyclobutane derivatives.[178–180] A substituted quinoxaline designated as compound 2 acts as a partial agonist at the GLP-1R, but it is particularly revealing that the efficacy of compound 2 as a cAMP-elevating agent is enhanced rather than reduced by GLP-1R antagonist exendin(9–39). This finding is consistent with the concept that allostery results from binding of compound 2 to a site on the GLP-1R that is not recognized by peptide-based agonists and antagonists.[181] Expanding on these findings, it is reported that novel substituted pyrimidines also act as GLP-1R agonists and that they do not compete with radiolabeled GLP-1 for binding to the GLP-1R.[182]

GLP-1R activation by the quinoxaline compound 3 is strongly influenced by mutations introduced into transmembrane α-helices 2 and 7, whereas such mutations do not alter the action of GLP-1.[183] These findings indicate that small molecule agonists activate or modulate the GLP-1R in a manner that is distinct from that of GLP-1. In fact, a quinoxaline (compound 2) and a pyrimidine (compound B) act in an additive manner to activate the GLP-1R under conditions in which the receptor is truncated to remove the N-terminal extracellular domain at which the C-terminus of GLP-1 binds.[184] Just as intriguing, GLP-1R-mediated signaling properties of ago-allosteric modulators are not identical, thereby suggesting that such agonists could be used in order to achieve signal transduction bias.[77,184–187]

9. DPP-IV INHIBITORS

DPP-IV encoded by the *DPP4* gene is a member of the prolyl oligopeptidase family of serine proteases, and it plays a role in the control of immune function and is a key determinant of incretin hormone action. DPP-IV exists as a soluble circulating form[188] or as a type II transmembrane serine exopeptidase.[189] Both forms of the enzyme catalyze the cleavage of dipeptides from the N-terminus of peptide substrates that contain on average 30-amino-acid residues and that have a proline or alanine residue in the penultimate position.[189] These substrates include chemokines (CCL5), neuropeptides (PYY and NPY), and hormones (GLP-1 and GIP).[190] Terminology exists in which DPP-IV is also known as adenosine deaminase complexing protein 2 (ADCP 2) or as the T-cell activation antigen CD26. DPP-IV is highly expressed on endothelial cells, differentiated epithelial cells, and lymphocytes. In the immune system, DPP-IV exists as an integral membrane glycoprotein in which it acts as a cofactor to control intracellular signaling pathways that are of importance to T-cell proliferation and T-cell activation.[191]

Crystallographic analysis combined with molecular modeling reveals that the 766-amino-acid residue DPP-IV contains an N-terminal β-propeller domain and a C-terminal α/β hydrolase domain that together form a cavity in which the enzyme's active site is located.[192,193] A distinguishing feature of DPP-IV is that the enzyme's α/β hydrolase domain contains a serine–aspartate–histidine catalytic triad, whereas the β-propeller domain contains two glutamate residues that are necessary for catalytic function and that align the substrate peptide so that only the penultimate proline or alanine residues may engage the active site. This structural feature of DPP-IV explains its substrate specificity in which it hydrolyzes peptides with N-terminal X-proline or X-alanine residues.[194]

Summarized in Table 2.2 are the pharmacological properties of small molecule DPP-IV inhibitors now in use for the treatment of T2DM. The xanthine class of DPP-IV inhibitors includes sitagliptin, linagliptin, and alogliptin, whereas vildagliptin and saxagliptin are members of the cyanopyrrolidine class of DPP-IV inhibitors. Inhibition of DPP-IV activity by sitagliptin is achieved by its noncovalent binding to the conserved glutamate residues 205 and 206 located within the enzyme's β-propeller, whereas saxagliptin binds not only to these glutamate residues but also to the serine residue located within the catalytic triad of the α/β hydroxylase domain.[190] In general, cyanopyrrolidines such as saxagliptin are competitive inhibitors rather than noncompetitive inhibitors of DPP-IV enzymatic activity since they form reversible covalent bonds with serine residue 630 located within the enzyme's active site.[190]

DPP-IV catalyzes the hydrolysis of GLP-1(7–36)amide to generate GLP-1(9–36)amide and the N-terminal histidine–alanine dipeptide. Therefore, DPP-IV inhibitors raise levels of endogenous GLP-1(7–36)amide, and it could be that at clinically relevant doses, this action of DPP-IV inhibitors produces a relatively selective increase of GLP-1(7–36)amide in the hepato-portal circulation or at the interface of L cells and vagal sensory nerve terminals. Since DPP-IV inhibitors suppress enzymatic production

Table 2.2 Pharmacological properties of DPP-IV inhibitors currently in use or under study for the treatment of T2DM

Inhibitor	Half-life	Route of administration	Plasma DPP-IV inhibition	Plasma GLP-1 increase	HbA1c reduction	Route of elimination	Refs.
Sitagliptin	11–13 h	PO 25–200 mg QD	80% with 50 mg	2 times	0.6–0.8%	Mostly renal	195–197
Vildagliptin	1.7–3 h	PO 25–200 mg QD or BID	80–90%	2–3 times	0.5–1.5%	Mostly renal	197–199
Saxagliptin	2.2–3.8 h 3.0–7.4 h	PO 2.5–50 mg QD	70%	1.5–2 times	0.5–0.9%	Mostly renal	23,197,200,201
Linagliptin	113–260 h	PO 0.5–10 mg QD	46% with 0.5 mg 78% with 2.5 mg 90% with 10 mg	2 times (0.5 mg) 3 times (2.5 mg) 4 times (10 mg)	0.4–0.8%	Hepatic (biliary excretion)	197,202,203
Alogliptin	12–21 h	PO 25–800 mg QD	74–97%	2–4 times	0.5–0.9%	Mostly renal	197,204,205

of GLP-1(9–36)amide, while also preventing release of the histidine–alanine dipeptide, it is of concern that these two metabolites might have important biological actions that would be missing in patients administered with DPP-IV inhibitors. In fact, GLP-1(9–36)amide exerts prosurvival actions in neurons and cardiomyocytes[154,206] while also suppressing hepatic glucose production in obese patients.[207,208] Furthermore, the histidine–alanine dipeptide is reported to influence glucose tolerance and insulin secretion in mice.[209]

10. GLP-1-BASED STRATEGIES FOR THE TREATMENT OF T2DM

A GLP-1-based strategy for the treatment of T2DM is indicated in view of the fact that GLP-1R agonists and DPP-IV inhibitors exert a beneficial constellation of physiological effects that include (1) glucose-dependent stimulation of insulin secretion, (2) suppression of glucagon secretion, (3) normalization of blood glucose without an attendant risk of hypoglycemia, (4) slowing of gastric emptying, (5) appetite suppression, and (6) weight loss. Potential additional benefits are actions to promote β-cell survival by slowing apoptosis or to promote β-cell regeneration by stimulating β-cell proliferation. Thus, it was originally anticipated that such a GLP-1-based therapy might lead to a long-term remission and possibly a cure for T2DM.[210] Since the notion of a GLP-1-based therapy has led to the term "incretin therapy," it is important to note that when considering the use of incretins for the treatment of T2DM, only GLP-1 is effective, whereas GIP is ineffective.[8,211]

Presently available GLP-1R agonists include exenatide (approved in the United States in 2005) and liraglutide (approved in the United States in 2010), both of which are administered by subcutaneous injection. Exenatide is approved for use twice a day, and liraglutide is approved for use once a day. An extended release (ER) formulation of exenatide is intended for use once a week. Additional long-acting formulations are under investigation including one depot preparation of exenatide that can be given once every 6 months. In contrast to exenatide and liraglutide, the DPP-IV inhibitors are orally administrable and are therefore a more convenient means by which to treat T2DM. Currently, in the United States, there are four approved drugs of this class. Sitagliptin was first approved in 2006, and since then, three additional DPP-IV inhibitors have been approved. They are saxagliptin, linagliptin, and alogliptin. In addition, vildagliptin and gemigliptin are available in other countries.

When considering the use of GLP-1R agonists or DPP-IV inhibitors for the treatment of T2DM, it is important to note that DPP-IV inhibitors raise levels of circulating GLP-1 by approximately twofold, whereas GLP-1R agonists exert a dose-dependent pharmacological effect that is considerably more potent since their circulating levels easily exceed endogenous GLP-1 levels by eightfold.[212] These pharmacological differences may explain why GLP-1R agonists are more effective inhibitors of gastric emptying, while also suppressing appetite and promoting weight loss. In fact, in some patients, the high potency of GLP-1R agonists can lead to adverse side effects of nausea and vomiting.[213]

GLP-1R agonists and DPP-IV inhibitors are approved for use in patients with T2DM, typically as adjuncts to diet and exercise and as either a monotherapy or a combination therapy with other antidiabetic medications.[214–217] Exenatide ER and liraglutide are not recommended as first-line therapies although they may be considered for monotherapy in patients who are unable to use other first-line therapies because of a lack of efficacy or due to contraindications such as allergic hypersensitivity, end-stage renal disease, and gastrointestinal diseases. DPP-IV inhibitors are also contraindicated in patients with hypersensitivity reactions such as urticaria, angioedema, or bronchial hypersensitivity. In addition to a history of serious hypersensitivity as a contraindication, exenatide ER and liraglutide are also contraindicated in patients with a personal or family history of medullary thyroid cancer or with a history of multiple endocrine neoplasia syndrome type 2 (MEN2). Prescribing information, warning labels, and precaution sections for exenatide and liraglutide or various DPP-IV inhibitors also list pancreatitis as a potential adverse side effect of their use.[214–217]

A review of the clinical DPP-IV literature concerning monotherapy for the treatment of T2DM indicates 25 randomized control trials (RCTs) in adult patients with trial durations of at least 12 weeks.[218] Sitagliptin and vildagliptin therapy results in an HbA1c reduction of 0.7% and 0.6%, respectively. In another review that includes 17 RCTs of 8 weeks minimum duration, monotherapy with GLP-1R agonists results in reductions of HbA1c of ca. 1%.[219] Although β-cell function improves with GLP-1R agonist treatment, it is interesting to note that a rapid deterioration of glucoregulation can occur after withdrawal of these medications. Thus, unlike the situation reported for mice administered with a GLP-1R agonist,[106] a "durable" effect of GLP-1R agonists is not so obvious in humans. This finding seems to argue that in humans, the primary effect of GLP-1R agonists is to exert an acute stimulatory effect on β-cell insulin secretion, rather than acting long term to alter β-cell gene expression.

GLP-1R agonists and DPP-IV inhibitors are also under study for use in combination with non–GLP-1-based medications such as insulin, sulfonylureas, the biguanide metformin, and the thiazolidinedione pioglitazone.[220–225] Especially noteworthy is the 2011 approval in the United States of exenatide as an add-on therapy to basal insulin analog glargine for patients with T2DM who are not achieving adequate glycemic control using glargine alone. Although GLP-1R agonists and DPP-IV inhibitors can be used in combination with sulfonylureas, there is an increased risk of hypoglycemia so that caution should be exercised and pre-emptive dose reduction should be implemented. In this regard, an attractive alternative therapy is based on the use of metformin in combination with a GLP-1R agonist or DPP-IV inhibitor. This combination therapy has a reduced risk hypoglycemia, yet it still promotes beneficial weight loss in patients with T2DM.

In the DURATION clinical trial series,[161,226–229] T2DM patients are reported to lose an average 2–4 kg body weight when treated with exenatide ER (2 mg per weekly as a single injection). The weight loss achieved with exenatide ER is similar to that observed in patients administered with non-ER exenatide twice daily (5–10 mcg per single injection). Importantly, body weight reduction is significantly larger for patients administered with exenatide ER in comparison to administered sitagliptin (-2.3 vs. -0.8 kg, respectively).[226] The LEAD (Liraglutide Effect and Action in Diabetes) trial also reveals significant body weight reduction with liraglutide mon-otherapy.[230] This weight loss is primarily due to reduced fat mass, mainly visceral adipose tissue.[231]

As summarized in Table 2.3, a GLP-1-based therapy for the treatment of T2DM is particularly attractive since it not only normalizes glycemia while reducing body weight but also improves cardiovascular function.[241,242] GLP-1R agonist treatment has positive effects on cardiovascular risk factors such as diabetes, hypertension, hyperlipidemia, and obesity. A pooled data analysis from six clinical trials investigating the outcomes of 6-month exenatide treatment in 2171 T2DM patients reveals significantly greater reductions in systolic blood pressure compared with placebo.[243] Mechanis-tically, such a reduction of blood pressure is consistent with the report that liraglutide exerts an action in the mouse atrial myocardium to stimulate the release of atrial natriuretic factor (ANF) that then acts to relax vascular smooth muscle while also promoting renal excretion of sodium ion.[244] GLP-1R agonist therapy also results in favorable changes in circulating lipids, which are another important cardiovascular risk factor. Meta-analysis demonstrates that liraglutide lowers blood levels of total cholesterol, low-density lipoproteins, free fatty acids, and triglycerides.[245]

Table 2.3 Cardiovascular actions of GLP-1-based therapeutics

GLP-1R agonist	Experimental/ clinical setting	Effect of GLP-1R agonist	Refs.
Exenatide	TG9 mice (murine DCM model)	Improvement of glucose tolerance; increase 2-deoxyglucose uptake and GLUT4 expression in myocardium	232
GLP-1	Dogs with pacing-induced DCM	Increase insulin sensitivity, basal and insulin-stimulated glucose extraction, and uptake in myocardium, and decrease plasma glucagon	233
GLP-1 GLP-1 (9–36)	Dogs with pacing-induced DCM	Both peptides increase insulin sensitivity and basal and insulin-stimulated glucose uptake in myocardium, and decrease plasma glucagon	234
Exenatide	Diabetic (STZ-induced) rats	Increase myocardial glucose uptake	235
Vildagliptin	Model of murine heart failure	Increase plasma GLP-1, improvement of glucose tolerance	236
Liraglutide	Mice on HFD	Decrease insulin resistance	237
GLP-1	Patients before, during, and after CABG	Decrease pre- and perioperative plasma glucose, decrease postoperative plasma glucagon, decrease postoperative insulin infusion required, decrease pharmacological or mechanical support to achieve hemodynamic stability in postoperative period	238
GLP-1	T2DM patients after CABG	Decrease postoperative insulin infusion required Decrease dobutamine infusion required	239
GLP-1	T1D patients	Decrease hyperglycemia- or hypoglycemia-induced oxidative stress, inflammation, and endothelial dysfunction	240

DCM, dilated cardiomyopathy; STZ, streptozotocin; HFD, high-fat diet; CABG, coronary artery bypass grafting.

11. IMPROVED GLUCOREGULATION AFTER BARIATRIC SURGERY

There is evidence that intestinally released GLP-1 might mediate the beneficial outcomes of bariatric surgery in which a Roux-en-Y gastric bypass (RYGB) leads to weight loss that is accompanied by elevated levels of plasma GLP-1 and improved glucoregulation in patients with T2DM.[246–252] This is a clinically important issue to address since RYGB surgery results in weight loss in 30–40% of obese patients, whereas improved glucoregulation is observed in about 80% of T2DM patients.[253–255] Unfortunately, our understanding how these beneficial outcomes of RYGB surgery are achieved is complicated by weaknesses in the experimental designs of published studies.[256] In fact, the role of elevated plasma GLP-1 as a determining factor in the remission of T2DM is disputed,[257] and it is instead reported that β-cells are rendered more sensitive to circulating GLP-1 after RYGB surgery.[258] Evidently, RYGB induces compensatory changes in β-cells that lead to improved blood glucose control.[258]

Additional clinical observations reveal that levels of blood glucose are quickly normalized after RYGB surgery, even before significant weight loss is achieved.[259] Though decreased caloric intake or reduced intestinal nutrient absorption is an obvious cause for the weight loss, there appear to be additional important factors that explain a remission of T2DM after RYGB surgery. This conclusion is supported by the following observations: (1) Remission occurs in the immediate postoperative period before any weight loss occurs, (2) remission is more pronounced after RYGB surgery as compared with outcomes achieved by dieting in order to achieve comparable weight loss, and (3) remission is more pronounced after RYGB surgery as compared with other forms of bariatric surgery (sleeve gastrectomy and gastric banding).

Differences in the outcomes achieved following RYGB or dieting are clearly evident since RYGB, but not dieting, leads to enhanced postprandial release of GLP-1, thereby restoring the incretin effect in patients with T2DM.[248] Under postoperative conditions of RYGB in T2DM, there is also a restoration of the missing first phase kinetic component of GSIS, and there is an accompanying improvement of oral glucose tolerance.[248] Unfortunately, the physiological basis for immediate or long-term endocrine and metabolic changes after RYGB is not fully elucidated. Changes in the rate of eating, gastric emptying, nutrient absorption and sensing, incretin hormone release, bile acid metabolism, and intestinal microbiota composition may all be important.[260–263]

Increased intestinal GLP-1 secretion after gastric bypass surgery appears to be sustained and can potentially have beneficial effects in terms of weight loss and long-term remission of T2DM. In addition to this surgery's acute stimulatory effect on insulin secretion, it could be that the sustained elevation of blood GLP-1 might regenerate β-cells. However, a potential drawback to surgery is that it is not yet clear whether a postoperative remission of T2DM is permanent or only temporary.[256,264–267] Furthermore, this gastric bypass surgery can lead to hyperinsulinemic hypoglycemia, thereby necessitating pancreatectomy.[268,269] Interestingly, the hyperinsulinemia in some patients who have had gastric bypass surgery does not appear to be secondary to an increase of β-cell mass, as might be expected if bypass surgery upregulates long-term actions of GLP-1 to stimulate β-cell proliferation. In one study of gastric bypass patients undergoing partial pancreatectomy to correct for hyperinsulinemia, histological analyses of pancreatic sections reveal no change in β-cell mass, proliferation, neogenesis, or apoptosis.[269] Thus, the nature of the adaptive change that underlies remission of T2DM after gastric bypass surgery remains to be determined.

12. SAFETY CONSIDERATIONS FOR GLP-1-BASED THERAPEUTICS

An ongoing controversy concerns whether the use of GLP-1-based therapeutics predisposes to unexpected side effects including inflammation of the pancreas (pancreatitis) or even pancreatic cancer in patients with T2DM.[270–274] Furthermore, postmortem histological analyses of pancreatic tissue from patients treated with a GLP-1R agonist or DPP-IV inhibitors provide evidence for an increased incidence of pancreatic exocrine cell dysplasia accompanied by hyperplasia of glucagon-secreting α-cells of the endocrine pancreas.[79] These findings have raised the specter that chronic GLP-1R activation in humans might lead to the appearance of exocrine cell adenocarcinomas or neuroendocrine tumors such as glucagonomas or insulinomas. Additional studies of rodents indicate that chronic stimulation of GLP-1 receptors on calcitonin-secreting C cells of the thyroid can lead to C-cell hyperplasia with eventual medullary thyroid cancer,[275] although this outcome is not measurable in nonhuman primates[276] and has yet to be demonstrated for humans. Countering these findings, it is argued that the benefits of GLP-1-based therapeutics outweigh their risks when considering their usefulness for the treatment of T2DM.[277,278] Currently, these safety concerns remain debated, and it is pointed out that in the published literature, there is no direct demonstration of causality linking GLP-1-based therapeutics to human pancreatitis, pancreatic cancer, or thyroid cancer. Despite this fact, the Food and Drug Administration acted in 2007 to issue a safety alert concerning the potential for pancreatitis in patients treated with exenatide. Furthermore, a black box warning is now provided with prescription information for both exenatide and liraglutide. In 2013, both the American Diabetes Association and the Endocrine Society called for independent review of findings relating to these potential adverse side effects of GLP-1R agonists and DPP-IV inhibitors.

13. CONCLUSION

Nearly 30 years of basic science and clinical research has culminated with the recognition that GLP-1-based therapies for the treatment of T2DM are highly effective. Unanticipated are the surprising beneficial cardiovascular and neuroprotective actions of this class of blood glucose-lowering agents. Since GLP-1R agonists also produce substantial weight loss in obese patients, it is clear that pharmacological GLP-1R activation can be particularly useful for treating or reversing the increasingly common metabolic syndrome of hyperglycemia, impaired cardiovascular function, excess weight, and neuropathology. Although safety concerns are increasingly debated, the general consensus at the present time is that additional clinical research is necessary in order to establish whether the use of GLP-1R agonists or DPP-IV inhibitors predisposes to pancreatitis or cancer. Looking to the future, it is anticipated that a new approach to drug development will be popularized in order to identify GLP-1R agonists that have a reduced propensity to promote cell growth while retaining their capacity to stimulate pancreatic insulin secretion. Particularly useful will be new approaches that allow oral delivery of GLP-1R agonists, either as synthetic small molecule compounds or as novel peptide conjugates.[279]

ACKNOWLEDGMENTS

G. G. H. and O. G. C. acknowledge the support of a Basic Science Award (7-12-BS-077) from the American Diabetes Association. All authors also acknowledge the institutional support of SUNY Upstate Medical University. All authors declare no conflict of interest concerning any of the concepts addressed in this review of the literature.

REFERENCES

1. Kieffer TJ, Habener JF. The glucagon-like peptides. *Endocr Rev.* 1999;20:876–913.
2. Holst JJ. The physiology of glucagon-like peptide-1. *Physiol Rev.* 2007;87:1409–1439.
3. Drucker DJ. The biology of incretin hormones. *Cell Metab.* 2006;3:153–165.
4. Burcelin R. The gut-brain axis: a major glucoregulatory player. *Diabetes Metab.* 2010;36(Suppl 3):S54–S58.
5. Drucker DJ, Nauck MA. The incretin system: glucagon-like peptide-1 agonists and dipeptidyl peptidase-4 inhibitors in type 2 diabetes. *Lancet.* 2006;368:1696–1705.

6. Nathan DM, Schreiber E, Fogel H, Mojsov S, Habener JF. Insulinotropic action of glucagonlike peptide-I-(7-37) in diabetic and nondiabetic subjects. *Diabetes Care.* 1992;15:270–276.

7. Gutniak M, Ørskov C, Holst JJ, Ahrén B, Efendić S. Antidiabetic effect of glucagon-like peptide-1 (7-36) amide in normal subjects and patients with diabetes mellitus. *N Engl J Med.* 1992;326:1316–1322.

8. Nauk MA, Heimesaat MM, Ørskov C, Holst JJ, Ebert R, Creutzfeldt W. Preserved incretin activity of glucagon-like peptide-1[7-36 amide] but not of synthetic human gastric inhibitory polypeptide in patients with type-2 diabetes mellitus. *J Clin Invest.* 1993;91:301–307.

9. Holz GG, Chepurny OG. Glucagon-like peptide-1synthetic analogs: new therapeutic agents for use in the treatment of diabetes mellitus. *Curr Med Chem.* 2003;10:2471–2483.

10. Lovshin JA, Drucker DJ. Incretin-based therapies for type 2 diabetes mellitus. *Nat Rev Endocrinol.* 2009;5:262–269.

11. Lund PK, Goodman RH, Dee PC, Habener JF. Pancreatic preproglucagon cDNA contains two glucagon-related coding sequences arranged in tandem. *Proc Natl Acad Sci U S A.* 1982;79:345–349.

12. Bell GI, Santerre RF, Mullenbach GT. Hamster preproglucagon contains the sequence of glucagon and two related peptides. *Nature.* 1983;302:716–718.

13. Mojsov S, Weir GC, Habener JF. Insulinotropin: glucagon-like peptide-1(7-37) co-encoded in the glucagon gene is a potent stimulator of insulin release in the perfused rat pancreas. *J Clin Invest.* 1987;79:616–619.

14. Kreymann B, Yiangou Y, Kanse S, Williams G, Ghatei MA, Bloom SR. Isolation and characterization of GLP-1 7-36 amide from rat intestine. Elevated levels in diabetic rats. *FEBS Lett.* 1988;242:167–170.

15. Kreymann B, Williams G, Ghatei MA, Bloom SR. Glucagon-like peptide-1 7-36: a physiological incretin in man. *Lancet.* 1987;2:1300–1304.

16. Mojsov S, Heinrich G, Wilson IB, Ravazzola M, Orci L, Habener JF. Preproglucagon gene expression in pancreas and intestine diversifies at the level of post-translational processing. *J Biol Chem.* 1986;261:11880–11889.

17. Habener JF, Stanojevic V. Alpha cells come of age. *Trends Endocrinol Metab.* 2013;24:153–163.

18. Tolhurst G, Reimann F, Gribble FM. Nutritional regulation of glucagon-like peptide-1 secretion. *J Physiol.* 2009;587:27–32.

19. Reimann F, Habib AM, Tolhurst G, Parker HE, Rogers GJ, Gribble FM. Glucose sensing in L cells: a primary cell study. *Cell Metab.* 2008;8:532–539.

20. Reimann F, Tolhurst G, Gribble FM. G-protein-coupled receptors in intestinal chemosensation. *Cell Metab.* 2012;15:421–431.

21. Kieffer TJ, McIntosh CH, Pederson RA. Degradation of glucose-dependent insulinotropic polypeptide and truncated glucagon-like peptide-1 in vitro and in vivo by dipeptidyl peptidase IV. *Endocrinology.* 1995;136:3585–3596.

22. Augustyns K, Bal G, Thonus G, et al. The unique properties of dipeptidyl-peptidase IV (DPP IV / CD26) and the therapeutic potential of DPP IV inhibitors. *Curr Med Chem.* 1999;6:311–327.

23. Golightly LK, Drayna CC, McDermott MT. Comparative clinical pharmacokinetics of dipeptidyl peptidase-4 inhibitors. *Clin Pharmacokinet.* 2012;51:501–514.

24. Shigeto M, Katsura M, Matsuda M, Ohkuma S, Kaku K. Low, but physiological, concentration of GLP-1 stimulates insulin secretion independent of the cAMP-dependent protein kinase pathway. *J Pharmacol Sci.* 2008;108:274–279.

25. Nielsen LL, Young AA, Parkes DG. Pharmacology of Exenatide (synthetic exendin-4): a potential therapeutic for improved glycemic control of type 2 diabetes. *Regul Pept.* 2004;117:77–88.

26. Eng J, Kleinman WA, Singh G, Raufman JP. Isolation and characterization of exendin-4, an exendin-3 analogue, from *Heloderma suspectum* venom. Further evidence for an exendin receptor on dispersed acini from guinea pig pancreas. *J Biol Chem.* 1992;267:7402–7405.

27. Drucker D, Easley C, Kirkpatrick P. Sitagliptin. *Nat Rev Drug Discov.* 2007;6:109–110.

28. Keating GM. Vildagliptin: a review of its use in type 2 diabetes mellitus. *Drugs.* 2010;70:2089–2112.

29. Dalle S, Burcelin R, Gourdy P. Specific actions of GLP-1 receptor agonists and DPP4 inhibitors for the treatment of pancreatic β-cell impairments in type 2 diabetes. *Cell Signal.* 2013;25:570–579.

30. Schmidt WE, Siegel EG, Creutzfeldt W. Glucagon-like peptide-1 but not glucagon-like peptide-2 stimulates insulin release from isolated rat pancreatic islets. *Diabetologia.* 1985;28:704–707.

31. Holst JJ, Ørskov C, Nielsen OV, Schwartz TW. Truncated glucagon-like peptide I, an insulin-releasing hormone from the distal gut. *FEBS Lett.* 1987;211:169–174.

32. Weir GC, Mojsov S, Heinrich G, Habener JF. Glucagon-like peptide-1(7-37) actions on endocrine pancreas. *Diabetes.* 1989;38:338–342.

33. Thorens B. Expression cloning of the pancreatic β cell receptor for the gluco-incretin hormone glucagon-like peptide 1. *Proc Natl Acad Sci U S A.* 1992;89:8641–8645.

34. Ørskov C, Jeppesen J, Matsbad S, Holst JJ. Proglucagon products in plasma of noninsulin-dependent diabetics and nondiabetic controls in the fasting state and after oral glucose and intravenous arginine. *J Clin Invest.* 1991;87:415–423.

35. Thorens B, Porret A, Bühler L, Deng SP, Morrel P, Widmann C. Cloning and functional expression of the human islet GLP-1 receptor. Demonstration that exendin-4 is an agonist and exendin-(9-39) an antagonist of the receptor. *Diabetes.* 1993;42:1678–1682.

36. Dillon JS, Tanizawa Y, Wheeler MB, et al. Cloning and functional expression of the human glucagon-like peptide-1 (GLP-1) receptor. *Endocrinology.* 1993;133:1907–1910.

37. Fehmann HC, Habener JF. Insulinotropic hormone glucagon-like peptide-1(7-37) stimulation of proinsulin gene expression and proinsulin biosynthesis in insulinoma beta TC-1 cells. *Endocrinology.* 1992;130:159–166.

38. Skoglund G, Hussain MA, Holz GG. Glucagon-like peptide-1 stimulates insulin gene promoter activity by protein kinase A-independent activation of the rat insulin I gene cAMP response element. *Diabetes.* 2000;49:1156–1164.

39. Chepurny OG, Hussain MA, Holz GG. Exendin-4 as a stimulator of rat insulin I gene promoter activity via bZIP/CRE interactions sensitive to serine/threonine protein kinase inhibitor Ro 31-8220. *Endocrinology.* 2002;143:2303–2313.

40. Hay CW, Sinclair EM, Bermano G, Durward E, Tadayyon M, Docherty K. Glucagon-like peptide-1 stimulates human insulin promoter activity in part through cAMP-responsive elements that lie upstream and downstream of the transcription start site. *J Endocrinol.* 2005;186:353–365.

41. Drucker DJ. Glucagon-like peptides: regulators of cell proliferation, differentiation, and apoptosis. *Mol Endocrinol.* 2003;17:161–171.

42. Buteau J. GLP-1 receptor signaling: effects on pancreatic β-cell proliferation and survival. *Diabetes Metab.* 2008;34(Suppl 2):S73–S77.

43. McIntosh CH, Widenmaier S, Kim SJ. Pleiotropic actions of the incretin hormones. *Vitam Horm.* 2010;84:21–79.

44. Portha B, Tourrel-Cuzin C, Movassat J. Activation of the GLP-1 receptor signaling pathway: a relevant strategy to repair a deficient beta-cell mass. *Exp Diabetes Res.* 2011;2011:376509. http://dx.doi.org/10.1155/2011/376509.

45. Yabe D, Seino S. Two incretin hormones GLP-1 and GIP: comparison of their actions on insulin secretion and β cell preservation. *Prog Biophys Mol Biol.* 2011;107:248–256.

46. Willard FS, Bueno AB, Sloop KW. Small molecule drug discovery at the glucagon-like peptide-1 receptor. *Exp Diabetes Res.* 2012;2012:709893.
47. Harmar AJ. Family-B G-protein-coupled receptors. *Genome Biol.* 2001;2 :reviews3013.1-reviews3013.10 (23 November 2001).
48. Furman B, Ong WK, Pyne NJ. Cyclic AMP signaling in pancreatic islets. *Adv Exp Med Biol.* 2010;654:281–304.
49. Leech CA, Chepurny OG, Holz GG. Epac2-dependent Rap1 activation and the control of islet insulin secretion by glucagon-like peptide-1. *Vitam Horm.* 2010;84:279–302.
50. Song WJ, Seshadri M, Ashraf U, et al. Snapin mediates incretin action and augments glucose-dependent insulin secretion. *Cell Metab.* 2011;13:308–319.
51. Dalle S, Quoyer J, Varin E, Costes S. Roles and regulation of the transcription factor CREB in pancreatic β -cells. *Curr Mol Pharmacol.* 2011;4:187–195.
52. Altarejos JY, Montminy M. CREB and the CRTC co-activators: sensors for hormonal and metabolic signals. *Nat Rev Mol Cell Biol.* 2011;12:141–151.
53. Jhala US, Canettieri G, Screaton RA, et al. cAMP promotes pancreatic β-cell survival via CREB-mediated induction of IRS2. *Genes Dev.* 2003;17:1575–1580.
54. Song WJ, Mondal P, Li Y, Lee SE, Hussain MA. Pancreatic beta-cell response to increased metabolic demand and to pharmacologic secretagogues requires EPAC2A. *Diabetes.* 2013;62:2796–2807.
55. Cheng K, Andrikopoulos S, Gunton JE. First phase insulin secretion and type 2 diabetes. *Curr Mol Med.* 2013;13:126–139.
56. Fehse F, Trautmann M, Holst JJ, et al. Exenatide augments first- and second-phase insulin secretion in response to intravenous glucose in subjects with type 2 diabetes. *J Clin Endocrinol Metab.* 2005;90:5991–5997.
57. Fukuda M, Williams KW, Gautron L, Elmquist JK. Induction of leptin resistance by activation of cAMP-Epac signaling. *Cell Metab.* 2011;13:331–339.
58. Friedrichsen BN, Neubauer N, Lee YC, et al. Stimulation of pancreatic beta-cell replication by incretins involves transcriptional induction of cyclin D1 via multiple signalling pathways. *J Endocrinol.* 2006;188:481–492.
59. Liu Z, Habener JF. Glucagon-like peptide-1 activation of TCF7L2-dependent Wnt signaling enhances pancreatic beta cell proliferation. *J Biol Chem.* 2008;283:8723–8735.
60. Shao W, Wang Z, Ip W, et al. GLP-1(28–36) improves β-cell mass and glucose disposal in streptozotocin induced diabetes mice and activates PKA-β-catenin signaling in beta-cells in vitro. *Am J Physiol Endocrinol Metab.* 2013;304:E1263–E1272.
61. Liu Z, Stanojevic V, Brindamour LJ, Habener JF. GLP1-derived nonapeptide GLP1 (28-36)amide protects pancreatic β-cells from glucolipotoxicity. *J Endocrinol.* 2012;213:143–154.
62. Guettier JM, Gautam D, Scarselli M, et al. A chemical-genetic approach to study G protein regulation of beta cell function in vivo. *Proc Natl Acad Sci U S A.* 2009;106:19197–19202.
63. Wang X, Zhou J, Doyle ME, Egan JM. Glucagon-like peptide-1 causes pancreatic duodenal homeobox-1 protein translocation from the cytoplasm to the nucleus of pancreatic beta-cells by a cyclic adenosine monophosphate/protein kinase A-dependent mechanism. *Endocrinology.* 2001;142:1820–1827.
64. Shao W, Yu Z, Fantus IG, Jin T. Cyclic AMP signaling stimulates proteasome degradation of thioredoxin interacting protein (TxNIP) in pancreatic beta-cells. *Cell Signal.* 2010;22:1240–1246.
65. Mukai E, Fujimoto S, Sato H, et al. Exendin-4 suppresses SRC activation and reactive oxygen species production in diabetic Goto-Kakizaki rat islets in an Epac-dependent manner. *Diabetes.* 2011;60:218–226.
66. Chen J, Couto FM, Minn AH, Shalev A. Exenatide inhibits beta-cell apoptosis by decreasing thioredoxin-interacting protein. *Biochem Biophys Res Commun.* 2006;346:1067–1074.

67. Bergwitz C, Gardella TJ, Flannery MR, et al. Full activation of chimeric receptors by hybrids between parathyroid hormone and calcitonin. Evidence for a common pattern of ligand-receptor interaction. *J Biol Chem*. 1996;271:26469–26472.
68. Miller LJ, Dong M, Harikumar KG, Gao F. Structural basis of natural ligand binding and activation of the Class II G-protein-coupled secretin receptor. *Biochem Soc Trans*. 2007;35:709–712.
69. Parthier C, Reedtz-Runge S, Rudolph R, Stubbs MT. Passing the baton in class B GPCRs: peptide hormone activation via helix induction? *Trends Biochem Sci*. 2009;34:303–310.
70. Donnelly D. The structure and function of the glucagon-like peptide-1 receptor and its ligands. *Br J Pharmacol*. 2011;166:27–41.
71. Chen Q, Pinon DI, Miller LJ, Dong M. Spatial approximations between residues 6 and 12 in the amino-terminal region of glucagon-like peptide-1 and its receptor. A region critical for biological activity. *J Biol Chem*. 2010;285:24508–24518.
72. Koole C, Wootten D, Simms J, Miller LJ, Christopoulos A, Sexton PM. Second extra-cellular loop of human glucagon-like peptide-1 receptor (GLP-1R) has a critical role in GLP-1 peptide binding and receptor activation. *J Biol Chem*. 2012;287:3642–3658.
73. Mathi SKI, Chan Y, Li X, Wheeler MB. Scanning of the glucagon-like peptide-1 receptor localizes G protein-activating determinants primarily to the N terminus of the third intracellular loop. *Mol Endocrinol*. 1997;11:424–432.
74. Dong M, Gao F, Pinon DI, Miller LJ. Insights into the structural basis of endogenous agonist activation of family B G protein-coupled receptors. *Mol Endocrinol*. 2008;22:1489–1499.
75. Harikumar KG, Wootten D, Pinon DI, et al. Glucagon-like peptide-1 receptor dimer-ization differentially regulates agonist signaling but does not affect small molecule allo-stery. *Proc Natl Acad Sci U S A*. 2012;109:18607–18612.
76. Knudsen LB, Kiel D, Teng M, et al. Small-molecule agonists for the glucagon-like pep-tide 1 receptor. *Proc Natl Acad Sci U S A*. 2007;104:937–942.
77. Koole C, Wootten D, Simms J, et al. Allosteric ligands of the glucagon-like peptide-1 receptor (GLP-1R) differentially modulate endogenous and exogenous peptide responses in a pathway-selective manner: implications for drug screening. *Mol Pharmacol*. 2010;78:456–465.
78. Willard FS, Wootten D, Showalter AD, et al. Small molecule allosteric modulation of the glucagon-like peptide-1 receptor enhances the insulinotropic effect of oxyntomodulin. *Mol Pharmacol*. 2012;82:1066–1073.
79. Butler AE, Campbell-Thompson M, Gurlo T, Dawson DW, Atkinson M, Butler PC. Marked expansion of exocrine and endocrine pancreas with incretin therapy in humans with increased exocrine pancreas dysplasia and the potential for glucagon-producing neuroendocrine tumors. *Diabetes*. 2013;62:2595–2604.
80. Gromada J, Holst JJ, Rorsman P. Cellular regulation of islet hormone secretion by the incretin hormone glucagon-like peptide-1. *Pflugers Arch*. 1998;435:583–594.
81. Holz GG. New insights concerning the glucose-dependent insulin secretagogue action of glucagon-like peptide-1 in pancreatic beta-cells. *Horm Metab Res*. 2004;36:787–794.
82. Gromada J, Brock B, Schmitz O, Rorsman P. Glucagon-like peptide-1: regulation of insulin secretion and therapeutic potential. *Basic Clin Pharmacol Toxicol*. 2004;95:252–262.
83. Rorsman P, Braun M. Regulation of insulin secretion in human pancreatic islets. *Annu Rev Physiol*. 2013;75:155–179.
84. Holz GG, Chepurny OG, Leech CA, Song WJ, Hussain MA. Molecular basis of cAMP signaling in pancreatic β-cells. In: *The Islets of Langerhans*. 2nd ed. Springer; 2013, in press.

85. Holz 4th GG, Kuhtreiber WM, Habener JF. Pancreatic beta-cells are rendered glucose-competent by the insulinotropic hormone glucagon-like peptide-1(7-37). *Nature.* 1993;361:362–365.
86. Holz GG, Leech CA, Heller RS, Castonguay M, Habener JF. cAMP-dependent mobilization of intracellular Ca^{2+} stores by activation of ryanodine receptors in pancreatic β-cells. A Ca^{2+} signaling system stimulated by the insulinotropic hormone glucagon-like peptide-1-(7-37). *J Biol Chem.* 1999;274:14147–14156.
87. Kang G, Chepurny OG, Holz GG. cAMP-regulated guanine nucleotide exchange factor II (Epac2) mediates Ca^{2+}-induced Ca^{2+} release in INS-1 pancreatic β-cells. *J Physiol.* 2001;536:375–385.
88. Kang G, Holz GG. Amplification of exocytosis by Ca^{2+}-induced Ca^{2+} release in INS-1 pancreatic β cells. *J Physiol.* 2003;546:175–189.
89. Kang G, Joseph JW, Chepurny OG, et al. Epac-selective cAMP analog 8-pCPT-2′-O-Me-cAMP as a stimulus for Ca^{2+}-induced Ca^{2+} release and exocytosis in pancreatic β-cells. *J Biol Chem.* 2003;278:8279–8285.
90. Kang G, Chepurny OG, Rindler MJ, et al. A cAMP and Ca^{2+} coincidence detector in support of Ca^{2+}-induced Ca^{2+} release in mouse pancreatic β cells. *J Physiol.* 2005;566:173–188.
91. Leech CA, Dzhura I, Chepurny OG, et al. Molecular physiology of glucagon-like peptide-1 insulin secretagogue action in pancreatic β cells. *Prog Biophys Mol Biol.* 2011;107:236–247.
92. Light PE, Manning Fox JE, Riedel MJ, Wheeler MB. Glucagon-like peptide-1 inhibits pancreatic ATP-sensitive potassium channels via a protein kinase A- and ADP-dependent mechanism. *Mol Endocrinol.* 2002;16:2135–2144.
93. Kang G, Chepurny OG, Malester B, et al. cAMP sensor Epac as a determinant of ATP-sensitive potassium channel activity in human pancreatic β cells and rat INS-1 cells. *J Physiol.* 2006;573:595–609.
94. Kang G, Leech CA, Chepurny OG, Coetzee WA, Holz GG. Role of the cAMP sensor Epac as a determinant of K_{ATP} channel ATP sensitivity in human pancreatic β cells and rat INS-1 cells. *J Physiol.* 2008;586:1307–1319.
95. Nakazaki M, Crane A, Hu M, et al. cAMP-activated protein kinase-independent potentiation of insulin secretion by cAMP is impaired in SUR1 null islets. *Diabetes.* 2002;51:3440–3449.
96. Shiota C, Larsson O, Shelton KD, et al. Sulfonylurea receptor type 1 knock-out mice have intact feeding-stimulated insulin secretion despite marked impairment in their response to glucose. *J Biol Chem.* 2002;277:37176–37183.
97. Miki T, Minami K, Shinozaki H, et al. Distinct effects of glucose-dependent insulinotropic polypeptide and glucagon-like peptide-1 on insulin secretion and gut motility. *Diabetes.* 2005;54:1056–1063.
98. Hugill A, Shimomura K, Ashcroft FM, Cox RD. A mutation in KCNJ11 causing human hyperinsulinism (Y12X) results in a glucose-intolerant phenotype in the mouse. *Diabetologia.* 2010;53:2352–2356.
99. Bourron O, Chebbi F, Halbron M, et al. Incretin effect of glucagon-like peptide-1 receptor agonist is preserved in presence of ABCC8/SUR1 mutation in β-cell. *Diabetes Care.* 2012;35:e76.
100. Winzell MS, Ahrén B. The high-fat diet-fed mouse: a model for studying mechanisms and treatment of impaired glucose tolerance and type 2 diabetes. *Diabetes.* 2004; 53(Suppl 3):S215–S219.
101. Bode HP, Moormann B, Dabew R, Göke B. Glucagon-like peptide 1 elevates cytosolic calcium in pancreatic beta-cells independently of protein kinase A. *Endocrinology.* 1999;140:3919–3927.

102. Holz GG. Epac: a new cAMP-binding protein in support of glucagon-like peptide-1 receptor-mediated signal transduction in the pancreatic β-cell. *Diabetes.* 2004;53:5–13.
103. Holz GG, Kang G, Harbeck M, Roe MW, Chepurny OG. Cell physiology of cAMP sensor Epac. *J Physiol.* 2006;577:5–15.
104. Holz GG, Chepurny OG, Leech CA. Epac2A makes a new impact in β-cell biology. *Diabetes.* 2013;62:2665–2666.
105. Ahlqvist L, Brown K, Ahrén B. Upregulated insulin secretion in insulin-resistant mice: evidence of increased islet GLP1 receptor levels and GPR119-activated GLP1 secretion. *Endocr Connect.* 2013;2:69–78.
106. Winzell MS, Ahrén B. Durable islet effects on insulin secretion and protein kinase A expression following exendin-4 treatment of high-fat diet-fed mice. *J Mol Endocrinol.* 2008;40:93–100.
107. Kaihara KA, Dickson LM, Jacobson DA, et al. β-cell-specific protein kinase A activation enhances the efficiency of glucose control by increasing acute-phase insulin secretion. *Diabetes.* 2013;62:1527–1536.
108. Yusta B, Baggio LL, Estall JL, et al. GLP-1 receptor activation improves β cell function and survival following induction of endoplasmic reticulum stress. *Cell Metab.* 2006;4:391–406.
109. Sonoda N, Imamura T, Yoshizaki T, Babendure JL, Lu JC, Olefsky JM. β-arrestin-1 mediates glucagon-like peptide-1 signaling to insulin secretion in cultured pancreatic β cells. *Proc Natl Acad Sci U S A.* 2008;105:6614–6619.
110. Dalle S, Ravier MA, Bertrand G. Emerging roles for β-arrestin in the control of the pancreatic β-cell function and mass: new therapeutic strategies and consequences for drug screening. *Cell Signal.* 2011;23:522–528.
111. Buteau J, Foisy S, Joly E, Prentki M. Glucagon-like peptide-1 induces pancreatic β-cell proliferation via transactivation of the epidermal growth factor receptor. *Diabetes.* 2003;52:124–132.
112. Quoyer J, Longuet C, Broca C, et al. GLP-1 mediates antiapoptotic effect by phosphorylating Bad through a β arrestin 1-mediated ERK1/2 activation in pancreatic β-cells. *J Biol Chem.* 2010;285:1989–2002.
113. Bastien-Dionne PO, Valenti L, Kon N, Gu W, Buteau J. Glucagon-like peptide-1 inhibits the sirtuin deacetylase SirT1 to stimulate pancreatic β-cell mass expansion. *Diabetes.* 2011;60:3217–3222.
114. Buteau J, Spatz ML, Accili D. Transcription factor FoxO1 mediates glucagon-like peptide-1 effects on pancreatic β-cell mass. *Diabetes.* 2006;55:1190–1196.
115. Talbot J, Joly E, Prentki M, Buteau J. β-Arrestin 1-mediated recruitment of c-Src underlies the proliferative action of glucagon-like peptide-1 in pancreatic β INS832/13 cells. *Mol Cell Endocrinol.* 2012;364:65–70.
116. Buteau J, Roduit R, Susini S, Prentki M. Glucagon-like peptide-1 promotes DNA synthesis, activates phosphatidylinositol 3-kinase and increases transcription factor pancreatic and duodenal homeobox gene 1 (PDX-1) DNA binding activity in beta (INS-1)-cells. *Diabetologia.* 1999;42:856–864.
117. Wang Q, Li L, Xu E, Wong V, Rhodes C, Brubaker PL. Glucagon-like peptide-1 regulates proliferation and apoptosis via activation of protein kinase B in pancreatic INS-1 beta cells. *Diabetologia.* 2004;47:478–487.
118. Buteau J, Foisy S, Rhodes CJ, Carpenter L, Biden TJ, Prentki M. Protein kinase Cζ activation mediates glucagon-like peptide-1-induced pancreatic β-cell proliferation. *Diabetes.* 2001;50:2237–2243.
119. Arnette D, Gibson TB, Lawrence MC, et al. Regulation of ERK1 and ERK2 by glucose and peptide hormones in pancreatic β cells. *J Biol Chem.* 2003;278:32517–32525.
120. Nakabayashi H, Nishizawa M, Nakagawa A, Takeda R, Niijima A. Vagal hepatopancreatic reflex effect evoked by intraportal appearance of tGLP-1. *Am J Physiol.* 1996;271:E808–E813.

121. Balkan B, Li X. Portal GLP-1 administration in rats augments the insulin response to glucose via neuronal mechanisms. *Am J Physiol Regul Integr Comp Physiol.* 2000;279:1449–1454.

122. Nakagawa A, Satake H, Nakabayashi H, et al. Receptor gene expression of glucagon-like peptide-1, but not glucose-dependent insulinotropic polypeptide, in rat nodose ganglion cells. *Auton Neurosci.* 2004;110:36–43.

123. Vahl TP, Tauchi M, Durler TS, et al. Glucagon-like peptide-1 (GLP-1) receptors expressed on nerve terminals in the portal vein mediate the effects of endogenous GLP-1 on glucose tolerance in rats. *Endocrinology.* 2007;148:4965–4973.

124. Bucinskaite V, Tolessa T, Pedersen J, et al. Receptor-mediated activation of gastric vagal afferents by glucagon-like peptide-1 in the rat. *Neurogastroenterol Motil.* 2009;21:978, e78.

125. Kakei M, Yada T, Nakagawa A, Nakabayashi H. Glucagon-like peptide-1 evokes action potentials and increases cytosolic Ca^{2+} in rat nodose ganglion neurons. *Auton Neurosci.* 2002;102:39–44.

126. Baraboi ED, St-Pierre DH, Shooner J, Timofeeva E, Richard D. Brain activation following peripheral administration of the GLP-1 receptor agonist exendin-4. *Am J Physiol Regul Integr Comp Physiol.* 2011;301:R1011–R1024.

127. Kastin AJ, Akerstrom V. Entry of exendin-4 into brain is rapid but may be limited at high doses. *Int J Obes Ralat Metab Disord.* 2003;27:313–318.

128. Göke R, Larsen PJ, Mikkelsen JD, Sheikh SP. Distribution of GLP-1 binding sites in the rat brain: evidence that exendin-4 is a ligand of brain GLP-1 binding sites. *Eur J Neurosci.* 1995;7:2294–3000.

129. Merchenthaler I, Lane M, Shurhrue P. Distribution of pre-pro-glucagon-like peptide-1 receptor messenger RNAs in the rat central nervous system. *J Comp Neurol.* 1999;403:261–280.

130. Plamboeck A, Veedfald S, Deacon CF, et al. The effect of exogenous GLP-1 on food intake is lost in male truncally vagotomized subjects with pyloroplasty. *Am J Physiol Gastrointest Liver Physiol.* 2013;304:G1117–G1127.

131. Llewellyn-Smith IJ, Reimann F, Gribble FM, Trapp S. Preproglucagon neurons project widely to autonomic control areas in the mouse brain. *Neuroscience.* 2011;180:111–121.

132. Llewellyn-Smith IJ, Gnanamanickam GJ, Reimann F, Gribble FM, Trapp S. Preproglucagon (PPG) neurons innervate neurochemically identified autonomic neurons in the mouse brainstem. *Neuroscience.* 2011;229:130–143.

133. Burcelin R, Serino M, Cabou C. A role of the gut-to-brain GLP-1-dependent axis in the control of metabolism. *Curr Opin Pharmacol.* 2009;9:744–752.

134. Trapp S, Hisadome K. Glucagon-like peptide 1 and the brain: central actions-central sources? *Auton Neurosci.* 2011;161:14–19.

135. Ghosal S, Myers B, Herman JP. Role of central glucagon-like peptide-1 in stress regulation. *Physiol Behav.* 2013, in press.

136. Lockie SH. Glucagon-like peptide-1 receptor in the brain: role in neuroendocrine control of energy metabolism and treatment target for obesity. *J Neuroendocrinol.* 2013; 25:597–604.

137. Wan S, Coleman FH, Travagli RA. Glucagon-like peptide-1 excites pancreas-projecting preganglionic vagal motoneurons. *Am J Physiol Gastrointest Liver Physiol.* 2007;292:G1474–G1482.

138. Wan S, Browning KN, Travagli RA. Glucagon-like peptide-1 modulates synaptic transmission to identified pancreas-projecting vagal motoneurons. *Peptides.* 2007;28:2184–2191.

139. Lamont BJ, Li Y, Kwan E, Brown TJ, Gaisano H, Drucker DJ. Pancreatic GLP-1 receptor activation is sufficient for incretin control of glucose metabolism in mice. *J Clin Invest.* 2012;122:388–402.

140. Barrera JG, Sandoval DA, D'Alessio DA, Seeley RJ. GLP-1 and energy balance: an integrated model of short-term and long-term control. *Nat Rev Endocrinol.* 2011;7:507–516.
141. Cabou C, Burcelin R. GLP-1, the gut-brain, and brain-periphery axes. *Rev Diabet Stud.* 2011;8:418–431.
142. D'Alessio DA, Sandoval DA, Seeley RJ. New ways in which GLP-1 can regulate glucose homeostasis. *J Clin Invest.* 2005;115:3406–3408.
143. Hayes MR. Neuronal and intracellular signaling pathways mediating GLP-1 energy balance and glycemic effects. *Physiol Behav.* 2012;106:413–416.
144. Ionut V, Liberty IF, Hucking K, et al. Exogenously imposed postprandial-like rises in systemic glucose and GLP-1 do not produce an incretin effect, suggesting an indirect mechanism of GLP-1 action. *Am J Physiol Endocrinol Metab.* 2006;291:E779–E785.
145. Thorens B. Central control of glucose homeostasis: the brain-endocrine pancreas axis. *Diabetes Metab.* 2010;36(Suppl 3):S45–S49.
146. Knauf C, Cani PD, Perrin C, et al. Brain glucagon-like peptide-1 increases insulin secretion and muscle insulin resistance to favor hepatic glycogen storage. *J Clin Invest.* 2005;115:3554–3563.
147. Sandoval DA, Bagnol D, Woods SC, D'Alessio DA, Seeley RJ. Arcuate glucagon-like peptide 1 receptors regulate glucose homeostasis but not food intake. *Diabetes.* 2008;57:2046–2054.
148. Sandoval DA. CNS GLP-1 regulation of peripheral glucose homeostasis. *Physiol Behav.* 2008;94:670–674.
149. Burdakov D, Luckman SM, Verkhratsky A. Glucose-sensing neurons of the hypothalamus. *Philos Trans R Soc Lond B Biol Sci.* 2005;360:2227–2235.
150. Spanswick D, Smith MA, Groppi VE, Logan SD, Ashford MLJ. Leptin inhibits hypothalamic neurons by activation of ATP-sensitive potassium channels. *Nature.* 1997;390:521–525.
151. Mirshamsi S, Laidlaw HA, Ning K, et al. Leptin and insulin stimulation of signaling pathways in arcuate nucleus neurons: PI3K dependent actin reorganization and K_{ATP} channel activation. *BMC Neurosci.* 2004;5:54.
152. Hölscher C. Potential role of glucagon-like peptide-1 (GLP-1) in neuroprotection. *CNS Drugs.* 2012;26:871–882.
153. Li Y, Duffy KB, Ottinger MA, et al. GLP-1 receptor stimulation reduces amyloid-β peptide accumulation and cytotoxicity in cellular and animal models of Alzheimer's disease. *J Alzheimers Dis.* 2010;19:1205–1219.
154. Ma T, Du X, Pick JE, Sui G, Brownlee M, Klann E. Glucagon-like peptide-1 cleavage product GLP-1(9-36) amide rescues synaptic plasticity and memory deficits in Alzheimer's disease model mice. *J Neurosci.* 2012;32:13701–13708.
155. Li Y, Perry T, Kindy MS, et al. GLP-1 receptor stimulation preserves primary cortical and dopaminergic neurons in cellular and rodent models of stroke and Parkinsonism. *Proc Natl Acad Sci U S A.* 2009;106:1285–1290.
156. Aviles-Olmos I, Dickson J, Kefalopoulou Z, et al. Exenatide and the treatment of patients with Parkinson's disease. *J Clin Invest.* 2013;123:2730–2736.
157. Gejl M, Egefjord L, Lerche S, et al. Glucagon-like peptide-1 decreases intracerebral glucose content by activating hexokinase and changing glucose clearance during hyperglycemia. *J Cereb Blood Flow Metab.* 2012;32:2146–2152.
158. Geil M, Lerche S, Egefjord L, et al. Glucagon-like peptide-1 raises blood-brain glucose transfer capacity and hexokinase activity in human brain. *Front Neuroenergetics.* 2013;5:2.
159. Moretto TJ, Milton DR, Ridge TD, et al. Efficacy and tolerability of exenatide monotherapy over 24 weeks in antidiabetic drug-naïve patients with type 2 diabetes: a randomized, double-blind, placebo-controlled, parallel-group study. *Clin Ther.* 2008;30:1448–1460.
160. Rosenstock J, Raccah D, Koranyi L, et al. Efficacy and safety of lixisenatide once daily versus exenatide twice daily in type 2 diabetes inadequately controlled on metformin:

a 24-week, randomized, open-label, active-controlled study (GetGoal-X). *Diabetes Care.* 2013;36:2945–2951.

161. Drucker DJ, Buse JD, Taylor K, et al. Exenatide once weekly versus twice daily for the treatment of type 2 diabetes: a randomized, open-label, non-inferiority study. *Lancet.* 2008;372:1240–1250.

162. Kim D, MacConell L, Zhuang D, et al. Effects of once-weekly dosing of a long-acting release formulation of exenatide on glucose control and body weight in subjects with type 2 diabetes. *Diabetes Care.* 2007;30:1487–1493.

163. Niswender K, Pi-Sunyer X, Buse J, et al. Weight change with liraglutide and comparator therapies: an analysis of seven phase 3 trials from the liraglutide diabetes development programme. *Diabetes Obes Metab.* 2013;15:42–54.

164. Buse JB, Rosenstock J, Sesti G, et al. Liraglutide once a day versus exenatide twice a day for type 2 diabetes: a 26-week randomized, parallel group, multinational, open-label trial (LEAD-6). *Lancet.* 2009;374:39–47.

165. Nauck M, Petrie JR, Sesti G, et al. The once-weekly human GLP-1 analogue semaglutide provides significant reductions in HbA1C and body weight in patients with type 2 diabetes (T2D). *Diabetologia.* 2012;55(Suppl 1):S7, EASD 48th Annual Meeting, Berlin, Germany.

166. St Onge EL, Miller SA. Albiglutide: a new GLP-1 analog for the treatment of the type 2 diabetes. *Exp Opin Biol Ther.* 2010;10:801–806.

167. Available from: http://us.gsk.com/html/media-news/pressreleases/2012/2012-pressrelease-1178461.htm.

168. Available from: http://us.gsk.com/html/media-news/pressreleases/2012/2012-pressrelease-1125277.htm.

169. Wang M, Matheson S, Picard J, Pezzullo J. Abstracts of 69th Scientific Sessions of the American Diabetes Association. New Orleans; 5–9 June 2009. Abstract 553-P.

170. Jimenez-Solem E, Rasmussen MH, Christensen M, Knop FK. Dulaglutide, a long-acting GLP-1 analog fused with an Fc antibody fragment for the potential treatment of type 2 diabetes. *Curr Opin Mol Ther.* 2010;12:790–797.

171. Grunberger G, Chang A, Garsia Soria G, Botros FT, Bsharat R, Milicevic Z. Monotherapy with the once-weekly GLP-1 analogue dulaglutide for 12 weeks in patients with type 2 diabetes: dose-dependent effects on glycaemic control in a randomized, double blind, placebo-controlled study. *Diabet Med.* 2012;29:1260–1267.

172. Available from: http://www.hanmi.co.kr/korea/research/120607_ADA_poster_HM11260C%20%28LAPS-Exendin-4%29.pdf.

173. Available from: http://www.diartispharma.com/content/newsandevents/releases/100212.htm.

174. Hoare SR. Allosteric modulators of class B G-protein-coupled receptors. *Curr Neuropharmacol.* 2007;5:168–179.

175. Bahekar RH, Jain MR, Gupta AA, et al. Synthesis and antidiabetic activity of 3,6,7-trisubstituted-2-(1H-imidazol-2-ylsulfanyl)quinoxalines and quinoxalin-2-yl isothioureas. *Arch Pharm.* 2007;340:359–366.

176. Teng M, Johnson MD, Thomas C, et al. Small molecule ago-allosteric modulators of the human glucagon-like peptide-1 (hGLP-1) receptor. *Bioorg Med Chem Lett.* 2007;17:5472–5478.

177. Irwin N, Flatt PR, Patterson S, Green BD. Insulin-releasing and metabolic effects of small molecule GLP-1 receptor agonist 6,7-dichloro-2-methylsulfonyl-3-N-tert-butylaminoquinoxaline. *Eur J Pharmacol.* 2010;628:268–273.

178. Su H, He M, Li H, et al. Boc5, a non-peptidic glucagon-like peptide-1 receptor agonist, invokes sustained glycemic control and weight loss in diabetic mice. *PLoS ONE.* 2008;3:e2892.

179. Chen D, Liao J, Zhou C, et al. A nonpeptidic agonist of glucagon-like peptide 1 receptors with efficacy in diabetic db/db mice. *Proc Natl Acad Sci U S A.* 2007;104:943–948.

180. Liu Q, Li N, Yuan Y, et al. Cyclobutane derivatives as novel nonpeptidic small molecule agonists of glucagon-like peptide-1 receptor. *J Med Chem*. 2012;55:250–267.
181. Coopman K, Huang Y, Johnston N, Bradley SJ, Wilkinson GF, Willars GB. Comparative effects of the endogenous agonist glucagon-like peptide-1 (GLP-1)-(7-36) amide and the small-molecule ago-allosteric agent "compound 2" at the GLP-1 receptor. *J Pharmacol Exp Ther*. 2010;334:795–808.
182. Sloop KW, Willard FS, Brenner MB, et al. Novel small molecule glucagon-like peptide-1 receptor agonist stimulates insulin secretion in rodents and from human islets. *Diabetes*. 2010;59:3099–3107.
183. Rye Underwood C, Möller Knudsen S, Schjellerup Wulff B, et al. Transmembrane α-helix 2 and 7 are important for small molecule-mediated activation of the GLP-1 receptor. *Pharmacology*. 2011;88:340–348.
184. Cheong YH, Kim MK, Son MH, Kaang BK. Two small molecule agonists of glucagon-like peptide-1 receptor modulate the receptor activation response differently. *Biochem Biophys Res Commun*. 2012;417:558–563.
185. Li N, Lu J, Willars GB. Allosteric modulation of the activity of the glucagon-like peptide-1 (GLP-1) metabolite GLP-1 9-36 amide at GLP-1 receptor. *PLoS One*. 2012;7:e47936.
186. Wootten D, Savage EE, Willard FS, et al. Differential activation and modulation of the glucagon-like peptide-1 receptor by small molecule ligands. *Mol Pharmacol*. 2013;83:822–834.
187. Koole C, Wootten D, Simms J, et al. Polymorphism and ligand dependent changes in human glucagon-like peptide-1 receptor (GLP-1R) function: allosteric rescue of loss of function mutation. *Mol Pharmacol*. 2011;80:486–497.
188. Ozeki N, Terasawa T, Naruse R, et al. Serum level of soluble CD26/dipeptidyl peptidase-4 (DPP-4) predicts the response to sitagliptin, a DPP-4 inhibitor, in patients with type 2 diabetes controlled inadequately by metformin and/or sulfonylurea. *Transl Res*. 2012;159:25–31.
189. Matteucci E, Giampietro O. Dipeptidyl peptidase-4 (CD26): knowing the function before inhibiting the enzyme. *Curr Med Chem*. 2009;16:2943–2951.
190. Kirby M, Yu DM, O'Connor S, Gorrell MD. Inhibitor selectivity in the clinical application of dipeptidyl peptidase-4 inhibition. *Clin Sci (Lond)*. 2009;118:31–41.
191. Ohnuma K, Dang NH, Morimoto C. Revisiting an old acquaintance: CD26 and its molecular mechanisms in T cell function. *Trends Immunol*. 2008;29:295–301.
192. Rasmussen HB, Branner S, Wiberg FC, Wagtmann N. Crystal structure of human dipeptidyl peptidase IV/CD26 in complex with a substrate analog. *Nat Struct Biol*. 2003;10:19–25.
193. Engel M, Hoffmann T, Wagner L, et al. The crystal structure of dipeptidyl peptidase IV (CD26) reveals its functional regulation and enzymatic mechanism. *Proc Natl Acad Sci U S A*. 2003;100:5063–5068.
194. Abbott CA, McCaughan GW, Gorrell MD. Two highly conserved glutamic acid residues in the predicted beta propeller domain of dipeptidyl peptidase IV are required for its enzyme activity. *FEBS Lett*. 1999;458:278–284.
195. Bergman AJ, Stevens C, Zhou Y, et al. Pharmacokinetic and pharmacodynamic properties of multiple oral doses of sitagliptin, a dipeptidyl peptidase-IV inhibitor: a double-blind, randomized, placebo-controlled study in healthy male volunteers. *Clin Ther*. 2006;28:55–72.
196. Herman GA, Stevens C, Van Dyck K, et al. Pharmacokinetics and pharmacodynamics of sitagliptin, an inhibitor of dipeptidyl peptidase IV, in healthy subjects: results from two randomized, double-blind, placebo-controlled studies with single oral doses. *Clin Pharmacol Ther*. 2005;78:675–688.

197. Aroda VR, Henry RR, Han J, et al. Efficacy of GLP-1 receptor agonists and DPP-4 inhibitors: meta-analysis and systematic review. *Clin Ther.* 2012;34:1247–1258.

198. He YL, Yamaguchi M, Ito H, Terao S, Sekiguchi K. Pharmacokinetics and pharmacodynamics of vildagliptin in Japanese patients with type 2 diabetes. *Int J Clin Pharmacol Ther.* 2010;48:582–595.

199. He YL, Valencia J, Zhang Y, et al. Hormonal and metabolic effects of morning or evening dosing of the dipeptidyl peptidase IV inhibitor vildagliptin in patients with type 2 diabetes. *Br J Clin Pharmacol.* 2010;70:34–42.

200. Boulton DW, Geraldes M. Safety, tolerability, pharmacokinetics and pharmacodynamics of once daily oral doses of saxagliptin for 2 weeks in type 2 diabetics and healthy subjects. *Diabetes.* 2007;56(Suppl 1):AI61, American Diabetes Association, 67th Scientific Sessions, Chicago, IL. Poster 606-P.

201. Kania DS, Gonzalvo JD, Weber ZA. Saxagliptin: a clinical review in the treatment of type 2 diabetes mellitus. *Clin Ther.* 2011;33:1005–1022.

202. Horie Y, Kanada S, Watada H, et al. Pharmacokinetic, pharmacodynamic, and tolerability profiles of the dipeptidyl peptidase-4 inhibitor linagliptin: a 4-week multicenter, randomized, double-blind, placebo-controlled phase IIa study in Japanese type 2 diabetes patients. *Clin Ther.* 2011;33:973–989.

203. Del Prato S, Barnett AH, Huisman H, Neubacher D, Woerle H-J, Dugi KA. Effect of linagliptin monotherapy on glycaemic control and markers of β-cell function in patients with inadequately controlled type 2 diabetes: a randomized controlled trial. *Diabetes Obes Metab.* 2011;13:258–267.

204. Covington P, Christopher R, Davenport M, et al. Pharmacokinetic, pharmacodynamic, and tolerability profiles of the dipeptidyl peptidase-4 inhibitor alogliptin: a randomized, double-blind, placebo-controlled, multiple-dose study in adult patients with type 2 diabetes. *Clin Ther.* 2008;30:499–512.

205. Christopher R, Covington P, Davenport M, et al. Pharmacokinetic, pharmacodynamic, and tolerability of single increasing doses of the dipeptidyl peptidase-4 inhibitor alogliptin in healthy male subjects. *Clin Ther.* 2008;30:513–527.

206. Ban K, Kim KH, Cho CK, et al. Glucagon-like peptide (GLP)-1(9-36)amide-mediated cytoprotection is blocked by exendin(9-39) yet does not require the known GLP-1 receptor. *Endocrinology.* 2010;151:1520–1531.

207. Elahi D, Egan JM, Shannon RP, et al. GLP-1 (9-36) amide, cleavage product of GLP-1 (7-36) amide, is a glucoregulatory peptide. *Obesity.* 2008;16:1501–1509.

208. Abu-Hamdah R, Rabiee A, Meneilly GS, Shannon RP, Andersen DK, Elahi D. Clinical review: the extrapancreatic effects of glucagon-like peptide-1 and related peptides. *J Clin Endocrinol Metab.* 2009;94:1843–1852.

209. Waget A, Cabou C, Masseboeuf M, et al. Physiological and pharmacological mechanisms through which the DPP-4 inhibitor sitagliptin regulates glycemia in mice. *Endocrinology.* 2011;152:3018–3029.

210. Holst JJ, Deacon C, Toft-Nielsen MB, Bjerre-Knudsen L. On the treatment of diabetes mellitus with glucagon-like peptide-1. *Ann N Y Acad Sci.* 1998;865:336–343.

211. Vilsbøll T, Knop FK, Krarup T, et al. The pathophysiology of diabetes involves a defective amplification of the late-phase insulin response to glucose by glucose-dependent insulinotropic polypeptide-regardless of etiology and phenotype. *J Clin Endocrinol Metab.* 2003;88:4897–4903.

212. Gentilella R, Bianchi C, Rossi A, Rotella CM. Exenatide: a review from pharmacology to clinical practice. *Diabetes Obes Metab.* 2009;11:544–556.

213. Fineman MS, Shen LZ, Taylor K, Kim DD, Baron AD. Effectiveness of progressive dose-escalation of exenatide (exendin-4) in reducing dose-limiting side effects in subjects with type 2 diabetes. *Diabetes Metab Res Rev.* 2004;20:411–417.

214. Bond A. Exenatide (Byetta) as a novel treatment option for type 2 diabetes mellitus. *Proc (Bayl Univ Med Cent)*. 2006;19:281–284.
215. Gough SC. Liraglutide: from clinical trials to clinical practice. *Diabetes Obes Metab*. 2012;14(Suppl 2):33–40.
216. White JR. Dipeptidyl peptidase-IV inhibitors: pharmacological profile and clinical use. *Clin Diabetes*. 2008;26:53–57.
217. Pathak R, Bridgeman MB. Dipeptidyl peptidase-4 (DPP-4) inhibitors in the management of diabetes. *Pharm Ther*. 2010;35:509–513.
218. Ritcher B, Bandeira-Echtler E, Bergerhoff K, Lerch CL. Dipeptidyl peptidase-4 (DPP-4) inhibitors for type 2 diabetes mellitus. *Cochrane Database Syst Rev*. 2008;2, CD006739.
219. Shyangdan DS, Royle P, Clar C, Sharma P, Waugh N, Snaith A. Glucagon-like peptide analogues for type 2 diabetes mellitus. *Cochrane Database Syst Rev*. 2011;10, CD006423.
220. Levin PA, Mersey JH, Zhou S, Bromberger LA. Clinical outcomes using long-term combination therapy with insulin glargine and exenatide in patients with type 2 diabetes mellitus. *Endocr Pract*. 2012;18:17–25.
221. Garber AJ. Liraglutide in oral antidiabetic drug combination therapy. *Diabetes Obes Metab*. 2012;14(Suppl 2):13–19.
222. St Onge EL, Miller S, Clements E. Sitagliptin/Metformin (janumet) as combination therapy in the treatment of type-2 diabetes mellitus. *Proc Natl Acad Sci U S A*. 2012;37:699–708.
223. Harashima SI, Ogura M, Tanaka D, et al. Sitagliptin add-on to low dosage sulphonylureas: efficacy and safety of combination therapy on glycaemic control and insulin secretion capacity in type 2 diabetes. *Int J Clin Pract*. 2012;66:465–476.
224. Williams-Herman D, Johnson J, Teng R, et al. Efficacy and safety of sitagliptin and metformin as initial combination therapy and as monotherapy over 2 years in patients with type 2 diabetes. *Diabetes Obes Metab*. 2010;12:442–451.
225. Jonas D, Van Scoyoc E, Gerrald K, et al. *Drug class review: newer diabetes medications, TZDs, and combinations: final original report [Internet]*. Portland, OR: Oregon Health & Science University; 2011, Available from: http://www.ncbi.nlm.nih.gov/books/NBK54209/. PMID: 21595121.
226. Bergenstal RM, Wysham C, Macconell L, et al. Efficacy and safety of exenatide once weekly versus sitagliptin or pioglitazone as an adjunct to metformin for treatment of type 2 diabetes (DURATION-2): a randomized trial. *Lancet*. 2010;376:431–439.
227. Diamant M, Van Gaal L, Stranks S, et al. Once weekly Exenatide compared with insulin glargine titrated to target in patients with type 2 diabetes (DURATION-3): an open-label randomized trial. *Lancet*. 2010;375:2234–2243.
228. Buse JB, Drucker DJ, Taylor KL, et al. DURATION-1: exenatide once weekly produces sustained glycemic control and weight loss over 52 weeks. *Diabetes Care*. 2010;33:1255–1261.
229. Blevins T, Pullman J, Malloy J, et al. DURATION-5: exenatide once weekly resulted in greater improvements in glycemic control compared with exenatide twice daily in patients with type 2 diabetes. *J Clin Endocrinol Metab*. 2011;96:1301–1310.
230. Nauk M, Frid A, Hermansen K, et al. Efficacy and safety comparison of Liraglutide, glimepiride, and placebo, all in combination with metformin, in type 2 diabetes: the LEAD (Liraglutide effect and action in diabetes)-2 study. *Diabetes Care*. 2009;32:84–90.
231. Jendle J, Nauk M, Matthews DR, et al. Weight loss with Liraglutide, a once-daily human glucagon-like peptide-1 analogue for type 2 diabetes treatment as monotherapy or added to metformin, is primarily as a result of a reduction in fat tissue. *Diabetes Obes Metab*. 2009;11:1163–1172.

232. Vyas AK, Yang K-C, Woo D, et al. Exenatide improves glucose homeostasis and prolongs survival in a murine model of dilated cardiomyopathy. *PLoS ONE*. 2011;6: e17178.

233. Nikolaidis LA, Elahi D, Hentosz T, et al. Recombinant glucagon-like peptide-1 increases myocardial glucose uptake and improves left ventricular performance in conscious dogs with pacing-induced dilated cardiomyopathy. *Circulation*. 2004;110:955–961.

234. Nikolaidis LA, Elahi D, Shen Y, Shennon RP. Active metabolite of GLP-1 mediates myocardial glucose uptake and improves left ventricular performance in conscious dogs with dilated cardiomyopathy. *Am J Physiol Heart Circ Physiol*. 2005;289:H2401–H2408.

235. Wang D, Luo P, Wang Y, et al. Glucagon-like peptide-1 protects against cardiac microvascular injury in diabetes via a cAMP/PKA/Rho-dependent mechanism. *Diabetes*. 2013;62:1697–1708.

236. Takahashi A, Asakura M, Ito S, et al. Dipeptidyl-peptidase IV inhibition improves pathophysiology of heart failure and increases survival rate in pressure-overloaded mice. *Am J Physiol Heart Circ Physiol*. 2013;304:H1361–H1369.

237. Noyan-Ashraf MH, Shikatani EA, Schuiki I, et al. A glucagon-like peptide-1 analog reverses the molecular pathology and cardiac dysfunction of a mouse model of obesity. *Circulation*. 2013;127:74–85.

238. Sokos GG, Bolukoglu H, German J, et al. Effect of glucagon-like peptide-1 (GLP-1) on glycemic control and left ventricular function in patients undergoing coronary artery bypass grafting. *Am J Cardiol*. 2007;100:824–829.

239. Müssig K, Onkü A, Lindauer P, et al. Effects of intravenous glucagon-like peptide-1 on glucose control and hemodynamics after coronary artery bypass surgery in patients with type 2 diabetes. *Am J Cardiol*. 2008;101:646–647.

240. Ceriello A, Novials A, Ortega E, et al. Glucagon-like peptide-1 reduces endothelial dysfunction, inflammation and oxidative stress induced by both hyperglycemia and hypoglycemia in type 1 diabetes. *Diabetes Care*. 2013;36:2346–2350.

241. Mundil D, Cameron-Vendrig A, Husain M. GLP-1 receptor agonists: a clinical perspective on cardiovascular effects. *Diab Vasc Dis Res*. 2012;9:95–108.

242. Okerson T, Chilton RJ. The cardiovascular effects of GLP-1 receptor agonists. *Cardiovasc Ther*. 2012;30:e146–e155.

243. Okerson T, Yan P, Stonehouse A, Brodows R. Effects of exenatide on systolic blood pressure in subjects with type 2 diabetes. *Am J Hypertens*. 2010;23:334–339.

244. Kim M, Platt MJ, Shibasaki T, et al. GLP-1 receptor activation and Epac2 link atrial natriuretic peptide secretion to control of blood pressure. *Nat Med*. 2013;19:567–575.

245. Plutzky J, Garber A, Toft AD, Poulter LR. Meta-analysis demonstrates that liraglutide, a once-daily human GLP-1 analogue, significantly reduces lipids and other markers of cardiovascular risks in type 2 diabetes. *Diabetologia*. 2009;52(Suppl 1):S299–S300.

246. Morinigo R, Moizé V, Musri M, et al. Glucagon-like peptide-1, peptide YY, hunger, and satiety after gastric bypass surgery in morbidly obese subjects. *J Clin Endocrinol Metab*. 2006;91:1735–1740.

247. Patriti A, Facchiano E, Gullà N, Aisa MC, Annetti C. Gut hormone profiles following bariatric surgery favor an anorectic state, facilitate weight loss, and improve metabolic parameters. *Ann Surg*. 2007;245:157–158.

248. Laferrère B, Heshka S, Wang K, et al. Incretin levels and effect are markedly enhanced 1 month after Roux-en-Y gastric bypass surgery in obese patients with type 2 diabetes. *Diabetes Care*. 2007;30:1709–1716.

249. Rodieux F, Giusti V, D'Alessio DA, Suter M, Tappy L. Effects of gastric bypass and gastric banding on glucose kinetics and gut hormone release. *Obesity*. 2008;16:298–305.

250. Laferrere B, Teixeira J, McGinty J, et al. Effect of weight loss by gastric bypass surgery versus hypocaloric diet on glucose and incretin levels in patients with type 2 diabetes. *J Clin Endocrinol Metab*. 2008;93:2479–2485.

251. Korner J, Bessler M, Inabnet W, Taveras C, Holst JJ. Exaggerated glucagon-like peptide-1 and blunted glucose-dependent insulinotropic peptide secretion are associated with Roux-en-Y gastric bypass but not adjustable gastric banding. *Surg Obes Relat Dis*. 2007;3:597–601.

252. Kashyap SR, Daud S, Kelly KR, et al. Acute effects of gastric bypass versus gastric restrictive surgery on beta-cell function and insulinotropic hormones in severely obese patients with type 2 diabetes. *Int J Obes*. 2010;34:462–471.

253. Schauer PR, Burguera B, Ikramuddin S, et al. Effect of laparoscopic Roux-en-Y gastric bypass on type 2 diabetes mellitus. *Ann Surg*. 2003;238:467–484.

254. le Roux CW, Welbourn R, Werling M, et al. Gut hormones as mediators of appetite and weight loss after Roux-en-Y gastric bypass. *Ann Surg*. 2007;246:780–785.

255. Buchwald H, Estok R, Fahrbach K, et al. Weight and type 2 diabetes after bariatric surgery: systematic review and meta-analysis. *Am J Med*. 2009;122:248–256.

256. Rhee NA, Vilsbøll T, Knop FK. Current evidence for a role of GLP-1in Roux-en-Y gastric bypass-induced remission of type 2 diabetes. *Diabetes Obes Metab*. 2012;14:291–298.

257. Jiménez A, Casamitjana R, Flores L, Delgado S, Lacy A, Vidal J. GLP-1 and the long-term outcome of type 2 diabetes mellitus after Roux-en-Y gastric bypass surgery in morbidly obese subjects. *Ann Surg*. 2013;257:894–899.

258. Salehi M, Prigeon RL, D'Alessio DA. Gastric bypass surgery enhances glucagon- like peptide 1-stimulated postprandial insulin secretion in humans. *Diabetes*. 2011;60:2308–2314.

259. Keidar A. Bariatric surgery for type 2 diabetes reversal: the risks. *Diabetes Care*. 2011; 34(Suppl 2):S361–S266.

260. Laferrere B. Do we really know why diabetes remits after gastric bypass surgery? *Endocrine*. 2011;40:162–167.

261. Bose M, Oliván B, Teixeira J, Pi-Sunyer FX, Laferrère B. Do incretins play a role in the remission of type 2 diabetes after gastric bypass surgery: what are evidence? *Obes Surg*. 2009;19:217–229.

262. Mason EE. The mechanism of surgical treatment of type 2 diabetes. *Obes Surg*. 2005;15:459–461.

263. Lee WJ, Hur KY, Lakadawala M, et al. Predicting success of metabolic surgery: age, body mass index, C-peptide, and duration score. *Surg Obes Relat Dis*. 2013;9:379–384.

264. Mingrone G, Panunzi S, De Gaetano A, et al. Bariatric surgery versus conventional medical therapy for type 2 diabetes. *N Engl J Med*. 2012;366:1577–1585.

265. Schauer PR, Kashyap SR, Wolski K, et al. Bariatric surgery versus intensive medical therapy in obese patients with diabetes. *N Engl J Med*. 2012;366:1567–1576.

266. Zimmet P, Alberti KG. Surgery or medical therapy for obese patients with type 2 diabetes? *N Engl J Med*. 2012;366:1635–1636.

267. Mason E. Gila Monster's guide to Surgery for obesity and diabetes. *J Am Coll Surg*. 2008;206:357–360.

268. Service, G. J . Hyperinsulinemic hypoglycemia with nesidioblastosis after gastric-bypass surgery. *N Engl J Med*. 2005;353:249–254.

269. Patti ME. Severe hypoglycemia post-gastric bypass requiring partial pancreatectomy : evidence for inappropriate insulin secretion and pancreatic islet hyperplasia. *Diabetologia*. 2005;48:2236–2240.

270. Butler PC, Dry S, Elashoff R. GLP-1-based therapy for diabetes: what you do not know can hurt you. *Diabetes Care*. 2010;33:453–455.

271. Elashoff M, Matveyenko AV, Gier B, Elashoff R, Butler PC. Pancreatitis, pancreatic, and thyroid cancer with glucagon-like peptide-1-based therapies. *Gastroenterology*. 2011;141:150–156.

272. Spranger J, Gundert-Remy U, Stammschulte T. GLP-1-based therapies: the dilemma of uncertainty. *Gastroenterology*. 2011;141:20–23.
273. Singh S, Chang HY, Richards TM, Weiner JP, Clark JM, Segal JB. Glucagonlike peptide 1-based therapies and risk of hospitalization for acute pancreatitis in type 2 diabetes mellitus: a population-based matched case-control study. *JAMA Intern Med*. 2013;173:534–539.
274. Butler PC, Elashoff M, Elashoff R, Gale EA. A critical analysis of the clinical use of incretin-based therapies: are the GLP-1 therapies safe? *Diabetes Care*. 2013;36:2118–2125.
275. Chiu WY, Shih SR, Tseng CH. A review on the association between glucagon-like peptide-1 receptor agonists and thyroid cancer. *Exp Diabetes Res*. 2012;2012:924168.
276. Bjerre Knudsen L, Madsen LW, Andersen S, et al. Glucagon-like peptide-1 receptor agonists activate rodent thyroid C-cells causing calcitonin release and C-cell proliferation. *Endocrinology*. 2010;151:1473–1486.
277. Drucker DJ, Sherman SI, Bergenstal RM, Buse JB. The safety of incretin-based therapies—review of the scientific evidence. *J Clin Endocrinol Metab*. 2011;96:2027–2031.
278. Nauck MA. A critical analysis of the clinical use of incretin-based therapies: the benefits by far outweigh the potential risks. *Diabetes Care*. 2013;36:2126–2132.
279. Clardy-James S, Chepurny OG, Leech CA, Holz GG, Doyle RP. Synthesis, characterization and pharmacodynamics of vitamin B12-conjugated glucagon-like peptide-1. *ChemMedChem*. 2013;8:582–586.

CHAPTER THREE

Physiology and Therapeutics of the Free Fatty Acid Receptor GPR40

Hui Huang[*], **Meng-Hong Dai**[*,†], **Ya-Xiong Tao**[*]

[*]Department of Anatomy, Physiology and Pharmacology, College of Veterinary Medicine, Auburn University, Auburn, Alabama, USA
[†]Department of Basic Veterinary Medicine, College of Veterinary Medicine, Huazhong Agricultural University, Wuhan, Hubei, China

Contents

Abstract

The G protein-coupled receptor 40 (GPR40) was deorphanized in 2003 as a receptor of medium- and long-chain free fatty acids (FFAs), now also called FFA receptor 1 (FFAR1). Studies have shown that GPR40 not only directly mediates FFA amplification of glucose-stimulated insulin secretion but also indirectly enhances insulin secretion by stimulating incretin release. Therefore, GPR40 has attracted considerable attention as a therapeutic drug target of type 2 diabetes mellitus, and numerous GPR40 ligands have been developed and investigated for their antidiabetic actions. Recently, one of these ligands, TAK-875, has been successfully tested in phase II clinical trials with reduced risk of hypoglycemia. This chapter will summarize studies on GPR40, including its molecular cloning and tissue distribution, physiology, pharmacology, and pathophysiology.

ABBREVIATIONS

AD atopic dermatitis
CCK cholecystokinin
CHS contact hypersensitivity
DHA docosahexaenoic acid
ERK1/2 extracellular signal-regulated kinase 1/2
FFA free fatty acid
FFAR free fatty acid receptor
GALR galanin receptor
GIP glucose-dependent insulinotropic polypeptide
GLP-1 glucagon-like peptide-1
GPCR G protein-coupled receptor
GPR40 G protein-coupled receptor 40
GSIS glucose-stimulated insulin secretion
HFD high-fat diet
hGPR40 human GPR40
$\mathbf{K_{ATP}}$ ATP-sensitive K^+ channels
LTCC L-type Ca^{2+} channel
PI3K phosphatidylinositol 3-kinase
PKB protein kinase B
PLC phospholipase C
PPAR peroxisome proliferator-activated receptor
PUFA polyunsaturated fatty acids
RNAi RNA interference
siRNA small interfering RNA
T2DM type 2 diabetes mellitus
TM transmembrane domain
TZD thiazolidinedione

1. INTRODUCTION

Free fatty acids (FFAs) are an essential energy source, especially during starvation, exercise, and pregnancy,[1] and they are also important modulators of various physiological responses. For pancreatic β-cells, FFAs are required for both basal and glucose-stimulated insulin secretion (GSIS).[2–4] The insulinotropic effect of FFAs depends on the chain length and the degree of saturation; the less saturated long-chain FFAs have a greater effect.[5]

However, chronically elevated plasma FFA concentrations, which are closely associated with obesity and type 2 diabetes mellitus (T2DM), may lead to insulin resistance in the skeletal muscle and liver and pancreatic β-cell dysfunction and apoptosis.[1,6,7]

The effects of FFAs on insulin secretion are traditionally believed to be related to the malonyl-CoA/long-chain acyl-CoA signaling network and glucose-responsive triglyceride/FFA cycling.[8] However, the deorphanization of G protein–coupled receptor 40 (GPR40) suggests the existence of a different mechanism. Activation of GPR40 in the presence of glucose increases cytosolic Ca^{2+} concentrations mainly through $G_{q/11}$–phospholipase C (PLC) pathway and eventually amplifies GSIS.[9] Therefore, GPR40 has become an attractive therapeutic target for T2DM, and numerous ligands of GPR40 have been developed.

GPR40, which is also known as FFA receptor 1 (FFAR1 or FFA1), was identified as an orphan receptor in the search for novel human galanin receptor (GALR) subtypes in 1997.[10] Using reverse pharmacology approaches measuring calcium transients, GPR40 were deorphanized and characterized as being activated by saturated and unsaturated medium- and long-chain fatty acids at physiologically relevant concentrations.[9,11] It was independently deorphanized using high-throughput reporter screening assay by a third group.[12] Of the saturated FFAs, pentadecanoic (C15) and palmitic (C16) acids are the most potent, whereas for the unsaturated FFAs, there is no major effect in chain length and/or degree of saturation on binding affinity or potency.[13] GPR40 is a promiscuous receptor that can be activated by several endogenous ligands, different from most of the typical GPCRs that are activated by one or a few endogenous ligands. In the primary sequence, GPR40 is highly related to other receptors within the same family, exhibiting high homology with GPR41 (FFAR3) and GPR43 (FFAR2),[10,14] whereas it has little homology with GPR119 and GPR120 (FFAR4).[15]

In this chapter, we will summarize the molecular cloning and tissue distribution, physiology, pharmacology, and pathophysiology of GPR40.

2. MOLECULAR CLONING AND TISSUE DISTRIBUTION OF GPR40

2.1. Molecular cloning

The gene encoding human (h) GPR40 was initially identified by Sawzdargo et al. as an intronless gene and located on chromosome 19q13.1 downstream of CD22.[10] In a search for new subtypes of GALRs, the new gene was identified using degenerate primers based on conserved sequences encoding transmembrane domains (TMs) of rat GALR1 and GALR2 and human GALR1.[10] The protein predicted from the gene sequence, composed of 300 amino acids, was 34% and 33% identical to hGPR41 and hGPR43, respectively.[10] Human GPR40 is a member, together with GPR41 and GPR43, of family A G protein-coupled receptors (GPCRs).[10]

Mouse *Gpr40* was cloned by hybridization of the ^{32}P-radiolabeled human *GPR40* coding region to a mouse genomic phage library under low-stringency conditions.[11] It contains an open reading frame of 300 amino acids, which had 83.0% identity to hGPR40. Rat *Gpr40* was cloned using primers designed in the regions of high homology between the human and mouse genes from rat pancreatic and brain cDNA.[11] It is 81.67% identical at the amino acid level to hGPR40.

To date, just three *GPR40* genes described in the text earlier have been cloned and published in peer-reviewed publications. The other putative sequences were predicted and submitted to NCBI. They originated from other primates including *Pan troglodytes*, *Macaca mulatta*, *Pan paniscus*, *Papio anubis*, and *Saimiri boliviensis boliviensis* and from other mammalian species including *Sus scrofa*, *Bos taurus*, *Canis lupus familiaris*, and *Felis catus*, as shown in Fig. 3.1A. Comparison of the putative GPR40 sequences of nonhuman primates with that of the hGPR40 shows that hGPR40 shares 99.67%, 96.33%, 99.67%, 97.00%, and 94.33% homology at the amino acid level with *P. troglodytes*, *M. mulatta*, *P. paniscus*, *P. anubis*, and *S. boliviensis boliviensis*, respectively (Fig. 3.1B). It suggests that there is high similarity among amino acids of human and monkey GPR40. The homologies of GPR40 across the other species using the NCBI BLASTP program are summarized in Fig. 3.1B. Phylogenetic analysis using MEGA5 indicates that hGPR40 has no obvious divergence from those of other mammalian species at the amino acid level (Fig. 3.1C).

In 1997, the structure of GPR40 was predicted by the Kyte–Doolittle hydropathy analysis.[10] The protein has seven TMs, which implies it is a GPCR (Fig. 3.1A).[10] In addition, the protein contains two cysteine residues (C79 and C170) in the TM3 and the second extracellular loop, respectively, which are predicted to form a disulfide bond.[10] This disulfide bond is a signature of family A GPCRs. GPR40 also displays several other characteristics of GPCRs, including one palmitoylation site (C289) in the C-terminal domain, two potential *N*-linked glycosylation sites (N155 and N165) in the second extracellular loop, three α-helical kink sites (P194, P239, and

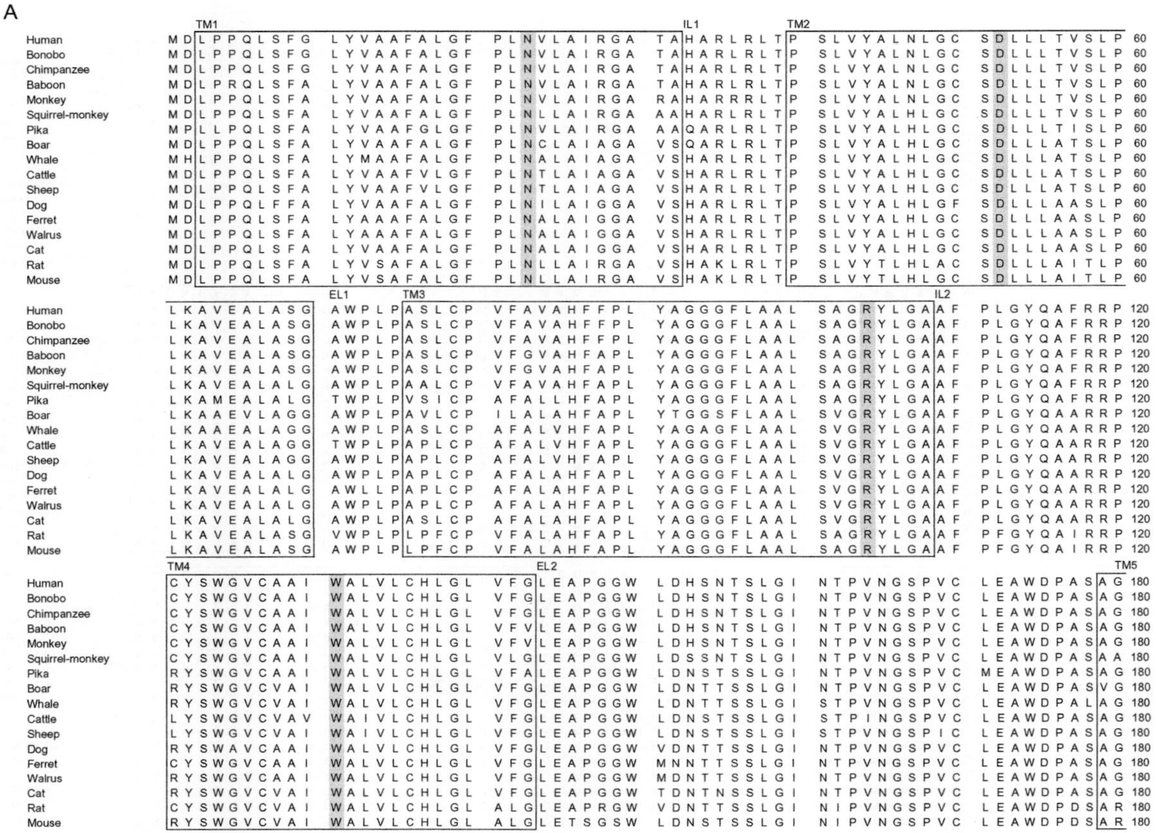

A

Figure 3.1—Cont'd

Figure 3.1 sequence alignment (continued). Multiple sequence alignment across species for transmembrane/loop regions IL3, TM6, EL3, and TM7.

Block 1 (IL3 – TM6), → 240

```
                        IL3                                    TM6
Human            P A R F S L S L L L L L F F L P L A I T A F  C Y Y V G C L R A L L A A  R S G L T H R R K L L  R A A W V A G G A L L L T T L L C V G P Y   240
Bonobo           P A R F S L S L L L L L F F F L P L L A I T A F  C Y Y V G C L R A L L A A  H S G L T H R R K L L  R A A W V A G G A L L L T T L L C V G P Y   240
Chimpanzee       P A R F S L S L L L L L F F F L P L L A I T A F  C Y Y V G C L R A L L A A  H S G L T H R R K L L  R A A W V A G G A L L L T T L L C V G P Y   240
Baboon           P A R F S L S L L L L L F F F L P L L A I T A F  C Y Y V G C L R A L L A A  R S G L T H R R K L L  R A A W V A G G A L L L T T L L C V G P Y   240
Monkey           P A R F S L S L L L L L F F F L P L L A I T A F  C Y Y V G C L R A L L A A  R S G L T H R R K L L  R A A W V A G G A L L L T T L L C V G P Y   240
Squirrel-monkey  P A R F S L S L L L L L F F F L P L L A I T A F  C Y Y V G C L R A L L A A  R S G L S H R R K L L  R A A W V A G G A L L L T T L L C V G P Y   240
Pika             P A R F S L S L L L L L F F F L P L L A L T A F  C Y Y V G C L R A L L A V  R S G L S H R R K L L  K A A W V A G G A L L L T T L L C L G P Y   240
Boar             P A R F S I S L L L L L F F F L P L L A L T A F  C Y Y V G C L R A L L A L  R S G L S H R R K L L  R A A W V A G G A L L L T T L L C L G P Y   240
Whale            P A R F S I S L L L L L F F F L P L L A L V T A F  C Y Y V G C L R A L L A L  H S G L S H R R R K L L  R A A W V A G G A L L L T T L L C L G P Y   240
Cattle           P A R F S I S L L L L L F F F L P L L A L T A F  C Y Y V G C L R A L L A L  R S G L S H R R K L L  K A A W V A G G A L L L T T L F C L G P Y   240
Sheep            P A R L S I S L L L L L F F F L P L L A L S T A F  C Y Y V G C L R A L L A L  R S G L S H R R K L L  R A A W V A G G A L L L T T L L C L G P Y   240
Dog              P A R L T L S L L L L L F F F L P L L A L V T A F  C Y Y V G C L R A L L A V  R S G L S H R K R K L L  R A A W V A G G A L L L T T L L C L G P Y   240
Ferret           P A R L S L S L L L L L F F F L P L L A L V T A F  C Y Y V G C L R A L L A V  R S G L S H R R K L L  R A A W V A G G A L L L T T L L C L G P Y   240
Walrus           P A R L S L S L L L L L F F F L P L L A L T A F  C Y Y V G C L R A L L A V  R S G L S H R R K L L  R A A W V A G G A L L L T T L L C L G P Y   240
Cat              P A R L S L S L L L L L F F F L P L L A L V T A F  C Y Y V G C L R A L L A V  R S G L S H K R K L L  R A A W V A G G A L L L T T L L C L G P Y   240
Rat              P A R L S F S I L L L L F F F L P L L A L V T A F  C Y Y V G C L R A L L A V  R S G L S H R K R K L L  R A A W V A G G A L L L T T L L C L G P Y   240
Mouse            P A R L S F S L L L L L F F F L P L L A L T A F  C Y Y V G C L R A L V A V  R S G L S H K R K L L  R A A W V A G G A L L L T T L L C L G P Y   240
```

Block 2 (EL3 – TM7), → 299/300/301

```
                     EL3                      TM7
Human            N A S N V A S F L Y P N L G G S W R K L L  G L I T G A W S V V  L N P L V T G Y L G  R G P G L K T V C A  A R T Q G G K S Q K   300
Bonobo           N A S N V A S F L Y P N L G G S W R K L L  G L I T G A W S V V  L N P L V T G Y L G  R G P G L K T V C A  A R T Q G G K S Q K   300
Chimpanzee       N A S N V A S F L Y P N L G G S W R K L L  G L I T G A W S V V  L N P L V T G Y L G  R G P G L K T V C A  A R T Q G G K S Q K   300
Baboon           N A S N V A G F L N P N L G G S W R K L L  G L I T G A W S V V  L N P L V T G Y L G  R G P G L K T V C A  A R T Q G G S T S Q K   300
Monkey           N A S N V A S F L N P D M G G G Y W R K L L  G L I T G A W S V V  L N P L V T G Y L G  K G P G L K T G C A  P R T Q G G S T S Q K   300
Squirrel-monkey  N A S N V A S F L N P N L G G S W R K L L  G L I T G A W S V V  L N P L V T G Y L G  R G P G L K T G C L  A R T Q G G T S Q K   300
Pika             N A S N V A G F L H P N S G G G Y W R K L L  G L I T G A W S V V  L N P L V T G Y L G  G G T G R G R G T T R G  A K T N G G A G A S Q K   300
Boar             N A S N V A G F L H P D I G G G W R K Q L  G L I T G A W S V V  L N P L V T G Y L G  G H P G A G R G T V C V  A K T K T G A T Q K   300
Whale            N A S N V A G F L H P N I G G G W R Q L L  G L I T G A W S V V  L N P L V T G Y L G  G R P G R G R V T V C V  A K T K G G T S Q K   299
Cattle           N A S N V A G F L R P N M G G G Y W R K L L  G L I T G A W S V V  L N P L V T G Y L G  G A T R R V T S V  A R T K G G T S Q K   299
Sheep            N A S N V A G F L H P D M G G G Y H W R K L L  G L I T G A W S V V  L N P L V T G Y L G  G G A G R A T I C V  V T S V  299
Walrus           N A S N V A A F L I N P D T G G G W R K L L  G L I T G A W S V V  L N P L V T G Y L G  G G A G R A T I C V  V T S V  299
Cat              N A S N V A F L I N P D L E G S W R K L L  G L I T G A W S V V  L N P L V T G Y L G  G R P G R G R G T I C V  A R T R G T  300
Rat              N A S N V A S F I N P D L G G S W R K L L  G L T G A W S V V  L N P L V T G Y L G  T G P G R G R G T I C V  T R T Q R G T I Q K   300
Mouse            N A S N V A S F I N P D L G G S W R K L L  G L T G A W S V V  L N P L V T G Y L G  T G P G R G R G T I C V  T R T Q R G T I Q K   300
```

Block 3 (C-terminal segment)

```
Human            -
Bonobo           -
Chimpanzee       -
Baboon           -
Monkey           -
Squirrel-monkey  -
Pika             -
Boar             -
Whale            -
Cattle           Q - - - - - - - - - - -                 301
Dog              -
Ferret           -
Walrus           Q Q P P L E R R G T                      P 320
Cat              Q Q P P L E R R G T  E Q G R P A A L A G   P 320
Rat              -
Mouse            -
```

Figure 3.1—Cont'd

B

	Human	Bonobo	Chimpanzee	Baboon	Monkey	Squirrel-monkey	Pika	Boar	Whale	Cattle	Sheep	Dog	Ferret	Walrus	Cat	Rat	Mouse
Human	100.0	99.7	99.7	97.0	96.3	94.3	87.7	85.3	85.0	84.3	85.0	85.6	86.0	86.6	87.6	81.7	83.0
Bonobo		100.0	100.0	97.3	96.7	94.0	88.0	85.0	84.7	84.7	84.7	85.3	85.6	86.3	87.3	82.0	82.7
Chimpanzee			100.0	97.3	96.7	94.0	88.0	85.0	84.7	84.7	84.7	85.3	85.6	86.3	87.3	82.0	82.7
Baboon				100.0	98.7	94.0	88.0	84.7	84.0	84.0	84.0	85.0	85.3	86.0	87.0	82.0	82.7
Monkey					100.0	93.7	87.7	84.7	84.0	84.0	84.0	85.0	85.3	86.0	87.0	82.0	82.7
Squirrel-monkey						100.0	88.0	85.3	84.0	84.0	84.0	87.0	87.0	88.3	88.6	84.0	85.3
Pika							100.0	86.0	85.3	85.3	85.0	86.3	86.3	87.6	87.3	81.0	82.3
Boar								100.0	91.0	88.0	89.0	90.3	90.0	91.3	91.6	82.3	83.7
Whale									100.0	91.0	91.3	89.3	89.3	91.3	91.6	81.0	82.3
Cattle										100.0	96.3	88.3	87.6	89.3	89.3	82.3	82.3
Sheep											100.0	88.6	88.6	89.6	89.7	81.7	83.3
Dog												100.0	94.3	95.3	94.7	83.3	84.0
Ferret													100.0	97.0	93.7	83.3	83.6
Walrus														100.0	96.3	84.6	85.6
Cat															100.0	84.3	85.3
Rat																100.0	95.7
Mouse																	100.0

Figure 3.1—Cont'd

C

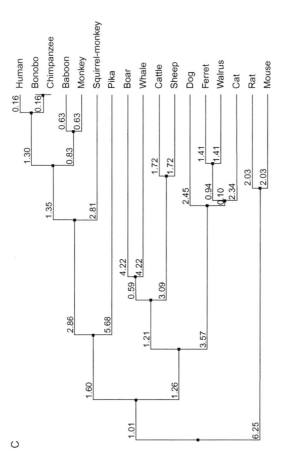

Figure 3.1 Sequence analysis of GPR40 in different species. (A) Alignment of the amino acid sequences between hGPR40 and GPR40 cloned or predicted from other species. The transmembrane domains are boxed. The most conserved residues in the transmembrane domains are shaded in gray. (B) Homology and (C) phylogenetic analysis of GPR40 amino acid sequences from different species. NCBI accession numbers for the proteins are as follows: human (*Homo sapiens*), NP_005294.1; bonobo (*Pan paniscus*), XP_003816203.1; chimpanzee (*Pan troglodytes*), XP_003816322.1; baboon (*Papio anubis*), XP_003915393.1; monkey (*Macaca mulatta*), XP_001094514.1; squirrel monkey (*Saimiri boliviensis boliviensis*), XP_003937408.1; pika (*Ochotona princeps*), XP_004595113.1; boar (*Sus scrofa*), NP_001265712.1; whale (*Orcinus orca*), XP_004284239.1; cattle (*Bos taurus*), XP_875595.1; sheep (*Ovis aries*), XP_004015691.1; dog (*Canis lupus familiaris*), XP_855486.1; ferret (*Mustela putorius furo*), XP_004810492.1; walrus (*Odobenus rosmarus divergens*), XP_004417152.1; cat (*Felis catus*), XP_003998032.1; rat (*Rattus norvegicus*), NP_695216.1; mouse (*Mus musculus*), NP_918946.2.

P273) in TMs, and five protein kinase C (PKC) phosphorylation sites (S212, T215, T287, T293, and S298) in the third intracellular loop and C-terminal tail (Fig. 3.2).[10]

2.2. Tissue distribution

GPR40 exhibits diverse tissue distribution. TaqMan RT-PCR showed that GPR40 is abundantly expressed in the brain and pancreas, with expression in rodent pancreas being localized to insulin-producing β-cells.[11] Others also showed that GPR40 is expressed in both human and rodent pancreatic islets of Langerhans[9,12,16,17] and in islet cell tumors.[18] However, no GPR40 mRNA is detected in mouse α-TC glucagonoma cells, human tissue extracts from the glucagonoma, and the acinar cell tumor specimen.[19]

GPR40 is also expressed ubiquitously in the human brain, with the highest expression in the substantia nigra and medulla oblongata by quantitative RT-PCR.[11] The expression of GPR40 is also detected in the monkey brain using a putative GPR40-specific antibody.[20] Ma et al.[20,21] further demonstrated expression of GPR40 in the subventricular and subgranular zones that form the neurogenic niche and in the spinal cord and pituitary gland. However, GPR40 mRNAs are not detectable in the human and rat brains by northern blotting and RT-PCR, respectively.[9,12] The study from Kotarsky et al. based on northern blotting revealed expression of GPR40 in the skeletal muscle, liver, and heart, inconsistent with findings from Briscoe et al. and Itoh et al. using quantitative RT-PCR in either human or rat tissues.[9,11] A recent study showed that in neonatal rat brain, GPR40 is not expressed in the neurons or glial cells; rather, it is highly expressed in microvessels.[22] Therefore, whether GPR40 is expressed in the brain and cells expressing the receptor in different species remains to be clarified by further studies.

GPR40 mRNA was initially described as being predominantly expressed in islet β-cells by TaqMan quantitative PCR.[11] Other studies also reported GPR40 expression in insulin-producing cell lines. Expression of GPR40 is highest in the mouse insulinoma cell line MIN6[23] and also present in mouse βTC-3, hamster HIT-T15, and rat insulinoma INS-1E cells.[9,11,12,24] GPR40 expression has also been reported in non-β-cells, including pancreatic α-cells,[25-27] monocytes,[11,26] enteroendocrine cells,[28-30] osteoclasts,[31-33] and human breast cancer cell lines MCF-7[34] and MDA-MB-231.[35] The roles of GPR40 in these cells will be summarized in the succeeding text.

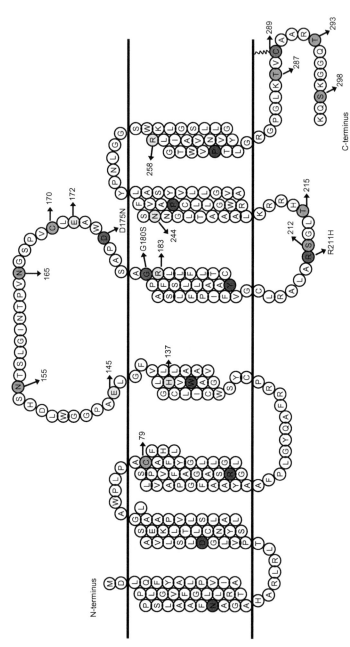

Figure 3.2 Two-dimensional illustration of hGPR40. The most conserved residues in the transmembrane domains are shaded in red. Residues that have naturally occurring mutations or polymorphisms are shaded in yellow. Residues that are important for the receptor activation are shaded in blue. Residues that are the potential sites for disulfide bond, palmitoylation, N-linked glycosylation, and PKC phosphorylation are shaded in gray. (See color plate.)

3. PHYSIOLOGY OF GPR40
3.1. Insulin secretion and glucolipotoxicity

The importance of GPR40 in acutely potentiating GSIS was first characterized in 2003. Itoh et al. showed that long-chain FFAs stimulate insulin secretion in the presence of glucose from mouse insulinoma MIN6 cells, and the stimulatory effect is eliminated when mouse *Gpr40* is knocked down by specific small interfering RNA (siRNA).[9] Subsequently, a number of studies confirmed this observation. Retrovirus-mediated RNA interference (RNAi) against rat *Gpr40* abolished palmitic acid's effect to amplify insulin secretion in rat insulinoma NIS-1E cells.[24] The GPR40-specific antagonist GW1100 abolished the effects of GW9508 or partially reduced that of linoleic acid on GSIS in MIN6 cells.[36] Antisense oligonucleotide[16] and antagonist ANT203[37] treatment of mouse islets inhibited linoleic acid- and palmitic acid-stimulated insulin secretion, respectively. The elimination of the acute potentiation of GSIS by palmitate was also observed in islets isolated from $Gpr40^{-/-}$ mice.[37–39]

The direct involvement of GPR40 in mediating FFA stimulation of GSIS has been further verified by *in vivo* studies. Latour et al. reported that insulin secretion stimulated by the intravenous injection of Intralipid is reduced by about 50% in $Gpr40^{-/-}$ mice, whereas insulin secretion stimulated by glucose is not altered.[39] Similar results were observed using selective β_3-adrenergic receptor agonist CL-316,243, which increases plasma FFAs; the increased plasma insulin levels of $Gpr40^{-/-}$ mice after administration of CL-316,243 are only 50% of that of $Gpr40^{+/+}$ mice.[40] In addition to impairing insulin secretion in response to FFAs, deletion of GPR40 also decreases insulin secretion stimulated by glucose and arginine by about 60% *in vivo*.[41]

Though it is well accepted that GPR40 mediates the acute effect of FFAs on insulin secretion, it is controversial whether GPR40 is also involved in the chronic effect of FFAs on pancreatic β-cell function. Steneberg et al. first showed that long-term (48 h) exposure to palmitic acid does not interfere with insulin secretion of islets isolated from $Gpr40^{-/-}$ mice with a mixed C57Bl6/SV129 background.[37,38] However, in this study, they also showed that $Gpr40^{-/-}$ mice fed on high-fat diet (HFD) develop less severe hyperinsulinemia, glucose intolerance, and insulin resistance than that of $Gpr40^{+/+}$ mice and that $Gpr40^{-/-}$ mice have no increase in stored lipids in the livers or in serum triglyceride levels.[38] In addition, transgenic overexpression of *Gpr40* under the *Ipf1/Pdx1* promoter mimicking chronic signaling of GPR40 results in pancreatic β-cell dysfunction and diabetes.[38] Another study performed by Brownlie et al. also showed that *Gpr40*$^{-/-}$ mice backcrossed onto C57Bl6/J from a 50:50 C57Bl6/129 background are protected from the effects of HFD on glucose tolerance and insulin sensitivity.[37] Based on these studies, the authors suggested that GPR40 antagonists should be applied for the prevention and treatment of T2DM.

However, a number of studies subsequently have provided evidences for the opposite conclusion, which emphasizes the importance of developing GPR40 agonists as therapeutic drug for T2DM. Islets isolated from *Gpr40*$^{-/-}$ mice were first shown not to be protected from FFA inhibition of insulin secretion upon a 72 h exposure.[39] Lan et al.[40] and Kebede et al.[42] then independently observed that *Gpr40*$^{-/-}$ mice in a C57Bl/6 background have similar hyperinsulinemia and glucose intolerance and that *Gpr40*$^{-/-}$ mice have similar increased liver triglyceride levels as the *Gpr40*$^{+/+}$ mice and are not protected from HFD-induced hepatic steatosis. Moreover, overexpression of human *GPR40* under the mouse insulin II promoter in diabetic mice increases insulin secretion and improves oral glucose tolerance.[43]

Studies using synthetic antagonists and agonists of GPR40 have provided further support that GPR40 agonists are beneficial for T2DM. Obese Zucker rats intraperitoneally treated with the GPR40 antagonist DC260126 for 8 weeks show no change in blood glucose and lipid levels, and no improvement in glucose tolerance, though they show increased insulin tolerance.[44] A number of studies have investigated the antidiabetic activity of GPR40 synthetic agonists.[45–53] Oral administration of AMG-837, a potent GPR40 agonist,[49] 30 min before an intraperitoneal glucose challenge in Zucker rats increases GSIS and decreases glucose excursions.[46] The effect is sustained following once daily dosing for 21 days.[46] Most importantly, another GPR40 agonist, TAK-875, has already been applied in 12-week phase II clinical trials by different groups.[54–57] In these studies, TAK-875 significantly improves glycemic control in patients with T2DM and causes much lower risk of hypoglycemia than glimepiride.[54–56]

3.2. Incretin secretion

FFAs stimulate the secretion of incretin hormones, and FFA receptors, such as GPR120 and GPR119, are involved in this physiological function.[58,59] GPR40 is also expressed in the enteroendocrine cells throughout the gastrointestinal tract, especially in the cells secreting glucagon-like peptide-1

(GLP-1) (L cells), glucose-dependent insulinotropic polypeptide (GIP) (K cells), and cholecystokinin (CCK) (I cells), which implies that GPR40 mediates FFA enhancement of GSIS not only through its direct function in the pancreatic β-cells but also through its indirect function in the regulation of incretin secretion.[28,30,40,60]

Edfalk et al. generated $Gpr40^{lacZ/lacZ}$ mice by replacing $Gpr40$ with the $lacZ$ coding frame and found that plasma insulin levels are decreased and glucose levels are increased after HFD at 60 min after oral glucose administration in $Gpr40^{lacZ/lacZ}$ mice.[28] Further analysis of plasma incretin concentrations found that, compared with $Gpr40^{+/+}$ mice, total GLP-1 and GIP concentrations are decreased after HFD in $Gpr40^{lacZ/lacZ}$ mice whereas they are not changed at 60 min after oral glucose administration.[28]

GPR40 synthetic agonist AM-1638 dose-dependently increases GLP-1 and GIP secretion from the cells isolated from fetal rat intestines.[50] AM-1638 was further found to increase insulin and GLP-1 levels 15 min after glucose challenge in NONcNZO10/LtJ diabetic mice.[50] The increase of GLP-1 levels elicited by AM-1638 is eliminated in $Gpr40^{-/-}$ mice.[50] These findings altogether suggest that GPR40 is essential for the secretion of GLP-1 and GIP in response to AM-1638.

CCK secretion from I cells is also directly regulated by GPR40. Liou et al. isolated I cells from mouse duodenal mucosa using fluorescence-activated cell sorting method.[30] They found that the increase of intracellular concentrations of Ca^{2+} stimulated by linoleic acid is reduced by 50% and the increase of CCK secretion induced by linolenic acid is eliminated in I cells isolated from $Gpr40^{-/-}$ mice.[30] Further *in vivo* studies showed that the increase of plasma CCK levels is significantly reduced in the $Gpr40^{-/-}$ mice after olive oil gavage.[30]

3.3. Glucagon secretion

Glucagon, a peptide hormone secreted by the pancreatic α-cells, elevates plasma glucose levels by promoting gluconeogenesis and glycogenolysis. Though the mechanism is still not fully understood, the release of glucagon requires elevated intracellular Ca^{2+} concentrations.[61] FFAs stimulate glucagon secretion and GPR40 has been shown, though still controversial, to be involved.

Previously, $GPR40$ expression in the pancreatic α-cells was detected neither in rodent islets[9,11] nor in human islets.[62] However, Flodgren et al. observed coexpression of GPR40 and glucagon in mouse islets.[25]

In response to linoleic acid stimulation, glucagon secretion is dose-dependently increased in hamster glucagonoma In–R1–G9 cells and in mouse islets with a greater secretion levels at lower glucose concentrations.[25] The effect is enhanced by *Gpr40* overexpression and attenuated by antisense knockdown of *Gpr40*.[25] Wang et al. also observed *Gpr40* expression in rat pancreatic islets and glucagon secretion in response to linoleic acid.[27] Further *in vivo* studies also found that the increase of plasma glucagon levels stimulated by CL-316,243 is attenuated in *Gpr40*$^{-/-}$ mice.[40]

However, Yashiro et al. used GPR40 synthetic agonist TAK-875 and found that TAK-875 has no effect on glucagon secretion in both rat and human islets.[61] Moreover, phase II clinical trials also reported that after 2 weeks of treatment, TAK-875 has no effect on fasting glucagon levels, but inhibits glucagon secretion during the oral glucose tolerance test.[54,57] These controversial results indicate that further investigation on whether GPR40 modulates glucagon secretion needs to be performed.

3.4. Inflammation

Recently, a study conducted by Fujita et al. showed that activation of GPR40 attenuates cutaneous immune inflammation.[63] Using the RT-PCR method, they detected *GPR40* expression in immortal human epidermal keratinocyte HaCaT cells and normal human epidermal keratinocyte NHEK cells.[63] CCL5 and CCL17 are important chemokines in the pathogenesis of allergic skin disease atopic dermatitis (AD) and CXCL10 is important in contact hypersensitivity (CHS) pathogenesis. Fujita et al. found that GPR40 agonist GW9508 dose-dependently suppresses cytokine-induced CCL5 and CCL17 mRNA and protein production in HaCaT cells and CCL5 and CXCL10 production in NHEK cells.[63] The suppressive effects of GW9508 in HaCaT cells are reversed by the treatment of low doses of pertussis toxin or siRNA against GPR40.[63]

Fujita et al. also detected *Gpr40* expression in keratinocytes of the mouse ear using RT-PCR and *in situ* hybridization methods.[63] They then generated AD and CHS mouse model induced by 2,4-dinitrofluorobenzene treatment. Administration of GW9508 topically to the ear significantly reduces ear swelling with reduced epidermal thickening and cell infiltration in the dermis.[63] Further analysis showed that GW9508 treatment suppresses *CCL5* and *CXCL10* expression in AD and CHS mouse models separately.[63] This study has identified the importance of GPR40 as a potential target for the treatment of allergic skin diseases.

3.5. Bone density

Different types of fatty acids exert different effects on the bone health; for example, omega-3 fatty acids inhibit bone turnover and increase bone mineral density, whereas omega-6 fatty acids promote bone loss.[64] The underlying mechanism, which might involve GPR120, peroxisome proliferator-activated receptor (PPAR), and Toll-like receptor, is still not fully understood.

Recently, *GPR40* expression has been detected in murine osteocytes, human osteoblasts, and human primary osteoclasts.[32] The use of thiazolidinediones (TZDs), the agonists of PPAR-γ, in the treatment of T2DM increases the risk of bone fracture by decreasing bone formation and increasing adiposity in the bone marrow.[65] Mieczkowska et al. found that osteocyte apoptosis induced by TZDs is attenuated by the siRNA treatment against *Gpr40,* which decreased the activation of extracellular signal-regulated kinase 1/2 (ERK1/2) and p38 in mouse osteocyte-like MLO-Y4 cells and mouse primary osteoblasts.[32] Silencing of *Gpr40* also attenuates the decrease of osteoclast formation induced by TZDs.[32] This study implies that GPR40 mediates TZD-induced osteocyte apoptosis and osteoclast formation.

In contrast, another study performed by Cornish et al. found that the GPR40/GPR120 agonist, GW9508, inhibits osteoclastogenesis in murine bone marrow cultures.[31] Consistent with this finding, Wauquier et al. also observed positive effects of GPR40 on bone density.[33] Wauquier et al. showed that *Gpr40*$^{-/-}$ mice displays osteoporotic phenotype.[33] GPR40 agonist GW9508 inhibits osteoclast differentiation by inhibiting the nuclear factor-κB signaling pathway in primary mouse bone marrow cultures and in mouse osteoclast precursor RAW264.7 cells.[33] The inhibitory effect of GW9508 is diminished in *Gpr40*$^{-/-}$ mouse primary cell cultures and in RAW264.7 cells treated with short hairpin RNA against GPR40.[33] In addition, administration of GW9508 prevents bone loss induced by ovariectomy *in vivo*.[33] These findings altogether indicate that development of GPR40 agonists for the treatment of T2DM may also prevent patients from osteoporosis. The controversial effects of TZDs and GW9508 on bone density via GPR40 may result from biased signaling of GPR40, the existence of which has been widely reported in many other GPCRs.[66]

3.6. Neurogenesis and pain control

Polyunsaturated fatty acids (PUFAs), such as docosahexaenoic acid (DHA) and arachidonic acids, are known to improve the cognitive function of humans[67,68] and rodents,[69,70] whereas PUFA deficiency may lead to cognitive impairment.[71] PUFA receptor GPR40 but not GPR120 is highly expressed in the central nervous system of humans, primates, and rodents,[11,20,72] implying that GPR40 might play a crucial role in neural function.

Ma et al. generated global cerebral ischemia monkey by clamping both the innominate and left subclavian arteries for 20 min.[21] They found that GPR40 protein expression and GPR40-positive cells in the subgranular zone, a well-known neurogenic niche, are increased significantly in the second week after ischemia.[21] Afterward, the transcription factor cAMP response element-binding protein, which is known to be involved in adult neurogenesis, learning, and memory, is shown to be coexpressed with GPR40 in the subgranular zone.[73] Furthermore, the differentiation of rat neural stem cells transfected with GPR40 into mature neuron is promoted by PUFA,[74] which indicates that it is GPR40 that mediates PUFA regulation of neurogenesis.

GPR40 may also have functions in the pain regulatory system.[72,75] It was reported that formalin-induced pain behavior of mice is remarkably attenuated by intracerebroventricular injection of GPR40 agonist DHA or GW9508.[75] Treatment of mice with the μ-opioid receptor antagonist β-funaltrexamine or anti-β-endorphin antiserum inhibits the attenuation of DHA on formalin-induced pain.[75] Most importantly, injection of DHA or GW9508 significantly enhances the protein expression of β-endorphin in the hypothalamus.[75] The antinociceptive effect of DHA and GW9508 suggests that the development of GPR40 agonists may also be important for pain control therapy.

3.7. Cell proliferation

It has been previously shown that unsaturated fatty acids such as oleic acids stimulate cell proliferation, whereas saturated fatty acids such as palmitic acids induce apoptosis in human breast cancer MDA–MB-231 cells.[76,77] The involvement of GPR40 in mediating fatty acid regulation of cell proliferation was first identified in 2004.[34] Yonezawa et al. found that GPR40 expression is increased during the beginning and end of cell proliferation in human breast cancer MCF-7 cells using RT-PCR.[34] They also found that activation of MCF-7 cells by oleate and linoleate results in increased intracellular Ca^{2+} concentrations, and this is partially sensitive to pertussis toxin treatment.[34]

Subsequently, the causal relationship between the proliferation of breast cancer cells stimulated by oleate and GPR40 was further studied.[35] Over-expression of *GPR40* in human breast cancer cells, MDA–MB-231, T47D, and MCF-7, amplifies cell proliferation induced by oleate, whereas knockdown of *GPR40* using siRNA inhibits cell proliferation in MDA–MB-231 cells.[35] It has been documented that the PLC, ERK1/2, Src, and phosphatidylinositol 3-kinase (PI3K)/protein kinase B (PKB/Akt) signaling pathways are associated with the proliferative signal mediated by GPR40.[35,78]

GPR40 has also been reported to mediate the proliferation of human bronchial epithelial cells induced by TZDs.[79] PPAR-γ synthetic agonist, but not endogenous agonist, stimulates the proliferation of human bronchial epithelial cells accompanied with enhanced Ca^{2+} mobilization.[79] Ca^{2+} mobilization and cell proliferation induced by TZDs are not affected by the treatment of PPAR-γ antagonist GW-9662 but are significantly inhibited by downregulation of *GPR40* using siRNA.[79] The earlier-mentioned data indicate that GPR40 plays a vital role in the human bronchial epithelial cell proliferation stimulated by TZDs.

3.8. Taste preference

In comparison with carbohydrates and proteins that are perceived by certain taste receptors, the detection of fat stimuli was previously considered to be dependent on olfactory and textural cues. Recently, however, studies suggest that long-chain FFAs released from triglycerides can be perceived by the taste system.[80] FFA receptors GPR40 and GPR120 may play a crucial role in sensing fatty acid.[58,81–83]

The sensing of fatty acids mediated by GPR40 and GPR120 was investigated in mice.[82] The expressions of both GPR40 and GPR120 are detected separately in type I and II taste bud cells of mice using immunohistochemistry.[82] During 48 h two-bottle preference tests, *Gpr40*$^{-/-}$ and *Gpr120*$^{-/-}$ mice show a reduced preference for linoleic acid and oleic acid, whereas they respond normally to other taste modalities.[82] *Gpr40*$^{-/-}$ and *Gpr120*$^{-/-}$ mice also display weaker glossopharyngeal taste nerve responses.[82] These findings indicate that GPR40 and GPR120 mediate the oral perception of fat.

However, unlike GPR120, the expression of *GPR40* in taste buds is not detected in rat[81] and, most importantly, not in humans[83] using RT-PCR. The authors suggested that the discrepancy in GPR40 expression in rodents and humans might be attributed to methodological differences or species differences.[82]

4. PHARMACOLOGY OF GPR40

4.1. Ligand binding and receptor activation

Saturated and unsaturated medium- and long-chain fatty acids are endogenous agonists for both human and rodent GPR40s. FFAs, including palmitic acid (C16), oleic acid (C18:1), elaidic acid (C18:1), linoleic acid (C18:2), linolenic acid (C18:3), arachidonic acid (C20:4), eicosapentaenoic acid (C20:5), and DHA (C22:6), were found to dose-dependently increase intracellular Ca^{2+} concentrations in CHO and HEK293 cells stably expressing GPR40.[9,11] The EC_{50} values of these FFAs are in micromolar range.[9,11] There is also one study reporting that short-chain fatty acids, butyric acid (C4), activate mouse GPR40 in *Xenopus* oocytes.[84] In addition, branched-chain fatty acids, including phytanic acid and pristanic acid, which are known to induce peroxisomal impairments, also activate GPR40 by inducing intracellular Ca^{2+} mobilization.[85] Berberine is known for its antidiabetic effects, and recently, it was also reported to be able to activate hGPR40 with an EC_{50} of 760 nM.[86] Very recently, it was shown that trans-arachidonic acids, arachidonic acid derivatives generated under nitrative stress, are also GPR40 agonists, inducing both intracellular Ca^{2+} mobilization and ERK1/2 phosphorylation.[22]

Ever since the deorphanization of GPR40 in 2003, tremendous efforts have been made on the development of GPR40 agonists. GW9508 is an agonist of GPR40 and GPR120 with an EC_{50} of 47.86 nM at hGPR40.[36] TAK-875 is a selective GPR40 agonist (EC_{50} for hGPR40, 14 nM) and has been successfully conducted in phase II clinical trials in T2DM patients[87] as described in the preceding text. Several other synthetic GPR40 agonists, such as AMG-837 (EC_{50} for hGPR40, 13 nM),[49] TUG-770 (EC_{50} for hGPR40, 6 nM),[53] and CNX-011-67 (EC_{50} for hGPR40, 0.24 nM),[88] are highly selective and orally bioavailable and therefore are good candidates for the treatment of T2DM. In addition, other synthetic ligands that are not specific for GPR40 have also been reported to activate the receptor, such as TZDs (compound C, EC_{50} at hGPR40, 10 nM),[89] and weight loss supplements conjugated linoleic acids.[90]

Based on the structure–activity relationship studies, several synthetic antagonists of GPR40 have also been developed. GW1100[36] and DC260126[91] were reported to dose-dependently inhibit agonist-stimulated increase of intracellular Ca^{2+} concentrations with IC_{50}s in micromolar range. Compound 15i is another potent antagonist of GPR40 with an IC_{50} of 20 nM.[92]

Several binding assays have been developed to directly measure interactions between GPR40 and its ligands. Hara et al.[93] set up a binding assay based on flow cytometry by immobilizing GPR40 on beads. Fluorescent-labeled ligand, C1-BODIPY-C12, bound GPR40 in a saturable and reversible way, whereas other GPR40 ligands can competitively bind with the receptor.[93] Radiolabeled binding assay has also been reported by labeling AMG-837 and AM-1638 with ^{3}H.[94]

Structure–function studies have identified several residues of GPR40 important for receptor activation. Residues H137, R183, N244, and R258 directly interact with the carboxylate groups of GW9508 or linoleic acid.[95,96] E145 and E172 interact with R183 and R258, respectively, by forming ionic interaction to constrain the receptor in an inactive form.[97] Alanine mutations of E145 and E172 result in potent constitutive activation of GPR40 (Fig. 3.2).[97]

However, mutations of R183 and R258 have no major effect on the function of other ligands, including DHA, AM-1638, and AM-8182.[94] Lin et al. found that GPR40 agonists display cooperation; AM-837 and AM-1638 positively cooperate with each other, whereas DHA and AM-8182 negatively or neutrally cooperate with AMG-837 and AM-1638 in binding assays, respectively.[94] This study implies that there are three allosteric binding sites on GPR40.[94] Interestingly, though DHA and AM-8182 negatively cooperate with AMG-837 in binding assays, they show positive cooperation in functional assays.[94] Xiong et al.[60] further showed that GPR40 agonists positively cooperate with each other to induce GLP-1 release in mice. Combination of allosteric agonists that have positive cooperation makes lower effective doses possible and therefore has promising prospects in the treatment of T2DM.

4.2. Signaling pathways

$G_{q/11}$–PLC mediates the major signaling pathway of GPR40.[9,11] Activation of GPR40 results in the cleavage of phospholipid phosphatidylinositol 4,5-bisphosphate (PIP2) to diacylglycerol (DAG) and inositol 1,4,5-trisphosphate (IP3). IP3 activates IP3 receptor on the endoplasmic reticulum membrane and therefore releases Ca^{2+} from the endoplasmic reticulum. DAG activates protein kinase D1, which induces F-actin depolymerization and therefore insulin secretion.[98] Insulin secretion mediated by GPR40 also involves the inhibition of ATP-sensitive K^+ channels (K_{ATP}), which results in the depolarization of cell membrane, and therefore extracellular Ca^{2+} influx through L-type Ca^{2+} channels (Fig. 3.3).[9] These studies have been previously reviewed by Mancini et al.[99] and will not be discussed in detail herein.

GPR40 has also been reported to activate other signaling pathways. Activation of hGPR40 inhibits cyclic AMP production, suggesting that GPR40 also couples to G_i and also stimulates ERK1/2 phosphorylation in CHO cells.[9] G_i was later reported to mediate GW9508-inhibited immune inflammation[63] and oleic acid–induced cell proliferation.[35] Besides G_i, PLC, ERK1/2, Src, and PI3K/Akt are also implicated in cell proliferation mediated by GPR40.[35] Activation of ERK1/2 and p38 by GPR40 has been reported to mediate TZD-induced osteocyte apoptosis.[32] The detailed signaling pathways of GPR40 have been discussed earlier in the text and illustrated in Fig. 3.3.

Similar to some other GPCRs, GPR40 displays high constitutive activities in ^{35}S-GTPγS binding assays.[100] However, Stoddart et al.[100] showed that this was an artifact attributed to the existence of endogenous agonists. The pseudomorphic constitutive activities can be eliminated with the

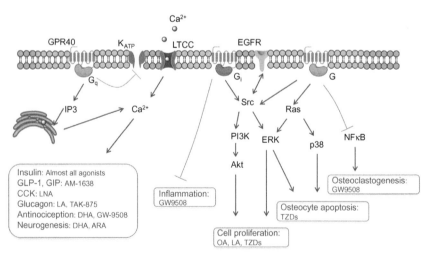

Figure 3.3 Ligands and mechanisms of GPR40-mediated physiological functions. Green arrows indicate stimulatory effects, whereas red arrows indicate inhibitory effects. Abbreviations in the figure: ARA, arachidonic acid; LA, linoleic acid; LNA, linolenic acid; OA, oleic acid; NF-κB, nuclear factor-κB; EGFR, epidermal growth factor receptor. (See color plate.)

treatment of fatty acid-free bovine serum albumin, which binds fatty acids, in cell lines both endogenously and exogenously expressing GPR40.[100]

5. PATHOPHYSIOLOGY OF GPR40

Up to now, human genomic studies have identified three nucleotide substitutions in GPR40 that result in coding variations, including one polymorphism, R211H, and two mutations, D175N and G180S (Fig. 3.2).

In 2005, Hamid et al.[101] studied the potential relationship between human *GPR40* gene variation and insulin resistance in T2DM patients. Using direct sequencing in Danish Caucasian subjects, they identified two single nucleotide substitutions in *GPR40* that result in R211H polymorphism located in the third intracellular loop and D175N mutation located in the second extracellular loop.[101] The D175N mutation was found in heterozygous form in both normal glucose tolerance subjects and T2DM patients. Further sequence analysis of the family members of one T2DM patient harboring the D175N mutation found three more members carrying the heterozygous mutation and observed no cosegregation of this mutation with T2DM.[101] The R211H polymorphism, detected in both T2DM patients and control subjects with a similar frequency (24%), was found to not associate with changes in insulin secretion or pancreatic β-cell function.[101] Functional studies by measuring inositol phosphate turnover showed that R211H GPR40 has similar signaling properties (EC_{50} and maximal response) as the wild-type receptor, whereas D175N mutant has decreased maximal signaling in response to eicosatriynoic acid by 39%, although the EC_{50} is similar to wild-type receptor.[101] No data on the expression and ligand binding properties of these receptors were reported.

However, the GPR40 R211H polymorphism was reported to affect the variation of insulin secretory capacity in other ethnic groups.[102] The allele frequency of H211 in Japanese was 78.4%.[102,103] By comparing clinical and metabolic parameters among Japanese men, Ogawa et al. found that histidine homozygotes have significantly higher serum insulin levels, homeostasis model of insulin resistance, and pancreatic β-cell function than arginine homozygotes do.[102] Therefore, the authors suggested that genetic variation of GPR40 might alter the insulin secretory capacity.[102]

Another *GPR40* mutation, G180S located in TM5, was identified in heterozygous form in the Sicilian population in Italy.[104] The frequency of G180S is 0.75%, which, interestingly, increases with the severity of obesity.[104] Subjects carrying *GPR40* G180S mutation have significantly decreased insulin secretion than those carrying wild-type *GPR40* during oral glucose tolerance test.[104] Functional studies showed that G180S mutants have similar expression levels (both total and cell surface) as the wild-type receptor but diminished increase of intracellular Ca^{2+} concentrations in response to oleic acid.[104] Whether this defect is due to ligand binding or G protein coupling/activation is not clear. Binding experiments were not performed.

6. CONCLUSIONS

GPCRs are versatile signaling molecules regulating almost all physiological processes, including energy homeostasis and glucose homeostasis.[105,106] They also prove to be important therapeutic targets. Drugs targeting GPCRs account for 30% of current pharmaceutical sales. GPR40 is a member of family A GPCRs. Since the deorphanization of GPR40 in 2003, various functions of GPR40 have been identified and characterized, including insulin secretion, incretin secretion, glucagon secretion, immune inflammation, bone density, neurogenesis, pain control, taste preference, and cell proliferation. Tremendous progress has been made on the development of GPR40 agonists in the treatment of T2DM, and most importantly, one of these ligands, TAK-875, has been successfully tried in phase II clinical trials. However, our understanding of some functions of GPR40 and the underlying mechanisms are still controversial and incomplete. Further studies are warranted to address these questions.

ACKNOWLEDGMENTS

Funding for our work on obesity and diabetes was provided by the American Diabetes Association Grant 1-12-BS212, the National Institutes of Health Grant R15-DK077213, Auburn University Intramural Grant Program, and the Interdisciplinary Grant and Animal Health and Diseases Research Program of the College of Veterinary Medicine at Auburn University (to Y.-X.T.).

REFERENCES

1. Boden G, Shulman GI. Free fatty acids in obesity and type 2 diabetes: defining their role in the development of insulin resistance and β-cell dysfunction. *Eur J Clin Invest.* 2002;32:14–23.
2. Crespin SR, Greenough WB, Steinberg D. Stimulation of insulin secretion by infusion of free fatty acids. *J Clin Invest.* 1969;48:1934–1943.
3. Stein DT, Esser V, Stevenson BE, et al. Essentiality of circulating fatty acids for glucose-stimulated insulin secretion in the fasted rat. *J Clin Invest.* 1996;97:2728–2735.
4. Boden G, Chen XH, Iqbal N. Acute lowering of plasma fatty acids lowers basal insulin secretion in diabetic and nondiabetic subjects. *Diabetes.* 1998;47:1609–1612.

5. Stein DT, Stevenson BE, Chester MW, et al. The insulinotropic potency of fatty acids is influenced profoundly by their chain length and degree of saturation. *J Clin Invest.* 1997;100:398–403.

6. Poitout V, Amyot J, Semache M, Zarrouki B, Hagman D, Fontes G. Glucolipotoxicity of the pancreatic beta cell. *Biochim Biophys Acta.* 2010;1801:289–298.

7. Bergman RN, Ader M. Free fatty acids and pathogenesis of type 2 diabetes mellitus. *Trends Endocrinol Metab.* 2000;11:351–356.

8. Nolan CJ, Madiraju MS, Delghingaro-Augusto V, Peyot ML, Prentki M. Fatty acid signaling in the β-cell and insulin secretion. *Diabetes.* 2006;55(Suppl 2):S16–S23.

9. Itoh Y, Kawamata Y, Harada M, et al. Free fatty acids regulate insulin secretion from pancreatic β cells through GPR40. *Nature.* 2003;422:173–176.

10. Sawzdargo M, George SR, Nguyen T, Xu S, Kolakowski LF, O'Dowd BF. A cluster of four novel human G protein-coupled receptor genes occurring in close proximity to CD22 gene on chromosome 19q13.1. *Biochem Biophys Res Commun.* 1997;239:543–547.

11. Briscoe CP, Tadayyon M, Andrews JL, et al. The orphan G protein-coupled receptor GPR40 is activated by medium and long chain fatty acids. *J Biol Chem.* 2003;278:11303–11311.

12. Kotarsky K, Nilsson NE, Olde B, Owman C. Progress in methodology. Improved reporter gene assays used to identify ligands acting on orphan seven-transmembrane receptors. *Pharmacol Toxicol.* 2003;93:249–258.

13. Vangaveti V, Shashidhar V, Jarrod G, Baune BT, Kennedy RL. Free fatty acid receptors: emerging targets for treatment of diabetes and its complications. *Ther Adv Endocrinol Metab.* 2010;1:165–175.

14. Brown AJ, Jupe S, Briscoe CP. A family of fatty acid binding receptors. *DNA Cell Biol.* 2005;24:54–61.

15. Fredriksson R, Hoglund PJ, Gloriam DEI, Lagerstrom MC, Schioth HB. Seven evolutionarily conserved human rhodopsin G protein-coupled receptors lacking close relatives. *FEBS Lett.* 2003;554:381–388.

16. Salehi A, Flodgren E, Nilsson NE, et al. Free fatty acid receptor 1 (FFA(1)R/GPR40) and its involvement in fatty-acid-stimulated insulin secretion. *Cell Tissue Res.* 2005;322:207–215.

17. Feng DD, Luo Z, Roh SG, et al. Reduction in voltage-gated K^+ currents in primary cultured rat pancreatic β-cells by linoleic acids. *Endocrinology.* 2006;147:674–682.

18. Tomita T, Masuzaki H, Noguchi M, et al. GPR40 gene expression in human pancreas and insulinoma. *Biochem Biophys Res Commun.* 2005;338:1788–1790.

19. Odori S, Hosoda K, Tomita T, et al. GPR119 expression in normal human tissues and islet cell tumors: evidence for its islet-gastrointestinal distribution, expression in pancreatic beta and alpha cells, and involvement in islet function. *Metabolism.* 2013;62:70–78.

20. Ma D, Tao B, Warashina S, et al. Expression of free fatty acid receptor GPR40 in the central nervous system of adult monkeys. *Neurosci Res.* 2007;58:394–401.

21. Ma D, Lu L, Boneva NB, et al. Expression of free fatty acid receptor GPR40 in the neurogenic niche of adult monkey hippocampus. *Hippocampus.* 2008;18:326–333.

22. Honore JC, Kooli A, Hamel D, et al. Fatty acid receptor Gpr40 mediates neuromicrovascular degeneration induced by transarachidonic acids in rodents. *Arterioscler Thromb Vasc Biol.* 2013;33:954–961.

23. Miyazaki JI, Araki K, Yamato E, et al. Establishment of a pancreatic β-cell line that retains glucose-inducible insulin-secretion: special reference to expression of glucose transporter isoforms. *Endocrinology.* 1990;127:126–132.

24. Shapiro H, Shachar S, Sekler I, Hershfinkel M, Walker MD. Role of GPR40 in fatty acid action on the β cell line INS-1E. *Biochem Biophys Res Commun.* 2005;335:97–104.

25. Flodgren E, Olde B, Meidute-Abaraviciene S, Winzell MS, Ahren B, Salehi A. GPR40 is expressed in glucagon producing cells and affects glucagon secretion. *Biochem Biophys Res Commun*. 2007;354:240–245.

26. Hirasawa A, Itsubo C, Sadakane K, et al. Production and characterization of a mono-clonal antibody against GPR40 (FFAR1; free fatty acid receptor 1). *Biochem Biophys Res Commun*. 2008;365:22–28.

27. Wang L, Zhao Y, Gui B, et al. Acute stimulation of glucagon secretion by linoleic acid results from GPR40 activation and $[Ca^{2+}]i$ increase in pancreatic islet α-cells. *J Endocrinol*. 2011;210:173–179.

28. Edfalk S, Steneberg P, Edlund H. Gpr40 is expressed in enteroendocrine cells and mediates free fatty acid stimulation of incretin secretion. *Diabetes*. 2008;57:2280–2287.

29. Parker HE, Habib AM, Rogers GJ, Gribble FM, Reimann F. Nutrient-dependent secretion of glucose-dependent insulinotropic polypeptide from primary murine K cells. *Diabetologia*. 2009;52:289–298.

30. Liou AP, Lu X, Sei Y, et al. The G-protein-coupled receptor GPR40 directly mediates long-chain fatty acid-induced secretion of cholecystokinin. *Gastroenterology*. 2011;140:903–912.

31. Cornish J, MacGibbon A, Lin JM, et al. Modulation of osteoclastogenesis by fatty acids. *Endocrinology*. 2008;149:5688–5695.

32. Mieczkowska A, Basle MF, Chappard D, Mabilleau G. Thiazolidinediones induce osteocyte apoptosis by a G protein-coupled receptor 40-dependent mechanism. *J Biol Chem*. 2012;287:23517–23526.

33. Wauquier F, Philippe C, Leotoing L, et al. The free fatty acid receptor G protein-coupled receptor 40 (GPR40) protects from bone loss through inhibition of osteoclast differentiation. *J Biol Chem*. 2013;288:6542–6551.

34. Yonezawa T, Katoh K, Obara Y. Existence of GPR40 functioning in a human breast cancer cell line, MCF-7. *Biochem Biophys Res Commun*. 2004;314:805–809.

35. Hardy S, St-Onge GG, Joly E, Langelier Y, Prentki M. Oleate promotes the prolifer-ation of breast cancer cells via the G protein-coupled receptor GPR40. *J Biol Chem*. 2005;280:13285–13291.

36. Briscoe CP, Peat AJ, McKeown SC, et al. Pharmacological regulation of insulin secre-tion in MIN6 cells through the fatty acid receptor GPR40: identification of agonist and antagonist small molecules. *Br J Pharmacol*. 2006;148:619–628.

37. Brownlie R, Mayers RM, Pierce JA, Marley AE, Smith DM. The long-chain fatty acid receptor, GPR40, and glucolipotoxicity: investigations using GPR40-knockout mice. *Biochem Soc Trans*. 2008;36:950–954.

38. Steneberg P, Rubins N, Bartoov-Shifman R, Walker MD, Edlund H. The FFA recep-tor GPR40 links hyperinsulinemia, hepatic steatosis, and impaired glucose homeostasis in mouse. *Cell Metab*. 2005;1:245–258.

39. Latour MG, Alquier T, Oseid E, et al. GPR40 is necessary but not sufficient for fatty acid stimulation of insulin secretion *in vivo*. *Diabetes*. 2007;56:1087–1094.

40. Lan H, Hoos LM, Liu L, et al. Lack of FFAR1/GPR40 does not protect mice from high-fat diet-induced metabolic disease. *Diabetes*. 2008;57:2999–3006.

41. Alquier T, Peyot ML, Latour MG, et al. Deletion of GPR40 impairs glucose-induced insulin secretion *in vivo* in mice without affecting intracellular fuel metabolism in islets. *Diabetes*. 2009;58:2607–2615.

42. Kebede M, Alquier T, Latour MG, Semache M, Tremblay C, Poitout V. The fatty acid receptor GPR40 plays a role in insulin secretion *in vivo* after high-fat feeding. *Diabetes*. 2008;57:2432–2437.

43. Nagasumi K, Esaki R, Iwachidow K, et al. Overexpression of GPR40 in pancreatic β-cells augments glucose-stimulated insulin secretion and improves glucose tolerance in normal and diabetic mice. *Diabetes*. 2009;58:1067–1076.

44. Zhang X, Yan G, Li Y, Zhu W, Wang H. DC260126, a small-molecule antagonist of GPR40, improves insulin tolerance but not glucose tolerance in obese Zucker rats. *Biomed Pharmacother*. 2010;64:647–651.

45. Doshi LS, Brahma MK, Sayyed SG, et al. Acute administration of GPR40 receptor agonist potentiates glucose-stimulated insulin secretion *in vivo* in the rat. *Metabolism*. 2009;58:333–343.

46. Lin DC, Zhang J, Zhuang R, et al. AMG 837: a novel GPR40/FFA1 agonist that enhances insulin secretion and lowers glucose levels in rodents. *PLoS One*. 2011;6: e27270.

47. Sasaki S, Kitamura S, Negoro N, et al. Design, synthesis, and biological activity of potent and orally available G protein-coupled receptor 40 agonists. *J Med Chem*. 2011;54:1365–1378.

48. Tsujihata Y, Ito R, Suzuki M, et al. TAK-875, an orally available G protein-coupled receptor 40/free fatty acid receptor 1 agonist, enhances glucose-dependent insulin secretion and improves both postprandial and fasting hyperglycemia in type 2 diabetic rats. *J Pharmacol Exp Ther*. 2011;339:228–237.

49. Houze JB, Zhu L, Sun Y, et al. AMG 837: a potent, orally bioavailable GPR40 agonist. *Bioorg Med Chem Lett*. 2012;22:1267–1270.

50. Luo J, Swaminath G, Brown SP, et al. A potent class of GPR40 full agonists engages the enteroinsular axis to promote glucose control in rodents. *PLoS One*. 2012;7: e46300.

51. Mikami S, Kitamura S, Negoro N, et al. Discovery of phenylpropanoic acid derivatives containing polar functionalities as potent and orally bioavailable G protein-coupled receptor 40 agonists for the treatment of type 2 diabetes. *J Med Chem*. 2012;55:3756–3776.

52. Negoro N, Sasaki S, Ito M, et al. Identification of fused-ring alkanoic acids with improved pharmacokinetic profiles that act as G protein-coupled receptor 40/free fatty acid receptor 1 agonists. *J Med Chem*. 2012;55:1538–1552.

53. Christiansen E, Hansen SV, Urban C, et al. Discovery of TUG-770: a highly potent free fatty acid receptor 1 (FFA1/GPR40) agonist for treatment of type 2 diabetes. *ACS Med Chem Lett*. 2013;4:441–445.

54. Burant CF, Viswanathan P, Marcinak J, et al. TAK-875 versus placebo or glimepiride in type 2 diabetes mellitus: a phase 2, randomised, double-blind, placebo-controlled trial. *Lancet*. 2012;379:1403–1411.

55. Naik H, Vakilynejad M, Wu J, et al. Safety, tolerability, pharmacokinetics, and pharmacodynamic properties of the GPR40 agonist TAK-875: results from a double-blind, placebo-controlled single oral dose rising study in healthy volunteers. *J Clin Pharmacol*. 2012;52:1007–1016.

56. Kaku K, Araki T, Yoshinaka R. Randomized, double-blind, dose-ranging study of TAK-875, a novel GPR40 agonist, in Japanese patients with inadequately controlled type 2 diabetes. *Diabetes Care*. 2013;36:245–250.

57. Araki T, Hirayama M, Hiroi S, Kaku K. GPR40-induced insulin secretion by the novel agonist TAK-875: first clinical findings in patients with type 2 diabetes. *Diabetes Obes Metab*. 2012;14:271–278.

58. Mo XL, Wei HK, Peng J, Tao YX. Free fatty acid receptor GPR120 and pathogenesis of obesity and type 2 diabetes mellitus. *Prog Mol Biol Transl Sci*. 2013;114:251–276.

59. Mo XL, Yang Z, Tao YX. Targeting GPR119 for the potential treatment of type 2 diabetes mellitus. *Prog Mol Biol Transl Sci*. 2014;121:95–131.

60. Xiong Y, Swaminath G, Cao Q, et al. Activation of FFA1 mediates GLP-1 secretion in mice. Evidence for allosterism at FFA1. *Mol Cell Endocrinol*. 2013;369:119–129.

61. Yashiro H, Tsujihata Y, Takeuchi K, Hazama M, Johnson PR, Rorsman P. The effects of TAK-875, a selective G protein-coupled receptor 40/free fatty acid 1 agonist, on

insulin and glucagon secretion in isolated rat and human islets. *J Pharmacol Exp Ther.* 2012;340:483–489.

62. Tomita T, Masuzaki H, Iwakura H, et al. Expression of the gene for a membrane-bound fatty acid receptor in the pancreas and islet cell tumours in humans: evidence for GPR40 expression in pancreatic beta cells and implications for insulin secretion. *Diabetologia.* 2006;49:962–968.

63. Fujita T, Matsuoka T, Honda T, Kabashima K, Hirata T, Narumiya S. A GPR40 agonist GW9508 suppresses CCL5, CCL17, and CXCL10 induction in keratinocytes and attenuates cutaneous immune inflammation. *J Invest Dermatol.* 2011;131:1660–1667.

64. Orchard TS, Pan XL, Cheek F, Ing SW, Jackson RD. A systematic review of omega-3 fatty acids and osteoporosis. *Br J Nutr.* 2012;107:S253–S260.

65. Grey A, Bolland M, Gamble G, et al. The peroxisome proliferator-activated receptor-γ agonist rosiglitazone decreases bone formation and bone mineral density in healthy postmenopausal women: a randomized, controlled trial. *J Clin Endocrinol Metab.* 2007;92:1305–1310.

66. Whalen EJ, Rajagopal S, Lefkowitz RJ. Therapeutic potential of β-arrestin- and G protein-biased agonists. *Trends Mol Med.* 2011;17:126–139.

67. McGahon BM, Martin DSD, Horrobin DF, Lynch MA. Age-related changes in synaptic function: analysis of the effect of dietary supplementation with ω-3 fatty acids. *Neuroscience.* 1999;94:305–314.

68. Yamashima T. A putative link of PUFA, GPR40 and adult-born hippocampal neurons for memory. *Prog Neurobiol.* 2008;84:105–115.

69. Okaichi Y, Ishikura Y, Akimoto K, et al. Arachidonic acid improves aged rats' spatial cognition. *Physiol Behav.* 2005;84:617–623.

70. He C, Qu X, Cui L, Wang J, Kang JX. Improved spatial learning performance of fat-1 mice is associated with enhanced neurogenesis and neuritogenesis by docosahexaenoic acid. *Proc Natl Acad Sci U S A.* 2009;106:11370–11375.

71. Conquer J, Tierney M, Zecevic J, Bettger W, Fisher R. Fatty acid analysis of blood plasma of patients with Alzheimer's disease, other types of dementia, and cognitive impairment. *Lipids.* 2000;35:1305–1312.

72. Nakamoto K, Nishinaka T, Matsumoto K, et al. Involvement of the long-chain fatty acid receptor GPR40 as a novel pain regulatory system. *Brain Res.* 2012;1432:74–83.

73. Boneva NB, Yamashima T. New insights into "GPR40-CREB interaction in adult neurogenesis" specific for primates. *Hippocampus.* 2012;22:896–905.

74. Ma D, Zhang M, Larsen CP, et al. DHA promotes the neuronal differentiation of rat neural stem cells transfected with GPR40 gene. *Brain Res.* 2010;1330:1–8.

75. Nakamoto K, Tokuyama S. A long chain fatty acid receptor GPR40 as a novel pain control system. *Nihon Shinkei Seishin Yakurigaku Zasshi.* 2012;32:233–237.

76. Hardy S, El-Assaad W, Przybytkowski E, Joly E, Prentki M, Langelier Y. Saturated fatty acid-induced apoptosis in MDA-MB-231 breast cancer cells—a role for cardiolipin. *J Biol Chem.* 2003;278:31861–31870.

77. Hardy S, Langelier Y, Prentki M. Oleate activates phosphatidylinositol 3-kinase and promotes proliferation and reduces apoptosis of MDA-MB-231 breast cancer cells, whereas palmitate has opposite effects. *Cancer Res.* 2000;60:6353–6358.

78. Soto-Guzman A, Robledo T, Lopez-Perez M, Salazar EP. Oleic acid induces ERK1/2 activation and AP-1 DNA binding activity through a mechanism involving Src kinase and EGFR transactivation in breast cancer cells. *Mol Cell Endocrinol.* 2008;294:81–91.

79. Gras D, Chanez P, Urbach V, Vachier I, Godard P, Bonnans C. Thiazolidinediones induce proliferation of human bronchial epithelial cells through the GPR40 receptor. *Am J Physiol Lung Cell Mol Physiol.* 2009;296:L970–L978.

80. Kawai T, Fushiki T. Importance of lipolysis in oral cavity for orosensory detection of fat. *Am J Physiol Regul Integr Comp Physiol.* 2003;285:R447–R454.

81. Matsumura S, Mizushige T, Yoneda T, et al. GPR expression in the rat taste bud relating to fatty acid sensing. *Biomed Res.* 2007;28:49–55.
82. Cartoni C, Yasumatsu K, Ohkuri T, et al. Taste preference for fatty acids is mediated by GPR40 and GPR120. *J Neurosci.* 2010;30:8376–8382.
83. Galindo MM, Voigt N, Stein J, et al. G protein-coupled receptors in human fat taste perception. *Chem Senses.* 2012;37:123–139.
84. Stewart G, Hira T, Higgins A, Smith CP, McLaughlin JT. Mouse GPR40 heterologously expressed in *Xenopus* oocytes is activated by short-, medium-, and long-chain fatty acids. *Am J Physiol Cell Physiol.* 2006;290:C785–C792.
85. Kruska N, Reiser G. Phytanic acid and pristanic acid, branched-chain fatty acids associated with Refsum disease and other inherited peroxisomal disorders, mediate intracellular Ca^{2+} signaling through activation of free fatty acid receptor GPR40. *Neurobiol Dis.* 2011;43:465–472.
86. Rayasam GV, Tulasi VK, Sundaram S, et al. Identification of berberine as a novel agonist of fatty acid receptor GPR40. *Phytother Res.* 2010;24:1260–1263.
87. Negoro N, Sasaki S, Mikami S, et al. Discovery of TAK-875: a potent, selective, and orally bioavailable GPR40 agonist. *ACS Med Chem Lett.* 2010;1:290–294.
88. Gowda N, Dandu A, Singh J, et al. Treatment with CNX-011-67, a novel GPR40 agonist, delays onset and progression of diabetes and improves beta cell preservation and function in male ZDF rats. *BMC Pharmacol Toxicol.* 2013;14:28.
89. Zhou C, Tang C, Chang E, et al. Discovery of 5-aryloxy-2,4-thiazolidinediones as potent GPR40 agonists. *Bioorg Med Chem Lett.* 2010;20:1298–1301.
90. Schmidt J, Liebscher K, Merten N, et al. Conjugated linoleic acids mediate insulin release through islet G protein-coupled receptor FFA1/GPR40. *J Biol Chem.* 2011;286:11890–11894.
91. Hu H, He LY, Gong Z, et al. A novel class of antagonists for the FFAs receptor GPR40. *Biochem Biophys Res Commun.* 2009;390:557–563.
92. Humphries PS, Benbow JW, Bonin PD, et al. Synthesis and SAR of 1,2,3,4-tetrahydroisoquinolin-1-ones as novel G-protein-coupled receptor 40 (GPR40) antagonists. *Bioorg Med Chem Lett.* 2009;19:2400–2403.
93. Hara T, Hirasawa A, Sun Q, et al. Flow cytometry-based binding assay for GPR40 (FFAR1; free fatty acid receptor 1). *Mol Pharmacol.* 2009;75:85–91.
94. Lin DC, Guo Q, Luo J, et al. Identification and pharmacological characterization of multiple allosteric binding sites on the free fatty acid 1 receptor. *Mol Pharmacol.* 2012;82:843–859.
95. Sum CS, Tikhonova IG, Neumann S, et al. Identification of residues important for agonist recognition and activation in GPR40. *J Biol Chem.* 2007;282:29248–29255.
96. Tikhonova IG, Sum CS, Neumann S, et al. Bidirectional, iterative approach to the structural delineation of the functional "chemoprint" in GPR40 for agonist recognition. *J Med Chem.* 2007;50:2981–2989.
97. Sum CS, Tikhonova IG, Costanzi S, Gershengorn MC. Two arginine-glutamate ionic locks near the extracellular surface of FFAR1 gate receptor activation. *J Biol Chem.* 2009;284:3529–3536.
98. Ferdaoussi M, Bergeron V, Zarrouki B, et al. G protein-coupled receptor (GPR)40-dependent potentiation of insulin secretion in mouse islets is mediated by protein kinase D1. *Diabetologia.* 2012;55:2682–2692.
99. Mancini AD, Poitout V. The fatty acid receptor FFA1/GPR40 a decade later: how much do we know? *Trends Endocrinol Metab.* 2013;24:398–407.
100. Stoddart LA, Brown AJ, Milligan G. Uncovering the pharmacology of the G protein-coupled receptor GPR40: high apparent constitutive activity in guanosine 5′-O-(3-[^{35}S]thio)triphosphate binding studies reflects binding of an endogenous agonist. *Mol Pharmacol.* 2007;71:994–1005.

101. Hamid YH, Vissing H, Holst B, et al. Studies of relationships between variation of the human G protein-coupled receptor 40 gene and type 2 diabetes and insulin release. *Diabet Med.* 2005;22:74–80.
102. Ogawa T, Hirose H, Miyashita K, Saito I, Saruta T. GPR40 gene Arg211His polymorphism may contribute to the variation of insulin secretory capacity in Japanese men. *Metabolism.* 2005;54:296–299.
103. Haga H, Yamada R, Ohnishi Y, Nakamura Y, Tanaka T. Gene-based SNP discovery as part of the Japanese Millennium Genome Project: identification of 190562 genetic variations in the human genome. *J Hum Genet.* 2002;47:605–610.
104. Vettor R, Granzotto M, De Stefani D, et al. Loss-of-function mutation of the GPR40 gene associates with abnormal stimulated insulin secretion by acting on intracellular calcium mobilization. *J Clin Endocrinol Metab.* 2008;93:3541–3550.
105. Tao YX, Yuan ZH, Xie J. G protein-coupled receptors as regulators of energy homeostasis. *Prog Mol Biol Transl Sci.* 2013;114:1–43.
106. Tao YX, Liang XF. G protein-coupled receptors as regulators of glucose homeostasis. *Prog Mol Biol Transl Sci.* 2014;121:1–21.

CHAPTER FOUR

Targeting GPR119 for the Potential Treatment of Type 2 Diabetes Mellitus

Xiu-Lei Mo[1], Zhao Yang[1], Ya-Xiong Tao

Department of Anatomy, Physiology and Pharmacology, College of Veterinary Medicine, Auburn University, Auburn, Alabama, USA
[1]Both the authors contributed equally to this work.

Contents

Abstract

G protein-coupled receptor 119 (GPR119) was initially identified as an orphan receptor through mining the human genome database. In 2005, GPR119 was deorphanized and shown to be a receptor for fatty acid metabolites, including some phospholipids and fatty acid amide derivatives. GPR119 regulates various physiological processes that improve glucose homeostasis, including glucose-dependent insulin secretion from pancreatic β-cells, gastrointestinal incretin hormone secretion, appetite control, epithelial electrolyte homeostasis, gastric emptying, and β-cell proliferation and cytoprotection. Therefore, GPR119, the sensing receptor for fatty acid metabolites, represents a novel drug target for the treatment of type 2 diabetes mellitus.

Progress in Molecular Biology and Translational Science, Volume 121
ISSN 1877-1173
http://dx.doi.org/10.1016/B978-0-12-800101-1.00004-1

95

ABBREVIATIONS

2-OG 2-oleoylglycerol
ECL extracellular loop
FFA free fatty acid
GIP glucose-dependent insulinotropic polypeptide
GLP-1 glucagon-like peptide 1
GPCR G protein-coupled receptor
GRK G protein-coupled receptor kinase
hGPR119 human GPR119
ICL intracellular loop
KO knockout
LPC lysophosphatidylcholine
OEA oleoylethanolamide
PKA protein kinase A
PKC protein kinase C
PPAR peroxisome proliferator-activated receptor
PYY peptide YY
T2DM type 2 diabetes mellitus
TRP transient receptor potential
WT wild-type

1. INTRODUCTION

Obesity and its closely associated comorbidities, such as type 2 diabetes mellitus (T2DM), have reached epidemic proportions in developed countries and are increasing at an alarming rate in developing countries, causing monumental health, societal, and economic problems.[1] The reason why obesity is associated with T2DM is not well understood.[2] Three criteria have been proposed for being a physiological link between obesity and T2DM.[3,4] These criteria are (1) that its circulation level is proportional to the adiposity level, (2) that increasing its circulating level induces insulin resistance, and (3) that decreasing its circulating level reduces insulin resistance. So far, only free fatty acids (FFAs) meet these three criteria in human subjects.[3,4]

Recently, multiple lines of investigations have demonstrated that FFAs not only are essential sources of energy and important metabolic substrates[5] but also serve as signaling molecules involved in various physiological mechanisms.[6] Nuclear receptors, including peroxisome proliferator-activated receptors (PPARs), were shown to act as FFA sensors and to regulate expression of genes that maintain homeostasis in various physiological conditions.[7] However, some FFA-mediated physiological effects were shown to be

PPAR independent and might be mediated by other mechanisms,[8,9] such as signaling through G protein-coupled receptors (GPCRs).

In the last decade, a series of family A, rhodopsin-like GPCRs, including GPR40, GPR41, GPR43, GPR84, GPR119, and GPR120, have been identified as receptors for FFAs and their derivatives through the GPCR deorphanization strategy.[10] GPR120 and GPR40 are receptors for long-chain FFAs. GPR84 is activated by medium-chain FFAs, while GPR43 and GPR41 are activated by short-chain FFAs.[11–13]

GPR119 is unique in that it is activated by a series of FFA metabolites.[14] This chapter reviews the studies on GPR119, from its cloning and tissue distribution to its physiological roles in regulating glucose-dependent insulin secretion, gastrointestinal hormone secretion, appetite and energy homeostasis, epithelial electrolyte homeostasis, gastric emptying, and β-cell proliferation and cytoprotection. We will then summarize the studies on the pharmacology of the receptor, including GPR119 agonist development, receptor activation, and signaling pathways and regulation.

2. MOLECULAR CLONING AND LOCALIZATION OF GPR119

2.1. Molecular cloning

GPR119 was initially identified by Fredriksson *et al.* as a novel member of class A, rhodopsin-like orphan GPCRs, that has no close primary sequence relative in the human genome,[10] although it was subsequently shown by Griffin to have some structural similarity with cannabinoid receptors using homology clustering analysis.[15] It belongs to the MECA (melanocortin, endothelial differentiation gene, cannabinoid, and adenosine) cluster of receptors. GPR119 has been independently studied and described in the literature using various synonyms including SNORF25,[16] GPCR2,[17] 19AJ,[18] OSGPR116,[15] HGPCR2,[17] MGC119957,[19] and glucose-dependent insulinotropic receptor.[20]

The gene for GPR119 is located on the short arm of X-chromosome (Xp26.1) containing a single exon (ENST00000276218) with no introns. Although the initial report of Fredriksson *et al.* suggested that a protein of 468 amino acids could be encoded by *GPR119* gene cloned from rat (accession number AY288429), it has now been widely accepted that in most species (including human), the GPR119 is encoded by a single transcript with an open reading frame of 1008 base pairs producing a protein of 335 amino acid residues,[16] and the overestimation in the initial report seems to result from the artificial additional C-terminal sequence.

GPR119 has been cloned from human, mouse, rat, and fugu.[10] As shown in Fig. 4.1A, GPR119 has also been predicted to exist in other mammalian species, including chimpanzee, rhesus monkey, olive baboon, rabbit, dog, horse, cow, sheep, pig, and opossum, and in several nonmammalian species, including chicken, zebra finch, green anole, and zebra fish, using the NCBI gene prediction method GNOMON. The homologies of GPR119 in different species are summarized in Fig. 4.1B. As shown in Fig. 4.1B, the mouse and rat GPR119s share 82.1% and 73.7% identity with human GPR119 (hGPR119), respectively. But there is only 39.2% identity between fugu and hGPR119s. Furthermore, hGPR119 was shown to be significantly divergent from nonmammalian species at amino acid level using phylogenetic analysis (Fig. 4.1C).

GPR119 is a member of family A, rhodopsin-like GPCRs, with seven transmembrane domains (TMs) connected by alternating extracellular loops (ECLs) and intracellular loops (ICLs). The N terminus is extracellular whereas the C terminus is intracellular. The hGPR119 shares some similar features with other rhodopsin-like family A GPCRs, including two signature motifs, DRY near the intracellular end of TM3 and NPxxY in TM7 that are highly conserved in this family. There seems to be no potential N-linked glycosylation site for hGPR119 since there is no consensus sequence for N-linked glycosylation, N-x-S/T, found in hGPR119. hGPR119 has extremely short N terminus with only one methionine residue, but relatively long C terminus (Fig. 4.2). Whether this receptor has other types of glycosylation, such as O-linked glycosylation, has not been reported yet.

The phosphorylation of GPCRs on the cytoplasmic tail or ICLs by G protein-coupled receptor kinases (GRKs) or second messenger-associated kinases, such as protein kinase A (PKA) and protein kinase C (PKC), plays an important role in cellular signal transduction. We identified several potential phosphorylation sites in GPR119 using online prediction tool.[21] There are four potential GRK phosphorylation sites within the C terminus. These sites are Thr327, Ser329, Ser330, and Ser331. In addition, there are six potential PKA/PKC phosphorylation sites. Four of these sites, Ser193, Ser196, Ser214, and Ser219, are within the third ICL, and the other two sites, Ser309 and Ser322, are within C terminus tail. The precise mechanism and sites of phosphorylation remain to be investigated.

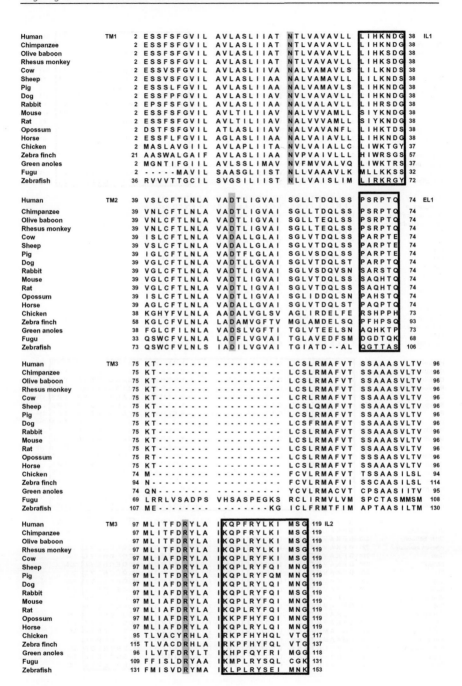

Figure 4.1—Cont'd

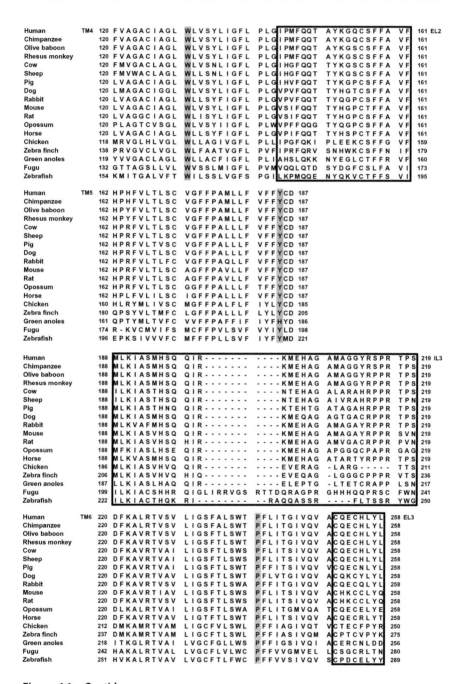

Figure 4.1—Cont'd

```
Human          TM7  259  V L E R Y L W L L G   V G N S L L N P L I   Y A Y W Q  283
Chimpanzee          259  V L E R Y L W L L G   V G N S L L N P L I   Y A Y W Q  283
Olive baboon        259  V L E R Y L W L L G   V G N S L L N P L I   Y A Y R Q  283
Rhesus monkey       259  V L E R Y L W L L G   V G N S L L N P L I   Y A Y R Q  283
Cow                 259  V L E Q Y L W L L G   V G N S L L N P L I   Y A Y W Q  283
Sheep               259  V L E R Y L W L L G   V G N S L L N P L I   Y A Y R Q  283
Pig                 259  V L E R Y L W L L G   V G N S L L N P L I   Y A Y R Q  283
Dog                 259  A L E R Y L W L L G   V G N S L L N P L I   Y A Y W Q  283
Rabbit              259  V L E K Y L W L L G   V G N S L L N P L I   Y A Y W Q  283
Mouse               259  V L E K Y L W V M G   V G N S L L N P L I   Y A Y W Q  283
Rat                 259  V L E S Y L W L L G   V G N S L L N P L I   Y A Y W Q  283
Opossum             259  V L E Q Y L W L L G   V S N S L L N P L I   Y A Y W Q  283
Horse               251  V L E S Y L W L L G   L C N S L L N P L L   Y S Y W Q  275
Chicken             276  V I E S F L W L L G   M G N S L L N P L L   Y S Y W Q  300
Zebra finch         257  L L E R Y L W L L G   V C N S L V N P L I   Y A Y R Q  281
Green anoles        281  V L Q N Y L W L L G   L S N S M I N P L V   Y A F W Q  305
Fugu                290  F L E N H L W L L G   T T N S L V N P L V   Y A C W Q  314
Zebrafish

Human    C-terminus  284  K E V R L Q L Y H M   A L G - V K K V L T   S F L L F L S - A R   N C G P E R P R E S  321
Chimpanzee          284  K E V R L Q L Y H M   A L G - V K K V L T   S F L L F L S - A R   N C G V E R P R E S  321
Olive baboon        284  K E V R L Q L Y H M   A L G - V K K A L T   S F L L F L S - A R   N G G V E R P R E S  321
Rhesus monkey       284  K E V R L Q L Y H M   A L G - V K K A L T   S F L L F L S - A R   N G G P E R P R E S  321
Cow                 284  K E V R Q Q F S Q M   A L A - M K K W L A   A C L L L L S - A R   D G G P E G R R E S  321
Sheep               284  K E V R Q Q F S Q M   A L A - M K K G L T   T C L L R L S - P R   D G G P E G R R E S  321
Pig                 284  K E V R R Q F S Q M   A L G - M K K G L A   A F L L L L S - A R   N G G P E R P R E S  321
Dog                 284  K E V Q Q Q L Y E M   V L G - V K K G F T   S C L L L L S - A R   A G R P E R P R E N  321
Rabbit              284  K E V R L Q L Y Q M   A L G - V K K G L T   A F L F L L V S A K   N G G P E R P R E T  322
Mouse               284  R E V R Q Q L Y H M   A L G - V K K F F T   S I L L L L P - A R   N R G P E R T R E S  321
Rat                 284  R E V R Q Q L C H M   A L G * A Q T L Y T   C D A Q T L Y - T S   S L V T G Q T E Q T  452
Opossum             284  K D V R M Q I Y Q M   A V G - V K K K F V   L L F F I L P - - R   D P G P G E P R E S  320
Horse               284  K E V R Q Q L Y Q M   A L G - V K K R L T   S F L L L L L - A R   D G A P E G P R E S  321
Chicken             276  K D V R R Q L S Q L   A A G - - - - - - V   K R K A L R - - - V   G K G C C F P S R G  306
Zebra finch         301  R D V Q L Q L S Q L   A A G - - - - - - V   K R R V L L H - - L   G N S R C F L G K D  332
Green anoles        282  K E V R L Q I C Q M   C V C - - - - - - M   K I K V F P L F H G   H S Q S H A P S R A  315
Fugu                306  K E V R L Q L A A M   F S C - F T G R L L   A A G T P S V A E R   R I L P S V V A V A  344
Zebrafish           315  R E V R D Q I C E L   F A Y - I K A G F C   R E R R S K G G D G   R T K D H L T V A H  353

Human               322  - - - - - - - - - -   - S C H I V T I S S   S E F D G - - - -   - - - - - - - - - -  335
Chimpanzee          322  - - - - - - - - - -   - S C H I V T I S S   - - - - - - - - - -                       335
Olive baboon        322  - - - - - - - - - -   - S C H I V T I S N   S E F D G - - - - -   - - - - - - - - - -  335
Rhesus monkey       322  - - - - - - - - - -   - S C H I V T I S N   S E F D G - - - - -   - - - - - - - - - -  335
Cow                 322  - - - - - - - - - -   - V R Y I T T M S H   S E L E G - - - - -   - - - - - - - - - -  335
Sheep               322  - - - - - - - - - -   - V R Y I T T M S H   S E L E G - - - - -   - - - - - - - - - -  335
Pig                 322  - - - - - - - - - -   - S C H I T T I S Q   S E L D G - - - - -   - - - - - - - - - -  335
Dog                 322  - - - - - - - - - -   - F C P I A T I S H   S Q L D G - - - - -   - - - - - - - - - -  335
Rabbit              323  - - - - - - - - - -   - S C H I V T I S H   S E L D G - - - - -   - - - - - - - - - -  336
Mouse               322  - - - - - - - - - -   - A Y H I V T I S H   P E L D G - - - - -   - - - - - - - - - -  335
Rat                 453  - - - - - - - - - -   - P L K R A N M S D   P L R T C R G - - -   - - - - - - - - - -  468
Opossum             321  - - - - - - - - - -   - S C H I V T I S H   P Q L - - - - - -   - - - - - - - - - -  332
Horse               322  - - - - - - - - - -   - S Y H I A T I S H   A Q L D G - - - - -   - - - - - - - - - -  335
Chicken             307  - - - - - - - - - -   - T K S I P T V S C   L Q L Q D - - - - -   - - - - - - - - - -  320
Zebra finch         333  - - - - - - - - - -   - T K A P P A V S C   L E L Q D - - - - -   - - - - - - - - - -  346
Green anoles        316  - - - - - - - - - -   - R P S V H I I S L   A H L E G - - - - -   - - - - - - - - - -  329
Fugu                345  F C V P N D A H I A   G S F F V P T Y E Q   G Q R S R G T T T D   Q H Q L S K G R G T   R R Q Q E P P E Q  393
Zebrafish           354  - - - - - - - - - A   D K F H D K T L A T   Q L N Y C G S V T -   - - - - - - - - - -   - - - - I P E -  376

Rat *            297  L L A D G S T Q P Q   I E T L K G K E E R   K K V G R K T L Y T   C D A Q T L Y T C D
                      A Q T L Y T C D A Q   T L Y T C D A C D T   Q T L Y T C D A Q T   L Y T C D A Q T L Y
                      T C D A Q T L Y T C   D A Q T L Y T C D A   Q T L Y T C D T Q T   L Y T C D A Q T L Y
                      T C D A Q T L Y T C   D  427
```

	Human	Chimpanzee	Oliver baboon	Rhesus monkey	Cow	Sheep	Pig	Dog	Rabbit	Mouse	Rat	Opossum	Horse	Chicken	Zebra finch	Green anole	Fugu	Zebrafish
Human	100.0	99.7	95.8	96.1	79.1	78.8	82.7	84.2	84.8	82.1	73.7	74.5	82.1	50.9	49.4	50.2	39.2	40.0
Chimpanzee		100.0	96.1	96.4	79.1	78.8	82.7	84.6	84.9	82.4	74.9	74.9	82.7	51.7	49.7	50.3	39.8	40.0
Olive baboon			100.0	99.7	79.4	79.1	83.6	84.8	85.7	82.1	74.6	73.0	82.7	50.5	50.2	50.5	38.7	39.1
Rhesus monkey				100.0	79.7	79.4	83.9	84.8	86.0	82.4	74.6	73.3	82.7	50.5	50.2	50.5	38.5	39.1
Cow					100.0	95.5	86.9	80.0	78.3	77.9	71.3	70.9	80.3	52.4	51.1	51.5	39.3	43.3
Sheep						100.0	84.5	80.0	78.3	77.9	72.8	70.9	80.0	51.1	50.2	51.8	39.4	43.0
Pig							100.0	83.0	84.2	80.0	72.5	71.8	82.4	51.7	50.5	52.9	38.3	40.6
Dog								100.0	84.2	83.3	76.4	74.2	85.1	50.2	51.5	48.8	40.1	41.8
Rabbit									100.0	82.7	75.2	74.6	83.9	49.9	50.5	49.9	38.3	40.2
Mouse										100.0	86.6	73.9	82.7	48.5	49.6	50.3	38.5	40.3
Rat											100.0	68.1	74.8	46.8	49.2	48.5	32.8	40.3
Opossum												100.0	73.6	52.3	51.5	51.2	39.2	40.1
Horse													100.0	51.8	51.2	49.7	38.7	40.9
Chicken														100.0	68.1	48.0	34.8	38.6
Zebra finch															100.0	46.7	33.8	36.5
Green anoles																100.0	35.7	39.5
Fugu																	100.0	42.9
Zebrafish																		100.0

Figure 4.1—Cont'd

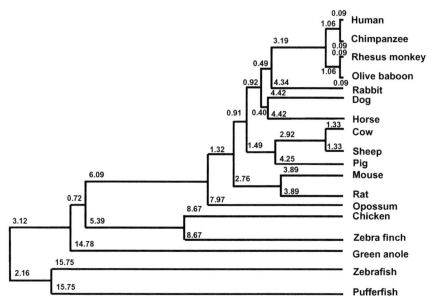

Figure 4.1 Comparison of GPR119 amino acid sequences between hGPR119 and GPR119s cloned or predicted from other species. (A) Alignment of the amino acid sequences between hGPR119 and GPR119s cloned or predicted from other species. The ECLs and ICLs are boxed. The most conserved residues in each TM are shaded in gray. (B) Homology and (C) phylogenetic analysis of GPR119 amino acid sequences from different species. NCBI reference numbers for the proteins are human, NP_848566.1; chimpanzee, XP_521262.2; mouse, NP_861416.1; rhesus monkey, XP_001093395.1; olive baboon, XP_003918318.1; dog, XP_549255.1; rabbit, XP_002720339.1; pig, XP_003135421.1; horse, XP_001500169.2; rat, NP_861435.1; cow, XP_002699578.1; sheep, XP_004022433; opossum, XP_001364948.1; chicken, XP_426248.1; zebra finch, XP_002195003.1; green anole, XP_003227136.1; zebra fish, XP_001337269.1; fugu, NP_001027835.1.

2.2. Tissue distribution

The distribution of GPR119 expression has been studied and evaluated in mammalian tissues by many researchers. By using quantitative real-time PCR or hybridization analysis, the pancreas and gastrointestinal tract are identified as major sites for GPR119 expression.[16,20,22–24]

Considerable evidence has demonstrated that GPR119 mRNA is abundantly expressed in pancreas in both human and rodents. Bonini et al.[16] suggested that the pancreas and gastrointestinal tract are the major sites of expression by studying the distribution of GPR119 mRNA in most of the rat and mouse tissues. Soga et al. reported strong *GPR119* gene

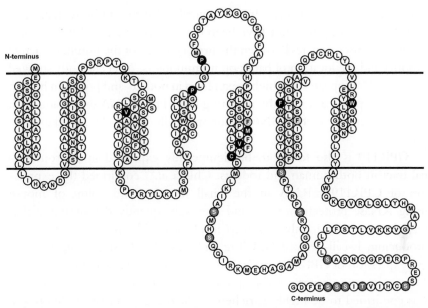

Figure 4.2 Two-dimensional illustration of hGPR119. The 10 amino acid residues that were predicted to be phosphorylated are indicated in *white* letter and shaded in *dark gray*. The three amino acid residues that were identified important for receptor basal activity are indicated in *black* letter and shaded in *light gray*. The eight amino acid residues that were predicted forming binding pocket for a synthetic agonist are indicated in *white* letter and shaded in *black*.

expression in rat pancreas[22] and this was subsequently confirmed by independent studies of different groups.[20,23] Odori et al.[24] detected the distribution of *GPR119* gene expression in normal human tissues, and they showed that GPR119 mRNA was most abundant in the pancreas with the level similar to those of GPR40 mRNA, which has been shown to be highly expressed in human pancreatic islets.

As for the specific distribution of GPR119 in the pancreatic islets, Chu et al.[20] concluded that β-cells are the main site for GPR119 expression in the rodent islet using *in situ* hybridization and immunofluorescent staining with polyclonal GPR119-specific antibody. This conclusion is supported by several observations of high GPR119 expression levels in β-cell-derived insulinoma cell lines including HIT-T15,[20,25] NIT-1[22,25], MIN6[26–28], BRIN-BD11[29], and INS-1E cells.[29] However, the immunofluorescence study by Sakamoto et al.[23] using a rabbit polyclonal antibody specific for GPR119 indicated the predominant GPR119 localization in pancreatic polypeptide cells of pancreatic islets while no immunoreactivity could be detected in islet α- or β-cells. It is noteworthy that the antibody developed by the latter group may not have been entirely specific for GPR119 since there were discrepancies between the predicted size of the immunoreactive product and the location of bands on immunoblots. A recent study showed that GPR119 appears to be highly expressed in both α- and β-cells in human and mouse pancreatic islets by detecting GPR119 mRNA expression with mouse pancreatic islets, MIN6 insulinoma cells, and α-TC glucagonoma cells.[24]

GPR119 is also found to be significantly expressed in gastrointestinal tract in both human and rodents. Chu et al.[20] confirmed the expression of GPR119 mRNA in the small intestine and colon of mouse using RNase protection assay, and they subsequently demonstrated that GPR119 was also highly expressed in glucagon-like peptide-1 (GLP-1)-producing, L-cell-derived GLUTag and STC-1 cells suggesting that the gut expression of GPR119 might correspond to enteroendocrine cells.[30,31] Moreover, using reverse transcription PCR, GPR119 mRNA transcript was confirmed to be highly enriched in glucose-dependent insulinotropic polypeptide (GIP)-containing K cells[32] and cholecystokinin-releasing I cells[33], both of which are located in the duodenal epithelia. As for the gastrointestinal distribution of GPR119 in human, the GPR119 mRNA was detected in duodenum, stomach, jejunum, ileum, and colon, but not in the esophagus or liver.[24] This finding seems to be consistent with the idea that GPR119 is expressed in enteroendocrine cells since human enteroendocrine cells are distributed throughout the gastrointestinal tract except the esophagus.[24]

At present, the expression of GPR119 in other tissues is still emerging and controversial. It was reported that GPR119 might be expressed in many areas of the rat brain, including cerebellum, cerebral cortex, choroid plexus, dorsal root ganglion, hippocampus, and hypothalamus.[34] However, Odori et al. reported that no significant GPR119 mRNA could be detected in the human hypothalamus, brain, or cerebrum in their study.[24] Moreover, it has been shown that GPR119 expression was detected in rat gastrocnemius muscle and cultured human primary skeletal muscle derived from obese and obese diabetic donors.[35] GPR119 mRNA expression was also observed in mouse C_2C_{12} myotubes.[35] However, the protein product in the mouse C_2C_{12} myotubes could not be confidently confirmed due to the unavailability of a reliable commercially available antibody against the mouse GPR119 protein. The potential physiological functions of GPR119 in these tissues remain to be elucidated.

3. PHYSIOLOGY OF GPR119

3.1. Glucose-dependent insulin secretion

The GPR119 was first deorphanized as a lysophosphatidylcholine (LPC) receptor enhancing glucose-dependent insulin secretion in both isolated rat pancreas and a cultured mouse pancreatic cell line NIT-1 cells.[22] Soga *et al.* demonstrated that in isolated rat pancreas, 18:1-LPC stimulates insulin secretion in glucose-dependent manner.[22] They further demonstrated that 18:1-LPC shows a dose-dependent stimulation of insulin secretion in NIT-1 cells.[22] Transfection with GPR119-specific siRNA or treatment with MDL12330A, an adenylate cyclase inhibitor, abolishes this LPC-induced insulin secretion in NIT-1 cells.[22] The molecular mechanism of regulating glucose-dependent insulin secretion through GPR119 activation is illustrated in Fig. 4.3 and will be elaborated in Section **4.2**.

The effect of GPR119 in mediating insulin secretion was further confirmed by several studies using other GPR119-specific endogenous[36] or synthesized agonists[25,27,28,37–47] *in vitro* and/or *in vivo*.

In 2007, Chu *et al.*[20] reported that AR231453, the most potent synthetic GPR119 agonist reported so far, stimulates glucose-dependent insulin secretion in HIT-T15 cells (a hamster islet cell line endogenously expressing

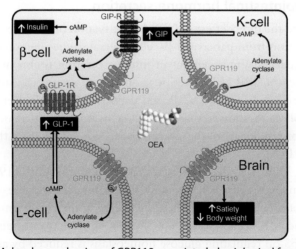

Figure 4.3 Molecular mechanism of GPR119-associated physiological functions, including regulation of insulin secretion, gastrointestinal hormone secretion, and energy homeostasis, upon GPR119 activation by the endogenous agonist OEA. (See color plate.)

GPR119) and RIN-5F cells (a rat islet cell line lacking endogenous GPR119) stably transfected with hGPR119, as well as in isolated rat and mouse islets. Moreover, oral administration of AR231453 significantly improves glucose tolerance in a dose-dependent manner in both normal and diabetic mice and rat models, most likely due to enhanced insulin secretion through activation of GPR119.[20] Indeed, the effect of AR231453 on glucose homeostasis is abolished in *Gpr119* knockout (KO) mice.[20]

The effect of GPR119 on glucose-dependent insulin release was questioned by several studies using *Gpr119* KO mice. Chu *et al.* reported that *Gpr119* KO mice have normal fed/fasted blood glucose level.[20] In 2009, Lan *et al.* also argued that the effects of LPC and oleoylethanolamide (OEA) on insulin secretion are not GPR119-specific using their independently generated *Gpr119* KO mice [48]. They reported that the isolated islets from both wild-type (WT) and *Gpr119* KO mice fed with chow or high-fat diet had normal glucose-dependent insulin secretion, and moreover, the effects of LPC and OEA on glucose-dependent insulin secretion were not significant.[48] Although the total insulin levels 30 min postglucose load were significantly decreased in *Gpr119* KO mice, most likely due to decreased GLP-1 secretion, it did not result in significantly decreased plasma glucose levels.

3.2. Gastrointestinal hormone secretion

In addition to the direct effect of GPR119 activation on glucose-dependent insulin secretion from pancreatic β-cells, several studies demonstrated that GPR119-mediated improvement in insulin secretion is also mediated by indirect effects of GPR119 activation in enteroendocrine cells, such as stimulating the release of GLP-1, GIP, and peptide YY (PYY). Very recently, Holz and colleagues showed that GPR119 activation also increases the gene expression of proglucagon, the precursor of GLP-1, therefore the biosynthesis of GLP-1.[49]

In 2008, Chu *et al.* demonstrated that activation of GPR119 improves glucose homeostasis through not only stimulating the glucose-dependent insulin secretion but also stimulating the release of incretin hormones, such as GLP-1 and GIP.[30] Early data from Chu *et al.* showed that the effectiveness of AR231453 on regulating glucose homeostasis is reduced by almost 50% when glucose is administered intraperitoneally, and they suggested that modulation of incretin-based mechanism might also be involved.[20] Indeed, they further demonstrated that AR231453 stimulates GLP-1 release both *in vitro* in GLUTag cells and *in vivo* in C57BL/6 mice.[30] The *in vivo* effects are markedly enhanced when GLP-1 is protected from degradation by dipeptidyl peptidase (DPP)-IV inhibitor.[30] Furthermore, it was shown that AR231453 significantly enhances GIP secretion *in vivo*, and the effect of AR231453 on GLP-1 and GIP release was abolished in *Gpr119* KO mice, demonstrating that this *in vivo* effect on enhancing incretin secretion is GPR119 mediated.[30] Lan *et al.* also reported that their independently generated *Gpr119* KO mice had significantly lower postprandial GLP-1, but not GIP, levels fed with low-fat diet (10% kcal as fat).[48]

Later, Lauffer *et al.* demonstrated that OEA stimulates GLP-1 secretion from both mouse and human intestinal L-cell lines, as well as from primary rat L cells *in vitro*.[31] This effect of OEA on enhancing GLP-1 secretion is GPR119 dependent, since *Gpr119* knockdown using siRNA led to a 45% reduction in the GLP-1 secretory response to OEA in mGLUTag cell.[31] Additionally, intraluminal application of OEA in rats induced a significant and persistent increase in bioactive GLP-1 levels over 1 h, supporting the *in vitro* findings and demonstrating a role for OEA as a GLP-1 secretagogue *in vivo*.[31] The effect of GPR119 in mediating incretin hormone secretion was further confirmed by several studies using other GPR119-specific endogenous[36,50] or synthesized agonists.[28,40,41,45–47,50–57]

In the initial study, Chu *et al.* reported that the effect of GPR119 agonist on stimulating GLP-1 release is glucose-dependent.[30] They demonstrated that the fold increase of plasma GLP-1 level upon GPR119 agonist stimulation significantly increased after glucose challenge compared with the basal glucose conditions.[30] Recently, Lan *et al.* demonstrated that, however, the GPR119 agonist-mediated GLP-1 release is not strictly glucose-dependent.[58] They showed that GPR119 agonist stimulates GLP-1 secretion both *in vitro* from GLUTag and primary intestinal cells and *in vivo* in C57BL/6 cells independent of glucose, although they were able to confirm that GPR119 stimulates insulin secretion glucose-dependently in MIN6 cells.[58] Therefore, whether GPR119-mediated GLP-1 secretion is glucose-dependent needs further investigation. A small clinical study in humans showed that administration of 2-oleoylglycerol (2-OG) (formed during fat digestion acting as GPR119 agonist) significantly increased plasma GLP-1 and GIP levels (but no increase in plasma glucagon and PYY levels).[52] It was suggested that because of the huge amount of 2-OG formed after ingestion of a meal containing fat, GLP-1 release stimulated by 2-OG might account for the well-known diet-induced GLP-1 release.[52] In an *in vitro* study of primary human intestinal cell cultures, treatment with

AR231453 does not increase GLP-1 release, whereas activation of G protein-coupled receptor 40, G protein-coupled receptor 120, and G protein-coupled bile acid receptor 1 each increases GLP-1 release by 1.5-fold, leading to the suggestion that GPR119 agonists may have weak potency in human L cells.[59] The molecular mechanisms regulating GLP-1 and GIP secretion upon GPR119 activation are illustrated in Fig. 4.3 and will be elaborated in Section 4.2.

Cox et al. reported that PYY is also required for GPR119-induced improvement in glucose tolerance and stimulation of circulating insulin levels.[60] First, they demonstrated that endogenous PYY selectively mediates the GPR119-mediated inhibitory effect on epithelial electrolyte secretion, suggesting the release of PYY upon GPR119 activation.[60] Furthermore, they demonstrated that the effect of PSN632408, a GPR119-specific agonist, on glucose tolerance was abolished in Pyy KO mice.[60] This observation was further supported by the finding that AR231453 activation of GPR119 results in significantly increased plasma levels of PYY in vivo.[40]

Taken together, GPR119 activation improves glucose homeostasis by stimulating insulin secretion by acting on pancreatic β-cells and by stimulating synthesis and secretion of endogenous antidiabetic hormones, the incretins (including GLP-1 and GIP).

3.3. Appetite and energy homeostasis

In 2006, Overton et al. showed that the endogenous anorectic lipid OEA is an endogenous GPR119 agonist that decreases food intake, but does not alter locomotor activity.[61] They demonstrated that the rats with acute intraperitoneal administration of OEA have significantly decreased food intake at 1 and 2 h postdosing and slight trend, although not statistically significant, toward lower cumulative food intake at later time points.[61] In addition to acting through activation of PPARα or the transient receptor potential (TRP) vanilloid receptor TRPV1, they suggested that the observed OEA-induced acute anorectic effect could also be GPR119-dependent, since PSN632408, a GPR119-specific agonist, also produces acute hypophagic effects when administered intraperitoneally or orally in rat.[61]

However, the results from the study of Gpr119 KO mice challenged the effects of GPR119-mediated hypophagic effect. Chu et al. reported that Gpr119 KO mice have normal size and body weight.[20] Lan et al.[48] also reported that under basal condition, Gpr119 KO mice have normal body weight and even lower body weight fed with low-fat diet. Moreover, they demonstrated that OEA is able to suppress food intake in both WT and *Gpr119* KO mice, indicating that OEA-induced hypophagic effect might not be GPR119-specific.[48] Therefore, whether GPR119 is involved in regulating appetite and energy homeostasis still needs further investigation.

3.4. Epithelial electrolyte homeostasis

In 2010, Cox *et al.* reported that GPR119 is critical in regulating gastrointestinal epithelial electrolyte homeostasis.[60] They demonstrated that PSN632408 induces the greatest inhibitory effects on epithelial electrolyte secretion in mouse descending colon and least in duodenal mucosa.[60] Moreover, this GPR119-mediated inhibitory effect was abolished in colonic mucosae in *Pyy* KO, *Npy* and *Pyy* double KO, and Y1 receptor antagonist-treated mice.[60] In human colonic mucosa, similar findings were observed that the mucosal GPR119 responses were blocked by the Y1 receptor antagonist.[60] These results suggested that the GPR119-mediated inhibitory effects were PYY and Y1 receptor mediated. In addition, they further demonstrated that the GPR119 inhibitory effect is glucose sensitive and partially mediated by ATP-sensitive potassium channel in both mouse and human colonic mucosae.[60] Later, Flock *et al.* reported that activation of GPR119 with AR231453 indeed results in significantly increased plasma levels of PYY *in vivo*.[40]

3.5. Gastric emptying

In 2011, Flock *et al.* demonstrated that GPR119 activation with AR231453 leads to significantly reduced gastric emptying in WT mice.[40] Moreover, GPR119 activation continued to reduce gastric emptying in both *Glp-1R* KO and in *Glp-1R/GipR* double KO mice, suggesting that AR231453 controls gastric emptying through a mechanism that does not require the GLP-1R or GIP receptor.[40] In addition, they demonstrated that the inhibitory actions of AR231453 on gastric emptying are indeed GPR119-specific, since the inhibitory effect was abolished in *Gpr119* KO mice, but retained in *Glp-2R* KO or Y2R antagonist-treated mice.[40]

3.6. β-Cell replication and cytoprotection

In 2011, Gao *et al.* demonstrated that several GPR119 agonists, including endogenous ligand OEA[50] and synthetic ligands PSN632408[50] and AR231453[51], stimulate β-cell replication both *in vitro* in cultured mouse islets and *in vivo* in mouse islet grafts. Furthermore, GPR119 agonist-treated diabetic recipient mice achieved significantly earlier normoglycemia, and thus suggesting that GPR119 agonists are potentially useful in improving islet graft function.[50,51] However, the specificity of GPR119 in mediating the effects of β-cell replication was not determined in these studies.

The proliferative effect of GPR119 was further supported by several other studies. Yoshida *et al.* reported that after 3 weeks of treatment with AS1535907 in *db/db* mice, the number of insulin- and proliferation cell nuclear antigen–positive cells and islet area is significantly higher than those in the vehicle-treated mice, suggesting the effects of GPR119 agonist on β-cell proliferation and protection.[44] Ansarullah *et al.* also demonstrated that 7-week treatment with PSN632408 alone or combined with DPP-IV inhibitor sitagliptin significantly stimulates β-cell proliferation and increases β-cell mass in diabetic C57BL/6 mice induced by streptozotocin.[57]

Subsequently, Stone *et al.* demonstrated that OEA indeed exerts cyto-protective effects in rat insulin-secreting cell lines, such as BRIN-BD11 or INS-1E cells.[29] However, cytoprotective effect was not reproduced using other synthetic GPR119 ligands.[29] Moreover, the cytoprotective response of OEA was lost during challenging with fatty acid amide hydrolase inhibitor suggesting that OEA *per se* is not the cytoprotective species, but that release of free oleate is required.[29]

Further studies are urgently needed to clarify some of the discrepancies. Whether GPR119 activation can preserve islet mass is of critical relevance in both type 1 and type 2 diabetes in demonstrating whether agonists for this receptor can slow progression of the diseases.

4. PHARMACOLOGY OF GPR119

4.1. Ligand development and receptor activation

Initially, GPR119 was classified as an orphan receptor, and efforts in deorphanizing GPR119 have identified two major classes of potential endogenous ligands: phospholipids and amide derivatives of fatty acids. LPC was first proposed as endogenous ligand for GPR119 that results in the stimulation of glucose-dependent insulin release and increased cAMP production in a mouse pancreatic β-cell line, NIT-1 cells endogenously expressing murine GPR119, and a rat hepatoma cell line, RH7777 cells stably expressing human GPR119.[22] OEA was also identified as GPR119 agonist that activates cAMP production in various cell lines.[52,56,61] Subsequently, several other endogenous ligands, as shown in Table 4.1, were identified as GPR119 agonist, including *N*-oleoyl-dopamine

Table 4.1 Endogenous and synthetic agonists, antagonist, and inverse agonists for GPR119

	Compound	Structure	EC50 (µM)	Ref.
Endogenous agonist				
1	Palmitoleoylethanolamine (POEA)		5	56
2	Linoleyl ethanolamine (LEA)		0.56–5	52,56
3	2-Oleoylglycerol (2-OG)		2.5–17	52,56
4	5-Hydroxy-eicosapentaenoic acid (5-HEPE)		NA	28
5	Palmitoylethanolamide (PEA)		0.84	52
6	N-oleoyl-dopamine (OLDA)		3.2	36

Continued

Table 4.1 Endogenous and synthetic agonists, antagonist, and inverse agonists for GPR119—cont'd

	Compound	Structure	EC50 (μM)	Ref.
7	Oleoylethanolamine (OEA)		0.2–5	50,52,56,61
8	1-Palmitoyl-lysophosphatidylcholine (16:0-LPC)		1.6	22
9	1-Stearoyl-lysophosphatidylcholine (18:0-LPC)		3.3	22
10	1-Oleoyl lysophosphatidylcholine (18:1-LPC)		1.5	22
11	Lysophosphatidylethanolamine (LPE)		5.7	22
12	Lysophosphatidylinositol (LPI)		5.7	22

Synthetic agonist

13	MW1219	0.96	47
14	GSK2041706A	NA	46
15	MBX2982	0.0039	62

Continued

Table 4.1 Endogenous and synthetic agonists, antagonist, and inverse agonists for GPR119—cont'd

Compound	Structure	EC50 (µM)	Ref.
16 Arena B3		0.29–0.40	58
17 Compound 34		0.024	63
18 Analog 12		0.0015	64

19	Quinazoline analog 12a		1	65
20	Quinazoline analog 12c		1	65
21	Quinazoline analog 12g		1	65

Continued

Table 4.1 Endogenous and synthetic agonists, antagonist, and inverse agonists for GPR119—cont'd

Compound	Structure	EC50 (μM)	Ref.
22 AS1669058		0.044	25
23 Compound 13		0.083	66
24 Compound 32i		0.022	67

25	Compound 42		0.006	55
26	Analog 2		0.051	54
27	GSK1292263		0.0077–0.0088	68,69

Continued

Table 4.1 Endogenous and synthetic agonists, antagonist, and inverse agonists for GPR119—cont'd

	Compound	Structure	EC50 (μM)	Ref.
28	Compound 16b		0.0083	70
29	Compound 24g		0.28	71
30	Compound 9t		1.2	72

| 31 | APD597, JNJ-38431055, Compound 16 | 0.046 | 41,53,73 |

| 32 | GSK1104252A | 0.05 | 45 |

| 33 | AS1535907 | 1.5 | 44 |

Continued

Table 4.1 Endogenous and synthetic agonists, antagonist, and inverse agonists for GPR119—cont'd

	Compound	Structure	EC50 (μM)	Ref.
34	Compound 36j		0.003	43
35	Compound 58		0.004	74
36	Compound 2		0.015	75
37	APD668, JNJ-28630368		0.0027	42

38	Compound 1		0.014	76
39	Compound 3		0.08	77
40	AS1907417		1.1	38
41	AS1535907		4.8	37

Continued

Table 4.1 Endogenous and synthetic agonists, antagonist, and inverse agonists for GPR119—cont'd

Compound	Structure	EC50 (µM)	Ref.
42 AS1269574		2.5	27
43 AR231453		0.00068–0.0047	20,40,51,78,79
44 PSN375963		8.4	61
45 PSN632408		5.6	50,61

Antagonist		
46 N-vanillylarachidonamide (Arvanil)	NA	56
Inverse agonist		
47 Compound 1	0.084	78
48 Compound 4e	0.05	78
49 Compound 4f	0.5	78

NA, not applicable.

(OLDA),[36] 5-hydroxy-eicosapentaenoic acid (5-HEPE),[28] and 2-OG and other 2-monoacylglycerols that are formed during fat digestion,[52] suggesting that GPR119 acts as a fat sensor.[52] However, the low potency (EC_{50} in micromolar range) and poor selectivity of these endogenous ligands, as well as acting as metabolism substrate, complicate their use in the characterization of the physiology and pharmacology of GPR119.

GPR119 agonists attracted the attention of the pharmaceutical industry at a very early stage, even before the elucidation of its biological and phys-iological importance, due to the unique selective expression of GPR119 in the endocrine pancreas. In the past few years, great efforts have been carried out by several pharmaceutical companies to discover potent and selective GPR119 agonists. The most potent agonist up to date is AR231453, the first-generation GPR119 agonist developed by Arena Pharmaceuticals, with EC_{50} reported ranging from 0.68[78] to 4.7[20] nM. Many other pharmaceutical companies, including Astellas Pharma Inc. (formerly Yamanouchi Pharma-ceutical), GlaxoSmithKline, Merck, Eli Lilly, Metabolex, Pfizer, Biovitrum, and Novartis, are pursuing GPR119 agonists as potential drug candidates for the treatment of T2DM and some of their disclosed compounds are listed in Table 4.1.

Several synthetic small molecule GPR119 agonists have entered clinical trials with high expectation due to their excellent preclinical performance. For example, in collaboration with Johnson & Johnson, Arena selected JNJ-38431055, also known as APD597, into clinical trial based on its good bal-ance between agonist potency, *in vivo* activity, good solubility, and reduced drug–drug interaction potential in preclinical animal study.[73] Unfortu-nately, however, it shows limited glucose lowering and incretin activity in T2DM patients.[53] Although single-dose administration decreases glucose excursion during an oral glucose tolerance test, treatment for 14 days does not lead to meaningful change in 24 h weighted mean glucose.[53] Therefore, recently, Johnson & Johnson returned their joint GPR119 project to Arena. Some other compounds, such as GSK1292263 (GlaxoSmithKline, Phase 2)[80] and MBX-2982 (Sanofi-Aventis/Metabolex, Phase 2),[62] were also in clinical trials. Although very limited information is yet available from these trials in scientific publications, it appears that the currently available GPR119 agonists sharing somewhat structural similarities as shown in Table 4.1 have provided disappointing results in Phase 2 clinical trials.

Computer modeling of GPR119, constructed from the agonist-bound human A_{2A} adenosine receptor X-ray structure, docked with compound 16b, a GPR119 agonist developed from fused pyrimidine derivatives, reveals a potential binding pocket formed by Val85, Pro140, Pro144, Met178, Val182, Cys186, Phe241, and Trp265 as shown in Fig. 4.2. Currently, however, no ligand binding assay has been set up to directly measure the binding of endogenous or synthetic ligands to the receptor, and thus the theoretically predicted binding pocket has not been verified experimentally. The establishment of a binding assay would be an important contribution to understanding the molecular pharmacology of the receptor, especially in elucidating the molecular mechanism of the ligand interaction with the receptor, and thus facilitating the development of new GPR119 agonists with novel chemical structures.

GPR119 has been shown to constitutively couple with $G\alpha s$, since several studies have shown that cells transfected with hGPR119 or mouse GPR119 have increased basal cAMP levels.[20,23,30,34,49,81,82] Mutagenesis study has demonstrated that Phe175[5.40], Phe234[6.44], and Trp238[6.48], as shown in Fig. 4.2, are crucial in contributing to the constitutive activity of GPR119.[81,82] It has been shown that alanine substitution of each of these three amino acid residues totally abolished the constitutive activity and impaired agonist-induced cAMP signaling in COS-7 cells transiently transfected with GPR119.[81,82] Altered constitutive activity of GPCRs has been shown to be associated with diseases[83,84]; however, the physiological relevance of constitutive activity in GPR119 has not been thoroughly examined. It was shown that the constitutive activity of hGPR119 could increase proglucagon gene expression through activation of PKA but not Epac.[49] Several GPR119 inverse agonists, as shown in Table 4.1, might be potential tools for future investigation of the physiological function of GPR119 constitutive activity.

4.2. Signaling pathways and regulation

The classical signaling pathway for the GPR119 is by coupling to $G\alpha s$ as shown in Fig 4.3. Soga et al. first demonstrated that GPR119 activation by LPC leads to dose-dependent increase of intracellular cAMP production in a mouse pancreatic β-cell line, NIT-1 cells that endogenously express murine GPR119, and a rat hepatoma cell line, RH7777 cells, stably expressing human GPR119.[22] Later, Overton et al. demonstrated that OEA also promotes a concentration-dependent increase in cAMP levels in HEK293 cells stably transfected with hGPR119 with greater potency than LPC.[61] The $G\alpha s$-coupled signaling pathway is further shown to be also activated in other cell lines, such as several insulinoma cell lines in regulating insulin secretion, enteroendocrine L cells in regulating GLP-1 secretion and K cells in regulating GIP secretion (Fig. 4.3), stimulated by various endogenous or synthetic GPR119 ligands. Cyclic AMP signaling pathway has been widely used to characterize the function of GPR119 and screen potential GPR119 agonists. There is no significant coupling of GPR119 to Gi or Gq.[20]

It has been well established that cAMP signaling pathway is closely associated with the physiological function of GPR119 in insulin and incretin secretion as shown in Fig. 4.3.[22,31,58] It was reported that LPC-induced insulin secretion from NIT-1 cells is markedly inhibited by using an adenylate cyclase inhibitor MDL12330A.[22] Some evidences show that OEA-induced [31] or Arena B3-induced [58] GLP-1 secretion from GLUTag cells is significantly decreased upon treatment with PKA inhibitor H89.

In addition to activation of Gαs-coupled signaling pathway, it has been suggested that depolarization of the cell might be involved in the regulation of GPR119 agonist-induced insulin and incretin secretion.[26,58] Ning et al. reported that OEA-induced insulin secretion is associated with not only increased cAMP but also the augmentation of glucose-stimulated increases of intracellular calcium concentration ($[Ca^{2+}]_i$).[26] It was shown that the addition of nitrendipine, an L-type calcium channel blocker, or diazoxide, an ATP-sensitive potassium channel activator, significantly reduces OEA-stimulated glucose-stimulated $[Ca^{2+}]_i$ and thus decreases the OEA-induced insulin secretion.[26] Lan et al. demonstrated recently that AR231453 induces a dose-dependent increase of $[Ca^{2+}]_i$.[58] Moreover, AR231453- or Arena B3-induced GLP-1 secretion from GLUTag cells is largely eliminated in the presence of nitrendipine.[58] However, some evidences questioned the role of $[Ca^{2+}]_i$ in GPR119 function. For example, it was reported that one GPR119 agonist, PSN632408, increased insulin release in a glucose-dependent manner but inhibited glucose-stimulated increase of $[Ca^{2+}]_i$. In contrast, another similar GPR119 agonist PSN375963 had no significant effect on insulin secretion but retained the ability to induce the increase of $[Ca^{2+}]_i$.[26] Different signaling pathways are activated by endogenous and synthetic agonists. Therefore, further studies are needed to clarify the contribution of $[Ca^{2+}]_i$ in GPR119 function.

5. CONCLUSIONS

Obesity is an established risk factor for T2DM, yet the links between obesity and T2DM are still mystery. Recent success with targeting GLP-1R for treating T2DM using peptide mimetics or DPP-IV inhibitor represents significant achievement in diabetes research and treatment; however, small molecule approaches to target GLP-1R have so far not been achieved. GPR119 has the potential to be an alternative GPCR target. GPR119, a receptor for fatty acid derivatives highly expressed on the β-cell, has several unique characteristics in improving glucose homeostasis, including stimulating glucose-stimulated insulin secretion, like the GLP-1R. In addition, GPR119 stimulates GLP-1 biosynthesis and secretion in intestinal enteroendocrine cells, further enhancing the insulinotropic activity of the β-cell. Therefore, GPR119 provides a novel route for investigating T2DM. However, the interaction of GPR119 with various endogenous fatty acid metabolites remains largely unknown yet. The signaling pathways involved in these processes for endogenous and synthetic agonists also remain to be elucidated completely. Therefore, more basic research regarding the molecular pharmacology and biology of GPR119 is essential before we can develop novel GPR119 agonists to target this receptor for T2DM treatment.

ACKNOWLEDGMENTS

Funding for our work on obesity and diabetes was provided by American Diabetes Association Grant 1-12-BS212, National Institutes of Health Grant R15-DK077213, Auburn University Intramural Grant Program, and Interdisciplinary Grant and Animal Health and Diseases Research Program of College of Veterinary Medicine at Auburn University (to Y.-X. T.).

REFERENCES

1. Tao YX, Yuan ZH, Xie J. G protein-coupled receptors as regulators of energy homeostasis. *Prog Mol Biol Transl Sci*. 2013;114:1–43.
2. Kahn SE, Hull RL, Utzschneider KM. Mechanisms linking obesity to insulin resistance and type 2 diabetes. *Nature*. 2006;444:840–846.
3. Boden G. Obesity and free fatty acids. *Endocrinol Metab Clin North Am*. 2008;37:635–646.
4. Boden G, Shulman GI. Free fatty acids in obesity and type 2 diabetes: defining their role in the development of insulin resistance and beta-cell dysfunction. *Eur J Clin Invest*. 2002;32:14–23.
5. Yaney GC, Corkey BE. Fatty acid metabolism and insulin secretion in pancreatic beta cells. *Diabetologia*. 2003;46:1297–1312.
6. Vangaveti V, Shashidhar V, Jarrod G, Baune BT, Kennedy RL. Free fatty acid receptors: emerging targets for treatment of diabetes and its complications. *Ther Adv Endocrinol Metab*. 2010;1:165–175.
7. Chawla A, Repa JJ, Evans RM, Mangelsdorf DJ. Nuclear receptors and lipid physiology: opening the X-files. *Science*. 2001;294:1866–1870.
8. Sauer LA, Dauchy RT, Blask DE. Mechanism for the antitumor and anticachectic effects of n-3 fatty acids. *Cancer Res*. 2000;60:5289–5295.

9. Louet JF, Chatelain F, Decaux JF, et al. Long-chain fatty acids regulate liver carnitine palmitoyltransferase I gene (L-CPT I) expression through a peroxisome-proliferator-activated receptor α (PPARα)-independent pathway. *Biochem J.* 2001;354:189–197.

10. Fredriksson R, Hoglund PJ, Gloriam DE, Lagerstrom MC, Schioth HB. Seven evolutionarily conserved human rhodopsin G protein-coupled receptors lacking close relatives. *FEBS Lett.* 2003;554:381–388.

11. Hirasawa A, Hara T, Katsuma S, Adachi T, Tsujimoto G. Free fatty acid receptors and drug discovery. *Biol Pharm Bull.* 2008;31:1847–1851.

12. Miyauchi S, Hirasawa A, Ichimura A, Hara T, Tsujimoto G. New frontiers in gut nutrient sensor research: free fatty acid sensing in the gastrointestinal tract. *J Pharmacol Sci.* 2010;112:19–24.

13. Mo XL, Wei HK, Peng J, Tao YX. Free fatty acid receptor GPR120 and pathogenesis of obesity and type 2 diabetes mellitus. *Prog Mol Biol Transl Sci.* 2013;114:251–276.

14. Ohishi T, Yoshida S. The therapeutic potential of GPR119 agonists for type 2 diabetes. *Expert Opin Investig Drugs.* 2012;21:321–328.

15. Griffin G. Methods for identification of modulators of OSGPR116 activity. *US 7083933*; 2006.

16. Bonini JA, Borowsky BE, Adham N, Boyle N, Thompson TO. DNA encoding SNORF25 receptor. *US 6221660*; 2001.

17. Takeda S, Kadowaki S, Haga T, Takaesu H, Mitaku S. Identification of G protein-coupled receptor genes from the human genome sequence. *FEBS Lett.* 2002;520:97–101.

18. Davey J. G-protein-coupled receptors: new approaches to maximise the impact of GPCRs in drug discovery. *Expert Opin Ther Targets.* 2004;8:165–170.

19. Oh DY, Kim K, Kwon HB, Seong JY. Cellular and molecular biology of orphan G protein-coupled receptors. *Int Rev Cytol.* 2006;252:163–218.

20. Chu ZL, Jones RM, He H, et al. A role for β-cell-expressed G protein-coupled receptor 119 in glycemic control by enhancing glucose-dependent insulin release. *Endocrinology.* 2007;148:2601–2609.

21. Xue Y, Ren J, Gao XJ, Jin CJ, Wen LP, Yao XB. GPS 2.0, a tool to predict kinase-specific phosphorylation sites in hierarchy. *Mol Cell Proteomics.* 2008;7:1598–1608.

22. Soga T, Ohishi T, Matsui T, et al. Lysophosphatidylcholine enhances glucose-dependent insulin secretion via an orphan G-protein-coupled receptor. *Biochem Biophys Res Commun.* 2005;326:744–751.

23. Sakamoto Y, Inoue H, Kawakami S, et al. Expression and distribution of Gpr119 in the pancreatic islets of mice and rats: predominant localization in pancreatic polypeptide-secreting PP-cells. *Biochem Biophys Res Commun.* 2006;351:474–480.

24. Odori S, Hosoda K, Tomita T, et al. GPR119 expression in normal human tissues and islet cell tumors: evidence for its islet-gastrointestinal distribution, expression in pancreatic beta and alpha cells, and involvement in islet function. *Metabolism.* 2013;62:70–78.

25. Oshima H, Yoshida S, Ohishi T, et al. Novel GPR119 agonist AS1669058 potentiates insulin secretion from rat islets and has potent anti-diabetic effects in ICR and diabetic *db/db* mice. *Life Sci.* 2013;92:167–173.

26. Ning Y, O'neill K, Lan H, et al. Endogenous and synthetic agonists of GPR119 differ in signalling pathways and their effects on insulin secretion in MIN6c4 insulinoma cells. *Br J Pharmacol.* 2008;155:1056–1065.

27. Yoshida S, Ohishi T, Matsui T, Shibasaki M. Identification of a novel GPR119 agonist, AS1269574, with in vitro and in vivo glucose-stimulated insulin secretion. *Biochem Biophys Res Commun.* 2010;400:437–441.

28. Kogure R, Toyama K, Hiyamuta S, Kojima I, Takeda S. 5-Hydroxy-eicosapentaenoic acid is an endogenous GPR119 agonist and enhances glucose-dependent insulin secretion. *Biochem Biophys Res Commun.* 2011;416:58–63.

29. Stone VM, Dhayal S, Smith DM, Lenaghan C, Brocklehurst KJ, Morgan NG. The cyto-protective effects of oleoylethanolamide in insulin-secreting cells do not require activation of GPR119. *Br J Pharmacol.* 2012;165:2758–2770.

30. Chu ZL, Carroll C, Alfonso J, et al. A role for intestinal endocrine cell-expressed g protein-coupled receptor 119 in glycemic control by enhancing glucagon-like Peptide-1 and glucose-dependent insulinotropic Peptide release. *Endocrinology.* 2008;149:2038–2047.

31. Lauffer LM, Iakoubov R, Brubaker PL. GPR119 is essential for oleoylethanolamide-induced glucagon-like peptide-1 secretion from the intestinal enteroendocrine L-cell. *Diabetes.* 2009;58:1058–1066.

32. Parker HE, Habib AM, Rogers GJ, Gribble FM, Reimann F. Nutrient-dependent secretion of glucose-dependent insulinotropic polypeptide from primary murine K cells. *Diabetologia.* 2009;52:289–298.

33. Sykaras AG, Demenis C, Case RM, Mclaughlin JT, Smith CP. Duodenal enter-oendocrine I-cells contain mRNA transcripts encoding key endocannabinoid and fatty acid receptors. *PLoS One.* 2012;7:e42373.

34. Overton HA, Fyfe MC, Reynet C. GPR119, a novel G protein-coupled receptor target for the treatment of type 2 diabetes and obesity. *Br J Pharmacol.* 2008;153(suppl 1):S76–S81.

35. Cornall LM, Mathai ML, Hryciw DH, et al. GPR119 regulates genetic markers of fatty acid oxidation in cultured skeletal muscle myotubes. *Mol Cell Endocrinol.* 2013;365:108–118.

36. Chu ZL, Carroll C, Chen R, et al. N-oleoyldopamine enhances glucose homeostasis through the activation of GPR119. *Mol Endocrinol.* 2010;24:161–170.

37. Yoshida S, Ohishi T, Matsui T, et al. Novel GPR119 agonist AS1535907 contributes to first-phase insulin secretion in rat perfused pancreas and diabetic *db/db* mice. *Biochem Biophys Res Commun.* 2010;402:280–285.

38. Yoshida S, Tanaka H, Oshima H, et al. AS1907417, a novel GPR119 agonist, as an insulinotropic and β-cell preservative agent for the treatment of type 2 diabetes. *Biochem Biophys Res Commun.* 2010;400:745–751.

39. Brocklehurst KJ, Broo A, Butlin RJ, et al. Discovery, optimisation and in vivo evaluation of novel GPR119 agonists. *Bioorg Med Chem Lett.* 2011;21:7310–7316.

40. Flock G, Holland D, Seino Y, Drucker DJ. GPR119 regulates murine glucose homeo-stasis through incretin receptor-dependent and independent mechanisms. *Endocrinology.* 2011;152:374–383.

41. Katz LB, Gambale JJ, Rothenberg PL, et al. Pharmacokinetics, pharmacodynamics, safety, and tolerability of JNJ-38431055, a novel GPR119 receptor agonist and potential antidiabetes agent, in healthy male subjects. *Clin Pharmacol Ther.* 2011;90:685–692.

42. Semple G, Ren A, Fioravanti B, et al. Discovery of fused bicyclic agonists of the orphan G-protein coupled receptor GPR119 with in vivo activity in rodent models of glucose control. *Bioorg Med Chem Lett.* 2011;21:3134–3141.

43. Xia Y, Chackalamannil S, Greenlee WJ, et al. Discovery of a nortropanol derivative as a potent and orally active GPR119 agonist for type 2 diabetes. *Bioorg Med Chem Lett.* 2011;21:3290–3296.

44. Yoshida S, Ohishi T, Matsui T, et al. The role of small molecule GPR119 agonist, AS1535907, in glucose-stimulated insulin secretion and pancreatic beta-cell function. *Diabetes Obes Metab.* 2011;13:34–41.

45. Katamreddy SR, Carpenter AJ, Ammala CE, et al. Discovery of 6,7-dihydro-5H-pyrrolo[2,3-a]pyrimidines as orally available G protein-coupled receptor 119 agonists. *J Med Chem.* 2012;55:10972–10994.

46. Ahlkvist L, Brown K, Ahren B. Upregulated insulin secretion in insulin-resistant mice: evidence of increased islet GLP1 receptor levels and GPR119-activated GLP1 secretion. *Endocr Connect.* 2013;2:69–78.

47. Zhang M, Feng Y, Wang J, et al. High-throughput screening for GPR119 modulators identifies a novel compound with anti-diabetic efficacy in *db/db* mice. *PLoS One.* 2013;8:e63861.
48. Lan H, Vassileva G, Corona A, et al. GPR119 is required for physiological regulation of glucagon-like peptide-1 secretion but not for metabolic homeostasis. *J Endocrinol.* 2009;201:219–230.
49. Chepurny OG, Bertinetti D, Diskar M, et al. Stimulation of proglucagon gene expression by human GPR119 in enteroendocrine L-cell line GLUTag. *Mol Endocrinol.* 2013;27:1267–1282.
50. Gao J, Tian L, Weng G, et al. Stimulating β-cell replication and improving islet graft function by GPR119 agonists. *Transpl Int.* 2011;24:1124–1134.
51. Gao J, Tian L, Weng G, O'brien TD, Luo J, Guo Z. Stimulating β-cell replication and improving islet graft function by AR231453, a gpr119 agonist. *Transplant Proc.* 2011;43:3217–3220.
52. Hansen KB, Rosenkilde MM, Knop FK, et al. 2-Oleoyl glycerol is a GPR119 agonist and signals GLP-1 release in humans. *J Clin Endocrinol Metab.* 2011;96:E1409–E1417.
53. Katz LB, Gambale JJ, Rothenberg PL, et al. Effects of JNJ-38431055, a novel GPR119 receptor agonist, in randomized, double-blind, placebo-controlled studies in subjects with type 2 diabetes. *Diabetes Obes Metab.* 2012;14:709–716.
54. Sakairi M, Kogami M, Torii M, et al. Synthesis and SAR studies of bicyclic amine series GPR119 agonists. *Bioorg Med Chem Lett.* 2012;22:5123–5128.
55. Scott JS, Birch AM, Brocklehurst KJ, et al. Use of small-molecule crystal structures to address solubility in a novel series of G protein coupled receptor 119 agonists: optimization of a lead and in vivo evaluation. *J Med Chem.* 2012;55:5361–5379.
56. Syed SK, Bui HH, Beavers LS, et al. Regulation of GPR119 receptor activity with endocannabinoid-like lipids. *Am J Physiol Endocrinol Metab.* 2012;303:E1469–E1478.
57. Ansarullah, Lu Y, Holstein M, DeRuyter B, Rabinovitch A, Guo Z. Stimulating β-cell regeneration by combining a GPR119 agonist with a DPP-IV inhibitor. *PLoS One.* 2013;8:e53345.
58. Lan H, Lin HV, Wang CF, et al. Agonists at GPR119 mediate secretion of GLP-1 from mouse enteroendocrine cells through glucose-independent pathways. *Br J Pharmacol.* 2012;165:2799–2807.
59. Habib AM, Richards P, Rogers GJ, Reimann F, Gribble FM. Co-localisation and secretion of glucagon-like peptide 1 and peptide YY from primary cultured human L cells. *Diabetologia.* 2013;56:1413–1416.
60. Cox HM, Tough IR, Woolston AM, et al. Peptide YY is critical for acylethanolamine receptor Gpr119-induced activation of gastrointestinal mucosal responses. *Cell Metab.* 2010;11:532–542.
61. Overton HA, Babbs AJ, Doel SM, et al. Deorphanization of a G protein-coupled receptor for oleoylethanolamide and its use in the discovery of small-molecule hypophagic agents. *Cell Metab.* 2006;3:167–175.
62. Roberts B, Gregoire FM, Karpf DB, et al. MBX-2982, a novel oral GPR119 agonist for the treatment of type 2 diabetes: results of single & multiple dose studies. In: *American Diabetes Association—69th Meeting 2009, New Orleans*; 2009.
63. Zhang JK, Li AR, Yu M, et al. Discovery and optimization of arylsulfonyl 3-(pyridin-2-yloxy)anilines as novel GPR119 agonists. *Bioorg Med Chem Lett.* 2013;23:3609–3613.
64. Yang Z, Fang Y, Pham TA, Lee J, Park H. Synthesis and biological evaluation of 5-nitropyrimidine analogs with azabicyclic substituents as GPR119 agonists. *Bioorg Med Chem Lett.* 2013;23:1519–1521.
65. Pham TA, Yang Z, Fang Y, Luo J, Lee J, Park H. Synthesis and biological evaluation of novel 2,4-disubstituted quinazoline analogues as GPR119 agonists. *Bioorg Med Chem.* 2013;21:1349–1356.

66. Futatsugi K, Mascitti V, Guimaraes CR, et al. From partial to full agonism: identification of a novel 2,4,5,6-tetrahydropyrrolo[3,4-c]pyrazole as a full agonist of the human GPR119 receptor. *Bioorg Med Chem Lett*. 2013;23:194–197.

67. Darout E, Robinson RP, Mcclure KF, et al. Design and synthesis of diazatricyclodecane agonists of the G-protein-coupled receptor 119. *J Med Chem*. 2013;56:301–319.

68. Polli JW, Hussey E, Bush M, et al. Evaluation of drug interactions of GSK1292263 (a GPR119 agonist) with statins: from in vitro data to clinical study design. *Xenobiotica*. 2012;43:498–508.

69. Zhu X, Huang D, Lan X, et al. The first pharmacophore model for potent G protein-coupled receptor 119 agonist. *Eur J Med Chem*. 2011;46:2901–2907.

70. Negoro K, Yonetoku Y, Moritomo A, et al. Synthesis and structure-activity relationship of fused-pyrimidine derivatives as a series of novel GPR119 agonists. *Bioorg Med Chem*. 2012;20:6442–6451.

71. Negoro K, Yonetoku Y, Misawa-Mukai H, et al. Discovery and biological evaluation of novel 4-amino-2-phenylpyrimidine derivatives as potent and orally active GPR119 agonists. *Bioorg Med Chem*. 2012;20:5235–5246.

72. Negoro K, Yonetoku Y, Maruyama T, Yoshida S, Takeuchi M, Ohta M. Synthesis and structure-activity relationship of 4-amino-2-phenylpyrimidine derivatives as a series of novel GPR119 agonists. *Bioorg Med Chem*. 2012;20:2369–2375.

73. Semple G, Lehmann J, Wong A, et al. Discovery of a second generation agonist of the orphan G-protein coupled receptor GPR119 with an improved profile. *Bioorg Med Chem Lett*. 2012;22:1750–1755.

74. Szewczyk JW, Acton J, Adams AD, et al. Design of potent and selective GPR119 agonists for type II diabetes. *Bioorg Med Chem Lett*. 2011;21:2665–2669.

75. Sharma R, Eng H, Walker GS, et al. Oxidative metabolism of a quinoxaline derivative by xanthine oxidase in rodent plasma. *Chem Res Toxicol*. 2011;24:2207–2216.

76. Mcclure KF, Darout E, Guimaraes CR, et al. Activation of the G-protein-coupled receptor 119: a conformation-based hypothesis for understanding agonist response. *J Med Chem*. 2011;54:1948–1952.

77. Mascitti V, Stevens BD, Choi C, et al. Design and evaluation of a 2-(2,3, 6-trifluorophenyl)acetamide derivative as an agonist of the GPR119 receptor. *Bioorg Med Chem Lett*. 2011;21:1306–1309.

78. Semple G, Fioravanti B, Pereira G, et al. Discovery of the first potent and orally efficacious agonist of the orphan G-protein coupled receptor 119. *J Med Chem*. 2008;51:5172–5175.

79. Ali S, Lamont BJ, Charron MJ, Drucker DJ. Dual elimination of the glucagon and GLP-1 receptors in mice reveals plasticity in the incretin axis. *J Clin Invest*. 2011;121:1917–1929.

80. Nunez D, Bush M, Collins D. Evaluation of GSK-1292263, a novel GPR119 agonist, in type 2 diabetes mellitus (T2DM): safety, tolerability, pharmacokinetics (PK) and pharmacodynamics (PD) of single and multiple doses. In: *American Diabetes Association—71th Meeting 2011, San Diego*; 2011.

81. Valentin-Hansen L, Holst B, Frimurer TM, Schwartz TW. PheVI:09 (Phe6.44) as a sliding microswitch in seven-transmembrane (7TM) G protein-coupled receptor activation. *J Biol Chem*. 2012;287:43516–43526.

82. Holst B, Nygaard R, Valentin-Hansen L, et al. A conserved aromatic lock for the tryptophan rotameric switch in TM-VI of seven-transmembrane receptors. *J Biol Chem*. 2010;285:3973–3985.

83. Tao YX. Constitutive activation of G protein-coupled receptors and diseases: insights into mechanisms of activation and therapeutics. *Pharmacol Ther*. 2008;120:129–148.

84. Seifert R, Wenzel-Seifert K. Constitutive activity of G-protein-coupled receptors: cause of disease and common property of wild-type receptors. *Naunyn Schmiedebergs Arch Pharmacol*. 2002;366:381–416.

CHAPTER FIVE

Enhanced Skeletal Muscle for Effective Glucose Homeostasis

Jinzeng Yang
Department of Human Nutrition, Food and Animal Sciences, University of Hawaii at Manoa, Honolulu, Hawaii, USA

Contents

Abstract

As the single largest organ in the body, the skeletal muscle is the major site of insulin-stimulated glucose uptake in the postprandial state. Skeletal muscles provide the physiological foundation for physical activities and fitness. Reduced muscle mass and strength is commonly associated with many chronic diseases, including obesity and insulin resistance. The complications of diabetes on skeletal muscle mass and physiology, resulting from either insulin deprivation or insulin resistance, may not be life-threatening, but accelerate the lost physiological functions of glucose homeostasis. The formation of skeletal muscle commences in the embryonic developmental stages at the time of mesoderm generation, where somites are the developmental milestone in musculoskeletal formation. Dramatic skeletal muscle growth occurs during adolescence as a result of muscle fiber hypertrophy since muscle fiber formation is mostly completed before birth. The rate of growth rapidly decelerates in the late stages of adulthood as adipose tissue gradually accumulates more fat when energy intake exceeds expenditure. Physiologically, the key to effective glucose homeostasis is the hormone insulin and insulin sensitivity of target tissues. Enhanced skeletal muscle, by either intrinsic mechanism or physical activity, offers great advantages and benefits in facilitating glucose regulation. One key protein factor named myostatin is a dominant inhibitor of muscle mass. Depression of myostatin by its propeptide or mutated receptor enhances muscle mass effectively. The muscle tissue utilizes a large portion of metabolic energy for its growth and maintenance. We demonstrated that transgenic overexpression of myostatin propeptide in mice fed with a high-fat diet enhanced muscle mass and circulating adiponectin, while the wild-type mice developed obesity and insulin resistance. Enhanced muscle growth has positive effects on fat metabolism through increasing adiponectin expression and its regulations. Molecular studies of the exercise-induced glucose uptake in skeletal muscle also provide insights on auxiliary substances that mimic the plastic adaptations of muscle to exercise so that the body may amplify the effects of exercise in contending physical activity limitations or inactivity. The recent results from the peroxisome proliferator-activated receptor γ coactivator 1α provide a promising therapeutic approach for future metabolic drug development. In summary, enhanced skeletal muscle and fundamental understanding of the biological process are critical for effective glucose homeostasis in metabolic disorders.

1. INTRODUCTION

Obesity and associated metabolic syndromes are consequences of energy intake exceeding what the body needs for normal physiological functions, breaking up the basic energy homeostasis for an extended period of time. Most troubling is the dramatic increase in juvenile obesity. The percentage of young people who are overweight has more than doubled since 1980. Type II diabetes and insulin resistance, which are typically associated with adult obesity, are expected to increase dramatically in children and adolescents. One of the main causes for such metabolic disorder prevalence in developed countries and some developing countries is the minimal physical efforts required for human life and work, which are diminishing the essential use and need of the musculoskeletal system of the body. Computer technologies are popular and widespread in almost every occupation and career, which certainly make work and life much easier in terms of physical demand. Dietary changes and profit-driving food industry are other important factors that influence what we eat and consume, which is not based on the physiology or nutritional requirement of our body, but on the appealing of the food commercial advertisements. Carbohydrates and meat products were expensive food for the generation of our grandparents as agricultural productions were inefficient and physical demand is high for most jobs. The nutrient compositions in most grains and meat products have not changed significantly, neither did the human genes within the couple of generations. However, the living environment and lifestyle, as well as the interactions between our genome and living environments, changed. The skeletal muscle is the largest organ in humans, providing the physiological foundation for physical activities and fitness, which is also known as "voluntary" muscles, playing a dynamic role in glucose uptake, storage, and utilization. Skeletal muscle and maintaining a healthy skeletal muscle physiology hold an important step to a healthy life in an industrialized society with more mental work and less physical activities. This chapter first begins with the basic introduction of skeletal muscle and glucose homeostasis and then proceeds to muscle formation and postnatal development, followed by an updated progress in enhanced skeletal muscle by myostatin and exercise, and implications for obesity and diabetes preventions.

2. SKELETAL MUSCLE AND GLUCOSE HOMEOSTASIS

Glucose, an essential energy source for all cells in the body, is initially absorbed into blood circulation across the intestine walls through food digestion. Like many other nutrients in the body, nutrients in the blood will be transported into various tissues to provide the nutrient requirements for the cells over a period of physiological actions. The body is able to regulate blood glucose level in a tight normal range of 70–100 mg/dl in 1–3 h after a meal by specific hormones. The postabsorptive glucose in the blood circulation is disposed by splanchnic tissues such as the liver and gastrointestinal tissue, as well as the main peripheral tissues—skeletal muscle and adipose tissues. Two important hormones, insulin and glucagon, produced in endocrine pancreas, play significant roles in the regulation of blood glucose homeostasis. The hormone insulin, produced from beta cells of pancreas in response to high blood glucose, facilitates the transport of glucose into cells to metabolize for adenosine triphosphate (ATP) production and/or to store as glycogen; therefore, the blood glucose concentration rapidly returns to preprandial level in a short period of time. The hormone glucagon, produced by the alpha cell of pancreas when the blood glucose is low, stimulates the liver to release glucose stored within its cells, therefore increasing blood glucose to a normal level. As the single largest organ in the body, the skeletal muscle is the major site of insulin-stimulated glucose uptake in the postprandial state in humans. Based on the euglycemic hyperinsulinemic condition, skeletal muscles dispose of ~80% of the glucose uptake. For example, at the state of physiological hyperinsulinemia (80–100 µU/ml), leg muscle glucose uptake increases linearly with time, reaching a plateau value of ~10 mg/kg leg weight per minute after 60 min.[1,2]

Glucose is transported into cell through specific proteins named glucose transporters. There are several types of glucose transporters located in the plasma membrane of the cells. Glucose transporters 1 and 3 (GLUT1 and GLUT3) are universally located in most cells of the body and have a high affinity to glucose, and their uptakes of blood glucose to cells remain at a constant and basal level for supporting cellular energy requirement. GLUT2 are mainly expressed and located in the plasma membranes of the hepato-cytes, pancreatic beta cells, and basolateral membrane of kidney proximal tubules. GLUT2 has low affinity to glucose, and the rate of glucose uptake by GLUT2 is proportional to the blood glucose levels. Therefore, the glu-cose transport and uptake by liver and pancreatic GLUT2 are mostly depen-dent on the glucose concentration in the blood. In healthy state, glucose is transported to the cell and then converted to glucose-6-phosphate by glucokinase, and the enzymatic reaction can be reversed by glucose-6-phosphatase to glucose in the liver tissue. The activity of GLUT2 is not reg-ulated by insulin, but hepatic glucokinase and glucose-6-phosphatase in the liver tissue are strongly regulated by insulin. Therefore, hepatic glucokinase is the first and rate-limiting step in glucose uptake and storage as glycogen in liver.[3] In a feedback regulation, glucokinase acts as a glucose sensor to control insulin secretion in the pancreatic beta cells. In muscle and adipose tissue, the most significant glucose transporter is GLUT4, which is insulin-sensitive, facilitating postprandial blood glucose uptake into muscle and adipose tissues. GLUT4 is expressed at cell surface and located within cyto-plasmic vesicles. When insulin binds to its receptors on the cell surface, the insulin–signaling cascade induces a rapid translocation of the GLUT4 from the storage vesicles to the plasma membrane by membrane fusion. GLUT4 is inserted to the plasma membrane and serves as important transporter for glu-cose in muscle and adipose tissue. After transportation to the cells, glucose is phosphorylated by hexokinase to form glucose-6-phosphate in the muscle and adipose tissue. Glucose-6-phosphate either enters the glycolytic path-way for ATP production or polymerizes to glycogen for glucose storage.

The signaling pathway of insulin action in the regulation of glucose uptake via GLUT4 in muscle and adipose tissue has been well defined. In skeletal muscle, the binding of insulin to its receptor leads to phosphoryla-tion of tyrosine kinases on the insulin receptor (IR), which phosphorylates and recruits different substrate adaptors such as the IRS (insulin receptor substrate) family of proteins. Tyrosine-phosphorylated IRS then displays binding sites for several signaling kinase. Tyrosine-phosphorylated IRS activates the p85 regulatory subunit of phosphatidylinositol (PI)

3-kinase (PI3K) and the p110 catalytic subunit, leading to an increase in phosphatidylinositol-3,4,5 triphosphate. This results in activation of downstream protein kinase B (also known as Akt) and phosphorylation of Akt substrate 160 (AS160), which facilitates the translocation of GLUT4 to the sarcolemma and subsequent entry of glucose into the cell.[4] The PI3K pathway plays a major role in insulin function. Activated Akt also induces glycogen synthesis through inhibition of GSK-3, protein synthesis via mTOR and downstream elements, and cell survival through inhibition of several proapoptotic agents. Maintaining the integrity of the IRS-1/PI3K/Akt pathway is essential for normal insulin-mediated glucose uptake in skeletal muscle.[5,6]

Obesity, metabolic syndrome, or subsequent type II diabetes is simply a disease of energy intake exceeding energy expenditure over a long period of time. Metabolic problems occur when adipose tissue is overloaded with abundant high-energy nutrients without subsequent physiological use. Increased triglycerides not only disrupt adipocyte endocrine functions but also cause dysfunctions of skeletal muscle (i.e., insulin resistance). The human body is largely composed of skeletal muscles and adipose tissues at the scale of the tissue mass, both of which are heavily involved in energy metabolisms. The skeletal muscle plays a major role in insulin-mediated glucose uptake through glucose transport to the muscle cell and glycogen synthesis in the cell. In the pathogenesis of type II diabetes, insulin resistance is the common metabolic defect, the loss of metabolic capability of insulin to drive glucose into its major target tissues—such as the skeletal muscle and liver. A loss of skeletal muscle mass is frequently associated with diabetes. Insulin deprivation in type I diabetes increases catabolism, resulting in serious muscle and adipose tissue weight loss without insulin treatment. The net muscle protein catabolism is due to a net increase in protein degradation rather than a decline in protein synthesis.[7] Similarly, an excessive loss of skeletal muscle mass is also widely observed in older adults with type II diabetes. Age-related loss of skeletal muscle mass is a physiological progression. However, the rapid decline in skeletal muscle loss in association with diabetes may lead to sarcopenia with decreased muscle strength, mobility limitations, physical disability, and eventually high mortality among the elderly.[8] Studies on well-functioning community-dwelling older adults aged 70–79 years reported that older adults with either diagnosed or undiagnosed type II diabetes showed excessive loss of appendicular lean mass and trunk fat mass compared with nondiabetic subjects. Thigh muscle cross-sectional area declined two times faster in older women with diabetes than their nondiabetic counterparts. Older adults (70–79 years old) with type II diabetes have an altered body composition with low skeletal muscle mass and strength compared with nondiabetic older adults.[9] As the declines in appendicular lean mass, an indicator of total skeletal muscle mass, are independent of weight changes over several years, the results suggest an excessive loss of skeletal muscle mass in older adults with type II diabetes. Older adults with type II diabetes lost their knee extensor strength more rapidly than their nondiabetic counterparts.[10,11]

The complications of diabetes on skeletal muscle mass and physiology, resulted from either insulin deprivation or insulin resistance, may not be immediately life-threatening, but accelerate the lost physiological function of glucose homeostasis. Similarly, accelerated loss of skeletal muscle mass and strength is found in other chronic and degenerative diseases such as heart failure, arthritis, autoimmune diseases, pulmonary diseases, chronic diseases, HIV infections, and cancer.[12] Muscle protein metabolism in type II diabetes is not consistent with the accelerated protein breakdown seen in type I diabetes in many studies from animal models and human subjects, although there are substantial alterations in glucose and lipid metabolisms resulting from defects in insulin functions in type II diabetes. Apparently, the muscle loss in type II diabetes is complicated by impaired muscle glucose uptake, mitochondrial ATP production, and reduced mitochondrial protein synthesis through the effects of insulin on AKT–mTOR pathway.[13] By real-time in vivo measurements of intracellular metabolites with magnetic resonance spectroscopy (MRS), researchers found that insulin resistance results from decreased glycogen synthesis due to reduced activities of glucose transporter GLUT4 in skeletal muscle.[14] Intracellular lipid induces the inhibition of IRS-1 tyrosine phosphorylation of insulin signaling.[14] Results from MRS studies with human subjects also confirmed mitochondrial dysfunction. The mitochondrial oxidative phosphorylation activity was reduced by ~40% in healthy, lean, elderly adults with insulin resistance compared with BMI- and physical activity-matched young adults. The reduced mitochondrial function predisposes to intramyocellular lipid accumulation in the elderly adults.[15] Research has also shown that the insulin-resistant offspring of parents with type II diabetes have impaired mitochondrial function as a result of reduced mitochondrial ATP synthesis in the offspring, which is associated with insulin resistance and increased intramyocellular lipid content.[16] Apparently, both diet-induced and inherited loss of mitochondrial function can lead to insulin resistance and fat accumulation in skeletal muscle.[14]

3. SKELETAL MUSCLE FORMATION

The skeletal muscle is the largest organ in humans, making up 40–50% of the total body mass in an average adult. Skeletal muscles are voluntary muscles, which are different from cardiac and smooth muscles. All the skeletal muscles are connected by motor neurons of the central nervous system so that controlled contractions and movements can be achieved. Together with the skeletal system, the muscular system in the body provides the physiological foundation for locomotion, posture, physical activities, and fitness.

The formation of skeletal muscle commences in the embryonic developmental stages in mammals. At the time of formation of the three primary germ layers known as endoderm, mesoderm, and ectoderm in the embryogenesis, the progenitor cells, which would generate muscle cells, accumulate to a structure known as somite in the mesoderm. Somites are formed along the anterior–posterior axis of the developing embryo in the pattern of pared structures, located in the two sides of the neural tube. There are four compartments in the developing somites, including sclerotome, myotome, dermatome, and syndetome. Somites mature according to a rostrocaudal gradient of differentiation, giving rise to different tissue types. Different compartments (i.e., cell blocks) of the somites eventually develop into vertebrae and the rib cartilage (sclerotome), skeletal muscles of the back, ribs and the limbs (myotome), dermis (dermatome), and tendons and blood vessels (syndetome). The myotomal compartments of somites will generate myoblast, a mononucleated muscle precursor cell. Two distinctive compartments present in the dermomyotome give rise to separate lineage of skeletal muscles during embryogenesis, that is, the dorsomedial epaxial domain forming the deep back and intercostal muscles and the lateral hypaxial domain forming the rest of the musculature of the body and limbs[17]; the hypaxial dermomyotome forms the ventrolateral extension of the dermomyotome for the body wall muscles, while some migratory precursor cells leave the dermomyotome and form more distant muscle masses of the wings and limbs.[17]

During embryonic development, myogenic precursor cells are formed from embryonic stem cell lineage in the somites. The proliferation and differentiations of these cells are programmed by intrinsic controlling mechanisms, as well as modified by neural and hormonal factors constituting the cell microenvironment. In the myogenic lineage, the cell type is referred to as myoblast before terminal differentiation to myotube and myofiber.

Interestingly, myoblasts are mononucleated cells, while a mature skeletal muscle cell or fiber is multinucleated. The development of myoblasts to a mature muscle fiber undergoes several transitional stages, including initial formation of mononucleated cell called myocyte, which is characterized by synthesis of myofibrillar proteins and metabolic enzymes for ATP synthesis; myotubes with centrally located multiple nuclei and innervations; and final formation of myofibers with cross striation, peripherally located multiple nuclei and contraction properties.

It has been well known for decades that myoblasts are rapidly dividing cells in culture, but cease the proliferation process or DNA synthesis once they fuse into a myotube.[18] It is therefore believed that myotube and myofiber cannot replicate its nuclear DNA; they do not have the capability of reentering cell cycle for dedifferentiation in normal muscle formation and development. Cell proliferations can only occur in the stage of formation of myoblasts or its preceding myogenic progenitors, but not in the terminally differentiated myofiber in mammals, implying that the myoblast differentiation is fundamentally irreversible. However, recent studies from urodele amphibians showed that dedifferentiation of myofiber plays an essential role in the regeneration of lost limbs and other musculature. The nuclei of amphibian myofiber reenter the cell cycle prior to dedifferentiation.[19] Mouse myotubes generated from myoblast cell line C2C12 can dedifferentiate when stimulated with appropriate signals and ectopic expression of the homeobox Msx1 transcription factor.[20] A portion of muscle cell isolated from the myotube developed to either smaller multinucleated myotubes or proliferating, mononucleated cells. Clonal populations of the myotube-derived mononucleated cells were induced to redifferentiate into cells expressing chondrogenic, adipogenic, myogenic, and osteogenic markers.[20,21] By using a Cre/Lox-β-galactosidase system to tag cells in muscle tissue, a recent study successfully demonstrated that multinuclear myofibers can dedifferentiate to mononuclear cells after traumatic injury *in vivo* and these mononuclear cells develop into a variety of different muscle cell lineages such as myoblasts, satellite cells, and muscle-derived stem cells.[22]

Prenatal myogenesis in many mammalian species is a biphasic process where two generations of muscle fibers are formed sequentially. For example, the primary fiber generation occurs from 35 to 55 days of gestation, followed by a second generation between 55 and 90 days of the 114 days of gestation in pigs. In the process of the secondary generation, myoblast cells use the primary fibers as a scaffold to develop new fibers.[23,24] There are only a small number of primary muscle fibers developed during the embryonic stage, while the secondary generation of myofibers makes up the majority of muscle fibers. The two stages of muscle fiber generation correspond well with the stage of embryo and fetus development, where the primary and secondary myoblast generations are referred as embryonic myoblast and fetal myoblast, respectively. As the formation of muscle fibers is accomplished before birth in most mammalian species, which intrinsically depends on the available myogenic progenitor cells and their proliferating activities, the secondary myogenesis is important for developing a full-scale skeletal muscle fiber number.[25] Postnatal growth of skeletal muscle is predominantly dependent upon the increase in muscle fiber size since there is not much net gain in the formation of new myofibers in skeletal musculature. Although the adult skeletal muscle is relatively stable with minimal DNA replication activities, injury or exercise-induced injuries can change muscle cellular and molecular activities dramatically, that is, muscle regenerations through extensive repair mechanisms, including generation of new myofiber and enlargement of the existing fibers.

4. MYOGENIC REGULATORY FACTORS

The development of myogenic progenitor cells and myoblast to myofiber is under intrinsic control mechanisms, known as the regulation of myogenic regulatory factors (MRFs). MRFs belong to a family of basic helix–loop–helix (bHLH) transcription factors, including MyoD, Myf5, myogenin, and MRF4. These proteins not only are characterized by the structural motif of bHLH with two α-helices connected by a loop but also contain a conserved E-box DNA-binding domain of the CANNTG sequence.[26] Expressions of MyoD and Myf5 are the first determination or specification factors of the myogenic lineage. Mice deficient in both MyoD and Myf5 lack skeletal muscle and fetuses can be delivered but die soon after birth. The putative muscle progenitor cells remain multipotential and can form nonmuscle tissues in the trunk and limbs of these mice. MyoD knockout embryo develops normal skeletal muscle because of the compensation by the overexpression of Myf5. The role of MyoD and Myf-5 is to induce the differentiation of myogenic precursor cells into myoblasts, and their functions are compensatory.[26] MyoD and Myf5 are widely used as the cell markers of proliferating myoblasts. The terminal differentiations of myoblast are through the withdrawal of the cell cycle in the proliferation process, where the late expressions of myogenin and MRF4 play important roles in the formation of myofiber. Myogenin is necessary for the fusion of myo-blasts into myotubes and expressions of muscle structural protein such as myosin heavy chain. There is no myofiber formation in myogenin-deficient embryos.[27,28] MRF4 mutation is viable with seemingly normal skeletal muscle, with the mutant mice having increased expression of myogenin.[29] It is believed that MRF4 functions late in the myogenic differentiation path-way, and its function can be substituted by myogenin.[30]

The expression of MRFs appears to be regulated by Pax3 and Pax7, members of the paired-box family of transcription factors. Pax3 and Pax7 are expressed in the early phase of embryonic development in der-momyotome of paraxial mesoderm. Based on the location of Pax3 expres-sion, its expression would contribute to myoblast formation since all myoblasts are derived from dermomyotome of paraxial mesoderm. *Pax3* and *Pax7* genes arose by duplication from a unique ancestral *Pax3/7* gene, with similar protein sequence and expression patterns. Forced expression of Pax3 using retroviral expression systems in chick tissue explants revealed that Pax3 activates the expression of MyoD, Myf5, and myogenin.[31] Studies have shown that Pax3 regulates MyoD and Myf5 expressions, somite seg-mentation, dermomyotome formation, and limb musculature develop-ment.[32] The Pax3/Pax7 double knockout mice were able to form primary myotome, but the subsequent phases of myogenesis are arrested,[33] but Pax7 mutant mice exhibit no overt muscle phenotype. Pax7 has shown to play an important role in the biology of satellites cells,[34] and it is necessary for the maintenance of adult satellite cells.[33] There-fore, it is believed that Pax7 attribute to the population of muscle progenitors present at the embryonic stage for continued muscle growth and buildup later in life.

5. POSTNATAL DEVELOPMENT OF SKELETAL MUSCLE

Muscle growth is determined by an increase in muscle fiber number (hyperplasia) as well as an increase in muscle fiber size (hypertrophy). The number of muscle fiber is determined before birth. Postnatal muscle growth mainly results from muscle hypertrophy through the increase in muscle fiber length and girth. The muscle fiber is formed by myoblast fusion, which needs continuous supply of nuclei. The enlargement of muscle tissue during postnatal growth is believed to be the consequence of increase in satellite cell activities, which fuse to adjacent muscle fibers to increase their sizes. After birth, the number of skeletal muscle fiber in most mammals and avian species does not increase due to the completion of embryonic proliferation of skeletal muscle cells.[35] For example, there is no significant change in postnatal fiber number in mice, rat, pig, and cattle.[35–38] However, other studies indicate that limited extent of muscle cell proliferation occurs after birth in some species. For example, the increase in muscle fiber number was observed shortly after birth in rodents.[39] Others argued that the increase in fiber number during the first days of postnatal life was a result of maturation and elongation of the existing myotubes rather than due to a production of new fibers.[40,41] Due to the limitation of fiber size increase, the growth potential for skeletal muscle is virtually determined by the number of fibers established at around the time of birth. This relationship of muscle fiber number and growth potential has been demonstrated in the enlarged muscles of double-muscled cattle, in genetically different sizes of animals, and in runts as compared with normal pigs.

Muscle fiber is a unique cell with multiple nuclei. Muscle growth is characterized by continuous recruitment of nuclei to the existing fiber through the period of postnatal growth. One of the main purposes of the recruitment of new nuclei is to maintain constant demand of protein synthesis. By adding nuclei to the existing myofiber, the cell will maintain appropriate nuclear to cytoplasmic ratio for efficient protein synthesis and utilization. The source of nuclei is strongly believed to come from muscle satellite cells, which are quiescent mononucleated cells, located between the sarcolemma and basement membrane of muscle fibers. Muscle satellite cells can account for about 30% of the nuclei associated with muscle growth during early postnatal growth. The contribution of satellite cells to muscle growth or buildup decreases as animals and humans age. In a healthy adult, satellite cells represent $\sim$2–7% of nuclei within skeletal muscle.[34] Satellite cells are normally in quiescent state in adult muscle, become activated to proliferate in response to injury, and give rise to more satellite cells and contribute to muscle growth or regeneration. Muscle satellite cell is functional as adult stem cell. A reserve pool of satellite cell is retained for future use after muscle regeneration by the asymmetrical division of satellite cells, in which one daughter cell returns to quiescence and replenish the satellite cell pool, while the other daughter cell enters the differentiation to form myofiber.

The MRF expressions during satellite cell activation, proliferation, and differentiation are analogous to their expressions manifested during the embryonic formation of myofiber. Quiescent satellite cells do not express MRFs. In mouse skeletal muscles, activated satellite cells (satellite cells entering the cell cycle) first express either Myf5 or MyoD following the coexpression of Myf5 and MyoD.[42] Myogenin and MRF4 are expressed in cells when they start differentiation.[43,44] Intrinsic activation of satellite cells is controlled by proximal signals from the myofiber microenvironment, inflammatory responses, and microvasculature.[45] Extrinsic cues have been shown to regulate satellite cell activation, including hepatocyte growth factor and neuronal nitric oxide synthase, fibroblast growth factors (FGFs), and FGF receptor 1.[45,46]

Skeletal muscle growth occurs dramatically during adolescence, along with physical skeletal development for the full functional capacity of musculoskeletal system. The rate of growth rapidly decelerates in late stages of adulthood as adipose tissue gradually accumulates more fat when energy intake exceeds expenditure in the body. As skeletal muscle mass and strength decline with age, known as sarcopenia, the satellite cell population decreases in absolute number as well as a percentage of nuclei in skeletal muscle tissue.[47] The reduction in satellite cells per myofiber with age is more prominent in females compared to males.[47] Not only do the number of satellite cells and the capability to contribute to muscle repair decline, but also the age-associated alterations of satellite cells to nonmyogenic lineages such as adipogenic and fibrogenic fate occur.[48–50]

Skeletal muscle fibers are heterogeneous in cell structure and functions, which is defined as muscle plasticity or malleability. The property of myofiber is based on the muscle types. The great variations of muscle fiber type are present in different muscles, as well as between individuals. Initially, muscle fibers are classified as three types, including type I fibers (also called slow–twitch oxidative fibers or red fibers), type IIa fibers (also known as fast-twitch oxidative glycolytic fibers or intermediate fibers), and type IIb fibers (also referred to as fast–twitch glycolytic fibers or white fibers). Type I fibers are characterized by a slow contraction and high resistance to fatigue as they have high mitochondrial and capillary density and high myoglobin content. These fibers use oxidative metabolism to generate ATP for muscle contractions, which are mostly used for aerobic, low–level force production such as walking and maintaining posture for routine daily living activities. Type IIb fibers initiate quick muscle contractions and have low resistance to fatigue. Their mitochondrial and myoglobin contents and capillary density are low. They mainly rely on anaerobic glycolysis to generate energy for contraction. Type IIb fiber-dominant muscles make up most body muscle mass and are used for anaerobic activities with a high force output such as sprinting and jumping. Type IIa fibers represent a transition between type I and type IIb fibers, and their structures have the characteristics of both type I and type IIb fibers. One of the muscle structural proteins named myosin heavy chain has been identified to have several isoforms in recent years, and their expression patterns highly represent the muscle fiber types previously classified by ATPase staining. The isoform of MHCI, MHCIIa, and MHCIIb corresponds to the type I, type IIa, and type IIb in ATPase staining, respectively. The fourth muscle fiber type is named as type IIx, which is the intermediate between types IIa and IIb in its structure and function, and expressed MHC IIx isoform. With advances in fiber typing by MHC isoform antibodies, individual fibers in adult muscles are identified more commonly as containing mixed or hybrid fiber types than purely one type of muscle fiber. Hybrid fiber types and fiber type switching or transition help explain the functional flexibility of skeletal muscle for different movements.

6. MYOSTATIN AND SKELETAL MUSCLE MASS

Myostatin, a member of transforming growth factor β (TGF-β) superfamily, acts as an inhibitor of myogenesis and skeletal muscle mass. Although scientists have studied and identified many genes and different molecular approaches to enhance skeletal muscle in animal models, myostatin is probably the single most significant gene, which has the most dramatic effects on muscle mass in a wide range of mammalian and avian species. Myostatin mutations have been associated with dramatic muscle mass in cattle, human, dogs, and sheep.[51–54] Although the phenotypes are predominantly on increased muscle mass, different mutations appear to result in variations of muscling, fiber compositions, and other phenotypes under different genetic backgrounds (Table 5.1). The initial study in myostatin knockout mice showed that myostatin mutant mice were hypermuscular with individual muscles weighing double that of wild-type mice.[55] When compared to wild-type littermates, myostatin-null mice are 30% heavier in body weight, which is the result of widespread increases in muscle mass. The increase in body weight observed at 3 months of age is attributed primarily to an increase in muscle weight because the myostatin-null mice, as a proportion of body weight, have less adipose tissue than the wild-type. At its widest point, the tibialis anterior muscle of the myostatin-null mice contains 86% more muscle fibers than the wild-type controls. In addition, the mean cross-sectional area of individual fibers is increased by 7% and 22% in tibialis anterior and gastrocnemius muscles, respectively. Furthermore, the protein/DNA ratio is increased in the myostatin-null mice when compared to the wild-type. These changes are observed in both male and female myostatin-null mice.[55]

Table 5.1 Myostatin mutations and their effects on protein and phenotypes

Breeds/ species	Location of mutation	Myostatin protein	Phenotypes
Belgian Blue (bovine)	11-Nucleotide deletion (exon 3)	Frameshift premature termination	Muscle mass ⇑ Intramuscular fat ⇓ Dystocia, small organ size
Piedmontese (bovine)	C to A (exon 1) G to A (exon 3) transition	Leu to Phe (ex1) Cys to Tyr (ex3) Loss of cystine knot structure	Muscle mass ⇑ Mild calving problem Normal birth weight
Whippet dog	2bp deletion (exon 3)	Cys to stop codon Loss of disulfide dimer	Muscle mass ⇑ Superior racing Unusual muscle body
Human	G to A transition (intron 1)	Insert 108 bp to exon 1 Premature termination	Muscle mass ⇑
Texel Sheep	G to A transition in 3′UTR	miRNA target site Translation inhibition	Muscle hypertrophy
Mouse	Exon 3 knockout	Null for mature myostatin	Muscle hyperplasia and hypertrophy

In beef cattle industry, double-muscled beef cattle breeds are known for their hypermuscularity, and the gene involved in muscling phenotype had been mapped to chromosome 2 and termed the mh locus.[56] At about the same time when myostatin gene was discovered, the gene at the mh locus was identified as the myostatin gene.[57,58] The human myostatin gene contains three exons, which encode for 125, 124, and 126 amino acids, respectively.[59] The promoter region of the myostatin gene has multiple response elements for glucocorticoids, androgens, monocyte-enhancing factor, peroxisome proliferator-activated receptor gamma, and nuclear factor κβ. Myostatin mRNA is primarily expressed in muscle and is detected as early as 9.5 days postcoitus in mice.[55] Myostatin expression is detected from 16 days in bovine embryo, and the total number of fibers is significantly increased during the fetal period in doubled muscle breeds.[56,58] Myostatin, however, has also been documented in the heart, the lactating mammary gland, brain, and ovarian tissue and at low levels in adipose tissue in mice.[60–62] Expression is influenced by several factors including strength training, disuse glucocorticoid treatment, obesity, and the administration of other myogenic factors such as MyoD.[61] Muscle-specific overexpression of myostatin results in reduced muscle mass in mice.[63]

Accumulating evidence indicates that myostatin initiates cell signaling by directly binding to the type II Ser/Thr kinase receptor, ActRIIB, that leads to the recruitment of an appropriate type I receptor and induction of phosphorylation of Smad2/Smad3 transcription factors.[64] Myostatin inhibits both myoblast cell proliferation and differentiation, reducing fiber number during muscle formation. McCroskery et al.[65] demonstrated that myostatin is expressed in satellite cell and adult myoblast cell using muscle tissue from normal and myostatin-null mice. Cyclin-dependent kinases (Cdks) regulate G1 phase transitions to S phase in a cell cycle. Myostatin is able to increase Cdk inhibitor p21 activity, therefore decreasing the Cdk levels, concurrently resulting in myoblast cell cycle arrest in the G1 phase. Myostatin also inhibits MyoD expression and activity via Smad3, which blocks myoblasts from differentiating into myotubes. Therefore, myostatin inhibits both myoblast cell proliferation and differentiation. Myostatin is also expressed in satellite cells and adult myoblasts. It negatively regulates the G1 to S progression of satellite cells to maintain their quiescent status. Muscles from myostatin-null mice have an increased number of satellite cells as well as a higher proportion of activated satellite cells than muscles of wild-type mice.[64,65]

Like other members of the TGF-β superfamily, the precursor peptide translated from myostatin mRNA is cleaved at the tetrapeptide (RSRR) site, generating two separate peptides: a~36 kDa N-terminal propeptide and a~12 kDa active C-terminal mature myostatin peptide. The biological activities of most TGF-β proteins are primarily regulated by posttranslational modifications in the form of latency-associated peptide complexes. Similar to TGF-β1, myostatin propeptide has an inhibitory effect on myostatin activity in vitro.[66,67] We and others have shown that transgenic expression of myostatin propeptide significantly improved muscle growth and mass.[68–70]

The transgenic mice with myostatin propeptide cDNA appear healthy, grow 17–30% faster, and produce 22–44% more muscle mass than their wild-type littermates.[68,69] Transgenic expression of myostatin inhibitors or overexpression of the propeptide of myostatin results in hyper-muscularity, though increased muscle mass in the propeptide overexpressing mice results from increased hypertrophy and not altered hyperplasia. Also noted is that the transgenic mice demonstrated much lower epididymal fat pad weights than their wild-type littermates.[68] Earlier analyses of transgenic and wild-type mice on a normal-fat or high-fat diet were conducted by our lab.[68,71] Similar to other studies, we found that the transgenic males were heavier than the wild-type animals on the normal diet. Both groups of transgenic mice consumed more kilocalories per day than their wild-type counterparts. Since the majority of the increased weight in transgenic animals was due to increased muscle mass, it indicates that these excess calories were partitioned into muscle mass instead of fat mass. Despite the presence of these other models, the original myostatin-null mouse continues to be the most commonly used model for research into myostatin function. The muscular phenotype resulted primarily from muscle hypertrophy. Individual major muscles of transgenic mice were 45–115% heavier than those of wild-type mice and maintained normal blood glucose, insulin sensitivity, and fat mass after a 2-month regimen with a high-fat diet (45% kcal fat). In contrast, high-fat diet induced wild-type mice with 170–214% more fat mass than transgenic mice and developed impaired glucose tolerance and insulin resistance.[69,71] Similarly, transgenic expression of mutated form of ActRIIB results in dramatic muscle mass.[70] Taken together, these findings from transgenic mice demonstrated successful and effective means of partial depression of myostatin function.

Myostatin is expressed in satellite cells and regulates satellite cell quiescence and self-renewal. Higher levels of myostatin have been observed in quiescent satellite cells and lower levels are observed when satellite cells are activated.[65] Additionally, higher levels of myostatin are observed during immobilization, resulting in suppression of satellite cell activation.[72] In vitro studies revealed that myostatin-null mice also exhibit greater proliferation and earlier differentiation of satellite cells.[72] Earlier studies suggest that myostatin maintains satellite cells in a quiescent state in adult muscle. However, recent studies indicate that there is no increase in satellite cells or the rate of satellite cell proliferation in the myostatin knockout mice compared to wild-type mice.[73] Muscle hypertrophy through myostatin inhibition is mainly caused by its effects on myofibers rather than on satellite cells. Age-induced loss in muscle mass and satellite cells also occurs in myostatin-null mice.[72] Therefore, the researchers believed that enhanced muscle mass in myostatin-null mice might not result from the activation of satellite cells in skeletal muscle. As satellite activation occurs mostly when muscle tissue is injured, the degree of satellite cell activation in normal growth and aging is not known; it is worth further investigation on satellite cell activation in myostatin genetically manipulated animals.

In the mean time, there are several studies that have demonstrated that enhanced muscle mass in myostatin-manipulated mice primarily resulted from myofiber protein metabolism. The effects of myostatin on myofiber protein metabolism were demonstrated by the observation that myostatin induces cachexia by activating the ubiquitin proteolytic system through a FOXO1-dependent mechanism.[74] A positive feedback mechanism between myostatin and FOXO1 pathways may amplify the atrophic response.[75] In our recent study, we have observed that levels of phosphorylated 4E-BP1 (Thr37/46) and p70S6k (Thr389), two key downstream effectors of the mTOR pathway in regulation of protein synthesis, were greater in the transgenic mice with depressed myostatin function compared with wild-type mice. This suggests that myostatin regulates muscle mass probably through the mTOR pathway.[76]

7. ENHANCED MUSCLE GROWTH FOR PREVENTING DIABETES

Active use and mobilization of energy resources by skeletal muscle through physical activity can be a very effective means in regulation of metabolic activities and energy balance. It will provide the most effective benefit for preventing and treating obesity and insulin resistance. It is recommended by many health organizations and health professionals that appropriate physical activity and balanced diet are the basic requirements for obesity and diabetes treatment and prevention. For example, the American Diabetes Association suggests regular exercising at least 3 days/weeks for at least 150 min/week, with no more than 2 consecutive days without physical activity. Enhanced physical activity modifies body composition and energy metabolism. A single session of physical exercise such as resistance exercise powerfully regulates fat and glucose metabolism and results in a remarkable improvement of insulin sensitivity by stimulating glucose transport. In addition, physical activities such as resistance exercise promote strength gains with better neuromuscular function and encompass muscle fiber hypertrophy, which are particularly important for the middle-aged and seniors due to aging-induced loss of muscle mass and endurance. It is important to realize that maintaining appropriate muscle mass at any age empowers exercise capacity, which will have significant treatment effects for most energy metabolic problems associated. Compared with medicine, enhanced muscle mass and associated exercise capacity have minimal side effects but add significant improvement of the drug treatment on diabetes because of the dynamic and active metabolic natures of the skeletal muscle and the positive correlations between muscle mass and energy expenditure of the body. One of the challenges with physical exercise for treating metabolic disease is patients with obesity and associated diabetes are often unable or reluctant to get into exercising activities, which may be caused be the mental effects of the disease, insufficient skeletal muscle, or lack of physical activity skills. A better understanding of enhanced muscle growth and buildup and its related regulation of glucose homeostasis may help develop new concepts and strategy for drug development and effective treatment targeting the skeletal muscle.

By using the myostatin propeptide transgenic mice, we have provided important insight about enhanced skeletal muscle for preventing obesity and type II diabetes. We and others have demonstrated that transgenic expression of myostatin propeptide cDNA sequence successfully depresses myostatin function and promoted muscularity phenotype. In our transgenic mice, generated by muscle-specific expression of the propeptide (the 5′-region 886 nucleotides) of myostatin, we observed significant muscling phenotypes: 20% faster growth rate and 44% more muscle mass than wild-type mice.[68] In comparison with myostatin knockout mice, the propeptide transgenic mice still produce myostatin in the muscle and maintain normal adipose tissues. To determine whether enhanced muscle growth minimizes the incidence of diet-induced obesity, we studied the response of the propeptide transgenic mice to a high-fat diet. We showed that the propeptide transgene not only enhanced muscle growth but also prevented dietary fat-induced obesity and insulin resistance. While wild-type mice on a high-fat diet for 2 months developed serious adiposity, impaired glucose tolerance, and insulin resistance, the transgenic mice are normal and healthy, accommodating a metabolic regulatory system that utilizes dietary fat for muscle growth and maintenance. Individual major muscles of transgenic mice were 45–115% heavier than those of wild-type mice and maintained normal blood glucose, insulin sensitivity, and fat mass after a 2-month regimen with a high-fat diet (45% kcal fat). In contrast, high-fat diet induced wild-type mice with 170–214% more fat mass than transgenic mice and developed impaired glucose tolerance and insulin resistance. Insulin signaling, analyzed by Akt phosphorylation, was elevated by 144% in transgenic mice over wild-type mice fed a high-fat diet. These findings from animal model clearly suggest that muscle development and buildup play a fundamental role in balancing the utilization and expenditure of body energy. The results support our hypothesis that well-developed skeletal muscles in early stage of life increase the flexibility of the body in utilizing dietary fat or other energy resource, so that the body would be better equipped to prevent obesity and its associated metabolic disorders in adulthood or later stages of life.[71]

Interestingly, the transgenic mice have increased adiponectin secretion.[77] Adiponectin is a 30 kDa adipocytokine hormone secreted primarily by the adipose tissue. It is a relatively abundant serum protein and accounts for between 0.01% and 0.03% of total serum protein in humans.[78] Increased serum levels of adiponectin are associated with increased insulin sensitivity, increased fatty acid oxidation, and decreased hepatic glucose production.[78] Obesity tends to increase the plasma concentrations of most proteins produced by adipose tissue because of the increase in total fat mass, but plasma levels of adiponectin were found to be much lower in obesity.[79] Significantly lower levels of adiponectin were also observed in fat tissues from obese mice and obese humans.[78,79] Adiponectin can ameliorate insulin resistance by decreasing hepatic glucose production through lowering activity of the gluconeogenic enzyme phosphoenolpyruvate carboxykinase.[80] Along with its ability to decrease the activity of hepatic gluconeogenic enzymes, adiponectin improves insulin sensitivity by decreasing the concentration of circulating triglycerides and free fatty acids.[81] To investigate the hormonal changes caused by high-fat diet and the propeptide transgene, we measured insulin and several adipocyte hormones.[71] The concentrations of insulin, leptin, and resistin were similar in transgenic mice fed either a normal or a high-fat diet; these levels were slightly higher but not significantly different from those seen in wild-type mice fed a normal diet. However, high-fat diet increased serum insulin, leptin, and resistin by 97%, 377%, and 32%, respectively, in wild-type mice compared to transgenic mice.[71] High-fat diet induced much higher levels of serum adiponectin concentration in the transgenic mice than that found in the wild-type mice (9.44 ± 0.79 vs. 7.90 ± 0.68 µg/ml, $P=0.008$), while transgenic mice fed a normal diet had significantly lower serum adiponectin levels (4.35 ± 0.60 µg/ml).

When we studied the gene expression of adiponectin in adipose tissue, myostatin propeptide transgenic mice fed either normal-fat or high-fat diet showed significant increased levels of adiponectin expression in epididymal fat pad. The relative expression level of adiponectin was the highest in the epididymal fat in the transgenic/high-fat group with a value of 3.10 ± 0.67, which is a 12-time increase over its control group—wild-type/high-fat diet.

The subcutaneous fat of the transgenic/normal-fat group also showed an increased level over its control group—transgenic/normal-fat. Within the epididymal fat pad, the wild-type/high-fat group was significantly lower in adiponectin expression than wild-type/normal-fat group. These results clearly suggest that an increased secretion of adiponectin may promote energy partition toward skeletal muscles, therefore promoting a beneficial interaction between muscle and adipose tissue.[77]

In addition, *ex vivo* delivery of a mutant form of myostatin propeptide also functionally improved dystrophic muscle in mice. By adeno-associated virus (AAV)-mediated expression of a mutated propeptide, researchers showed a boost in muscle mass and an increase in absolute force in calpain 3-deficient mice. Myostatin inhibition by its propeptide overexpression could be effective for therapeutic treatment of atrophic disorder.[82] A similar experiment with mdx mice, a murine model of Duchenne muscular dystrophy, was conducted by adeno-associated virus serotype 8 delivery of myostatin propeptide. There was a significant increase in skeletal muscle mass in the form of fiber hypertrophy after the propeptide virus injection. A grip force test and an *in vitro* tetanic contractile force test showed improved muscle strength.[82]

Myostatin sequence is highly conserved among mammalian species. Mouse, pig, human, and chicken myostatins are 100% identical in the amino acid sequence of the mature peptide. Pig is becoming a valuable model animal for biomedical research. We have created a mutated form of porcine myostatin propeptide at the cleavage site of metalloproteinase BMP-1/TLD family, which strongly blocked myostatin activity in A204 cell assay. Administration of this mutated propeptide to mice significantly increased skeletal muscle growth. Enhanced muscle growth by two injections of the propeptide in neonatal stages can last 4 weeks after injection and primarily resulted from fiber hypertrophy. These results suggest that depression of myostatin activity by BMP-1/TLD proteinase-resistant propeptide can be an effective way to promote muscle growth.[83] The result is consistent with early experimental data with murine myostatin propeptide obtained by Wolfman *et al.*,[84] where the concentration of murine propeptide at 500 pM was used for suppressing 1 pM myostatin. However, the result of porcine myostatin propeptide from this study may appear more effective as the concentration of the propeptide at 8 pM (100 ng/ml) depressed the activity of 2 pM (20 ng/ml) myostatin. Mice injected with the mutated propeptide showed a significant increase ($P < 0.05$) in body weight from littermate controls as early as 1 week after the first injection (at the age of 11 days) in females and 1 week after the second injection in males (18 days of age).

After the second injection, male mice injected with mutated propeptide were 12–15% heavier than their controls from the age of 25 to 57 days ($P < 0.05$), whereas female mice injected with the propeptide were 11–15% heavier than their littermate controls ($P < 0.05$) from the age of 25 to 57 days. The effect of the injections of mutated propeptide on muscle growth was maintained for at least 4 weeks from the age of 25 to 53 days.[83] Taken together, these findings indicate that suppressing myostatin activity by its propeptide in native or mutated form can be an effective strategy to enhance muscle growth. The size and amount of skeletal muscle certainly are important factors for muscle strength. A therapeutic strategy based on myostatin propeptide for muscle dystrophy or any other muscle wasting appears effective in small animals. Further testing and *ex vivo* delivery to a large animal model such as pigs would provide initial evidence for application to human therapy.

With the technological advancements of gene expression and protein detection, skeletal muscle and adipose tissues have also been known as endocrine glands by constitutively secreting proteins that regulate metabolic and cellular events. Although myostatin is mainly produced in the skeletal muscle, its actions appear on skeletal muscle via paracrine signaling, as well as in an endocrine way of regulation. The level of myostatin protein is much higher in skeletal muscle and plasma in obese women compared with healthy women. The levels of increased myostatin are significantly correlated with the severity of insulin resistance.[85] In Sprague–Dawley rats, the levels of myostatin protein increased progressively at 6, 12, and 27 months of age compared with young animals.[86] In streptozotocin-induced diabetes, myostatin expressions are well corresponded with the diabetic syndrome and the effects of insulin treatment. Streptozotocin treatment induces diabetes and body weight loss. It also increases muscle myostatin expression due to the loss of skeletal muscle. While insulin treatment reversed the hyperglycemia and body weight loss, myostatin expression was also recovered.[87] Studies with myostatin injection found that increased myostatin expression is inversely correlated with insulin sensitivity. It was believed that myostatin might directly regulate skeletal muscle glucose uptake.[61] Earlier, myostatin was hypothesized to act as a chalone to limit muscle size,[88] whereby inhibition or complete deficiency of myostatin would remove the limiting effects on muscle mass, resulting in enhanced skeletal muscle tissue in an intrinsic control mechanism. As the size of muscle tissue is dynamic with great plasticity in response to various physiological and biochemical stimuli, myostatin expression may represent a significant contribution to the deterioration or loss of muscle mass in diabetes. As the effects of muscle mass on insulin sensitivity are predominantly significant, the direct effects of myostatin on muscle glucose uptake hardly distinguish from its indirect effects simply due to its primary genetic regulations on muscle mass.

The mechanism by which myostatin function is disrupted by its propeptide is likely the result of formation of latency-associated complex. Results from both *in vitro* and *in vivo* support the mechanism of myostatin latency during posttranslational processing. The myostatin precursor protein can be cleaved by furin proteases *in vitro*, producing N-terminal propeptide and C-terminal mature peptide or myostatin. The propeptide remains noncovalently bound to the mature C-terminal dimer in a latent state,[89–91] which also appears to circulate in the blood.[91] In transgenic mice with mutated BMP-1/TLD family of metalloproteases, which can block the cleavage pathway, myostatin function is partially inhibited as the muscle mass is increased, along with increased circulating myostatin levels.[92] In the myostatin propeptide transgenic mice in my laboratory, a large, latent form of myostatin complex was detected in skeletal muscle (unpublished). The molar ratio between propeptide and mature myostatin in serum is approximately 1:1, and no data are available about the molar ratio in muscle. As the propeptide and mature myostatin are generated from the same precursor peptide, we assumed their molar ratio is close to 1:1 in muscle tissue. A transgenic overexpression of the propeptide may promote forming of more myostatin-latent complex, therefore resulting in less biologically active mature myostatin in muscle or even in circulation. In the propeptide transgenic mice, transgene mRNA was not detected in the adipose tissue, excluding its possible direct effects on adipose tissue. We believe that enhanced muscling induced by the transgene during the early-growing stage significantly changed nutrient partitions in the body. Energy demand for muscle growth and maintenance in the transgenic mice was so prominent that less energy was available for fat accumulation and, consequently, they maintained normal adipose tissue and high insulin sensitivity.[71] Interestingly, when we did muscle histology in the old transgenic animals, the observation of the myofiber histology indicated more nuclei were localized in the central and basal lamina of the myofibers of the transgenic mice at 1 year old. Long and stretched nuclei were also noted in some fibers, suggesting active fiber fusion in these muscles at 1 year old. In contrast, muscle histology from wild-type littermates did not show such changes in nuclei distributions. The number of nuclei per fiber in both basal and central lamina of the myofiber was significantly higher in transgenic mice than in wild-type littermates by 58.3% and 45.8%, respectively ($P < 0.01$). These results provide evidences that transgenic expression of myostatin propeptide supported continuous muscle buildup in adult skeletal muscle tissue.[93]

To understand how the enhanced muscle growth by myostatin propeptide is regulated at the whole genome level, we employed microarray analysis. The results from microarray analysis of global gene expression profile, supported by qRT-PCR assays and biochemical analysis, provide distinct muscle growth pathways that integrate low protein degradation and ATP synthesis for efficient muscle buildup and energy utilization. Skeletal muscle buildup is maintained by high-level expressions of myogenin, Cdk inhibitor P21, follistatin-like factor (Fstl), Rho-associated kinase (Rock1), and ECM components such as procollagen, fibronectin, and biglycan. In the meantime, there are decreased levels of protein degradation and mitochondrial ATP synthesis in the transgenic mice, which suggests efficient energy utilization for muscle buildup. Although the profile change and patterns of muscle gene expressions were caused and maintained by the manipulation of a single gene, namely, myostatin, clearly, the genes associated with myofiber fusions and energy utilizations were dynamically changed when expressed at the mRNA levels. In the current myostatin propeptide transgenic mouse model, decreased protein degradation and mitochondrial ATP synthesis may represent an important metabolic type that significantly enhances muscle growth in adult stages. Since adult muscle buildup is complicated by age-induced muscle atrophy, any regulatory and metabolic pathway or mechanism that can maintain skeletal muscle mass and growth would provide important directions for future studies. We have begun to define more specific mechanisms of myogenic initiation, maintenance of myogenic states, and mitochondrial energy production in adult stages.[93] In addition, further studies with this model may shed light on its potential application to the treatment of muscle dystrophy and cachexia by depressing myostatin activity.

Physiologically, the key to effective glucose homeostasis is insulin hormone and its sensitivity on its target tissues. Glucose transport in skeletal muscle is stimulated by at least two distinct pathways. Insulin-stimulated IRS-1 and PI3 kinase primarily regulate glucose transport in response to the postprandial increased level of blood glucose. Muscle contraction through physical exercise or resistance training initiates insulin-independent glucose uptake. For example, the 5'AMP (adenosine monophosphate)-activated protein kinase participates in the muscle contraction-activated glucose transport and uptake. Muscle contraction causes depolarization of

T-tubules and then calcium (Ca^{2+}) release from the sarcoplasmic reticulum, which triggers actin and myosin interaction. The energy demand of contraction increases the ratio of AMP/ATP, which stimulates AMP-activated protein kinase (AMPK). Therefore, AMPK serves as the sensing enzyme for the energy levels such as the AMP to ATP and creatine-to-phosphocreatine ratios via a complex mechanism of allosteric modification and phosphorylation.[94,95] Although how AMPK signaling regulates GLUT4 remains to be elucidated, AMPK is activated in both rat and human muscles in response to muscle contraction.[96,97] AMPK is activated chemically by 5′AMP and aminoimidazole-4-carboxamide-riboside. Once activated, AMPK is thought to phosphorylate proteins involved in triggering fatty acid oxidation and glucose uptake via increasing GLUT4 and hexokinase and uncoupling protein 3 and mitochondrial oxidative enzymes. Besides AMPK pathways, muscle contraction-stimulated glucose transport may use other intermediates such as Ca^{2+}/calmodulin, nitric oxide, bradykinin, and AKT substrate AS160. The signaling pathway of the muscle contraction-stimulated glucose uptake appears also interconnected with insulin pathway via intermediates. In the skeletal muscle, the actin cytoskeleton-regulating GTPase, Rac1, is initially defined for insulin-dependent GLUT4 translocation. A recent study showed that Rac1 activation by muscle contraction (i.e., exercise) in mice and human is AMPK-independent and Rac1 inhibition reduces contraction-stimulated glucose uptake in mouse muscles.[98]

The fact that increased physical activity is a simple and effective pathway to prevent and treat metabolic disorders is widely accepted by health professionals. However, it is not implemented by most patients. Reduced exercise or continued physical inactivity may be due to chronic subclinical inflammation associated with the obesity or metabolic syndrome. It is generally agreed that physical activity is not desirable during inflammation by the body and the mind.[99–101] Researchers have suggested to develop auxiliary substances that mimic the plastic adaptations to exercise so that the body will amplify the effects of exercise in contending physical activity limitations or inactivity.[101] One of the interesting key regulators of skeletal muscle contraction is the peroxisome proliferator-activated receptor γ coactivator 1α (PGC-1α). PGC-1α acts as a transcriptional coactivator that induces mitochondrial biogenesis, promotes angiogenesis, and increases peak oxygen consumption and fatigue resistance.[102,103] Elevated expression of PGC-1α in skeletal muscle increases endurance performance by switching fast glycolytic fibers toward slow oxidative fibers and increases metabolic flexibility.[104] Skeletal muscle-specific overexpression of PGC-1α results in diet-induced insulin resistance in sedentary conditions.[103] However, the elevated PGC-1α in combination with exercise preferentially improves glucose homeostasis, increases Krebs cycle activity, and reduces the levels of acylcarnitines and sphingosine. Moreover, patterns of lipid partitioning are altered in favor of enhanced insulin sensitivity in response to combined PGC-1α and exercise.[105] The results provide a promising therapeutic approach for metabolic drug development by addressing active skeletal muscle combined with exercise for prevention and treatment of metabolic disorders.

8. CONCLUDING REMARKS

Glucose homeostasis is fundamental for life and long-term health. As the most effective organ for insulin-stimulated glucose uptake in the body, the skeletal muscle is the key element for maintaining effective glucose homeostasis in many chronic diseases. Preservation of muscle mass in obesity, insulin resistance, or type II diabetes is important for effective treatment and prevention of these diseases. Enhanced skeletal muscle by intrinsic biology control or physical activities provides effective approaches to regulate metabolic activities and energy balance. Studies from myostatin, a dominant inhibitor of muscle mass, have produced significant and interesting results in understanding and promoting skeletal muscle growth. The results in transgenic mice indicate that depression of myostatin function by its propeptide produces dramatic muscle mass at the growth stage and less fat at older ages. A high level of myostatin has been found in skeletal muscle and plasma in severely obese patients. Plasma myostatin is correlated with the severity of insulin resistance. Enhancing muscle mass by depressing myostatin activity can be an effective way to reverse diabetic pathogenesis. Experiments from the myostatin propeptide transgenic mice and high-fat feeding trial further support the flexibility of skeletal muscle growth and buildup while myostatin was modified. There have not been strong experimental data to directly link myostatin function and insulin sensitivity in skeletal muscles. Due to the size of the skeletal muscle tissue, there are great challenges in reversing age- or disease-induced skeletal muscle loss by physiological or medical manipulations. Exercise or resistance training offers simple and practical approaches to enhance skeletal muscle buildup and function, which can stimulate insulin-independent glucose uptake. Research results from muscle formation and regeneration provide exciting scientific progress in the fundamental understanding of muscle tissue. Studies in exercise-induced muscle regenerations, in combination with dominant genes such as myostatin manipulation and diabetes-induced muscle loss, will further broaden the basic understanding of adult muscle growth and maintenances and translate the molecular mechanism into developing therapeutic approaches for the rejuvenation of skeletal muscle with effective glucose homeostasis.

ACKNOWLEDGMENTS

This work is supported by grants from Alan M. Krassner Fund and Ingeborg v. F. McKee Fund of the Hawaii Community Foundation (Grant ID #13P-60310), the USDA National Institute for Food and Agriculture Project HAW244-R administered by the College of Tropical Agriculture and Human Resources, University of Hawaii at Manoa, and Wuhan "3551" Talent Scheme. Thanks for technical assistance and support from Baoping Zhao, Zicong Li, Xiang Huang, Liang Wu, Wenhui Pi, Ahmed Hussein, Awat Yousif, Xin Zhang, and Sharon Agacid.

REFERENCES

1. Thiebaud D, Jacot E, DeFronzo RA, Maeder E, Jequier E, Felber JP. The effect of graded doses of insulin on total glucose uptake, glucose oxidation, and glucose storage in man. *Diabetes*. 1982;31:957–963.
2. DeFronzo RA, Jacot E, Jequier E, Maeder E, Wahren J, Felber JP. The effect of insulin on the disposal of intravenous glucose: results from indirect calorimetry and hepatic and femoral venous catheterization. *Diabetes*. 1981;30:1000–1007.
3. Wilson JE. Isozymes of mammalian hexokinase: structure, subcellular localization and metabolic function. *J Exp Biol*. 2003;206:2049–2057.
4. DeFronzo RA, Tripathy D. Skeletal muscle insulin resistance is the primary defect in type 2 diabetes. *Diabetes Care*. 2009;32:S157–S163.
5. Taniguchi CM, Emanuelli B, Kahn CR. Critical nodes in signalling pathways: insights into insulin action. *Nat Rev Mol Cell Biol*. 2006;7:85–96.
6. Krook A, Bjornholm M, Galuska D, et al. Characterization of signal transduction and glucose transport in skeletal muscle from type 2 diabetic patients. *Diabetes*. 2000;49:284–292.
7. Koopman R, van Loon LJC. Aging, exercise, and muscle protein metabolism. *J Appl Physiol*. 2009;106:2040–2048.
8. Davison KK, Ford ES, Cogswell ME, Dietz WH. Percentage of body fat and body mass index are associated with mobility limitations in people aged 70 and older from NHANES III. *J Am Geriatr Soc*. 2002;50:1802–1809.
9. Park SW, Goodpaster BH, Strotmeyer ES, et al. Decreased muscle strength and quality in older adults with type 2 diabetes: the health, aging, and body composition study. *Diabetes*. 2006;55:1813–1818.
10. Park SW, Goodpaster BH, Strotmeyer ES, et al. Accelerated loss of skeletal muscle strength in older adults with type 2 diabetes: the health, aging, and body composition study. *Diabetes Care*. 2007;30:1507–1512.
11. Park SW, Goodpaster BH, Lee JS, et al. Excessive loss of skeletal muscle mass in older adults with type 2 diabetes. *Diabetes Care*. 2009;32:1993–1997.

12. International Working Group on Sarcopenia . Sarcopenia: an undiagnosed condition in older adults. Current consensus definition: prevalence, etiology, and consequences. *J Am Med Dir Assoc.* 2009;12:249–256.

13. Schiaffino S, Mammucari C. Regulation of skeletal muscle growth by the IGF1-Akt/PKB pathway: insights from genetic models. *Skelet Muscle.* 2011;1:4.

14. Petersen KF, Shulman GI. Etiology of insulin resistance. *Am J Med.* 2006;119:S10–S16.

15. Petersen KF, Befroy D, Dufour S, et al. Mitochondrial dysfunction in the elderly (possible role in insulin resistance). *Science.* 2003;300:1140–1142.

16. Petersen KF, Dufour S, Befroy D, et al. Impaired mitochondrial activity in the insulin-resistant offspring of patients with type 2 diabetes. *N Engl J Med.* 2004;350:664–671.

17. Bentzinger CF, Wang YX, Rudnicki MA. Building muscle: molecular regulation of myogenesis. *Cold Spring Harb Perspect Biol.* 2012;4:a008342.

18. Konigsberg IR. Clonal analysis of myogenesis. *Science.* 1963;140:1273–1284.

19. Echeverri K, Clarke JD, Tanaka EM. In vivo imaging indicates muscle fiber dedifferentiation is a major contributor to the regenerating tail blastema. *Dev Biol.* 2001;236:151–164.

20. Odelberg SJ, Kollhoff A, Keating MT. Dedifferentiation of mammalian myotubes induced by msx1. *Cell.* 2000;103:1099–1109.

21. Odelberg SJ. Unraveling the molecular basis for regenerative cellular plasticity. *PLoS Biol.* 2004;2:1068–1071.

22. Mu X, Peng H, Pan H, Huard J, Li Y. Study of muscle cell dedifferentiation after skeletal muscle injury of mice with a Cre-Lox system. *PLoS One.* 2011;6(2):e16699.

23. Lefaucheur L, Edom F, Ecolan P, Butler-Browne GS. Pattern of muscle fiber type formation in the pig. *Dev Dyn.* 1995;203:27–41.

24. Lefaucheur L, Hoffman RK, Gerrard DE, Okamura CS, Rubinstein N, Kelly A. Evidence for three adult fast myosin heavy chain isoforms in type II skeletal muscle fibers in pigs. *J Anim Sci.* 1998;76:1584–1593.

25. Yan X, Zhu MJ, Dodson MV, Du M. Developmental programming of fetal skeletal muscle and adipose tissue development. *J Genomics.* 2012;1:29–38.

26. Perry RL, Rudnick MA. Molecular mechanisms regulating myogenic determination and differentiation. *Front Biosci.* 2000;5:D750–D767.

27. Hasty P, Bradley A, Morris JH, et al. Muscle deficiency and neonatal death in mice with a targeted mutation in the myogenin gene. *Nature.* 1993;364:501–506.

28. Nabeshima Y, Hanaoka K, Hayasaka M, et al. Myogenin gene disruption results in perinatal lethality because of severe muscle defect. *Nature.* 1993;364:532–534.

29. Zhang W, Behringer RR, Olson EN. Inactivation of the myogenic bHLH gene MRF4 results in up-regulation of myogenin and rib anomalies. *Genes Dev.* 1995;9:1388–1399.

30. Braun T, Arnold HH. Inactivation of Myf-6 and Myf-5 genes in mice leads to alterations in skeletal muscle development. *EMBO J.* 1995;14:1176–1186.

31. Maroto M, Reshef R, Munsterberg AE, Koester S, Goulding M, Lassar AB. Ectopic Pax-3 activates MyoD and Myf-5 expression in embryonic mesoderm and neural tissue. *Cell.* 1997;89:139–148.

32. Messina G, Cossu G. The origin of embryonic and fetal myoblasts: a role of Pax3 and Pax7. *Genes Dev.* 2009;23:902–905.

33. Relaix F, Rocancourt D, Mansouri A, Buckingham M. Divergent functions of murine Pax3 and Pax7 in limb muscle development. *Genes Dev.* 2004;18:1088–1105.

34. Hawke TJ, Garry DJ. Myogenic satellite cells: physiology to molecular biology. *J Appl Physiol.* 2001;91:534–551.

35. Rowe RW, Goldspink G. Muscle fibre growth in five different muscles in both sexes of mice. II. Dystrophic mice. *J Anat.* 1969;104(Pt 3):531–538.

36. Rosenblatt JD, Woods RI. Hypertrophy of rat extensor digitorum longus muscle injected with bupivacaine. A sequential histochemical, immunohistochemical, histological and morphometric study. *J Anat*. 1992;181:11–27.
37. Fiedler I, Rehfeldt C, Dietl G, Ender K. Phenotypic and genetic parameters of muscle fiber number and size. *J Anim Sci*. 1997;75(suppl 1):165 (Abstract).
38. Wegner J, Albrecht E, Fiedler I, Teuscher F, Papstein HJ, Ender K. Growth- and breed-related changes of muscle fiber characteristics in cattle. *J Anim Sci*. 2000;78:1485–1496.
39. Summers P, Medrano JF. Morphometric analysis of skeletal muscle growth in the high growth mouse. *Growth Dev Aging*. 1994;58:135–148.
40. Ontell M, Kozeka K. The organogenesis of murine striated muscle: a cytoarchitectural study. *Am J Anat*. 1984;171:133–148.
41. Ontell M, Kozeka K. The organogenesis of the mouse extensor digitorum longus striated muscle: a quantitative study. *Am J Anat*. 1984;171:149–161.
42. Seale P, Rudnicki MA. A new look at the origin, function, and "stem-cell" status of muscle satellite cells. *Dev Biol*. 2000;218:115–124.
43. Cornelison DD, Wold BJ. Single-cell analysis of regulatory gene expression in quiescent and activated mouse skeletal muscle satellite cells. *Dev Biol*. 1997;191:270–283.
44. Cornelison DD. Context matters: in vivo and in vitro influences on muscle satellite cell activity. *J Cell Biochem*. 2008;105:663–669.
45. Yablonka-Reuveni Z. The skeletal muscle satellite cell: still young and fascinating at 50. *J Histochem Cytochem*. 2011;59:1041–1059.
46. Yablonka-Reuveni Z, Day K, Vine A, Shefer G. Defining the transcriptional signature of skeletal muscle stem cells. *J Anim Sci*. 2008;86:E207–E216.
47. Snow MH. Myogenic cell formation in regenerating rat skeletal muscle injured by mincing. I. A fine structural study. *Anat Rec*. 1977;188:181–200.
48. Snow MH. Myogenic cell formation in regenerating rat skeletal muscle injured by mincing. II. An autoradiographic study. *Anat Rec*. 1977;188:201–218.
49. Day K, Shefer G, Shearer A, Yablonka-Reuveni Z. The depletion of skeletal muscle satellite cells with age is concomitant with reduced capacity of single progenitors to produce reserve progeny. *Dev Biol*. 2010;340:330–343.
50. Relaix F, Zammit PS. Satellite cells are essential for skeletal muscle regeneration: the cell on the edge returns centre stage. *Development*. 2012;139:2845–2856.
51. McPherron AC, Lee S-J. Double muscling in cattle due to mutations in the myostatin gene. *Proc Natl Acad Sci USA*. 1997;94:12457–12461.
52. Mosher DS, Quignon P, Bustamante CD, et al. A mutation in the myostatin gene increases muscle mass and enhances racing 79 performance in heterozygote dogs. *PLoS Genet*. 2007;3:0779–0786.
53. Clop A, Marcq F, Takeda H, et al. A mutation creating a potential illegitimate micro-RNA target site in the myostatin gene affects muscularity in sheep. *Nat Genet*. 2006;38:813–818.
54. Schuelke M, Wagner K, Stolz L, et al. Myostatin mutation associated with gross muscle hypertrophy in a child. *N Engl J Med*. 2004;350:2682–2688.
55. McPherron AC, Lawler AM, Lee S-J. Regulation of skeletal muscle mass in mice by a new TGF-ß superfamily member. *Nature*. 1997;387:83–90.
56. Hanset R, Michaux C. On the genetic determinism of muscular hypertrophy in the Belgian White and Blue cattle breed. *Genet Sel Evol*. 1985;17:359–368.
57. Kambadur R, Sharma M, Smith TP, Bass JJ. Mutations in myostatin (GDF8) in double-muscled Belgian Blue and Piedmontese cattle. *Genome Res*. 1997;7:910–916.
58. Grobet L, Martin LJ, Poncelet D, et al. A deletion in the bovine myostatin gene causes the double-muscled phenotype in cattle. *Nat Genet*. 1997;17:71–74.

59. Gonzalez-Cadavid NF, Taylor WE, Yarasheski K, et al. Organization of the human myostatin gene and expression in healthy men and HIV-infected men with muscle wasting. *Proc Natl Acad Sci USA*. 1998;95:14938–14943.
60. Deveaux V, Cassar-Malek I, Picard B. Comparison of contractile characteristics of muscle from Holstein and double-muscled Belgian Blue fetuses. *Comp Biochem Physiol*. 2001;131A:21–29.
61. Rodgers BD, Garikipati DK. Clinical, agricultural, and evolutionary biology of myostatin: a comparative review. *Endocr Rev*. 2008;29:513–534.
62. Allen DL, Hittel DS, McPherron AC. Expression and function of myostatin in obesity, diabetes, and exercise adaptation. *Med Sci Sports Exerc*. 2011;43:1828–1835.
63. Reisz-Porszasz BS, Artaza JN, Shen R, et al. Lower skeletal muscle mass in male transgenic mice with muscle-specific over-expression of myostatin. *Am J Physiol Endocrinol Metab*. 2003;285:E876–E888.
64. Rebbapragada A, Benchabane H, Wrana JL, Celeste AJ, Attisano L. Myostatin signals through a transforming growth factor β-like signaling pathway to block adipogenesis. *Mol Cell Biol*. 2003;23:7230–7242.
65. McCroskery S, Thomas M, Maxwell L, Sharma M, Kambadur R. Myostatin negatively regulates satellite cell activation and self-renewal. *J Cell Biol*. 2003;162:1135–1147.
66. McPherron AC, Huynh TV, Lee S-J. Redundancy of myostatin and growth/differentiation factor 11 function. *BMC Dev Biol*. 2009;9:24–32.
67. Thies RS, Chen T, Davies MV, et al. GDF-8 propeptide binds to GDF-8 and antagonizes biological activity by inhibiting GDF-8 receptor binding. *Growth Factors*. 2001;18:251–259.
68. Yang J, Ratovitski T, Brady JP, Solomon MB, Wells KD, Wall RJ. Expression of myostatin pro domain results in muscular transgenic mice. *Mol Reprod Dev*. 2001;60:351–361.
69. Yang J, Zhao B. Postnatal expression of myostatin propeptide cDNA maintained high muscle growth and normal adipose tissue mass in transgenic mice fed a high-fat diet. *Mol Reprod Dev*. 2006;73:462–469.
70. Lee SJ, McPherron AC. Regulation of myostatin activity and muscle growth. *Proc Natl Acad Sci USA*. 2001;98:9306–9311.
71. Zhao B, Wall RJ, Yang J. Transgenic expression of myostatin propeptide prevents diet-induced obesity and insulin resistance. *Biochem Biophys Res Commun*. 2005;337:248–255.
72. Wagner KR, Liu X, Chang X, Allen RE. Muscle regeneration in the prolonged absence of myostatin. *Proc Natl Acad Sci USA*. 2005;102:2519–2524.
73. Lee SJ, Huynh TV, Lee YS, et al. Role of satellite cells versus myofibers in muscle hypertrophy induced by inhibition of the myostatin/activin signaling pathway. *Proc Natl Acad Sci USA*. 2012;109:2353–2360.
74. Wang Q, McPherron AC. Myostatin inhibition induces muscle fibre hypertrophy prior to satellite cell activation. *J Physiol*. 2012;590:2151–2165.
75. McFarlane E, Plummer M, Thomas A, et al. Myostatin induces cachexia by activating the ubiquitin proteolytic system through an NF-kappaB-independent, FoxO1-dependent mechanism. *J Cell Physiol*. 2006;209:501–510.
76. Kim KH, Kim YS, Yang J. The muscle-hypertrophic effect of clenbuterol is additive to the hypertrophic effect of myostatin suppression. *Muscle Nerve*. 2011;43:700–709.
77. Suzuki ST, Zhao B, Yang J. Enhanced muscle by myostatin propeptide increases adipose tissue adiponectin, PPAR-α and PPAR-γ expressions. *Biochem Biophys Res Commun*. 2008;369:767–773.
78. Berg H, Combs TP, Du X, Brownlee M, Scherer PE. The adipocyte-secreted protein Acrp30 enhances hepatic insulin action. *Nat Med*. 2002;7:947–953.

79. Hu P Liang, Spiegelman BM. AdipoQ is a novel adipose-specific gene dysregulated in obesity. *J Biol Chem.* 1996;271:10697–10702.
80. Combs TP, Berg AH, Obici S, Scherer PE, Rossetti L. Endogenous glucose production is inhibited by the adipose-derived protein Acrp30. *J Clin Invest.* 2001;108:1875–1880.
81. Muoio M, Way JM, Tanner CJ, et al. Peroxisome proliferator-activated receptor-alpha regulates fatty acid utilization in primary human skeletal muscle cells. *Diabetes.* 2002;51:901–908.
82. Qiao J, Li J, Jiang X, Zhu B, Wang J Li, Xiao X. Myostatin propeptide gene delivery by adeno-associated virus serotype 8 vectors enhances muscle growth and ameliorates dystrophic phenotypes in mdx mice. *Hum Gene Ther.* 2008;19:241–247.
83. Li Z, Zhao B, Kim YS, Hu CY, Yang J. Administration of a mutated myostatin pro-peptide to neonatal mice significantly enhances skeletal muscle growth. *Mol Reprod Dev.* 2009;77:76–81.
84. Wolfman NM, McPherron AC, Pappano WN, et al. Activation of latent myostatin by the BMP-1/tolloid family of metalloproteinases. *Proc Natl Acad Sci USA.* 2003;100:15842–15846.
85. Hittel S, Berggren JR, Shearer J, Boyle K, Houmard JA. Increased secretion and expression of myostatin in skeletal muscle from extremely obese women. *Diabetes.* 2009;58:30–39.
86. Baumann P, Ibebunjo C, Grasser WA, Paralkar VM. Myostatin expression in age and denervation-induced skeletal muscle atrophy. *J Musculoskelet Neuronal Interact.* 2003;3:8–13.
87. Chen Y, Cao L, Ye J, Zhu D. Upregulation of myostatin gene expression in streptozotocin-induced type 1 diabetes mice is attenuated by insulin. *Biochem Biophys Res Commun.* 2009;388:112–121.
88. Lee SJ. Regulation of muscle mass by myostatin. *Annu Rev Cell Dev Biol.* 2004;20:61–71.
89. Harrison CA, Al-Musawi SL, Walton KL. Prodomains regulate the synthesis, extracel-lular localisation and activity of TGF-β superfamily ligands. *Growth Factors.* 2011;29:174–180.
90. Thies RS, Chen T, Davies MV, et al. GDF-8 propeptide binds to GDF-8 and antag-onizes biological activity by inhibiting GDF-8 receptor binding. *Growth Factors.* 2001;18:251–260.
91. Hill JJ, Davies MV, Pearson AA, et al. The myostatin propeptide and the follistatin-related gene are inhibitory binding proteins of myostatin in normal serum. *J Biol Chem.* 2002;277:40735–40740.
92. Lee SJ. Genetic analysis of the role of proteolysis in the activation of latent myostatin. *PLoS One.* 2008;3:e1628.
93. Zhao B, Li EJ, Wall RJ, Yang J. Coordinated patterns of gene expressions for adult muscle build-up in transgenic mice expressing myostatin propeptide. *BMC Genomics.* 2009;10:305–341.
94. Goodyear LJ, Kahn BB. Exercise, glucose transport, and insulin sensitivity. *Annu Rev Med.* 1998;49:235–261.
95. Jessen N, Goodyear LJ. Contraction signaling to glucose transport in skeletal muscle. *J Appl Physiol.* 2005;99:330–351.
96. Winder WW, Hardie DG. AMP-activated protein kinase, a metabolic master switch: possible roles in type 2 diabetes. *Am J Physiol Endocrinol Metab.* 1999;277:E1–E10.
97. Winder WW, Wilson HA, Hardie DG, et al. Phosphorylation of rat muscle acetyl-CoA carboxylase by AMP-activated protein kinase and protein kinase A. *J Appl Physiol.* 1997;82:219–225.

98. Sylow L, Jensen TE, Kleinert M, et al. Rac1 signaling is required for insulin-stimulated glucose uptake and is dysregulated in insulin-resistant murine and human skeletal muscle. *Diabetes*. 2013;62:1865–1875.

99. Pahor M, Kritchevsky S. Research hypotheses on muscle wasting, aging, loss of function and disability. *J Nutr Health Aging*. 1998;2:97–100.

100. Roubenoff R, Hughes VA. Sarcopenia: current concepts. *J Gerontol A Biol Sci Med Sci*. 2000;55:M716–M724.

101. Matsakas A, Narkar VA. Endurance exercise mimetics in skeletal muscle. *Curr Sports Med Rep*. 2010;9:227–232.

102. Handschin C. The biology of PGC-1α and its therapeutic potential. *Trends Pharmacol Sci*. 2009;30:322–329.

103. Calvo JA, Daniels TG, Wang X, et al. Muscle-specific expression of PPARgamma coactivator-1alpha improves exercise performance and increases peak oxygen uptake. *J Appl Physiol*. 2008;104:1304–1309.

104. Lin J, Wu H, Tarr PT, et al. Transcriptional co-activator PGC-1 alpha drives the formation of slow-twitch muscle fibres. *Nature*. 2002;418:797–801.

105. Summermatter S, Shui G, Maag D, Santos G, Wenk MR, Handschin C. PGC-1α improves glucose homeostasis in skeletal muscle in an activity-dependent manner. *Diabetes*. 2013;62:85–95.

CHAPTER SIX

Factors Affecting Insulin-Regulated Hepatic Gene Expression

Hong-Ping Guan*, Guoxun Chen†

*Department of Diabetes, Merck Research Laboratories, Kenilworth, New Jersey, USA
†Department of Nutrition, University of Tennessee at Knoxville, Knoxville, Tennessee, USA

Contents

Abstract

Obesity has become a major concern of public health. A common feature of obesity and related metabolic disorders such as noninsulin-dependent diabetes mellitus is insulin resistance, wherein a given amount of insulin produces less than normal physiological responses. Insulin controls hepatic glucose and fatty acid metabolism, at least in part, via the regulation of gene expression. When the liver is insulin-sensitive, insulin can stimulate the expression of genes for fatty acid synthesis and suppress those for gluconeogenesis. When the liver becomes insulin-resistant, the insulin-mediated suppression of gluconeogenic gene expression is lost, whereas the induction of fatty acid synthetic gene expression remains intact. In the past two decades, the mechanisms of insulin-regulated hepatic gene expression have been studied extensively and many components of insulin signal transduction pathways have been identified. Factors that alter these pathways, and the insulin-regulated hepatic gene expression, have been revealed and the underlying mechanisms have been proposed. This chapter summarizes the recent grogresses in our understanding of the effects of dietary factors, drugs, bioactive compounds, hormones, and cytokines on insulin-regulated hepatic gene expression. Given the large amount of information and progresses regarding the roles of insulin, this chapter focuses on findings in the liver and hepatocytes and not those described for other tissues and cells. Typical insulin-regulated hepatic genes, such as insulin-induced glucokinase and sterol regulatory element-binding protein-1c and insulin-suppressed cytosolic phosphoenolpyruvate carboxyl kinase and insulin-like growth factor-binding protein 1, are used as examples to discuss the mechanisms such as insulin regulatory element-mediated transcriptional regulation. We also propose the potential mechanisms by which these factors affect insulin-regulated hepatic gene expression and discuss potential future directions of the area of research.

ABBREVIATIONS

ACC acetyl-CoA carboxylase
AMPK AMP-activated protein kinase
BBR berberine
CBP CREB-binding protein
C/EBP CCAAT/enhancer-binding protein
CREB cAMP regulatory element-binding protein
FAS fatty acid synthase
FFA free fatty acid
FoxO1 forkhead box protein O1
G6Pase glucose-6-phosphatase catalytic subunit
GCGR glucagon receptor
GK glucokinase
GH growth hormone
GLUT2 glucose transporter 2
GRB2 growth factor receptor-bound protein 2
GRE glucocorticoid-responsive element
GSIS glucose-stimulated insulin secretion
HNF4 hepatocyte nuclear factor 4
IGFBP-1 insulin-like growth factor-binding protein 1
IRE insulin-response element
IRE-BP1 insulin-responsive element-binding protein 1
IRS insulin receptor substrate
LDL low-density lipoprotein
LPKR pyruvate kinase
LXR liver X receptor
MAPK mitogen-activated protein kinase

mTORC1 mammalian target of rapamycin complex 1
NAD nicotinamide adenine dinucleotide
NIDDM noninsulin-dependent diabetes mellitus
PEPCK-C phosphoenolpyruvate carboxyl kinase—cytosolic form
PEPCK-M phosphoenolpyruvate carboxyl kinase—mitochondrial form
PGC-1α peroxisome proliferator-activated receptor coactivator 1α
PI3K phosphatidylinositol 3-kinase
PKB/Akt protein kinase B
PPRE peroxisome proliferator-activated receptor element
PTPs protein tyrosine phosphatases
RA retinoic acid
RAR retinoic acid receptor
RXR retinoid X receptor
SCD-1 stearoyl-CoA desaturase 1
SREBP-1c sterol regulatory element-binding protein-1c
SHIP2 SH2-containing inositol 5′-phosphatase 2
SOCS3 suppressor of cytokine signaling 3
SRC1 steroid receptor coactivator 1
T3 triiodothyronine
TAT tyrosine transaminase
TNFα tumor necrosis factor α
VAD vitamin A deficiency

1. INTRODUCTION

In a multicellular organism, cells and organs are constantly communicating with each other to access the statuses. The rise or fall of serum concentrations of functional molecules causes corresponding changes in a specific organ or the whole body. The delicate balance becomes essential for the very healthy status of an individual. The lack of signaling molecules, their receptors, or the sensitivity of their signal transduction cascades can cause diseases in humans and animals. Since its discovery, insulin has been linked to the development of diabetes mellitus, due to lack of the hormone itself or function it exerts.[1] It has been known for a long time that dietary factors influence insulin activity.[2] As a key hormone maintaining fuel homeostasis, insulin regulates expression of hepatic genes involved in glucose and fatty acid metabolism.[3,4] This process can be altered not only by hormones but also by environmental factors including nutrients, phytochemicals, and pharmaceutical drugs. This chapter is aimed to summarize the recent progresses in the study of factors that affect insulin-regulated gene expression in hepatocytes in the liver.

2. INSULIN SECRETION, ITS SIGNALING PATHWAYS, AND ITS REGULATED HEPATIC GENE EXPRESSION

2.1. Insulin secretion

2.1.1 Factors affecting insulin secretion

The link of the pancreas to diabetes mellitus was discovered when Oskar Minkowski removed the organ from dogs.[5] The idea of Frederick Banting to degenerate the exocrine portion and preserve the endocrine portion of the pancreas did not come out until 30 years later, which led him and his colleagues to successfully isolate insulin from the pancreas for treatment of diabetes.[1] The differential amounts of insulin needed to control plasma glucose of diabetic patients led Himsworth to set up elegant experiments to test the responses of glucose to insulin injection, which caused him to conclude that there were two types of diabetes, one with deficiency of insulin and the other with loss of insulin actions. He even correctly predicted that noninsulin-dependent diabetes mellitus (NIDDM) patients had higher plasma insulin levels. This was two decades before the radio-immunoassay (RIA) of insulin became available.[6] The subsequent employ-ment of RIA techniques allowed analysis of insulin secretion in responses to different secretagogues. This led scientists to believe that the eventual failure of pancreatic β-cells was responsible for the overt development of NIDDM.

Glucose has been considered one of the most important natural secreta-gogues of insulin secretion. The entry and metabolism of glucose are con-trolled by glucose transporter 2 (GLUT2) and glucokinase in pancreatic β-cells, respectively. Both have high K_m values to glucose. Glucose metab-olism results in a rise of ATP/ADP ratio, which can inhibit ATP-sensitive K^+ (K_{ATP}) channels, causing plasma membrane depolarization and activat-ing exocytosis of insulin granules in pancreatic β-cells.[7] The metabolism of glucose inside pancreatic β-cells raises the intracellular ATP level, which leads to K_{ATP} closure. The rise of intracellular potassium level results in depolarization of plasma membrane, which causes the opening of voltage-gated calcium channel of β-cell. The elevation of cytosolic calcium level triggers the fusion of insulin granules with plasma membrane and release of insulin to extracellular environment. This glucose-stimulated insulin secretion (GSIS) assures that insulin is released only when the plasma glucose level rises, a physiological response to maintain glucose homeostasis. Insulin secretion can be stimulated by other physiologically relevant secretagogues as well, such as arginine,[8] which directly causes depolarization rather than through change of intracellular ATP levels.

Nutrients have profound effects on insulin secretion. Long–chain fatty acids (FA), probably in the forms of long-chain acyl-CoAs, potentiate GSIS for short term.[9,10] In isolated pancreas, GSIS can be modulated by FA metabolism. In pancreas isolated from fasted rats, GSIS was impaired, which returned to normal when FAs were provided in the secretion medium. The potentiated GSIS was only observed in medium containing long-chain FA, but not short–chain FAs. This phenomenon indicates the roles of elevated plasma FA levels in the regulation of GSIS. Depletion of lipid stores in pancreatic β-cells diminished GSIS, which was restored when FAs were added to the medium with glucose.[11] However, long-term exposure of islets and insulin-secreting cells to long–chain FAs increased the content of intracellular lipids and reduced the function of β-cells, a phenomenon that has been attributed to lipotoxicity of FAs in pancreatic β-cells.[12]

GSIS is also regulated by vitamin A (VA) status in rats. GSIS was impaired in pancreas from VA-deficient (VAD) rats and restored in those isolated from the rats with repletion of VA using retinyl palmitate.[13] Interestingly, VA either inhibited (100 μM) or potentiated (100 nM) GSIS in rat islets.[14] The active metabolic product of VA, all-*trans* retinoic acid (RA), has been shown to potentiate GSIS in insulin-secreting cells.[15,16] This all-*trans* RA-mediated increase of GSIS in INS-1 cells was attributed to the induction of transglutaminase activity.[17] In cultured islets treated with RA for 24 h, GSIS was also increased, presumably through the stimulation of glucokinase gene expression and glucokinase protein (GK) activity.[18]

2.1.2 Metabolic diseases and insulin resistance

The development of specialized cells or mechanisms to store excessive energy and nutrients is evolutionarily advantageous for an individual. The stored energy and nutrients not only allow the individual to face an uncertain environmental challenge but also signal the organism to be prepared for proliferation. A good example is the obese gene (*ob*). An individual with *ob* gene mutation developed massive obesity.[19] The *ob* gene product, leptin, an adipokine highly expressed in and secreted from adipose cells, has been shown to regulate food intake, body composition, energy status, fuel metabolism, and even readiness for reproduction in females.[20–23]

The rise of obese population in the United States has become a concern of public health.[24] The high obesity prevalence in the United States population[25] indicates an increase of the number of patients with NIDDM and other metabolic diseases.[26] Genetic mutations, such as mutations of leptin and its receptor, can cause obesity.[27] Genes responsible for development of NIDDM have also been suggested.[28] However, it seems unlikely that human genome has experienced a drastic evolution process during the past half century. Therefore, the monogenic mutations may not contribute profoundly to the current epidemic of obesity. This leads to the hypothesis that the interactions of genetic and environmental/nutritional factors may play a significant role in the development of obesity and its associated metabolic diseases.

The long evolutionary process has equipped human and animal bodies with a variety of regulatory mechanisms to control the interactions between genes and environmental/nutritional stimuli that may play a role in the current epidemic of obesity.[29,30] It is not surprising that overconsumption of nutrients, such as fructose in the sweetened soft drinks, has been considered as a contributing factor to the epidemic of obesity.[31] It is worthy to note that diets contain not only energy derived from macronutrients but also micronutrients and other factors, which may regulate and affect metabolic homeostasis. Interestingly, despite the obvious link between nutrition and metabolic diseases, how individual micronutrients in conjunction with macronutrients contribute to the development of obesity is still an open question.

One characteristic of human obesity and NIDDM is insulin resistance. Reduction of body weight (BW) in obese and diabetic patients improves insulin sensitivity[32] and corrects abnormalities of NIDDM.[33] Dietary manipulation has been considered a potent factor affecting insulin action. For example, young, healthy, male medical students consuming high-fat diet (HFD) for 2 days developed glucose intolerance,[34] indicating insulin resistance after short-term dietary interference. Rats fed with an HFD (50% fat calories) for 8 weeks showed impaired glucose uptake due to diminished insulin-stimulated GLUT4 translocation in the skeletal muscle.[35] All these observations suggest that a specific dietary factor(s) associated with high-fat feeding contributes to leptin and insulin resistance. Free fatty acids (FFAs) have been considered the factor responsible for HFD-induced leptin and insulin resistance.[36–38]

The liver plays an essential role in the regulation of glucose and lipid homeostasis in response to the hormonal and nutritional stimuli. In fed state, the liver oxidizes glucose for energy and produces glycogen for the storage of glucose. It also converts excessive acetyl-CoA into triglyceride (TG) for the storage in adipose tissues. In fasting state, the liver breaks down glycogen and uses metabolic intermediates to generate glucose via gluconeogenesis for use in the rest of the body. It also oxidizes FA for its own energy needs and produces ketone bodies to be released for fuel in other organs such as the brain. All these hepatic processes are coordinately regulated by hormonal and nutritional stimuli in which insulin plays an important role.[39] This is attributed, at least in part, to the insulin-regulated expression of genes involved in hepatic glucose and lipid metabolism.[3,4]

2.1.3 Hepatic insulin resistance

In the liver, insulin regulates the expression of a variety of genes responsible for glycolysis, glycogenesis, lipogenesis, and gluconeogenesis.[40] With the development of insulin resistance in obesity and NIDDM, profound changes of hepatic lipid and glucose metabolism can be observed.[39] If the liver is insulin-sensitive, insulin stimulates lipogenesis and inhibits gluconeogenesis. For example, to regulate hepatic glucose homeostasis, insulin increases the expression of glucokinase gene *(Gck)*,[41,42] the enzyme responsible for the first step of hepatic glycolysis. It suppresses the expression of the cytosolic phosphoenolpyruvate carboxykinase gene *(Pck1)*[43] and glucose-6-phosphatase catalytic subunit gene *(G6pc)*,[40] which controls the first and last steps of gluconeogenesis, respectively. For hepatic lipid metabolism, insulin increases the mRNA level of sterol regulatory element-binding protein-1c gene *(Srebp-1c)*,[44] a family member of SREBPs that regulate the hepatic cholesterol and FA biosynthesis and their homeostasis.[45] In the liver of *Srebp-1c* knockout mice,[46] the fasting–refeeding cycle no longer appropriately regulates the expression levels of critical lipogenic genes such as fatty acid synthase *(Fas)* and stearoyl-CoA desaturase 1 *(Scd1)*.[47,48] When the liver is insulin-resistant, insulin loses its inhibition on gluconeogenesis. However, the induction of insulin on lipogenesis remains intact. The elevated FFA levels in the insulin-resistant state stimulate more insulin secretion from β-cell, which further induces lipogenesis and results in a vicious cycle in the liver.[5,49] The coexistence of the hepatic insulin resistance and sensitivity at the gene expression level has been observed in rodent obese and diabetic models.[3,39] However, the molecular mechanism of this coexistence of insulin sensitivity and resistance in different pathways has not been revealed.[49]

By using genetic approach, Moon *et al.* knocked out hepatic SREBP cleavage-activating protein gene *(Scap)* in *ob/ob* mice, an escort protein necessary for generating nuclear isoforms of SREBPs. Deficiency of hepatic *Scap* reversed liver lipogenesis and steatosis in *ob/ob* mice. Interestingly, the plasma levels of glucose and insulin were comparable between control *ob/ob* mice and *ob/ob* mice with liver-specific *Scap* knockout, indicating a persistence of insulin resistance despite correction of liver steatosis. Hepatic expression levels of *Pck1* and *G6pc* were not different between control *ob/ob* mice and *ob/ob* mice with liver-specific *Scap* knockout.[50] This finding was somewhat surprising in that the correction of liver steatosis did not seem to affect insulin sensitivity significantly and thus raised question on whether suppressing lipogenesis in the liver will lead to any beneficial effect on insulin sensitivity and glucose tolerance in human.

2.2. Insulin signaling pathways

2.2.1 Insulin receptor and its substrates

Insulin molecule consists of two chains, A-chain and B-chain, which are linked by one intrachain linkage within the A-chain and two interchain disulfide bonds between A- and B-chains.[51] The insulin receptor is a transmembrane and heterotetramer tyrosine kinase receptor, consisting of two α- and two β-subunits linked by disulfide bonds.[52] There are two insulin receptor isoforms arising from the alternative splicing of exon 11 resulting in either the IR-B or IR-A isoform.[53] The structure of insulin receptor ectodomain suggests that ligand-binding regions are in juxtaposition.[54] Insulin binding leads to domain rearrangements, transphosphorylation of the intracellular β-subunit tyrosine kinase,[55,56] and initiation of the signal transduction.[54] Insulin engages the insulin receptor carboxy-terminal α-chain (aCT) segment, which displaces the B-chain C-terminus away from the hormone core, and initiates the conformational switch for signal transduction.[57] The signaling of insulin binding to its receptor is transduced by multiple components in a complex network containing cascades of kinases.[58,59]

The first intracellular substrate protein phosphorylated by insulin receptor β-subunit was identified from Fao rat hepatoma cells labeled with radioactive ^{32}P-orthophosphate and stimulated with insulin.[60] Subsequently, several insulin receptor substrates (IRSs) were identified and their roles in mediating insulin actions in different tissues have been demonstrated by using genetic deletions.[61] In addition to tyrosine phosphorylation, serine and threonine sites on IRS-1 and 2 can be phosphorylated, which may play positive or negative roles in mediating insulin sensitivity.[62]

2.2.2 Activation of protein kinase B

The transduction of insulin signals depends on protein–protein interactions. A detailed analysis of protein interaction with peptides of IRS-1 and IRS-2 shows that phosphorylation of their tyrosine residues offers docking sites for various proteins involved in signal transduction.[63] The binding of Src homology 2 (SH2) domain proteins, such as the regulatory subunits of phosphatidylinositol 3-kinase (PI3K), growth factor receptor-bound protein 2 (GRB2), and protein tyrosine phosphatase-2 (SHP-2), plays significant roles in mediating insulin signal transduction.[62] The insulin signaling events are further amplified via PI3K/protein kinase B (PKB/Akt) pathway and GRB2/mitogen-activated protein kinase (MAPK) pathway.[64–66]

The product of *PKB/Akt* gene was identified as a serine-threonine kinase[67] after the genome of AKT8 retrovirus was cloned and analyzed.[68] It was independently cloned and termed rac (*related to the A and C* kinases) due to its similarities to protein kinases A and C.[69] PKB/Akt has been shown to phosphorylate transcription factors mediating insulin action. For example, the members of winged helix–forkhead family of transcription factors share a similar DNA-binding domain and have broad functions.[70] The transcriptional activity of forkhead box protein O1 (FoxO1), previously called FKHR for forkhead in human rhabdomyosarcoma,[71] is regulated by the serine/threonine phosphorylation by PKB/Akt.[72,73] The phosphorylated FoxO1 is excluded from the nucleus and thus activation of its downstream genes is prevented. In addition to phosphorylation of transcription factors, PKB/Akt can phosphorylate transcriptional coactivators, such as peroxisome proliferator-activated receptor coactivator 1α (PGC-1α), which inhibits its roles in gluconeogenesis and fatty acid oxidation.[74]

2.2.3 Activation of mitogen-activated protein kinase

Insulin has been shown to induce extracellular-signal-regulated kinases (ERK) 1/2 activation in many types of cells since its identification. The activated ERK1/2 can be translocated to nucleus where they can affect gene expression. Although ERK1/2 mediate cell growth and differentiation, their roles in mediating insulin-regulated metabolism have been unclear.[64,66,75]

2.2.4 The activation of atypical protein kinase C

Another serine/threonine kinase that can be activated by PI3K is the atypical protein kinase C (aPKC).[62] aPKC has been studied regarding their roles in regulation of hepatic insulin signal transduction and sensitivity. Mice lacking the hepatic expression of two of the three isoforms of the PI3K regulatory subunits (L-p85DKO), *Pik3r1* and *Pi3k3r2*, were generated via albumin Cre-recombinase-mediated deletion of *Pik3r1* on the background of

Pik3r2-null allele.[76] The L-p85DKO mice had reduced liver/body weight ratio and increased white adipose tissue/body weight ratio. They also had reductions in serum levels of TG, cholesterol, and FFA and elevations in serum levels of glucose, insulin, leptin, and adiponectin. The hepatic mRNA level of *Srebp-1c* was decreased in L-p85DKO, which can be increased after reintroduction of PKCλ, but not PKB/Akt, via recombinant adenovirus. On the other hand, the hepatic mRNA levels of *Pck1* and *G6pc* were increased in L-p85DKO, whose expression can be suppressed largely by PKB/Akt overexpression, to a small extent by PKCλ overexpression. These results led the authors to conclude that insulin–activated PI3K pathway may regulate hepatic glucose and lipid metabolism via PKB/Akt and atypical protein PKCλ/ζ, respectively.[76] The expression of PKCλ, but not PKCζ, was observed to play a role in insulin–stimulated glucose uptake in 3T3-L1 cells.[77]

The effects of aPKC on hepatic and muscular insulin activities have been studied in *ob/ob* mice and Goto–Kakizaki rats (a nonobese Wistar substrain that develops NIDDM).[78] It has been observed that Goto–Kakizaki rats had higher hepatic expression of *Srebp-1c* than their controls after overnight fasting, which also lost responses to insulin–induced expression.[78] In addition, the refeeding-induced aPKC, but not PKB/Akt activity, was still intact. When the hepatic activities of aPKC in Goto–Kakizaki rats were reduced by administering an adenovirus encoding kinase-inactive (KI) *Prkcz* mRNA for 5 days, the hepatic expression of *Srebp-1c* mRNA and mature form of SREBP-1c protein and serum TG and glucose levels were all reduced, whereas insulin–induced activity of PKB/Akt was restored. The same inhibition of aPKC activity in *ob/ob* mice for 5 days resulted in reduction of hepatic expression of *Srebp-1c* mRNA and induction of hepatic activity of PKB/Akt.[78] In male C57Bl/6 mice fed with an HFD for 3–4 weeks, the same inhibition reduced the expression levels of *Srebp-1c*, *Fas*, and *Acc* mRNA levels in *ad libitum* or after an overnight fasting.[79] These were associated with reduction of serum levels of TG, cholesterol, and glucose in the mice fed with the HFD. On the other hand, treatment of adenovirus expressing Cre recombinase to mice with the floxed PKCλ genes caused the reduction of *Srebp-1c* mRNA and mature SREBP-1c protein in fed state.[79]

Crossing the mice that have exon 5 of PKCλ flanked by loxP sequences and those expressing Cre recombinase under the control of the albumin gene promoter resulted in mice with liver-specific knockout of PKCλ without affecting on the expression levels of PKCλ in other tissues and the expression of PKCζ, a closely related isoform.[14] The hepatic mRNA levels of *Srebp-1c* and *Fas*, but not *Gck* and *Pck1*, in PKCλ knockout mice were reduced in both fasting and refeeding conditions but still elevated in response to the treatment of T0901317, indicating the impairment of signal transduction rather than general transcription of *Srebp-1c* gene.[14] In comparison to that in primary hepatocytes without any adenovirus infection, the adenovirus-mediated overexpression of the dominant-negative PKCλ attenuated insulin-induced *Srebp-1c* mRNA levels, whereas overexpression of the wild-type PKCλ resulted in elevation of *Srebp-1c* expression in both basal and insulin-stimulated conditions.[14]

When the differences of HFD-induced insulin resistance were compared between two mouse strains, obesity-sensitive C57BL/6 J (B6) and obesity-resistant 129S6/Sv,[80] the elevation of PKCδ expression in the liver starting from birth was a feature of prodiabetic mice. The elevated mRNA levels of other PKCs were also observed in *ob/ob* mice as compared to their lean controls. Importantly, the hepatic levels of PKCδ mRNA are also increased in obese human subjects. Mice with PKCδ deletion had improvement of insulin sensitivity and reduction of hepatic mRNA levels of *Srebp-1c* and *Pck1* genes, which were all reversed with hepatic overexpression of PKCδ mediated by recombinant adenovirus.[81]

Interestingly, PKCε, but not other isoforms of PKC, was activated in hepatic steatosis and insulin resistance in rats fed with an HFD. PKCε was found to be associated with insulin receptor β-subunit. The treatment of antisense oligonucleotide in rats resulted in reduction of PKCε expression in both the liver and white adipose tissue, which led to improvement of whole-body insulin sensitivity and hepatic insulin signal transduction in rats on HFD.[82]

It becomes obvious that alteration of the expression levels of different aPKC isoforms can be observed in the liver of mice with insulin resistance caused by genetic and nutritional factors.[81] The unanswered question is which one plays which role under different physiological conditions. Another conclusion that can be made is that the alteration of aPKC expression may depend on species as the isoforms changed by HFD feeding in rats are PKCε but not others.[82] Lastly, in cultured cell lines, PKCλ has been shown to be responsible for mediating the receptor tyrosine kinase signal transduction pathways activated by receptors of epidermal growth factor (EGF) and platelet-derived growth factor (PDGF).[83] This raises an extremely important question, what is the specificity of insulin signal transduction pathway? If PKB/Akt and aPKC are shared by insulin and other growth factors, the important signaling events that differentiate insulin from other growth factors still stay unrevealed.

2.2.5 Phosphatase involved in insulin signaling

The human liver expresses several protein tyrosine phosphatases (PTPs).[84] Okadaic acid, an inhibitor of protein phosphatase type 2A and type 1, inhibits glycogen synthesis and insulin-stimulated *Gck* expression in primary rat hepatocytes, indicating that dephosphorylation mechanism may also participate in mediating insulin signaling.[85]

The knockout of PTP-1B elevated and extended tyrosine phosphorylation of insulin receptor β-subunit in the liver, and increased insulin sensitivity in the whole body.[86] The liver-specific knockout of PTP-1B in mice caused increase of insulin sensitivity and reduction of hepatic expression of *Pck1* and *Srebp-1c*.[87] The expression of dominant-negative mutants of SHP-2, a tyrosine-specific protein phosphatase, inhibited insulin-stimulated signaling events in CHO cells, indicating its positive effects on insulin actions.[88] Overexpression of SHP-2 caused insulin resistance in mice and attenuation of insulin-stimulated PI3K and PKB/Akt kinase activities in the liver.[89] Mice overexpressing a catalytically defective SHP-1/PTP showed improvement of insulin sensitivity comparing with their wild-type controls and had enhanced insulin receptor signaling to IRS–PI3K–Akt in the liver.[90]

Suppression of the PTP-LAR expression in McA-RH7777 rat hepatoma cells increased insulin-stimulated phosphorylation of its downstream signaling cascade components.[91] However, the phosphorylation levels of other growth factor receptors were also increased, suggesting the general role of PTP-LAR.[84] The PTP-LAR knockout mice had lower plasma glucose and insulin levels, indicating an improvement of insulin sensitivity. However, insulin-activated PI3K was attenuated, suggesting insulin resistance in these mice.[92]

Mice with heterozygous knockout of T-cell protein tyrosine phosphatase fed with an HFD were more insulin-sensitive than the wild-type controls, and insulin-stimulated phosphorylation of IR and PKB/Akt in their hepatocytes was enhanced as well.[93] Another protein tyrosine phosphatase, PTP-MEG2 (or PTPN9), reduced insulin-stimulated phosphorylation of PKB/Akt in mouse liver when it was overexpressed by recombinant adenovirus.[94] Hepatic overexpression of a dominant-negative SH2-containing inositol 5′-phosphatase 2 (SHIP2) decreased blood glucose level and affected the expression of certain insulin-regulated genes in *db/db* mice.[95]

2.3. Insulin-regulated hepatic gene expression

This chapter is focused on the mechanisms of insulin-regulated gene expression in the liver and hepatocytes, but not in other tissues. It is intended to cover the genes involved in metabolism of carbohydrates, fatty acids, and proteins. Typical genes whose transcription levels are directly regulated positively or negatively by insulin treatment are discussed in this section.

2.3.1 Sterol regulatory element-binding protein-1c gene

It has been known that insulin was needed for the hepatic utilization of glucose for lipogenesis and CO_2 production.[96] When rats were induced to develop insulin-dependent diabetes via Alloxan, the conversion of glucose into FAs was impaired in liver slices,[97,98] whole liver, and adipose tissues,[99] which were back to normal after insulin treatment. The hepatic FA synthesis rose sharply and continuously for more than 20 h after rats were fasted for 48 h and then refed with a high-carbohydrate diet.[100] The amount and activity of FAS, the enzyme responsible for fatty acid synthesis, were increased in normal rats refed with a high-carbohydrate diet after a fasting and in diabetic rats after insulin injection.[101] However, in insulin-deficient rats, the induction of FAS after refeeding was lost until insulin was injected, demonstrating the important role of insulin in support of hepatic fatty acid synthesis.[102]

This induction of FAS activity is attributed to the induction of its transcripts.[103,104] It is clear now that a transcription factor named SREBP-1c plays an essential role in this induction.[48] Overexpression of SREBP-1c protein has been shown to mimic insulin effects on gene expression in hepatocytes.[105,106] Whole-body deletion of Srebp-1 gene (deletion of both Srebp-1a and 1c) resulted in impaired induction of hepatic lipogenic genes in response to the fasting and refeeding cycle in mice.[107] When Srebp-1c was specifically deleted in the mouse liver, the induction of lipogenic genes in response to the cycle of fasting and refeeding, and treatment of liver X receptor (LXR) agonist T0901317, was impaired, indicating the essential role of SREBP-1c in mediating hepatic lipogenesis.[46] Insulin specifically induced the expression of Srebp-1c mRNA in the liver of diabetic rats induced by the treatment of streptozotocin (STZ).[44] However, in liver-specific insulin receptor knockout mice, the induction of hepatic Srebp-1c by refeeding largely remained unchanged.[108] This suggests that nutrients-derived signals may activate downstream of insulin receptor signaling pathway to regulate Srebp-1c expression and hepatic lipogenesis.[108]

2.3.2 Glucokinase gene (Gck)

To be utilized, glucose must be phosphorylated into glucose-6-phosphate (G6P), a reaction catalyzed by hexokinases found in different organisms ranging from bacteria to humans. In vertebrates, four isozymes, hexokinase I (A), II (B), III (C), and IV (D), have been identified to be with different tissue distributions, intracellular locations, and kinetic characteristics. Mammalian hexokinase IV (D) is also known as GK (ATP: D-hexose 6-phosphotransferase, EC 2.7.1.1).[41,42,109,110] Mutations that alter its enzymatic activity (K_m and/or V_{max}) have been associated with maturity-onset diabetes of the young.[111] Two features of GK, low affinity for glucose (higher K_m) and unresponsiveness to allosteric inhibition mediated by physiological concentrations of G6P, differentiate from other members of hexokinase family.[109,110,112,113] GK activity is present in pancreatic islets and liver, which are, respectively, expressed as β-cell and liver isoforms with similar kinetic properties.[114,115] Whole-body and tissue-specific deletion of Gck demonstrated that either pancreatic β-cell or hepatic GK activity was essential for glucose homeostasis.[116,117] Altered glucose metabolism has been observed in transgenic mice with hepatic GK overexpression[118–121] and in rats with GK overexpressed by recombinant adenovirus.[122]

In short-term condition, hepatic GK activity is subjected to regulation by binding to glucokinase regulatory protein,[123] phosphorylation by protein kinase A,[124] and interaction with a cytosolic glucokinase-associated phosphatase.[125] Recently, GK activation induced by allosteric activators or small molecules has been shown to modulate insulin secretion and hepatic glucose metabolism.[126,127] The long-term regulation of GK activity is through change in Gck mRNA level, which is differentially controlled by an upstream promoter in pancreatic β-cells and a downstream promoter in hepatocytes.[42,114,128,129] This leads to generation of different Gck mRNA molecules with unique 5′ sequences and thus distinct GK proteins in the liver and pancreatic β-cells. Hepatic expression of Gck is affected by fasting and refeeding, and it is induced by insulin and suppressed by glucagon in rat hepatocytes.[80,130,131] In contrast, Gck expression level in β-cells does not respond to fasting and refeeding, but its activity is regulated by plasma concentration of glucose directly.[128,132]

2.3.3 The cytosolic form of phosphoenolpyruvate carboxyl kinase gene (Pck1)

The plasma glucose homeostasis is balanced by its dynamic usage and production. Gluconeogenesis, the production of glucose from noncarbohydrate precursors, is not simply a reversed process of glycolysis. The first rate-limiting enzyme of this process is phosphoenolpyruvate carboxyl kinase (PEPCK, EC4.1.1.32), which catalyzes the generation of phosphoenolpyruvate from oxaloacetate. Two isoforms, the cytosolic form (PEPCK-C) whose expression level is regulated by dietary and hormonal conditions and the mitochondrial form (PEPCK-M) whose expression is kept constant, have been identified in the liver. Expression and activity of PEPCK have been found in the liver, kidney, adipose tissues, and some other tissues.[43,133,134] It has been proposed that PEPCK activity controls gluconeogenesis in the liver and kidney and supports glyceroneogenesis in white adipose tissue.[135] Transgenic mice and rats overexpressing the *Pck1* gene developed hyperglycemia and hyperinsulinemia.[136–138] Mice with homozygous deletion of *Pck1* gene died within 2–3 days after birth with severe hypoglycemia,[139] and changes in lipid and amino acid metabolism were observed prior to death.[139–141] Although mice with liver-specific *Pck* deletion maintained euglycemia after 24 h fasting, their lipid metabolism was impaired,[139,141] suggesting that hepatic *Pck1* expression contributes to both glucose and lipid homeostasis. By using an *ex vivo* liver perfusion system, the contribution of PEPCK to hepatic gluconeogenesis was studied in mouse models with varying protein contents of *Pck1*. Livers with a 90% reduction in PEPCK content only reduced gluconeogenic flux by 40%, which indicated a lower than expected capacity for PEPCK protein content to control gluconeogenesis and it suggested that PEPCK flux and expression must coordinate with hepatic energy metabolism to control gluconeogenesis.[142]

Allosteric regulation of PEPCK-C activity has not been reported in mammalian cells. However, PEPCK-C activity can be controlled by regulation of *Pck1* gene expression under different physiological conditions.[43,134] In the liver, fasting, diabetes, and carbohydrate-free and high-fat diet increase, whereas refeeding, insulin treatment, and high-carbohydrate diet decrease, *Pck1* gene expression. In H4IIE hepatoma cells, insulin suppressed the expression of PEPCK in a PI3K-dependent but MAPK-independent manner.[143]

2.3.4 Insulin-like growth factor-binding protein 1 gene (Igfbp1)

Insulin-like growth factor-binding proteins (IGFBPs) are a family of protein with high binding affinity to insulin-like growth factor 1 (IGF-1) that regulates clearance of IGFBPs. This system has been considered a marker for protein nutrition and amino acid homeostasis. It has been shown that the plasma IGFBP-1 level is elevated in states like fasting, malnutrition, and diabetes.[144,145] Transgenic mice overexpressing IGFBP1 developed insulin resistance.[146] Interestingly, it was reported that refeeding caused a decline of hepatic *Igfbp1* mRNA levels in rats. However, the decline of *Igfbp1* mRNA was achieved through administration of growth hormone (GH), but not insulin, in the liver of fasted rats.[147] Insulin rapidly inhibited hepatic *Igfbp1* gene expression in the liver of STZ-induced diabetic rats[148] and HepG2 cells.[149]

2.4. Insulin-responsive elements
2.4.1 Insulin-responsive elements in insulin-induced genes
2.4.1.1 Insulin-responsive elements in *Srebp-1c* promoter

Insulin-responsive elements (IREs) in the *Srebp-1c* promoter have been identified as two LXR-binding elements (LXRE) and one sterol regulatory element (SRE).[150] Other transcription factor-binding sites around this region were also considered to be involved in insulin-induced *Srebp-1c* transcription.[151,152] The same LXREs in human *SREBP-1c* promoter also mediate insulin-stimulated transcription.[153] This implies that insulin regulates the expression of its responsive genes after it stimulates the synthesis of endogenous agonists for nuclear receptor activation.

When we analyzed the effects of a lipophilic extract (LE) from rat livers on insulin-regulated gene expression, the LE that contained retinol and retinal synergized with insulin to induce hepatic *Gck* and *Srebp-1c* expression.[154] It has been reported that SREBP-1c mediated the retinoid-dependent increase in *Fas* promoter activity in HepG2.[155] A peroxisome proliferator-activated receptor (PPAR)-binding element (PPRE) was identified in the promoter of human *SREBP-1c* gene, and PPARα agonist synergizes with insulin to induce lipogenesis in human hepatocytes.[153] In hepatocytes, glucagon blocked insulin-induced *Srebp-1c* transcription[3] and SREBP-1c protein processing.[156]

2.4.1.2 Insulin-responsive element in *Gck* promoter

The hepatic *Gck* promoter region controlling liver-specific expression has been identified in the region from -1003 to -707 relative to the transcription initiation site in primary rat hepatocytes. However, the same region seems to act as a place for the suppression of *Gck* expression in FTO-2B hepatoma cells.[157]

SREBP-1c, an insulin-responsive gene by itself,[44] has been considered as the major mediator of insulin-induced *Gck* expression in hepatocytes.[106,107,158] This conclusion was made based on the fact that recombinant adenovirus-mediated overexpression of mature form of SREBP-1c induced *Gck* mRNA.[105,106] Therefore, the SRE can be considered as an IRE.

However, studies analyzing endogenous *Srebp-1c* mRNA expression in response to insulin treatment seemed to paint a different story that insulin-induced *Gck* expression occurs in an SREBP-1c-independent manner.[159,160] In primary hepatocytes with knockdown of SREBP-1c, insulin-induced *Gck* expression was not affected. This conclusion was supported by results obtained from mice bearing *Srebp-1c* deletion. The mutant mice had the same hepatic *Gck* expression in response to the cycle of fasting and refeeding as their wild-type controls.[46] More importantly, the insulin-induced *Gck* mRNA expression occurred much quicker than insulin-induced *Srebp-1c*, with peak time at 3 and 9 h, respectively.[161] As shown in Fig. 6.1A, comparing to the levels at time 0, mRNA level of

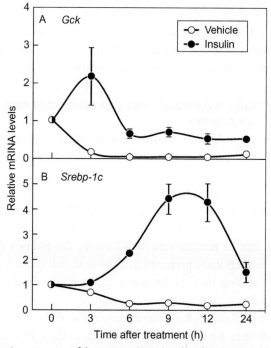

Figure 6.1 The time course of hepatic mRNA levels of *Gck* (A) and *Srebp-1c* (B) in response to insulin treatment. Primary rat hepatocytes were isolated, pretreated as described in Ref. 160 and incubated in medium without or with 1 nM insulin. Total RNA was extracted at 0, 3, 6, 9, 12, and 24 h after the incubation and subjected to real-time PCR analysis. The levels of indicated transcripts at time 0 were arbitrarily assigned a value of 1. Data are presented as fold induction (mean ± SD, $n = 3$).

Gck dropped rapidly in the absence of insulin. The addition of 1 nM insulin only transiently induced *Gck* mRNA at 3 h posttreatment. Here, the main function of insulin is to transiently induce and then to keep *Gck* mRNA at a level relatively higher than basal level without insulin in the medium. For *Srebp-1c*, it is a different scenario as shown in Fig. 6.1B. The *Srebp-1c* mRNA level in hepatocytes without insulin did not reach the bottom at 3 but 6 h posttreatment. Comparing with time 0, expression level of *Srebp-1c* remained unchanged at 3 h, started to increase at 6 h, and peaked at 9/12 h and stayed elevated at 24 h. The mRNA level of *Gck* started to drop before insulin-induced *Srebp-1c* expression reached the peak. It seems that the factor responsible for *Gck* induction begins to fade before the factor for *Srebp-1c* expression starts to act.

Unfortunately, attempts to identify insulin-responsive element in the *Gck* promoter have not been successful by studying reporter gene constructs. A 5.5 kb promoter of liver *Gck* gene failed to respond to insulin treatment in rat primary hepatocytes.[162] This may be due to IRE of *Gck* being localized upstream of this 5.5-kb promoter fragment or unresponsiveness of the current reporter gene system for mediating insulin response. Further studies are needed to identify the insulin-responsive element in *Gck* promoter.

2.4.2 IRE for insulin-suppressed gene expression and roles of Forkhead proteins

The first IRE responsible for insulin-inhibited *Pck1* expression was identified after a series of promoter reporter gene constructs were stably transfected into H4IIE rat hepatoma cells.[163] It demonstrated that disrupting the interaction between DNA and protein at the IRE site impaired insulin-mediated regulation of *Pck1* expression. However, the binding pattern of nuclear extracts from the liver of STZ-induced diabetic rats was not different from that of normal rats. In addition, the pattern did not differ when nuclear extracts were prepared from H4IIE cells treated with or without insulin, suggesting that insulin might not completely alter the binding pattern of the IRE.[163]

The level of *Pck1* mRNA is determined by the rates of transcription and mRNA degradation. To evaluate the relevance of these IRE sites to its regulation, chimeric *Pck1* reporter gene constructs controlling either transcription of bovine growth hormone (PEPCK-bGH)[164] or expression of a portion of chicken PEPCK cDNA[165,166] were generated and used to create a series of transgenic mice containing different length and mutant sites of *Pck1* promoter. The length and sites for developmental regulation, tissue-specific expression, and hormonal responses were evaluated by a series of studies.[164,165,167–171] The promoter sequences downstream of -540 relative to the start codon were shown to be required for liver-specific expression.[164] However, disruption of the IRE site did not alter the inhibition of high-carbohydrate feeding on *Pck1* expression, suggesting that additional mechanism might exist to control the response of *Pck1* to high carbohydrate.[169]

One copy of IRE with a T(G/A)TTT(T/G)(G/T) core sequence was found in the promoter of apolipoprotein C3 (*ApoC3*) gene, which is responsible for insulin-mediated suppression of its hepatic expression in HepG2 cells.[172] Two copies of the same IRE showing dyad symmetry were identified in promoters of human *IGFBP1* and rat *Igfbp1* gene that mediated insulin action in HepG2[173] and H4IIE cells,[174] respectively. Three copies of IRE in the promoter of *G6pc*, another insulin-suppressed gene, were identified in H4IIE cells.[175] Five copies of the similar IRE in the promoter of rat cytosolic aspartate aminotransferase gene were observed, two of which were responsible for insulin-mediated inhibition of its gene expression in H4IIE cells.[176]

The proteins bound to this kind of IRE have been designated winged helix–forkhead transcription factors. FKHR (forkhead in rhabdomyosarcoma) or FoxO1 was first identified as a transcription factor fused with PAX3 transcription DNA-binding domain in the solid tumor alveolar rhabdomyosarcoma.[177] The activity of FKHR was negatively regulated by PKB/Akt in human embryonic kidney 293 cells through phosphorylation.[178] The phosphorylation can be induced by insulin, which can be blocked by Wortmannin treatment in SV40-transformed mouse hepatocytes.[179] The insulin-suppressed expression of *Igfbp1* was mediated by phosphorylating FKHR at Ser-256, one of three residues phosphorylated upon insulin treatment.[180] It has been shown that FoxO1 directly interacted with PGC-1α to induce the expression of gluconeogenic genes, such as *Pck1* and *G6pc*. This interaction was disrupted upon insulin treatment, a process attributed to insulin-inhibited gluconeogenesis.[181]

Activation of PKB/Akt also caused exclusion of FoxO3/FKHRL1 from the nucleus, which promotes the survival of 293 cells.[182] When FoxO3/FKHRL1 was overexpressed in H4IIE rat hepatoma cells, it induced *Pck1* and *Igfbp1* gene expression, which can be suppressed by insulin. However, the FoxO3/FKHRL1 binding to mutated IREs of the *Pck1* and *Igfbp1* genes did not match the response of these mutants to insulin-mediated suppression, indicating that other proteins interacting with the IREs might mediate insulin action in those cells.[183]

Another member of forkhead family, Foxa2/HNF3β,[184] has been shown to be excluded from nucleus in the same manner after insulin stimulation and to be responsible for the control of β-oxidation and ketogenesis in the liver.[185] Overexpression of a nonphosphorylatable mutant, Foxa2T156A, resulted in improvement of glucose and lipid homeostasis in the liver of obese and diabetic mice.[185] When the expression level of Foxoa2 was increased in hepatocytes under the control of transthyretin (TTR) promoter, reduction of glycogen content and increased expression of genes for β-oxidation were observed in the liver of those transgenic mice.[186] All these studies set up the tune for insulin-suppressed gene expression.

In mice with liver-specific FoxO1 deletion, the hepatic expression of *Igfbp1*, but not *Pck1*, was reduced. The expression of *Pck1* was only affected when the hepatic expressions of FoxO1/3/4 were all specifically deleted.[187] It suggests that all three of them are able to mediate *Pck1* expression. In addition, it will be interesting to determine how this reduction results in the elevation of insulin sensitivity or loss of response to fasting.

2.4.3 Insulin-response element-binding protein 1

Insulin-stimulated IGFBP-3 expression has been thought to be mediated by insulin-response element-binding protein 1 (IRE-BP1), which was identified via yeast one-hybrid system and phosphorylated by PKB/Akt.[188] IRE-BP1 was located to promoters of insulin-responsive genes such as *Igfbp1* and *Igfbp3*. In the liver of insulin-resistant animals, IRE-BP1 was excluded from the nucleus.[189]

3. EFFECTS OF HORMONE AND NUTRIENT ON INSULIN-REGULATED HEPATIC GENE EXPRESSION

3.1. Hormones affecting insulin-regulated hepatic gene expression

3.1.1 Glucagon and its receptor

In an attempt to optimize methods to concentrate and purify insulin, a substance precipitated from beef pancreas by acetone and solubilized in 95% alcohol was identified to have hyperglycemic effect and named *glucagon*.[190] Later on, it was purified and crystallized,[191,192] and the sequence of this 29-amino acid peptide hormone was determined.[193] The development of glucagon RIA allowed accurate and sensitive measurement of plasma glucagon levels.[194] The realization of diabetic human subjects with higher plasma glucagon levels[195] led to the recognition that the pathology of type I diabetes was the lack of insulin and excessive activity of glucagon.[196] For roles of glucagon in the development of diabetes, please see the recent review by Unger and Cherrington.[197]

The glucagon receptor (GCGR) is a 62 kDa glycoprotein dimerized with intermolecular disulfide bonds. Glucagon binds closely to the carboxyl terminus of its receptor.[198–200] In purified rat hepatocyte membrane, glucagon activates adenylyl cyclase to generate cAMP,[201] a phenomenon that allows the assessment of glucagon activity in cells. Glucagon stimulation caused increase of intracellular levels of cAMP and calcium.[202]

Mice with deletion of glucagon receptor gene ($Gcgr^{-/-}$) had reduction of glucose and elevation of glucagon and somatostatin levels. They had normal body weight, food intake, energy expenditure, and insulin levels but reduced adiposity and leptin levels.[203] Reduction of hepatic $Gcgr$ expression by using antisense oligonucleotide suppressed hepatic glucose production and improved pancreatic β-cell functions.[204] The $Gcgr^{-/-}$ mice had improved insulin sensitivity[205] and were resistant to STZ-induced diabetes[206] and HFD-induced obesity.[207] Islets isolated from $Gcgr^{-/-}$ mice had impaired insulin secretion in response to glucose and other secretagogues.[205] In addition, $Gcgr$ seems to be essential for regulation of FA oxidation, but not for that of FA synthesis.[208]

3.1.2 Leptin

Since the cloning of *ob* gene,[19] the adipocyte–derived adipokines have been recognized as factors other than FAs with regulatory roles in metabolism.[38] Leptin, the product of *ob* gene, activates several signaling cascades via leptin receptors (OB-R) present in a variety of cells.[209] The expression of hepatic insulin–regulated genes can be affected by leptin treatment.

The expression of leptin receptor was observed in human liver and hepatoma cell lines.[210] In several hepatoma cell lines, the insulin–activated IRS-1 phosphorylation was attenuated in the presence of leptin via reduction of IRS-1 association with components in insulin signaling cascade.[210] On the other hand, in differentiated hepatoma Fao cells expressing both long and short forms of OB-R, leptin pretreatment enhanced insulin–induced tyrosine phosphorylation of IRS-1.[211] In addition to activating ERK1/2 and MAPK as insulin, leptin also activates Janus kinase 2 (Jak2) and p90RSK in the same cells.[212] The insulin-mediated phosphorylation of IRS1 was enhanced in HepG2 cells treated with 5 μM RA for 24 h and then 100 ng/ml leptin for 3 h. In KK-A^y mice, Am80 (a retinoic acid receptor

(RAR) agonist) induced hepatic leptin receptor mRNA expression and improved insulin sensitivity.[213]

In mice, leptin infusion for 6 h increased the hepatic abundance of *Gck* mRNA and decreased that of *Pck1* mRNA.[214] However, leptin treatment also attenuated insulin-suppressed *Pck1* expression in HepG2 cells, demonstrating that leptin antagonized insulin action.[210] In HepG2 cells, insulin-induced *Scd1* was impaired in the presence of leptin.[212] Both leptin and insulin treatments decreased the expression levels of angiopoietin-like protein 3 in rat hepatoma H4IIEC3 and human hepatoma HepG2 cells.[215] In HepG2 cells, insulin increased the protein expression levels of aquaglyceroporins (AQP) 3, 7, and 9, whereas leptin stimulated only AQP 3 but suppressed AQP 7 and 9.[216] However, the effects of leptin on insulin-mediated induction of AQP 7 and 9 were not investigated. It is worthy to note that both insulin and leptin signals are mediated through PKB/Akt and mammalian target of rapamycin complex (mTORC) pathways in the same cells.[216]

3.1.3 Tumor necrosis factor alpha, adiponectin, growth hormone, and others

In H4IIE cells, tumor necrosis factor α (TNFα) treatment blocked expression of some insulin-regulated genes.[217] In HepG2 cells, TNFα attenuated insulin-suppressed *Pck1* expression via maintaining the nuclear localization of Foxa2 protein.[218]

Adiponectin-suppressed *Srebp-1c* expression in the liver of *db/db* mice and Fao hepatoma cells via its receptor AdipoR1 and activation of AMP-activated protein kinase (AMPK).[219] In isolated rat primary hepatocytes, adiponectin potentiated insulin-suppressed glucose production.[220] In H4IIE hepatoma cells, adiponectin suppressed the expression of *G6pc* and *Pck1* without any additive effect with insulin.[221] The adiponectin-suppressed hepatic glucose production was impaired in liver-specific AMPKα2 knockout mice.[222]

In HepG2 cells, resistin induced *Pck1* and *G6pc* expression and attenuated insulin-stimulated phosphorylation of PKB/Akt depending on suppressor of cytokine signaling 3 (SOCS3).[223] The hepatic expression of *Pck1* and *G6pc* in the liver-specific SOCS3 knockout mice fed with chow diet, but not HFD, was reduced after hyperinsulinemic–euglycemic clamps, suggesting an increase of insulin sensitivity. HFD feeding resulted in higher weight gain and higher hepatic expression of lipogenic genes in SOCS3 knockout mice than in control.[224]

GH also affects hepatic glucose and lipid metabolism.[225] Both insulin and GH induced IGF-1 expression in primary rat hepatocytes, and there was either an additive[226] or no additive[227] effect between them. It was reported that GOAT–ghrelin–GH axis might play essential role to stimulate gluconeogenesis in starvation after calorie restriction.[228]

Recently, aP2, an intracellular lipid chaperone and biomarker of adipocyte differentiation, was reported to be an adipokine that was elevated in fasting state and obesity in mice and humans. Interestingly, aP2 could somehow stimulate hepatic gluconeogenesis and neutralization of aP2 with an antibody-suppressed hepatic glucose production and improved glucose tolerance. Although the exact mechanism was not known, hepatic mRNA levels of both *Pck1* and *G6pc* were significantly decreased.[229]

3.2. Vitamin A effects on insulin-regulated gene expression

VA (retinol) and molecules with similar physiological activities are essential micronutrient for vision, embryogenesis, immunity, and differentiation.[230] The active metabolite of VA, RA, regulates gene expression through activation of two families of nuclear receptors,[231] RARs, and retinoid X receptors (RXRs).[232,233] Upon binding of ligand to them RAR/RXR hetero- or RXR/RXR homodimer associates with the RA-responsive elements (RARE) as the promoters activates the transcription of RA-responsive genes.[234,235]

More recently, retinoids have been proposed to play roles in glucose and lipid metabolism and maintain energy homeostasis.[236,237] The use of isotretinoin (13-*cis* RA) caused elevation of plasma TG levels in human subjects.[238–240] This increase in plasma TG level was also observed in rats.[241–243] It has been shown that VA status affected the obesity development in Zucker fatty (ZF) rats and reduced plasma TG levels in Zucker lean (ZL) and ZF rats.[244] The hepatic expression of *Srebp1-c* was reduced in VAD animals.[244] RA synergized with insulin to induce the expression of *Srebp1-c* in primary rat hepatocytes via the previously identified LXREs responsible for insulin-induced expression, indicating that the LXREs are also RAREs.[161] The proximal one of the two previously identified RAREs in *Pck1* promoter[245–247] mediated the RA effect in primary hepatocytes. RAR activation can induce *Pck1* expression and attenuate insulin-mediated suppression of its expression in rat primary hepatocytes.[248] Altogether, these observations demonstrate the convergence of retinoids and insulin signals in the regulation of a particular gene involved in glucose and lipid metabolism.

3.3. Hormonal and nutritional effects on hepatic *Gck, Srebp-1c, Pck1*, and *Igfbp-1* gene expression

3.3.1 Hormonal and nutritional effects on Gck expression

Multiple transcriptional factor-binding sites have been identified in the liver *Gck* promoter. These include putative binding sites of hepatocyte nuclear factor-6,[249] O_2-responsive elements,[250,251] PPRE,[252] and two SREs.[253] In neonatal rat hepatocytes, *Gck* mRNA was induced by both insulin and triiodothyronine (T3).[254] Biotin induced *Gck* transcription and GK activity in fasted rat liver and rat hepatocytes in culture.[255,256] Mice fed with synthetic agonist of LXR, an oxysterol receptor, had higher level of liver *Gck* mRNA than control.[257] Recently, it has been shown that PPAR, LXR, and SREBP bound to *Gck* promoter and acted together with RXR and small heterodimer partner (SHP) to mediate insulin action in Alexander cells, a human hepatoma cell line.[258]

It has been reported that RA induced *Gck* expression in rat hepatocytes without any additive effects on insulin-mediated induction.[259,260] Recently, we have reported that retinoids synergized with insulin to induce *Gck* expression in rat primary hepatocytes.[154]

3.3.2 Hormonal and nutritional effects on Pck1 expression

Glucagon, glucocorticoids, thyroid hormone, and RA induced hepatic *Pck1* expression, while insulin and glucose inhibited it.[40] Insulin played a dominant role in suppressing *Pck1* expression induced by other hormones or nutrients in hepatocytes.[40,43,261] The transcriptional regulation of *Pck1* has been studied extensively.[43,134] DNase I footprinting assays were performed to determine the protein binding sites at the proximal region of rat *Pck1* gene promoter.[262,263] Two cAMP-responsive elements (CRE 1 and 2) and six additional binding sites (P1 to 6) that interacted with proteins in rat liver nuclear extracts were identified.[43]

Transient transfection of reporter genes has been used to identify the regulatory elements at 5′ promoter region of the *Pck1* gene in cultured cells, largely in hepatoma cell lines.[43,134] Multiple regulatory elements including distal PPRE[264] and proximal CRE1[265] sites have been identified. The transcription factors that bound to these elements and regulate *Pck1* expression include hepatocyte nuclear factor 4 (HNF4),[266] glucocorticoid receptor (GR),[265,267–270] RAR,[171,245–247] thyroid hormone receptor (TR),[245,271] LXR,[257] chicken ovalbumin upstream transcription factor (COUP-TF),[266] PPAR,[264] forkhead family of transcription factors,[183] CCAAT/enhancer-binding protein (C/EBP) and cAMP regulatory element-binding protein (CREB),[272] and SREBP-1c.[273,274] Their coactivators, such as CREB-binding protein (CBP),[275] steroid receptor coactivator 1 (SRC1),[276] and PGC-1α,[277] also played critical roles in regulating *Pck1* gene expression.

It is worth noting that the majority of these transcription factors also play important roles in controlling hepatic lipid metabolism. For instance, over-expression of the active form of SREBP-1c stimulated lipogenesis and abolished *Pck1* gene expression in cultured rat hepatocytes.[105,273] Rodents fed with T0901317, an agonist for LXR, had increased lipogenic gene expression and reduced *Pck1* expression.[257,278]

3.3.3 Hormonal and nutritional effects on Igfbp1 expression

Insulin-suppressed *Igfbp1* expression was blocked by inhibitors of PI3K (Wortmannin) and mTORC1 (rapamycin) in rat primary hepatocytes.[279] In HepG2 cells, activation of PI3K and PKB/Akt, but not mTORC1, was responsible for insulin-inhibited *Igfbp1* expression.[280] In H4IIE cells, the glucocorticoid-responsive element (GRE) in *Igfbp1* promoter was also considered to be IRE.[174] The insulin-suppressed *Igfbp1* expression was mediated by phosphorylation of FKHR at Ser-256, one of three residues phosphorylated after insulin treatment.[180]

In addition, inflammation caused induction of hepatic *Igfbp1* mRNA expression mediated by cytokines.[281] Interestingly, bovine insulin did not induce salmon hepatic *Igfbp1* gene expression, whereas glucagon and dexamethasone (DEX) did.[282] Somatostatin analog, octreotide, induced *Igfbp1* mRNA expression in the liver without affecting plasma insulin level[283] and in HepG2 cells in the presence of insulin.[284]

3.3.4 The interaction of glucagon and insulin on tyrosine transaminase expression

As counterregulatory hormones maintaining glucose homeostasis, insulin and glucagon exert opposite effects on hepatic glucose output. Insulin-regulated hepatic gene expression, however, is not always antagonized by glucagon. In adrenalectomized rats, both glucagon and insulin induced the hepatic activity of tyrosine transaminase (TAT).[285,286] However, combined treatment of these two hormones resulted in a level of enzyme activity lower than that induced by the individual ones, suggesting additive effect of insulin and glucagon.[286] Cycloheximide, an inhibitor of protein synthesis, synergized with insulin to induce TAT mRNA levels in the liver of adrenalectomized rats.[287] In cultured fetal rat liver, either insulin or glucagon induced the activity of TAT after a 4-h treatment. There was no additive effect between them. In the same lysate, however, glucagon-induced PEPCK activity was inhibited by insulin.[288] A single injection of cAMP induced hepatic TAT mRNA within 20 min, which did not persist at 2 h after treatment.[289] In rat Fao hepatoma cell line, insulin induced TAT mRNA levels at as early as 15 min and did not affect DEX-induced TAT expression.[290] In H4IIE cells, insulin blocked this cAMP-mediated induction of TAT expression. In addition, insulin exerted a biphasic regulation of TAT transcription with suppression in the first 1–2 h followed by induction at around 4 h.[291] In KRC-7 hepatoma cells, insulin reduced TAT mRNA level but induced TAT enzyme activity.[292] In cultured fetal hepatocytes, DEX, cAMP, and insulin all induce TAT mRNA level after a 32 h treatment. DEX synergized with cAMP to enhance this induction even further, whereas insulin attenuated this synergy.[293]

4. DRUGS AND PHYTOCHEMICALS AFFECTING HEPATIC INSULIN ACTION

4.1. Insulin signaling pathway modulators

4.1.1 Rapamycin

Rapamycin is an inhibitor of mTOR, a serine-threonine protein kinase that forms a complex with regulatory-associated protein (Raptor), mammalian lethal with SEC13 protein 8, PRAS40, and DEPTOR. The mTORC1 signaling pathway plays an important role for cells to sense stimuli from nutrients, growth factors, energy status, oxygen, and stress. Insulin-induced expression of *Srebp-1c* mRNA in epithelial cells was mediated by PI3K/Akt/mTORC1 signaling pathway.[294] By dissecting the signaling pathways with inhibitors of PI3K, PKB/Akt, mTORC1, and S6K, Li *et al.* confirmed that rapamycin specifically blocked insulin-induced *Srebp-1c* expression, but had no effect on insulin-suppressed *Pck1* expression in rat primary hepatocytes. This observation suggests that mTORC1 plays an essential role in the insulin-induced hepatic lipogenesis but not insulin-suppressed gluconeogenesis.[295] It is still not clear how mTORC1 regulates transcription of *Srebp-1c*; the mTORC1 substrate p70 S6-kinase may not play a role in this process.[295,296] Interestingly, the insulin-stimulated SREBP-1c protein processing also requires mTORC1. In contrast to the regulation of transcription, mTORC1 regulates protein processing of SREBP-1c in a p70 S6-kinase-dependent manner, demonstrating the complexity of the insulin-stimulated lipogenesis.[156]

4.1.2 Wortmannin

Wortmannin is a natural product that potently inhibits PI3K with anti-inflammatory and immunosuppressant effects *in vivo*.[297] It has been used extensively in the study of insulin-regulated events in a variety of cells. It was also used to study the activation of PI3K by other growth factors, such as platelet-derived growth factor.[298] Wortmannin was reported to block insulin-stimulated expression of lipogenic genes, such as *Srebp-1c* and *Acc*.[299] It also blocked insulin-stimulated glycogen synthase, S6K, Akt, and GSK-3 activities.[300–302] Interestingly, it has been reported that Wortmannin also inhibited the activity of smooth muscle myosin light-chain kinase.[303]

4.1.3 C2 ceramide

C2 ceramide, a ceramide analog, is a sphingolipid that can inhibit insulin-mediated PKB/Akt activation after the production of $3'$-phosphoinositides.[304,305] In primary hepatocytes, C2 ceramide inhibited insulin-induced mRNA levels of *Gck* and *Srebp-1c*.[306]

4.1.4 Benfluorex

Benfluorex is an amphetamine derivative related to fenfluramine and dex-fenfluramine and has been used in patients with hypertriglyceridemia and type II diabetes mellitus in Europe, Asia, and South Africa. It had also been used off-label as a slimming aid and then withdrawn from the European market in 2010 due to its association with heart valve lesions. In rat primary hepatocytes, benfluorex and its metabolite were shown to decrease mRNA levels of *Pck1* and *G6pc* and increase mRNA levels of *Gck* and liver type pyruvate kinase *(Lpkr)*.[307]

4.2. Phytochemicals

4.2.1 Resveratrol

Resveratrol is believed to be the main component attributed to the antiatherosclerotic benefits of red wine consumption. So far, resveratrol has been reported to have a variety of health-promoting activities, such as antiaging,[308] anti-inflammatory,[309] and antidiabetic effects,[310] in addition to its potent antiviral[311] and antineoplastic[312] actions. It has been shown to enhance coronary relaxation, reduce ventricular arrhythmias, inhibit platelet aggregation and smooth muscle proliferation, improve glucose metabolism, and reduce low-density lipoprotein (LDL) while elevating high-density lipoprotein (HDL).[313]

The exact molecular mechanism by which resveratrol causes these beneficial effects has not been fully revealed. The sirtuin system, termed silent information regulator genes (SIR), is thought to be responsible for mediating effects of resveratrol. There are seven sirtuin genes in mammals, SIRT 1 to 7, which are highly conserved in different species. They are widely expressed in a variety of tissues and these genes encode for nicotinamide adenine dinucleotide (NAD)-dependent deacetylase or ADP–ribosyl transferase. These proteins are found in multiple subcellular locations, such as the nucleus (SIRT-1, -2, -6, and -7), cytoplasm (SIRT-1 and -2), and mitochondria (SIRT 3, -4, and -5). Resveratrol modulates these favorable effects via activation of sirtuin proteins.[314]

Effect of resveratrol on gluconeogenesis is believed to be mediated by SIRT1, although the effects are controversial from study to study. Liver-specific knockout of SIRT1 resulted in decreased glucose production. In primary hepatocytes, resveratrol attenuated insulin-repressed mRNA levels of Pck1 and G6pc. This effect was mediated by deacetylation of FoxO1 by SIRT1 in the nucleus.[315] In another study, SIRT1 was demonstrated to repress transcription of Pck1 gene by deacetylating HNF4α.[316]

4.2.2 Metformin

Metformin has been used as a treatment of type II diabetes for decades. It is also an effective agent to decrease the risk of development of the disease.[317] In addition to lowering the blood glucose levels, metformin also inhibits adipose tissue lipolysis, reduces circulating FFAs, and diminishes VLDL production.[318,319] It is effective in decreasing diabetes–related death, myocardial infarction, and stroke according to the results of the United Kingdom Prospective Diabetes Study.[320] Metformin inhibits the activity of the mitochondrial respiratory chain complex I, and its antidiabetic effect is believed to be mediated by inhibiting hepatic gluconeogenesis and increasing glucose utilization in peripheral tissues.[321] Metformin treatment has been shown to suppress Srebp-1c mRNA and protein levels, reduce mRNA levels of lipogenic genes, and induce fatty acid oxidation via activation of AMPK in hepatocytes.[322] The molecular mechanism of metformin had been revisited by using genetically manipulated mice and cells. Metformin inhibited expression of G6pc, while Pck1 was unaffected in wild-type, AMPK-deficient, and LKB1-deficient hepatocytes, suggesting that metformin inhibited hepatic gluconeogenesis in an LKB1- and AMPK-independent manner via a decrease in hepatic energy state.[323,324] More recently, the mechanism of metformin on hepatic gluconeogenesis has been linked to inhibition of cAMP production and activation of PKA and, in turn, suppression of glucagon-induced glucose output.[325]

4.2.3 Berberine

Berberine (BBR) is an isoquinoline alkaloid of the protoberberine type, which is found in the root, rhizome, and stem bark of many plant species such as *Coptis chinensis* Franch, a famous traditional Chinese medicine. It shows significant antimicrobial activity against microorganisms. Therefore, it has been used conventionally as an antidiarrheal, bitter stomachic, and antimalarial.[326] Recently, it has been shown that BBR has antidiabetic effects in experimental animals and clinical diabetic patients.[327–330] BBR treatment inhibited hepatic gluconeogenesis in rats of an insulin-insufficient type II diabetic model. This effect was attributed to BBR-mediated inhibition of *Pck1* and *G6pc* expression in an insulin-independent manner.[331] BBR increased *Gck* expression and suppressed *Pck1* and *G6pc* expression in hepatocytes of Zucker rats in the absence of insulin.[332] BBR was also reported to affect expression of key genes involved in adipocyte differentiation and fatty acid metabolism, such as *Srebp-1c*.[333]

4.2.4 Quercetin

Quercetin is a naturally occurring flavonoid with ability to modulate glucose and lipid metabolism by activating AMPK.[334,335] It has been reported that quercetin increased energy expenditure transiently and decrease plasma levels of inflammation markers in mice fed with high-fat diet.[336] It enhanced exercise endurance in mice, which was related to increased mitochondrial biogenesis.[337] Recent study combining transcriptomic and metabolomic profiling suggested that quercetin and resveratrol combo treatment in mice fed with an HFD resulted in significant restoration of gene sets in functional pathways of glucose/lipid metabolism, liver function, cardiovascular system, and inflammation/immunity that were induced by HFD feeding.[338]

4.3. Plant extracts

Numerous plant extracts have been shown, reproducibly or nonreproducibly, to affect glucose metabolism and thus effectively improve diabetic symptoms such as decreasing plasma glucose and enhancing insulin sensitivity. Due to the complexity and impurity of plant extracts in most cases, it is hard to make a firm conclusion on whether and how each individual plant extract affects transcription of gluconeogenic genes. Some effects might well be due to *in vitro* artifacts or were related to toxicity of plant extracts that led to decrease in food intake and body weight loss and thus changes in glycemia and insulin sensitivity in preclinical studies. The key is to identify the exact active phytochemical that exerts the antiglycemic effect observed *in vitro* and *in vivo*. Here, we will discuss a few examples in the field and briefly review the effect of phytochemicals on gluconeogenic gene expression.

Fenugreek seed is an herbal medicine widely used as a galactagogue to induce lactation in nursing mothers to increase inadequate breast milk supply. Early study in type I diabetic patients indicated that fenugreek diet significantly reduced fasting blood glucose levels and improved glucose tolerance and dyslipidemia.[339] Fenugreek seed treatment decreased plasma glucose levels, suppressed *Pck1*, and increased *Lpkr* mRNA levels in the liver of diabetic rats.[340] The active component in fenugreek seed with anti-diabetic activity is still not known, whereas 4-hydroxyleucine extracted from fenugreek has been demonstrated to affect glucose-induced insulin secretion.[341,342]

Ginkgo biloba extract is a potent oxygen radical scavenger. It was reported that *G. biloba* extract protected the oxidative muscle fibers of STZ-induced diabetic rats.[343] Treatment of STZ rats with *G. biloba* extract for 30 days suppressed the progression of diabetes and reduced plasma glucose levels.[344] *G. biloba* extract induced transcription of IRS-2 in L-02 cells, an immortalized human hepatocyte cell line.[338,344]

Syzygium aromaticum ethanol extract (SAE) is traditionally used as an anodyne, carminative, and anthelmintic in Asian countries. Its effects on obesity have been investigated recently. Treatment of SAE caused reduction of body weight, liver weight, and fat mass in HFD-fed mice. It decreased mRNA level of *Srebp-1c*, *Fas*, *CD36*, and *Pparg* in the liver.[345]

Kuding tea is a traditional Chinese beverage and its extract has been shown to prevent symptoms of metabolic disorders in HFD-fed mice such as body weight gain, fat mass, and plasma lipid levels. Liver mRNA levels of *Lxra*, *Lxrb*, *Abca1*, and *Srebp-1c* were decreased after 2-week treatment of kuding tea ethanol extract.[346]

Paeoniae radix is the root of a Chinese peony, *Paeonia lactiflora* Pallas, and its ethanol extract (PR-Et) has been shown to stimulate glucose uptake and suppress hyperglycemia in rodents. In STZ rats and *db/db* mice, PR-Et treatment dramatically decreased plasma levels of glucose and insulin and suppressed hepatic mRNA level of *Pck1*. PR-Et also inhibited mRNA level of *Pck1* induced by DEX in H4IIE hepatoma cells without indication of cell toxicity.[347]

5. PERSPECTIVES AND FUTURE DIRECTIONS

Insulin-regulated transcription in hepatocytes is a dynamic and complex event. The same signaling events initiated after binding of insulin to its receptor and activation of receptor tyrosine kinase result in up- or down-regulation of gene expression. Even for the genes upregulated by insulin, the extent of induction and time curve might be very different from gene to gene. The sequential events may or may not have direct connections. For example, it may be unlikely that insulin-induced SREBP-1c protein level is responsible for insulin-stimulated *Gck* expression in the short term as indicated in Fig. 6.1. However, overexpression of SREBP-1c does induce *Gck* expression in primary rat hepatocytes.[105,106] It seems that more than one transcription factors are mediating insulin-induced or suppressed expression of a particular gene. Each one of them may play a specific role within a particular time frame for the insulin-regulated gene expression one direction or another. Whether this is true or not deserves further investigation.

The challenge facing the studies of insulin-regulated gene expression is the system available to detect the activation or suppression of the transcription. Using the insulin-induced *Gck* and *Srebp-1c* expression as examples, large differences of endogenous mRNA levels in response to insulin treatments are very easy to be detected in primary hepatocytes. Reporter gene constructs have been successfully used to identify the IREs responsible for insulin-induced *Srebp-1c* transcription.[150] However, a reporter gene construct containing a 5.5-kb promoter fragment of hepatic *Gck* promoter, which included regions for liver-specific expression, still failed to respond to insulin in primary rat hepatocytes[55] and (unpublished observation from the corresponding author's lab). Whether the lack of chromatin structure in plasmid DNA for reporter gene assay causes unresponsiveness of the reporter constructs to insulin stimulation or an IRE further upstream of the studied region is responsible for liver-specific expression remains to be determined. It also remains to be investigated whether chromatin immunoprecipitation assay can be used to reveal the dynamic associations of major transcription factors with the promoter.

The insulin signals generated on cell membrane are transduced to nucleus for the regulation of gene transcription. As shown in Fig. 6.2, this is accomplished after several steps and involved in components of multiple signaling cascades. The early events, such as insulin receptor phosphorylation status, can be altered by certain factors, such as PTPs, which reduce

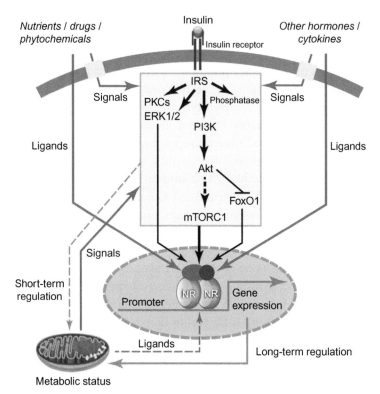

Figure 6.2 Schematic diagram of potential effects of diet/drug/phytochemicals, hormones/cytokines, and metabolic status on hepatic gene expression regulated by insulin. The binding of insulin to its receptor initiates a signal transduction cascade including activation of a series of protein kinases and phosphatases. These signal transducers immediately regulate nutrient metabolism via regulation of the activities of metabolic enzymes in mitochondria and cytosol in short term. The same signals also change the expression of genes involved in nutrient metabolism and, in turn, control the activities of enzymes in the long term. Dietary bioactive components/drugs and other hormones/cytokines exert similar effects through interacting with their receptors on the surface of hepatocytes or entering the cell to directly interact with intracellular proteins or transcription factors. When they bind receptors on the cell surface, signal transduction cascades are initiated, which may interfere or synergize with insulin signal transduction pathway to regulate its downstream effects. When these factors enter the nucleus, they can regulate the activities of transcriptional machinery that may have already been activated or inhibited by insulin signaling pathway. In hepatocytes, these factors can also modulate the activities of metabolic pathways by affecting flux of metabolic intermediates in mitochondria and thus changing metabolic status of the cell, which may curtail or enhance insulin actions and result in alteration of insulin-regulated events indirectly. (See color plate.)

insulin action. Additionally, it is worth noting that insulin shares many of these components of signaling cascades with other hormones or cytokines. For example, ERK1/2 and PKB/Akt, components activated upon insulin stimulation, can be activated by hepatic growth factor and epidermal growth factor treatments in rat primary hepatocytes.[348] This observation provides another mechanism for the modulation of insulin-regulated gene expression. Moreover, the activities of transcription complex on the promoter can be modulated by binding of ligands or affected by cell metabolic status and activation states of other transcription factors. Therefore, hormonal, nutritional, and dietary factors can affect insulin-regulated gene expression at multiple points. Finally, a given stimulus may use more than one of these mechanisms to regulate the final formation of the transcription machineries assembled at the target genes' promoter. Future studies are needed to define how each step is regulated by using sample genes such as *Gck*, *Srebp-1c*, *Pck1*, or *Igfbp1* to reveal the complex regulatory systems. The data obtained will certainly provide feasible approaches to combat metabolic diseases.

ACKNOWLEDGMENTS

The corresponding author thanks the following agencies for supporting his research projects: research grant from Allen Foundation Inc. (to G. C.), start-up fund from the University of Tennessee at Knoxville (to G. C.), and Scientist Development Grant from American Heart Association (09SDG2140003, to G. C.).

REFERENCES

1. Bliss M. *The Discovery of Insulin*. 2nd ed. University of Chicago Press; Chicago, 2007.
2. Himsworth HP. Dietetic factors influencing the glucose tolerance and the activity of insulin. *J Physiol*. 1934;81:29–48.
3. Shimomura I, Matsuda M, Hammer RE, Bashmakov Y, Brown MS, Goldstein JL. Decreased IRS-2 and increased SREBP-1c lead to mixed insulin resistance and sensitivity in livers of lipodystrophic and *ob/ob* mice. *Mol Cell*. 2000;6:77–86.
4. Spiegelman BM, Flier JS. Obesity and the regulation of energy balance. *Cell*. 2001;104:531–543.
5. McGarry JD. What if Minkowski had been ageusic? An alternative angle on diabetes. *Science*. 1992;258:766–770.
6. Yalow RS, Berson SA. Immunoassay of endogenous plasma insulin in man. *J Clin Invest*. 1960;39:1157–1175.
7. Jensen MV, Joseph JW, Ronnebaum SM, Burgess SC, Sherry AD, Newgard CB. Metabolic cycling in control of glucose-stimulated insulin secretion. *Am J Physiol Endocrinol Metab*. 2008;295:E1287–E1297.
8. Floyd JC, Fajans SS, Conn JW, Knopf RF, Rull J. Stimulation of insulin secretion by amino acids. *J Clin Invest*. 1966;45:1487–1502.
9. Stein DT, Stevenson BE, Chester MW, et al. The insulinotropic potency of fatty acids is influenced profoundly by their chain length and degree of saturation. *J Clin Invest*. 1997;100:398–403.

10. Stein DT, Esser V, Stevenson BE, et al. Essentiality of circulating fatty acids for glucose-stimulated insulin secretion in the fasted rat. *J Clin Invest.* 1996;97:2728–2735.

11. Koyama K, Chen G, Wang MY, et al. Beta-cell function in normal rats made chronically hyperleptinemic by adenovirus-leptin gene therapy. *Diabetes.* 1997;46:1276–1280.

12. Milburn JL, Hirose H, Lee YH, et al. Pancreatic β-cells in obesity. Evidence for induction of functional, morphologic, and metabolic abnormalities by increased long chain fatty acids. *J Biol Chem.* 1995;270:1295–1299.

13. Chertow BS, Blaner WS, Baranetsky NG, et al. Effects of vitamin A deficiency and repletion on rat insulin secretion in vivo and in vitro from isolated islets. *J Clin Invest.* 1987;79:163–169.

14. Matsumoto M, Ogawa W, Akimoto K, et al. PKCλ in liver mediates insulin-induced SREBP-1c expression and determines both hepatic lipid content and overall insulin sensitivity. *J Clin Invest.* 2003;112:935–944.

15. Blumentrath J, Neye H, Verspohl EJ. Effects of retinoids and thiazolidinediones on proliferation, insulin release, insulin mRNA, GLUT 2 transporter protein and mRNA of INS-1 cells. *Cell Biochem Funct.* 2001;19:159–169.

16. Chertow BS, Goking NQ, Driscoll HK, Primerano DA, Matthews KA. Effects of all-trans-retinoic acid (ATRA) and retinoic acid receptor (RAR) expression on secretion, growth, and apoptosis of insulin-secreting RINm5F cells. *Pancreas.* 1997;15:122–131.

17. Driscoll HK, Adkins CD, Chertow TE, Cordle MB, Matthews KA, Chertow BS. Vitamin A stimulation of insulin secretion: effects on transglutaminase mRNA and activity using rat islets and insulin-secreting cells. *Pancreas.* 1997;15:69–77.

18. Cabrera-Valladares G, German MS, Matschinsky FM, Wang J, Fernandez-Mejia C. Effect of retinoic acid on glucokinase activity and gene expression and on insulin secretion in primary cultures of pancreatic islets. *Endocrinology.* 1999;140:3091–3096.

19. Zhang Y, Proenca R, Maffei M, Barone M, Leopold L, Friedman JM. Positional cloning of the mouse obese gene and its human homologue. *Nature.* 1994;372:425–432.

20. Campfield LA, Smith FJ, Guisez Y, Devos R, Burn P. Recombinant mouse OB protein: evidence for a peripheral signal linking adiposity and central neural networks. *Science.* 1995;269:546–549.

21. Considine RV, Sinha MK, Heiman ML, et al. Serum immunoreactive-leptin concentrations in normal-weight and obese humans. *N Engl J Med.* 1996;334:292–295.

22. Halaas JL, Gajiwala KS, Maffei M, et al. Weight-reducing effects of the plasma protein encoded by the Obese gene. *Science.* 1995;269:543–546.

23. Paz-Filho G, Wong ML, Licinio J. Ten years of leptin replacement therapy. *Obes Rev.* 2011;12:e315–e323.

24. Haslam DW, James WP. Obesity. *Lancet.* 2005;366:1197–1209.

25. Yanovski SZ, Yanovski JA. Obesity prevalence in the United States: up, down, or sideways? *N Engl J Med.* 2011;364:987–989.

26. Schulze MB, Hu FB. Primary prevention of diabetes: what can be done and how much can be prevented? *Annu Rev Pub Health.* 2005;26:445–467.

27. Friedman JM. Leptin at 14 y of age: an ongoing story. *Am J Clin Nutr.* 2009;89:973S–979S.

28. Gaulton KJ, Willer CJ, Li Y, et al. Comprehensive association study of Type 2 diabetes and related quantitative traits with 222 candidate genes. *Diabetes.* 2008;57:3136–3144.

29. Popkin BM, Gordon-Larsen P. The nutrition transition: worldwide obesity dynamics and their determinants. *Int J Obes Relat Metab Disord.* 2004;28:S2–S9.

30. Wells JCK. Thrift: a guide to thrifty genes, thrifty phenotypes and thrifty norms. *Int J Obes.* 2009;33:1331–1338.

31. Bray GA, Nielsen SJ, Popkin BM. Consumption of high-fructose corn syrup in beverages may play a role in the epidemic of obesity. *Am J Clin Nutr.* 2004;79:537–543.

32. Reaven GM. The insulin resistance syndrome: definition and dietary approaches to treatment. *Annu Rev Nutr.* 2005;25:391–406.
33. Dixon JB, O'Brien PE, Playfair J, et al. Adjustable gastric banding and conventional therapy for Type 2 diabetes: a randomized controlled trial. *JAMA.* 2008;299:316–323.
34. Sweeney JS. Dietary factors that influence the dextrose tolerance test: a preliminary study. *Arch Intern Med.* 1927;40:818–830.
35. Hansen PA, Han DH, Marshall BA, et al. A high-fat diet impairs stimulation of glucose transport in muscle. *J Biol Chem.* 1998;273:26157–26163.
36. McGarry JD. Glucose-fatty acid interactions in health and disease. *Am J Clin Nutr.* 1998;67:500S–504S.
37. Myers Jr MG, Leibel RL, Seeley RJ, Schwartz MW. Obesity and leptin resistance: distinguishing cause from effect. *Trends Endocrinol Metab.* 2010;21:643–651.
38. Yu YH, Ginsberg HN. Adipocyte signaling and lipid homeostasis. *Circ Res.* 2005;96:1042–1052.
39. McGarry JD. Banting Lecture 2001: dysregulation of fatty acid metabolism in the etiology of Type 2 diabetes. *Diabetes.* 2002;51:7–18.
40. O'Brien RM, Granner DK. Regulation of gene expression by insulin. *Physiol Rev.* 1996;76:1109–1161.
41. Iynedjian PB. Mammalian glucokinase and its gene. *Biochem J.* 1993;293:1–13.
42. Magnuson MA, Andreone TL, Printz RL, Koch S, Granner DK. Rat glucokinase gene: structure and regulation by insulin. *Proc Natl Acad Sci USA.* 1989; 86:4838–4842.
43. Hanson RW, Reshef L. Regulation of phosphoenolpyruvate carboxykinase (GTP) gene expression. *Annu Rev Biochem.* 1997;66:581–611.
44. Shimomura I, Bashmakov Y, Ikemoto S, Horton JD, Brown MS, Goldstein JL. Insulin selectively increases SREBP-1c mRNA in the livers of rats with streptozotocin-induced diabetes. *Proc Natl Acad Sci USA.* 1999;96:13656–13661.
45. Horton JD, Goldstein JL, Brown MS. SREBPs: activators of the complete program of cholesterol and fatty acid synthesis in the liver. *J Clin Invest.* 2002;109:1125–1131.
46. Liang G, Yang J, Horton JD, Hammer RE, Goldstein JL, Brown MS. Diminished hepatic response to fasting/refeeding and liver X receptor agonists in mice with selective deficiency of sterol regulatory element-binding protein-1c. *J Biol Chem.* 2002;277:9520–9528.
47. Flowers MT, Ntambi JM. Stearoyl-CoA desaturase and its relation to high-carbohydrate diets and obesity. *Biochim Biophys Acta.* 2009;1791:85–91.
48. Horton JD. Physiology: unfolding lipid metabolism. *Science.* 2008;320:1433–1434.
49. Brown MS, Goldstein JL. Selective versus total insulin resistance: a pathogenic paradox. *Cell Metab.* 2008;7:95–96.
50. Moon YA, Liang G, Xie X, et al. The Scap/SREBP pathway is essential for developing diabetic fatty liver and carbohydrate-induced hypertriglyceridemia in animals. *Cell Metab.* 2012;15:240–246.
51. Adams MJ, Blundell TL, Dodson EJ, et al. Structure of rhombohedral 2 zinc insulin crystals. *Nature.* 1969;224:491–495.
52. De Meyts P. Insulin and its receptor: structure, function and evolution. *Bioessays.* 2004;26:1351–1362.
53. Denley A, Wallace JC, Cosgrove LJ, Forbes BE. The insulin receptor isoform exon 11- (IR-A) in cancer and other diseases: a review. *Horm Metab Res.* 2003;35:778–785.
54. McKern NM, Lawrence MC, Streltsov VA, et al. Structure of the insulin receptor ectodomain reveals a folded-over conformation. *Nature.* 2006;443:218–221.
55. Kasuga M, Karlsson FA, Kahn CR. Insulin stimulates the phosphorylation of the 95,000-dalton subunit of its own receptor. *Science.* 1982;215:185–187.

56. Petruzzelli LM, Ganguly S, Smith CJ, Cobb MH, Rubin CS, Rosen OM. Insulin activates a tyrosine-specific protein kinase in extracts of 3T3-L1 adipocytes and human placenta. *Proc Natl Acad Sci USA.* 1982;79:6792–6796.
57. Menting JG, Whittaker J, Margetts MB, et al. How insulin engages its primary binding site on the insulin receptor. *Nature.* 2013;493:241–245.
58. Cohen P. The twentieth century struggle to decipher insulin signalling. *Nat Rev Mol Cell Biol.* 2006;7:867–873.
59. Taniguchi CM, Emanuelli B, Kahn CR. Critical nodes in signalling pathways: insights into insulin action. *Nat Rev Mol Cell Biol.* 2006;7:85–96.
60. White MF, Maron R, Kahn CR. Insulin rapidly stimulates tyrosine phosphorylation of a Mr-185,000 protein in intact cells. *Nature.* 1985;318:183–186.
61. White MF. IRS proteins and the common path to diabetes. *Am J Physiol Endocrinol Metab.* 2002;283:E413–E422.
62. Copps KD, White MF. Regulation of insulin sensitivity by serine/threonine phosphorylation of insulin receptor substrate proteins IRS1 and IRS2. *Diabetologia.* 2012;55:2565–2582.
63. Hanke S, Mann M. The phosphotyrosine interactome of the insulin receptor family and its substrates IRS-1 and IRS-2. *Mol Cell Proteomics.* 2008;8:519–534.
64. Avruch J. MAP kinase pathways: the first twenty years. *Biochim Biophys Acta.* 2007;1773:1150–1160.
65. Manning BD, Cantley LC. AKT/PKB signaling: navigating downstream. *Cell.* 2007;129:1261–1274.
66. Osborne JK, Zaganjor E, Cobb MH. Signal control through Raf: in sickness and in health. *Cell Res.* 2012;22:14–22.
67. Bellacosa A, Staal SP, Testa JR, Tsichlis PN. A retroviral oncogene, akt, encoding a serine-threonine kinase containing an SH2-like region. *Science.* 1991;254:274.
68. Staal SP. Molecular cloning of the akt oncogene and its human homologues AKT1 and AKT2: amplification of AKT1 in a primary human gastric adenocarcinoma. *Proc Natl Acad Sci USA.* 1987;84:5034–5037.
69. Jones PF, Jakubowicz T, Pitossi FJ, Maurer F, Hemmings BA. Molecular cloning and identification of a serine/threonine protein kinase of the second-messenger subfamily. *Proc Natl Acad Sci USA.* 1991;88:4171–4175.
70. Kaestner KH, Knöchel W, Martínez DE. Unified nomenclature for the winged helix/forkhead transcription factors. *Genes Dev.* 2000;14:142–146.
71. Fredericks WJ, Galili N, Mukhopadhyay S, et al. The PAX3-FKHR fusion protein created by the t(2;13) translocation in alveolar rhabdomyosarcomas is a more potent transcriptional activator than PAX3. *Mol Cell Biol.* 1995;15:1522–1535.
72. Biggs WH, Meisenhelder J, Hunter T, Cavenee WK, Arden KC. Protein kinase B/Akt-mediated phosphorylation promotes nuclear exclusion of the winged helix transcription factor FKHR1. *Proc Natl Acad Sci USA.* 1999;96:7421–7426.
73. Rena G, Guo S, Cichy SC, Unterman TG, Cohen P. Phosphorylation of the transcription factor forkhead family member FKHR by protein kinase B. *J Biol Chem.* 1999;274:17179–17183.
74. Li X, Monks B, Ge Q, Birnbaum MJ. Akt/PKB regulates hepatic metabolism by directly inhibiting PGC-1α transcription coactivator. *Nature.* 2007;447: 1012–1016.
75. Siddle K. Molecular basis of signaling specificity of insulin and IGF receptors: neglected corners and recent advances. *Front Endocrinol.* 2012;3:34, http://dx.doi.org/10.3389/fendo.2012.00034.
76. Taniguchi CM, Kondo T, Sajan M, et al. Divergent regulation of hepatic glucose and lipid metabolism by phosphoinositide 3-kinase via Akt and PKC[lambda]/[zeta]. *Cell Metab.* 2006;3:343–353.

77. Kotani K, Ogawa W, Matsumoto M, et al. Requirement of atypical protein kinase Cλ for insulin stimulation of glucose uptake but not for Akt activation in 3T3-L1 adipocytes. *Mol Cell Biol.* 1998;18:6971–6982.

78. Sajan MP, Standaert ML, Rivas J, et al. Role of atypical protein kinase C in activation of sterol regulatory element binding protein-1c and nuclear factor kappa B (NFκB) in liver of rodents used as a model of diabetes, and relationships to hyperlipidaemia and insulin resistance. *Diabetologia.* 2009;52:1197–1207.

79. Sajan MP, Standaert ML, Nimal S, et al. The critical role of atypical protein kinase C in activating hepatic SREBP-1c and NFκB in obesity. *J Lipid Res.* 2009;50:1133–1145.

80. Iynedjian PB, Gjinovci A, Renold AE. Stimulation by insulin of glucokinase gene transcription in liver of diabetic rats. *J Biol Chem.* 1988;263:740–744.

81. Bezy O, Tran TT, ki J, et al. PKCδ regulates hepatic insulin sensitivity and hepatosteatosis in mice and humans. *J Clin Invest.* 2011;121:2504–2517.

82. Samuel VT, Liu ZX, Wang A, et al. Inhibition of protein kinase Cε prevents hepatic insulin resistance in nonalcoholic fatty liver disease. *J Clin Invest.* 2007; 117:739–745.

83. Akimoto K, Takahashi R, Moriya S, et al. EGF or PDGF receptors activate atypical PKClambda through phosphatidylinositol 3-kinase. *EMBO J.* 1996;15:788–798.

84. Norris K, Norris F, Kono DH, et al. Expression of protein-tyrosine phosphatases in the major insulin target tissues. *FEBS Lett.* 1997;415:243–248.

85. Agius L, Peak M. Interactions of okadaic acid with insulin action in hepatocytes: role of protein phosphatases in insulin action. *Biochim Biophys Acta.* 1991;1095:243–248.

86. Elchebly M, Payette P, Michaliszyn E, et al. Increased insulin sensitivity and obesity resistance in mice lacking the protein tyrosine phosphatase-1B gene. *Science.* 1999;283:1544–1548.

87. Delibegovic M, Zimmer D, Kauffman C, et al. Liver-specific deletion of protein-tyrosine phosphatase 1B (PTP1B) improves metabolic syndrome and attenuates diet-induced endoplasmic reticulum stress. *Diabetes.* 2009;58:590–599.

88. Yamauchi K, Milarski KL, Saltiel AR, Pessin JE. Protein-tyrosine-phosphatase SHPTP2 is a required positive effector for insulin downstream signaling. *Proc Natl Acad Sci USA.* 1995;92:664–668.

89. Maegawa H, Hasegawa M, Sugai S, et al. Expression of a dominant negative SHP-2 in transgenic mice induces insulin resistance. *J Biol Chem.* 1999;274:30236–30243.

90. Dubois MJ, Bergeron S, Kim HJ, et al. The SHP-1 protein tyrosine phosphatase negatively modulates glucose homeostasis. *Nat Med.* 2006;12:549–556.

91. Kulas DT, Goldstein BJ, Mooney RA. The transmembrane protein-tyrosine phosphatase LAR modulates signaling by multiple receptor tyrosine kinases. *J Biol Chem.* 1996;271:748–754.

92. Ren JM, Li PM, Zhang WR, et al. Transgenic mice deficient in the LAR protein-tyrosine phosphatase exhibit profound defects in glucose homeostasis. *Diabetes.* 1998;47:493–497.

93. Fukushima A, Loh K, Galic S, et al. T-cell protein tyrosine phosphatase attenuates STAT3 and insulin signaling in the liver to regulate gluconeogenesis. *Diabetes.* 2010;59:1906–1914.

94. Cho CY, Koo SH, Wang Y, et al. Identification of the tyrosine phosphatase PTP-MEG2 as an antagonist of hepatic insulin signaling. *Cell Metab.* 2006;3:367–378.

95. Fukui K, Wada T, Kagawa S, et al. Impact of the liver-specific expression of SHIP2 (SH2-containing inositol 5'-phosphatase 2) on insulin signaling and glucose metabolism in mice. *Diabetes.* 2005;54:1958–1967.

96. Chernick SS, Chaikoff IL. Insulin and hepatic utilization of glucose for lipogenesis. *J Biol Chem.* 1950;186:535–542.

97. Gibson DM, Hubbard DD. Incorporation of malonyl CoA into fatty acids by liver in starvation and alloxan-diabetes. *Biochem Biophys Res Commun.* 1960;3:531–535.

98. Williams WR, Hill R, Chaikoff IL. Portal venous injection of insulin in the diabetic rat: time of induction of changes in hepatic lipogenesis, cholesterogenesis, and glycogenesis. *J Lipid Res.* 1960;1:236–240.

99. Hausberger FX, Milstein SW, Rutman RJ. The influence of insulin on glucose utilization in adipose and hepatic tissues in vitro. *J Biol Chem.* 1954;208:431–438.

100. Allmann DW, Hubbard DD, Gibson DM. Fatty acid synthesis during fat-free refeeding of starved rats. *J Lipid Res.* 1965;6:63–74.

101. Burton DN, Collins JM, Kennan AL, Porter JW. The effects of nutritional and hormonal factors on the fatty acid synthetase level of rat liver. *J Biol Chem.* 1969;244:4510–4516.

102. Lakshmanan MR, Nepokroeff CM, Porter JW. Control of the synthesis of fatty-acid synthetase in rat liver by insulin, glucagon, and adenosine 3':5' cyclic monophosphate. *Proc Natl Acad Sci USA.* 1972;69:3516–3519.

103. Paulauskis JD, Sul HS. Hormonal regulation of mouse fatty acid synthase gene transcription in liver. *J Biol Chem.* 1989;264:574–577.

104. Porter J, Swenson T. Induction of fatty acid synthetase and acetyl-CoA carboxylase by isolated rat liver cells. *Mol Cell Biochem.* 1983;53–54:307–325.

105. Becard D, Hainault I, Azzout-Marniche D, Bertry-Coussot L, Ferre P, Foufelle F. Adenovirus-mediated overexpression of sterol regulatory element binding protein-1c mimics insulin effects on hepatic gene expression and glucose homeostasis in diabetic mice. *Diabetes.* 2001;50:2425–2430.

106. Foretz M, Guichard C, Ferre P, Foufelle F. Sterol regulatory element binding protein-1c is a major mediator of insulin action on the hepatic expression of glucokinase and lipogenesis-related genes. *Proc Natl Acad Sci USA.* 1999;96:12737–12742.

107. Shimano H, Yahagi N, Amemiya-Kudo M, et al. Sterol regulatory element-binding protein-1 as a key transcription factor for nutritional induction of lipogenic enzyme genes. *J Biol Chem.* 1999;274:35832–35839.

108. Haas J, Miao J, Chanda D, et al. Hepatic insulin signaling is required for obesity-dependent expression of SREBP-1c mRNA but not for feeding-dependent expression. *Cell Metab.* 2012;15:873–884.

109. Cardenas ML, Cornish-Bowden A, Ureta T. Evolution and regulatory role of the hexokinases. *Biochim Biophys Acta.* 1998;1401:242–264.

110. Matschinsky FM, Magnuson MA, Zelent D, et al. The network of glucokinase-expressing cells in glucose homeostasis and the potential of glucokinase activators for diabetes therapy. *Diabetes.* 2006;55:1–12.

111. Froguel P, Zouali H, Vionnet N, et al. Familial hyperglycemia due to mutations in glucokinase—definition of a subtype of diabetes mellitus. *N Engl J Med.* 1993;328:697–702.

112. Gidh-Jain M, Takeda J, Xu LZ, et al. Glucokinase mutations associated with non-insulin-dependent (Type 2) diabetes mellitus have decreased enzymatic activity: implications for structure/function relationships. *Proc Natl Acad Sci USA.* 1993;90:1932–1936.

113. Vinuela E, Salas M, Sols A. Glucokinase and hexokinase in liver in relation to glycogen synthesis. *J Biol Chem.* 1963;238:C1175–C1177.

114. Iynedjian PB, Mobius G, Seitz HJ, Wollheim CB, Renold AE. Tissue-specific expression of glucokinase: identification of the gene product in liver and pancreatic islets. *Proc Natl Acad Sci USA.* 1986;83:1998–2001.

115. Liang Y, Jetton TL, Zimmerman EC, Najafi H, Matschinsky FM, Magnuson MA. Effects of alternate RNA splicing on glucokinase isoform activities in the pancreatic islet, liver, and pituitary. *J Biol Chem.* 1991;266:6999–7007.

116. Grupe A, Hultgren B, Ryan A, Ma YH, Bauer M, Stewart TA. Transgenic knockouts reveal a critical requirement for pancreatic [beta] cell glucokinase in maintaining glucose homeostasis. *Cell.* 1995;83:69–78.
117. Postic C, Shiota M, Niswender KD, et al. Dual roles for glucokinase in glucose homeostasis as determined by liver and pancreatic beta cell-specific gene knock-outs using Cre recombinase. *J Biol Chem.* 1999;274:305–315.
118. Ferre T, Pujol A, Riu E, Bosch F, Valera A. Correction of diabetic alterations by glucokinase. *Proc Natl Acad Sci USA.* 1996;93:7225–7230.
119. Ferre T, Riu E, Bosch F, Valera A. Evidence from transgenic mice that glucokinase is rate limiting for glucose utilization in the liver. *FASEB J.* 1996;10:1213–1218.
120. Ferre T, Riu E, Franckhauser S, Agudo J, Bosch F. Long-term overexpression of glucokinase in the liver of transgenic mice leads to insulin resistance. *Diabetologia.* 2003;46:1662–1668.
121. Hariharan N, Farrelly D, Hagan D, et al. Expression of human hepatic glucokinase in transgenic mice liver results in decreased glucose levels and reduced body weight. *Diabetes.* 1997;46:11–16.
122. O'Doherty RM, Lehman DL, Telemaque-Potts S, Newgard CB. Metabolic impact of glucokinase overexpression in liver: lowering of blood glucose in fed rats is accompanied by hyperlipidemia. *Diabetes.* 1999;48:2022–2027.
123. Van Schaftingen E, Detheux M, Veiga da Cunha M. Short-term control of glucokinase activity: role of a regulatory protein. *FASEB J.* 1994;8:414–419.
124. Ekman P, Nilsson E. Phosphorylation of glucokinase from rat liver in vitro by protein kinase A with a concomitant decrease of its activity. *Arch Biochem Biophys.* 1988;261:275–282.
125. Munoz-Alonso MJ, Guillemain G, Kassis N, Girard J, Burnol AF, Leturque A. A novel cytosolic dual specificity phosphatase, interacting with glucokinase, increases glucose phosphorylation rate. *J Biol Chem.* 2000;275:32406–32412.
126. Brocklehurst KJ, Payne VA, Davies RA, et al. Stimulation of hepatocyte glucose metabolism by novel small molecule glucokinase activators. *Diabetes.* 2004;53:535–541.
127. Grimsby J, Sarabu R, Corbett WL, et al. Allosteric activators of glucokinase: potential role in diabetes therapy. *Science.* 2003;301:370–373.
128. Iynedjian PB, Pilot PR, Nouspikel T, et al. Differential expression and regulation of the glucokinase gene in liver and islets of Langerhans. *Proc Natl Acad Sci USA.* 1989;86:7838–7842.
129. Magnuson MA. Tissue-specific regulation of glucokinase gene expression. *J Cell Biochem.* 1992;48:115–121.
130. Iynedjian PB, Jotterand D, Nouspikel T, Asfari M, Pilot PR. Transcriptional induction of glucokinase gene by insulin in cultured liver cells and its repression by the glucagon-cAMP system. *J Biol Chem.* 1989;264:21824–21829.
131. Sibrowski W, Seitz HJ. Rapid action of insulin and cyclic AMP in the regulation of functional messenger RNA coding for glucokinase in rat liver. *J Biol Chem.* 1984;259:343–346.
132. Bedoya FJ, Matschinsky FM, Shimizu T, O'Neil JJ, Appel MC. Differential regulation of glucokinase activity in pancreatic islets and liver of the rat. *J Biol Chem.* 1986;261:10760–10764.
133. Hanson RW, Garber AJ. Phosphoenolpyruvate carboxykinase. I. Its role in gluconeogenesis. *Am J Clin Nutr.* 1972;25:1010–1021.
134. O'Brien RM, Printz RL, Halmi N, Tiesinga JJ, Granner D. Structural and functional analysis of the human phosphoenolpyruvate carboxykinase gene promoter. *Biochim Biophys Acta.* 1995;1264:284–288.
135. Hanson RW, Reshef L. Glyceroneogenesis revisited. *Biochimie.* 2003;85:1199–1205.

136. Rosella G, Zajac JD, Baker L, et al. Impaired glucose tolerance and increased weight gain in transgenic rats overexpressing a non-insulin-responsive phosphoenolpyruvate carboxykinase gene. *Mol Endocrinol*. 1995;9:1396–1404.

137. Sun Y, Liu S, Ferguson S, et al. Phosphoenolpyruvate carboxykinase overexpression selectively attenuates insulin signaling and hepatic insulin sensitivity in transgenic mice. *J Biol Chem*. 2002;277:23301–23307.

138. Valera A, Pujol A, Pelegrin M, Bosch F. Transgenic mice overexpressing phosphoenol-pyruvate carboxykinase develop non-insulin-dependent diabetes mellitus. *Proc Natl Acad Sci USA*. 1994;91:9151–9154.

139. She P, Shiota M, Shelton KD, Chalkley R, Postic C, Magnuson MA. Phosphoenol-pyruvate carboxykinase is necessary for the integration of hepatic energy metabolism. *Mol Cell Biol*. 2000;20:6508–6517.

140. Hakimi P, Johnson M, Yang J, et al. Phosphoenolpyruvate carboxykinase and the crit-ical role of cataplerosis in the control of hepatic metabolism. *Nutr Metab*. 2005;2:33.

141. She P, Burgess SC, Shiota M, et al. Mechanisms by which liver-specific PEPCK knock-out mice preserve euglycemia during starvation. *Diabetes*. 2003;52:1649–1654.

142. Burgess SC, He T, Yan Z, et al. Cytosolic phosphoenolpyruvate carboxykinase does not solely control the rate of hepatic gluconeogenesis in the intact mouse liver. *Cell Metab*. 2007;5:313–320.

143. Sutherland C, Waltner-Law M, Gnudi L, Kahn BB, Granner DK. Activation of the Ras mitogen-activated protein kinase-ribosomal protein kinase pathway is not required for the repression of phosphoenolpyruvate carboxykinase gene transcription by insulin. *J Biol Chem*. 1998;273:3198–3204.

144. Fafournoux P, Bruhat A, Jousse C. Amino acid regulation of gene expression. *Biochem J*. 2000;351:1–12.

145. Noguchi T. Protein nutrition and insulin-like growth factor system. *Br J Nutr*. 2000; 84(Suppl 2):S221–S224.

146. Rajkumar K, Krsek M, Dheen ST, Murphy LJ. Impaired glucose homeostasis in insulin-like growth factor binding protein-1 transgenic mice. *J Clin Invest*. 1996;98:1818–1825.

147. Murphy LJ, Seneviratne C, Moreira P, Reid RE. Enhanced expression of insulin-like growth factor-binding protein-I in the fasted rat: the effects of insulin and growth hor-mone administration. *Endocrinology*. 1991;128:689–696.

148. Ooi GT, Tseng LY, Tran MQ, Rechler MM. Insulin rapidly decreases insulin-like growth factor-binding protein-1 gene transcription in streptozotocin-diabetic rats. *Mol Endocrinol*. 1992;6:2219–2228.

149. Powell DR, Suwanichkul A, Cubbage ML, DePaolis LA, Snuggs MB, Lee PD. Insulin inhibits transcription of the human gene for insulin-like growth factor-binding protein-1. *J Biol Chem*. 1991;266:18868–18876.

150. Chen G, Liang G, Ou J, Goldstein JL, Brown MS. Central role for liver X receptor in insulin-mediated activation of Srebp-1c transcription and stimulation of fatty acid syn-thesis in liver. *Proc Natl Acad Sci USA*. 2004;101:11245–11250.

151. Cagen LM, Deng X, Wilcox HG, Park EA, Raghow R, Elam MB. Insulin activates the rat sterol-regulatory-element-binding protein 1c (SREBP-1c) promoter through the combinatorial actions of SREBP, LXR, Sp-1 and NF-Y cis-acting elements. *Biochem J*. 2005;385:207–216.

152. Deng X, Yellaturu C, Cagen L, et al. Expression of the rat sterol regulatory element-binding protein-1c gene in response to insulin is mediated by increased transactivating capacity of specificity protein 1 (Sp1). *J Biol Chem*. 2007;282:17517–17529.

153. Fernández-Alvarez A, Alvarez MS, Gonzalez R, Cucarella C, Muntané J, Casado M. Human SREBP1c expression in liver is directly regulated by peroxisome proliferator-activated receptor α (PPARα). *J Biol Chem*. 2011;286:21466–21477.

154. Chen G, Zhang Y, Lu D, Li N, Ross AC. Retinoids synergize with insulin to induce hepatic Gck expression. *Biochem J.* 2009;419:645–653.
155. Roder K, Zhang L, Schweizer M. SREBP-1c mediates the retinoid-dependent increase in fatty acid synthase promoter activity in HepG2. *FEBS Lett.* 2007;581:2715–2720.
156. Owen JL, Zhang Y, Bae SH, et al. Insulin stimulation of SREBP-1c processing in transgenic rat hepatocytes requires p70 S6-kinase. *Proc Natl Acad Sci USA.* 2012;109:16184–16189.
157. Iynedjian PB, Marie S, Wang H, Gjinovci A, Nazaryan K. Liver-specific enhancer of the glucokinase gene. *J Biol Chem.* 1996;271:29113–29120.
158. Azzout-Marniche D, Becard D, Guichard C, Foretz M, Ferre P, Foufelle F. Insulin effects on sterol regulatory-element-binding protein-1c (SREBP-1c) transcriptional activity in rat hepatocytes. *Biochem J.* 2000;350(Pt 2):389–393.
159. Gregori C, Guillet-Deniau I, Girard J, Decaux JF, Pichard AL. Insulin regulation of glucokinase gene expression: evidence against a role for sterol regulatory element binding protein 1 in primary hepatocytes. *FEBS Lett.* 2006;580:410–414.
160. Hansmannel F, Mordier S, Iynedjian PB. Insulin induction of glucokinase and fatty acid synthase in hepatocytes: analysis of the roles of sterol-regulatory-element-binding protein-1c and liver X receptor. *Biochem J.* 2006;399:275–283.
161. Li R, Chen W, Li Y, Zhang Y, Chen G. Retinoids synergized with insulin to induce Srebp-1c expression and activated its promoter via the two liver X receptor binding sites that mediate insulin action. *Biochem Biophys Res Commun.* 2011;406:268–272.
162. Noguchi T, Takenaka M, Yamada K, Matsuda T, Hashimoto M, Tanaka T. Characterization of the 5' flanking region of rat glucokinase gene. *Biochem Biophys Res Commun.* 1989;164:1247–1252.
163. O'Brien RM, Lucas PC, Forest CD, Magnuson MA, Granner DK. Identification of a sequence in the PEPCK gene that mediates a negative effect of insulin on transcription. *Science.* 1990;249:533–537.
164. Eisenberger CL, Nechushtan H, Cohen H, Shani M, Reshef L. Differential regulation of the rat phosphoenolpyruvate carboxykinase gene expression in several tissues of transgenic mice. *Mol Cell Biol.* 1992;12:1396–1403.
165. Mcgrane MM, Yun JS, Moorman AF, et al. Metabolic effects of developmental, tissue-, and cell-specific expression of a chimeric phosphoenolpyruvate carboxykinase (GTP)/bovine growth hormone gene in transgenic mice. *J Biol Chem.* 1990;265:22371–22379.
166. McGrane MM. Vitamin A, regulation of gene expression: molecular mechanism of a prototype gene. *J Nutr Biochem.* 2007;18:497–508.
167. Friedman JE, Yun JS, Patel YM, Mcgrane MM, Hanson RW. Glucocorticoids regulate the induction of phosphoenolpyruvate carboxykinase (GTP) gene transcription during diabetes. *J Biol Chem.* 1993;268:12952–12957.
168. Mcgrane MM, de Vente J, Yun J, et al. Tissue-specific expression and dietary regulation of a chimeric phosphoenolpyruvate carboxykinase/bovine growth hormone gene in transgenic mice. *J Biol Chem.* 1988;263:11443–11451.
169. Patel YM, Yun JS, Liu J, Mcgrane MM, Hanson RW. An analysis of regulatory elements in the phosphoenolpyruvate carboxykinase (GTP) gene which are responsible for its tissue-specific expression and metabolic control in transgenic mice. *J Biol Chem.* 1994;269:5619–5628.
170. Shin DJ, Tao AY, Mcgrane MM. Effects of vitamin A deficiency and retinoic acid treatment on expression of a phosphoenolpyruvate carboxykinase-bovine growth hormone gene in transgenic mice. *Biochem Biophys Res Commun.* 1995;213:706–714.
171. Shin DJ, Odom DP, Scribner KB, Ghoshal S, McGrane MM. Retinoid regulation of the phosphoenolpyruvate carboxykinase gene in liver. *Mol Cell Endocrinol.* 2002;195:39–54.

172. Chen M, Breslow JL, Li W, Leff T. Transcriptional regulation of the apoC-III gene by insulin in diabetic mice: correlation with changes in plasma triglyceride levels. *J Lipid Res*. 1994;35:1918–1924.

173. Suwanickul A, Morris SL, Powell DR. Identification of an insulin-responsive element in the promoter of the human gene for insulin-like growth factor binding protein-1. *J Biol Chem*. 1993;268:17063–17068.

174. Suh DS, Ooi GT, Rechler MM. Identification of cis-elements mediating the stimulation of rat insulin-like growth factor-binding protein-1 promoter activity by dexamethasone, cyclic adenosine 3',5'-monophosphate, and phorbol esters, and inhibition by insulin. *Mol Endocrinol*. 1994;8:794–805.

175. Streeper RS, Svitek CA, Chapman S, Greenbaum LE, Taub R, O'Brien RM. A multicomponent insulin response sequence mediates a strong repression of mouse glucose-6-phosphatase gene transcription by insulin. *J Biol Chem*. 1997;272:11698–11701.

176. Beurton F, Bandyopadhyay U, Dieumegard B, Barouki R, Aggerbeck M. Delineation of the insulin-responsive sequence in the rat cytosolic aspartate aminotransferase gene: binding sites for hepatocyte nuclear factor-3 and nuclear factor I. *Biochem J*. 1999;343:687–695.

177. Galili N, Davis RJ, Fredericks WJ, et al. Fusion of a fork head domain gene to PAX3 in the solid tumour alveolar rhabdomyosarcoma. *Nat Genet*. 1993;5:230–235.

178. Tang ED, Nuñez G, Barr FG, Guan KL. Negative regulation of the Forkhead transcription factor FKHR by Akt. *J Biol Chem*. 1999;274:16741–16746.

179. Nakae J, Park BC, Accili D. Insulin stimulates phosphorylation of the Forkhead transcription factor FKHR on serine 253 through a wortmannin-sensitive pathway. *J Biol Chem*. 1999;274:15982–15985.

180. Guo S, Rena G, Cichy S, He X, Cohen P, Unterman T. Phosphorylation of serine 256 by protein kinase B disrupts transactivation by FKHR and mediates effects of insulin on insulin-like growth factor-binding protein-1 promoter activity through a conserved insulin response sequence. *J Biol Chem*. 1999;274:17184–17192.

181. Puigserver P, Rhee J, Donovan J, et al. Insulin-regulated hepatic gluconeogenesis through FOXO1-PGC-1[alpha] interaction. *Nature*. 2003;423(6939):550–555.

182. Brunet A, Bonni A, Zigmond MJ, et al. Akt promotes cell survival by phosphorylating and inhibiting a Forkhead transcription factor. *Cell*. 1999;96:857–868.

183. Hall RK, Yamasaki T, Kucera T, Waltner-Law M, O'Brien R, Granner DK. Regulation of phosphoenolpyruvate carboxykinase and insulin-like growth factor-binding protein-1 gene expression by insulin. The role of Winged helix/Forkhead proteins. *J Biol Chem*. 2000;275:30169–30175.

184. Wolfrum C, Besser D, Luca E, Stoffel M. Insulin regulates the activity of forkhead transcription factor Hnf-3β/Foxa-2 by Akt-mediated phosphorylation and nuclear/cytosolic localization. *Proc Natl Acad Sci USA*. 2003;100:11624–11629.

185. Wolfrum C, Asilmaz E, Luca E, Friedman JM, Stoffel M. Foxa2 regulates lipid metabolism and ketogenesis in the liver during fasting and in diabetes. *Nature*. 2004;432:1027–1032.

186. Hughes DE, Stolz DB, Yu S, et al. Elevated hepatocyte levels of the Forkhead box A2 (HNF-3β) transcription factor cause postnatal steatosis and mitochondrial damage. *Hepatology*. 2003;37:1414–1424.

187. Haeusler RA, Kaestner KH, Accili D. FoxOs function synergistically to promote glucose production. *J Biol Chem*. 2010;285:35245–35248.

188. Villafuerte BC, Phillips LS, Rane MJ, Zhao W. Insulin-response element-binding protein 1: a novel Akt substrate involved in transcriptional action of insulin. *J Biol Chem*. 2004;279:36650–36659.

189. Chahal J, Chen CC, Rane MJ, et al. Regulation of insulin-response element binding protein-1 in obesity and diabetes: potential role in impaired insulin-induced gene transcription. *Endocrinology*. 2008;149:4829–4836.

190. Kimball CP, Murlin JR. Aqueous extracts of pancreas: III. Some precipitation reactions of insulin. *J Biol Chem*. 1923;58:337–346.

191. Staub A, Sinn L, Behrens OK. Purification and crystallization of hyperglycemic glycogenolytic factor (HGF). *Science*. 1953;117:628–629.

192. Staub A, Sinn L, Behrens OK. Purification and crystallization of glucagon. *J Biol Chem*. 1955;214:619–632.

193. Bromer WW, Sinn LG, Staub A, Behrens OK. The amino acid sequence of glucagon. *Diabetes*. 1957;6:234–238.

194. Unger RH, Eisentraut AM, McCall MS, Madison LL. Glucagon antibodies and an immunoassay for glucagon. *J Clin Invest*. 1961;40:1280–1289.

195. Aguilar-Parada E, Eisentraut AM, Unger RH. Pancreatic glucagon secretion in normal and diabetic subjects. *Am J Med Sci*. 1969;257:415–419.

196. Unger RH. The Banting Memorial Lecture 1975: diabetes and the alpha cell. *Diabetes*. 1976;25:136–151.

197. Unger RH, Cherrington AD. Glucagonocentric restructuring of diabetes: a pathophysiologic and therapeutic makeover. *J Clin Invest*. 2012;122:4–12.

198. Iwanij V, Hur KC. Direct cross-linking of 125I-labeled glucagon to its membrane receptor by UV irradiation. *Proc Natl Acad Sci USA*. 1985;82:325–329.

199. Iwanij V, Vincent AC. Characterization of the glucagon receptor and its functional domains using monoclonal antibodies. *J Biol Chem*. 1990;265:21302–21308.

200. Iyengar R, Herberg JT. Structural analysis of the hepatic glucagon receptor. Identification of a guanine nucleotide-sensitive hormone-binding region. *J Biol Chem*. 1984;259:5222–5229.

201. Pohl SL, Birnbaumer L, Rodbell M. Glucagon-sensitive adenyl cylase in plasma membrane of hepatic parenchymal cells. *Science*. 1969;164:566–567.

202. Jelinek LJ, Lok S, Rosenberg GB, et al. Expression cloning and signaling properties of the rat glucagon receptor. *Science*. 1993;259:1614–1616.

203. Gelling RW, Du XQ, Dichmann DS, et al. Lower blood glucose, hyperglucagonemia, and pancreatic α cell hyperplasia in glucagon receptor knockout mice. *Proc Natl Acad Sci USA*. 2003;100:1438–1443.

204. Sloop KW, Cao JX-C, Siesky AM, et al. Hepatic and glucagon-like peptide-1-mediated reversal of diabetes by glucagon receptor antisense oligonucleotide inhibitors. *J Clin Invest*. 2004;113:1571–1581.

205. Sørensen H, Winzell MS, Brand CL, et al. Glucagon receptor knockout mice display increased insulin sensitivity and impaired β-cell function. *Diabetes*. 2006;55:3463–3469.

206. Lee Y, Wang MY, Du XQ, Charron MJ, Unger RH. Glucagon receptor knockout prevents insulin-deficient Type 1 diabetes in mice. *Diabetes*. 2011;60:391–397.

207. Conarello SL, Jiang G, Mu J, et al. Glucagon receptor knockout mice are resistant to diet-induced obesity and streptozotocin-mediated beta cell loss and hyperglycaemia. *Diabetologia*. 2007;50:142–150.

208. Longuet C, Sinclair EM, Maida A, et al. The glucagon receptor is required for the adaptive metabolic response to fasting. *Cell Metab*. 2008;8:359–371.

209. Myers MG, Cowley MA, Münzberg H. Mechanisms of leptin action and leptin resistance. *Annu Rev Physiol*. 2008;70:537–556.

210. Cohen B, Novick D, Rubinstein M. Modulation of insulin activities by leptin. *Science*. 1996;274:1185–1188.

211. Szanto I, Kahn CR. Selective interaction between leptin and insulin signaling pathways in a hepatic cell line. *Proc Natl Acad Sci USA*. 2000;97:2355–2360.

212. Mauvoisin D, Prévost M, Ducheix S, Arnaud MP, Mounier C. Key role of the ERK1/2 MAPK pathway in the transcriptional regulation of the stearoyl-CoA desaturase (SCD1) gene expression in response to leptin. *Mol Cell Endocrinol*. 2010;319:116–128.
213. Tsuchiya H, Ikeda Y, Ebata Y, et al. Retinoids ameliorate insulin resistance in a leptin-dependent manner in mice. *Hepatology*. 2012;56:1319–1330.
214. Rossetti L, Massillon D, Barzilai N, et al. Short term effects of leptin on hepatic gluconeogenesis and in vivo insulin action. *J Biol Chem*. 1997;272:27758–27763.
215. Shimamura M, Matsuda M, Ando Y, et al. Leptin and insulin down-regulate angiopoietin-like protein 3, a plasma triglyceride-increasing factor. *Biochem Biophys Res Commun*. 2004;322:1080–1085.
216. Rodríguez A, Catalán V, Gómez-Ambrosi J, et al. Insulin- and leptin-mediated control of aquaglyceroporins in human adipocytes and hepatocytes is mediated via the PI3K/Akt/mTOR signaling cascade. *J Clin Endocrinol Metab*. 2011;96:E586–E597.
217. Solomon SS, Odunusi O, Carrigan D, et al. TNF-α inhibits insulin action in liver and adipose tissue: a model of metabolic syndrome. *Horm Metab Res*. 2010;42:115–121.
218. Pandey AK, Bhardwaj V, Datta M. Tumour necrosis factor-α attenuates insulin action on phosphoenolpyruvate carboxykinase gene expression and gluconeogenesis by altering the cellular localization of Foxa2 in HepG2 cells. *FEBS J*. 2009;276:3757–3769.
219. Awazawa M, Ueki K, Inabe K, et al. Adiponectin suppresses hepatic SREBP1c expression in an AdipoR1/LKB1/AMPK dependent pathway. *Biochem Biophys Res Commun*. 2009;382:51–56.
220. Berg AH, Combs TP, Du X, Brownlee M, Scherer PE. The adipocyte-secreted protein Acrp30 enhances hepatic insulin action. *Nat Med*. 2001;7:947–953.
221. Zhou H, Song X, Briggs M, et al. Adiponectin represses gluconeogenesis independent of insulin in hepatocytes. *Biochem Biophys Res Commun*. 2005;338:793–799.
222. Andreelli F, Foretz M, Knauf C, et al. Liver adenosine monophosphate-activated kinase-α2 catalytic subunit is a key target for the control of hepatic glucose production by adiponectin and leptin but not insulin. *Endocrinology*. 2006;147:2432–2441.
223. Luo Z, Zhang Y, Li F, et al. Resistin induces insulin resistance by both AMPK-dependent and AMPK-independent mechanisms in HepG2 cells. *Endocrine*. 2009;36:60–69.
224. Sachithanandan N, Fam BC, Fynch S, et al. Liver-specific suppressor of cytokine signaling-3 deletion in mice enhances hepatic insulin sensitivity and lipogenesis resulting in fatty liver and obesity. *Hepatology*. 2010;52:1632–1642.
225. Vijayakumar A, Novosyadlyy R, Wu Y, Yakar S, LeRoith D. Biological effects of growth hormone on carbohydrate and lipid metabolism. *Growth Horm IGF Res*. 2010;20:1–7.
226. Boni-Schnetzler M, Schmid C, Meier PJ, Froesch ER. Insulin regulates insulin-like growth factor I mRNA in rat hepatocytes. *Am J Physiol Endocrinol Metab*. 1991;260:E846–E851.
227. Johnson TR, Blossey BK, Denko CW, Ilan J. Expression of insulin-like growth factor I in cultured rat hepatocytes: effects of insulin and growth hormone. *Mol Endocrinol*. 1989;3:580–587.
228. Zhao TJ, Liang G, Li RL, et al. Ghrelin O-acyltransferase (GOAT) is essential for growth hormone-mediated survival of calorie-restricted mice. *Proc Natl Acad Sci USA*. 2010;107:7467–7472.
229. Cao H, Sekiya M, Ertunc ME, et al. Adipocyte lipid chaperone aP2 is a secreted adipokine regulating hepatic glucose production. *Cell Metab*. 2013;17:768–778.
230. Blomhoff R, Blomhoff HK. Overview of retinoid metabolism and function. *J Neurobiol*. 2006;66:606–630.
231. Evans RM. The nuclear receptor superfamily: a rosetta stone for physiology. *Mol Endocrinol*. 2005;19:1429–1438.

232. Napoli JL. Physiological insights into all-trans-retinoic acid biosynthesis. *Biochim Biophys Acta*. 1821;2011:152–167.
233. Wolf G. Tissue-specific increases in endogenous all-trans retinoic acid: possible contributing factor in ethanol toxicity. *Nutr Rev*. 2010;68:689–692.
234. Chambon P. A decade of molecular biology of retinoic acid receptors. *FASEB J*. 1996;10:940–954.
235. Mangelsdorf DJ, Evans RM. The RXR heterodimers and orphan receptors. *Cell*. 1995;83:841–850.
236. Shirakami Y, Lee SA, Clugston RD, Blaner WS. Hepatic metabolism of retinoids and disease associations. *Biochim Biophys Acta*. 1821;2012:124–136.
237. Zhao S, Li R, Li Y, Chen W, Zhang Y, Chen G. Roles of vitamin A status and retinoids in glucose and fatty acid metabolism. *Biochem Cell Biol*. 2012;90:1–11.
238. Koistinen HA, Remitz A, Gylling H, Miettinen TA, Koivisto VA, Ebeling P. Dyslipidemia and a reversible decrease in insulin sensitivity induced by therapy with 13-cis-retinoic acid. *Diabetes Metab Res Rev*. 2001;17:391–395.
239. Lyons S, Laker MF, Marsden JR, Manuel R, Shuster SAM. Effect of oral 13-cis-retinoic acid on serum lipids. *Br J Dermatol*. 1982;107:591–595.
240. Marsden JR, Trinick TR, Laker MF, Shuster S. Effects of isotretinoin on serum lipids and lipoproteins, liver and thyroid function. *Clin Chim Acta*. 1984;143:243–251.
241. Gerber LE, Erdman Jr J. Retinoic acid and hypertriglyceridemia. *Ann N Y Acad Sci*. 1981;359:391–392.
242. Gerber LE, Erdman JW. Effect of retinoic acid and retinyl acetate feeding upon lipid metabolism in adrenalectomized rats. *J Nutr*. 1979;109:580–589.
243. Gerber LE, Erdman Jr JW. Comparative effects of all-trans and 13-cis retinoic acid administration on serum and liver lipids in rats. *J Nutr*. 1980;110:343–351.
244. Zhang Y, Li R, Li Y, Chen W, Zhao S, Chen G. Vitamin A status affects obesity development and hepatic expression of key genes for fuel metabolism in Zucker fatty rats. *Biochem Cell Biol*. 2012;90:548–557.
245. Lucas PC, Forman BM, Samuels HH, Granner DK. Specificity of a retinoic acid response element in the phosphoenolpyruvate carboxykinase gene promoter: consequences of both retinoic acid and thyroid hormone receptor binding. *Mol Cell Biol*. 1991;11:5164–5170.
246. Lucas PC, O'Brien RM, Mitchell JA, et al. A retinoic acid response element is part of a pleiotropic domain in the phosphoenolpyruvate carboxykinase gene. *Proc Natl Acad Sci USA*. 1991;88:2184–2188.
247. Scott DK, Mitchell JA, Granner DK. Identification and characterization of a second retinoic acid response element in the phosphoenolpyruvate carboxykinase gene promoter. *J Biol Chem*. 1996;271:6260–6264.
248. Zhang Y, Li R, Chen W, Li Y, Chen G. Retinoids induced *Pck1* expression and attenuated insulin-mediated suppression of its expression via activation of retinoic acid receptor in primary rat hepatocytes. *Mol Cell Biochem*. 2011;355:1–8.
249. Lannoy V, Decaux J, Pierreux C, Lemaigre F, Rousseau G. Liver glucokinase gene expression is controlled by the onecut transcription factor hepatocyte nuclear factor-6. *Diabetologia*. 2002;45:1136–1141.
250. Roth U, Jungermann K, Kietzmann T. Activation of glucokinase gene expression by hepatic nuclear factor 4alpha in primary hepatocytes. *Biochem J*. 2002;365(Pt 1): 223–228.
251. Roth U, Jungermann K, Kietzmann T. Modulation of glucokinase expression by hypoxia-inducible factor 1 and upstream stimulatory factor 2 in primary rat hepatocytes. *Biol Chem*. 2004;385:239–247.
252. Sy Kim, Hi Kim, Park SK, et al. Liver glucokinase can be activated by peroxisome proliferator-activated receptor-{gamma}. *Diabetes*. 2004;53:S66–S70.

253. Sy Kim, Hi Kim, Kim TH, et al. SREBP-1c mediates the insulin-dependent hepatic glucokinase expression. *J Biol Chem.* 2004;279:30823–30829.
254. Narkewicz MR, Iynedjian PB, Ferre P, Girard J. Insulin and tri-iodothyronine induce glucokinase mRNA in primary cultures of neonatal rat hepatocytes. *Biochem J.* 1990;271:585–589.
255. Chauhan J, Dakshinamurti K. Transcriptional regulation of the glucokinase gene by biotin in starved rats. *J Biol Chem.* 1991;266:10035–10038.
256. Spence JT, Koudelka AP. Effects of biotin upon the intracellular level of cGMP and the activity of glucokinase in cultured rat hepatocytes. *J Biol Chem.* 1984;259:6393–6396.
257. Laffitte BA, Chao LC, Li J, et al. Activation of liver X receptor improves glucose tolerance through coordinate regulation of glucose metabolism in liver and adipose tissue. *Proc Natl Acad Sci USA.* 2003;100:5419–5424.
258. Kim TH, Kim H, Park JM, et al. Interrelationship between liver X receptor α, sterol regulatory element-binding protein-1c, peroxisome proliferator-activated receptor γ, and small heterodimer partner in the transcriptional regulation of glucokinase gene expression in liver. *J Biol Chem.* 2009;284:15071–15083.
259. Cabrera-Valladares G, Matschinsky FM, Wang J, Fernandez-Mejia C. Effect of retinoic acid on glucokinase activity and gene expression in neonatal and adult cultured hepatocytes. *Life Sci.* 2001;68:2813–2824.
260. Decaux JF, Juanes M, Bossard P, Girard J. Effects of triiodothyronine and retinoic acid on glucokinase gene expression in neonatal rat hepatocytes. *Mol Cell Endocrinol.* 1997;130:61–67.
261. Sasaki K, Cripe TP, Koch SR, et al. Multihormonal regulation of phosphoenolpyruvate carboxykinase gene transcription. The dominant role of insulin. *J Biol Chem.* 1984;259:15242–15251.
262. Benvenisty N, Reshef L. Developmental acquisition of DNase I sensitivity of the phosphoenolpyruvate carboxykinase (GTP) gene in rat liver. *Proc Natl Acad Sci USA.* 1987;84:1132–1136.
263. Roesler WJ, Vandenbark GR, Hanson RW. Identification of multiple protein binding domains in the promoter- regulatory region of the phosphoenolpyruvate carboxykinase (GTP) gene. *J Biol Chem.* 1989;264:9657–9664.
264. Glorian M, Duplus E, Beale EG, Scott DK, Granner DK, Forest C. A single element in the phosphoenolpyruvate carboxykinase gene mediates thiazolidinedione action specifically in adipocytes. *Biochimie.* 2001;83:933–943.
265. Imai E, Miner JN, Mitchell JA, Yamamoto KR, Granner DK. Glucocorticoid receptor-cAMP response element-binding protein interaction and the response of the phosphoenolpyruvate carboxykinase gene to glucocorticoids. *J Biol Chem.* 1993;268:5353–5356.
266. Hall RK, Sladek FM, Granner DK. The orphan receptors COUP-TF and HNF-4 serve as accessory factors required for induction of phosphoenolpyruvate carboxykinase gene transcription by glucocorticoids. *Proc Natl Acad Sci USA.* 1995;92:412–416.
267. Kucera T, Waltner-Law M, Scott DK, Prasad R, Granner DK. A point mutation of the AF2 transactivation domain of the glucocorticoid receptor disrupts its interaction with steroid receptor coactivator 1. *J Biol Chem.* 2002;277:26098–26102.
268. Nyirenda MJ, Lindsay RS, Kenyon CJ, Burchell A, Seckl JR. Glucocorticoid exposure in late gestation permanently programs rat hepatic phosphoenolpyruvate carboxykinase and glucocorticoid receptor expression and causes glucose intolerance in adult offspring. *J Clin Invest.* 1998;101:2174–2181.
269. Scott DK, Stromstedt PE, Wang JC, Granner DK. Further characterization of the glucocorticoid response unit in the phosphoenolpyruvate carboxykinase gene. The role of the glucocorticoid receptor-binding sites. *Mol Endocrinol.* 1998;12:482–491.

270. Sugiyama T, Scott DK, Wang JC, Granner DK. Structural requirements of the gluco-corticoid and retinoic acid response units in the phosphoenolpyruvate carboxykinase gene promoter. *Mol Endocrinol.* 1998;12:1487–1498.

271. Giralt M, Park EA, Gurney AL, Liu JS, Hakimi P, Hanson RW. Identification of a thyroid hormone response element in the phosphoenolpyruvate carboxykinase (GTP) gene. Evidence for synergistic interaction between thyroid hormone and cAMP cis-regulatory elements. *J Biol Chem.* 1991;266:21991–21996.

272. Park EA, Gurney AL, Nizielski SE, et al. Relative roles of CCAAT/enhancer-binding protein beta and cAMP regulatory element-binding protein in controlling transcription of the gene for phosphoenolpyruvate carboxykinase (GTP). *J Biol Chem.* 1993;268:613–619.

273. Chakravarty K, Leahy P, Becard D, et al. Sterol regulatory element-binding protein-1c mimics the negative effect of insulin on phosphoenolpyruvate carboxykinase (GTP) gene transcription. *J Biol Chem.* 2001;276:34816–34823.

274. Chakravarty K, Wu SY, Chiang CM, Samols D, Hanson RW. SREBP-1c and Sp1 interact to regulate transcription of the gene for phosphoenolpyruvate carboxykinase (GTP) in the liver. *J Biol Chem.* 2004;279:15385–15395.

275. Leahy P, Crawford DR, Grossman G, Gronostajski RM, Hanson RW. CREB binding protein coordinates the function of multiple transcription factors including nuclear fac-tor I to regulate phosphoenolpyruvate carboxykinase (GTP) gene transcription. *J Biol Chem.* 1999;274:8813–8822.

276. Stafford JM, Waltner-Law M, Granner DK. Role of accessory factors and steroid recep-tor coactivator 1 in the regulation of phosphoenolpyruvate carboxykinase gene tran-scription by glucocorticoids. *J Biol Chem.* 2001;276:3811–3819.

277. Yoon JC, Puigserver P, Chen G, et al. Control of hepatic gluconeogenesis through the transcriptional coactivator PGC-1. *Nature.* 2001;413:131–138.

278. Cao G, Liang Y, Broderick CL, et al. Antidiabetic action of a liver X receptor agonist mediated by inhibition of hepatic gluconeogenesis. *J Biol Chem.* 2003;278:1131–1136.

279. Band CJ, Posner BI. Phosphatidylinositol 3'-kinase and p70^{s6k} are required for insulin but not bisperoxovanadium 1,10-phenanthroline (bpV(phen)) inhibition of insulin-like growth factor binding protein gene expression: evidence for MEK-independent activation of mitogen-activated protein kinase by bpV(phen). *J Biol Chem.* 1997;272:138–145.

280. Cichy SB, Uddin S, Danilkovich A, Guo S, Klippel A, Unterman TG. Protein kinase B/Akt mediates effects of insulin on hepatic insulin-like growth factor-binding protein-1 gene expression through a conserved insulin response sequence. *J Biol Chem.* 1998;273:6482–6487.

281. Novosyadlyy R, Lelbach A, Sheikh N, et al. Temporal and spatial expression of IGF-I and IGFBP-1 during acute-phase response induced by localized inflammation in rats. *Growth Horm IGF Res.* 2009;19:51–60.

282. Pierce AL, Shimizu M, Felli L, Swanson P, Dickhoff WW. Metabolic hormones reg-ulate insulin-like growth factor binding protein-1 mRNA levels in primary cultured salmon hepatocytes; lack of inhibition by insulin. *J Endocrinol.* 2006;191:379–386.

283. Flyvbjerg A, Schuller AGP, van Neck JW, Groffen C, Ørskov H, Drop SLS. Stimu-lation of hepatic insulin-like growth factor-binding protein-1 and -3 gene expression by octreotide in rats. *J Endocrinol.* 1995;147:545–551.

284. Ren SG, Ezzat S, Melmed S, Braunstein GD. Somatostatin analog induces insulin-like growth factor binding protein-1 (IGFBP-1) expression in human hepatoma cells. *Endo-crinology.* 1992;131:2479–2481.

285. Holten D, Kenney FT. Regulation of tyrosine α-ketoglutarate transaminase in rat liver: VI. Induction by pancreatic hormones. *J Biol Chem.* 1967;242:4372–4377.

286. Wicks WD, Kenney FT, Lee KL. Induction of hepatic enzyme synthesis in vivo by adenosine 3',5'-monophosphate. *J Biol Chem*. 1969;244:6008–6013.
287. Lee KL, Isham KR, Johnson A, Kenney FT. Insulin enhances transcription of the tyrosine aminotransferase gene in rat liver. *Arch Biochem Biophys*. 1986;248:597–603.
288. Wicks WD. Induction of hepatic enzymes by adenosine 3',5'-monophosphate in organ culture. *J Biol Chem*. 1969;244:3941–3950.
289. Hashimoto S, Schmid W, Schütz G. Transcriptional activation of the rat liver tyrosine aminotransferase gene by cAMP. *Proc Natl Acad Sci USA*. 1984;81:6637–6641.
290. Crettaz M, Muller-Wieland D, Kahn CR. Transcriptional and posttranscriptional regulation of tyrosine aminotransferase by insulin in rat hepatoma cells. *Biochemistry*. 1988;27:495–500.
291. Messina JL, Chatterjee AK, Strapko HT, Weinstock RS. Short- and long-term effects of insulin on tyrosine aminotransferase gene expression. *Arch Biochem Biophys*. 1992;298:56–62.
292. Moore PS, Koontz JW. Regulation of tyrosine aminotransferase by insulin and cyclic AMP: similar effects on activity but opposite effects on transcription. *Mol Endocrinol*. 1989;3:1724–1732.
293. Cake MH, Ho KKW, Shelly L, Milward E, Yeoh GCT. Insulin antagonism of dexamethasone induction of tyrosine aminotransferase in cultured fetal hepatocytes. *Eur J Biochem*. 1989;182:429–435.
294. Porstmann T, Santos CR, Griffiths B, et al. SREBP activity is regulated by mTORC1 and contributes to Akt-dependent cell growth. *Cell Metab*. 2008;8:224–236.
295. Li S, Brown MS, Goldstein JS. Bifurcation of insulin signaling pathway in rat liver: mTORC1 required for stimulation of lipogenesis, but not inhibition of gluconeogenesis. *Proc Natl Acad Sci USA*. 2010;107:3441–3446.
296. Li S, Ogawa W, Emi A, et al. Role of S6K1 in regulation of SREBP1c expression in the liver. *Biochem Biophys Res Commun*. 2011;412:197–202.
297. Arcaro A, Wymann MP. Wortmannin is a potent phosphatidylinositol 3-kinase inhibitor: the role of phosphatidylinositol 3,4,5-trisphosphate in neutrophil responses. *Biochem J*. 1993;296:297–301.
298. Wymann M, Arcaro A. Platelet-derived growth factor-induced phosphatidylinositol 3-kinase activation mediates actin rearrangements in fibroblasts. *Biochem J*. 1994;298:517–520.
299. Cifuentes M, Albala C, Rojas CV. Differences in lipogenesis and lipolysis in obese and non-obese adult human adipocytes. *Biol Res*. 2008;41197–41204.
300. Tang X, Wang L, Proud CG, Downes CP. Muscarinic receptor-mediated activation of p70 S6 kinase 1 (S6K1) in 1321 N1 astrocytoma cells: permissive role of phosphoinositide 3-kinase. *Biochem J*. 2003;374:137–143.
301. Hinault C, Mothe-Satney I, Gautier N, Van Obberghen E. Amino acids require glucose to enhance, through phosphoinositide-dependent protein kinase 1, the insulin-activated protein kinase B cascade in insulin-resistant rat adipocytes. *Diabetologia*. 2006;49:1017–1026.
302. Jensen J, Brennesvik EO, Lai YC, Shepherd PR. GSK-3beta regulation in skeletal muscles by adrenaline and insulin: evidence that PKA and PKB regulate different pools of GSK-3. *Cell Signal*. 2007;19:204–210.
303. Nakanishi S, Kakita S, Takahashi I, et al. Wortmannin, a microbial product inhibitor of myosin light chain kinase. *J Biol Chem*. 1992;267:2157–2163.
304. Hajduch E, Balendran A, Batty IH, et al. Ceramide impairs the insulin-dependent membrane recruitment of protein kinase B leading to a loss in downstream signalling in L6 skeletal muscle cells. *Diabetologia*. 2001;44:173–183.
305. Teruel T, Hernandez R, Lorenzo M. Ceramide mediates insulin resistance by tumor necrosis factor-alpha in brown adipocytes by maintaining Akt in an inactive dephosphorylated state. *Diabetes*. 2001;50:2563–2571.

306. Ribaux PG, Iynedjian PB. Analysis of the role of protein kinase B (cAKT) in insulin-dependent induction of glucokinase and sterol regulatory element-binding protein 1 (SREBP1) mRNAs in hepatocytes. *Biochem J.* 2003;376:697–705.

307. Kohl C, Ravel D, Girard J, Pegorier JP. Effects of benfluorex on fatty acid and glucose metabolism in isolated rat hepatocytes: from metabolic fluxes to gene expression. *Diabetes.* 2002;51:2363–2368.

308. Howitz KT, Bitterman KJ, Cohen HY, et al. Small molecule activators of sirtuins extend *Saccharomyces cerevisiae* lifespan. *Nature.* 2003;425:191–196.

309. Khanduja KL, Bhardwaj A, Kaushik G. Resveratrol inhibits N-nitrosodiethylamine-induced ornithine decarboxylase and cyclooxygenase in mice. *J Nutr Sci Vitaminol (Tokyo).* 2004;50:61–65.

310. Baur JA, Pearson KJ, Price NL, et al. Resveratrol improves health and survival of mice on a high-calorie diet. *Nature.* 2006;444:337–342.

311. Docherty JJ, Fu MM, Hah JM, Sweet TJ, Faith SA, Booth T. Effect of resveratrol on herpes simplex virus vaginal infection in the mouse. *Antiviral Res.* 2005;67: 155–162.

312. Jang M, Cai L, Udeani GO, et al. Cancer chemopreventive activity of resveratrol, a natural product derived from grapes. *Science.* 1997;275:218–220.

313. Carluccio MA, Ancora MA, Massaro M, et al. Homocysteine induces VCAM-1 gene expression through NF-kappaB and NAD(P)H oxidase activation: protective role of Mediterranean diet polyphenolic antioxidants. *Am J Physiol Heart Circ Physiol.* 2007;293:H2344–H2354.

314. Yamamoto H, Schoonjans K, Auwerx J. Sirtuin functions in health and disease. *Mol Endocrinol.* 2007;21:1745–1755.

315. Park JM, Kim TH, Bae JS, Kim MY, Kim KS, Ahn YH. Role of resveratrol in FOXO1-mediated gluconeogenic gene expression in the liver. *Biochem Biophys Res Commun.* 2010;403:329–334.

316. Yang J, Kong X, Martins-Santos ME, et al. Activation of SIRT1 by resveratrol represses transcription of the gene for the cytosolic form of phosphoenolpyruvate carboxykinase (GTP) by deacetylating hepatic nuclear factor 4alpha. *J Biol Chem.* 2009;284:27042–27053.

317. Knowler WC, Barrett-Connor E, Fowler SE, et al. Reduction in the incidence of type 2 diabetes with lifestyle intervention or metformin. *N Engl J Med.* 2002;346:393–403.

318. Bailey CJ, Turner RC. Metformin. *N Engl J Med.* 1996;334:574–579.

319. Wiernsperger NF, Bailey CJ. The antihyperglycaemic effect of metformin: therapeutic and cellular mechanisms. *Drugs.* 1999;58(Suppl 1):31–39.

320. Prospective UK. Diabetes Study (UKPDS) Group. Effect of intensive blood-glucose control with metformin on complications in overweight patients with type 2 diabetes (UKPDS 34). UK Prospective Diabetes Study (UKPDS) Group. *Lancet.* 1998;352:854–865.

321. Owen MR, Doran E, Halestrap AP. Evidence that metformin exerts its anti-diabetic effects through inhibition of complex 1 of the mitochondrial respiratory chain. *Biochem J.* 2000;348:607–614.

322. Zhou G, Myers R, Li Y, et al. Role of AMP-activated protein kinase in mechanism of metformin action. *J Clin Invest.* 2001;108:1167–1174.

323. Foretz M, Hebrard S, Leclerc J, et al. Metformin inhibits hepatic gluconeogenesis in mice independently of the LKB1/AMPK pathway via a decrease in hepatic energy state. *J Clin Invest.* 2010;120:2355–2369.

324. Miller RA, Birnbaum MJ. An energetic tale of AMPK-independent effects of metformin. *J Clin Invest.* 2010;120:2267–2270.

325. Miller RA, Chu Q, Xie J, Foretz M, Viollet B, Birnbaum MJ. Biguanides suppress hepatic glucagon signalling by decreasing production of cyclic AMP. *Nature.* 2013;494:256–260.

326. Gibbs PJ, Seddon KR. Berberine. *Altern Med Rev.* 2000;5:175–177.
327. Lee YS, Kim WS, Kim KH, et al. Berberine, a natural plant product, activates AMP-activated protein kinase with beneficial metabolic effects in diabetic and insulin-resistant states. *Diabetes.* 2006;55:2256–2264.
328. Leng SH, Lu FE, Xu LJ. Therapeutic effects of berberine in impaired glucose tolerance rats and its influence on insulin secretion. *Acta Pharmacol Sin.* 2004;25:496–502.
329. Yin J, Xing H, Ye J. Efficacy of berberine in patients with type 2 diabetes mellitus. *Metabolism.* 2008;57:712–717.
330. Zhang Y, Li X, Zou D, et al. Treatment of type 2 diabetes and dyslipidemia with the natural plant alkaloid berberine. *J Clin Endocrinol Metab.* 2008;93:2559–2565.
331. Xia X, Yan J, Shen Y, et al. Berberine improves glucose metabolism in diabetic rats by inhibition of hepatic gluconeogenesis. *PLoS One.* 2011;6:e16556.
332. Ge Y, Zhang Y, Li R, Chen W, Li Y, Chen G. Berberine regulated Gck, G6pc, Pck1 and Srebp-1c expression and activated AMP-activated protein kinase in primary rat hepatocytes. *Int J Biol Sci.* 2011;7:673–684.
333. Li GS, Liu XH, Zhu H, et al. Berberine-improved visceral white adipose tissue insulin resistance associated with altered sterol regulatory element-binding proteins, liver X receptors, and peroxisome proliferator-activated receptors transcriptional programs in diabetic hamsters. *Biol Pharm Bull.* 2011;34:644–654.
334. Davis JM, Murphy EA, Carmichael MD. Effects of the dietary flavonoid quercetin upon performance and health. *Curr Sports Med Rep.* 2009;8:206–213.
335. Hardie DG. Sensing of energy and nutrients by AMP-activated protein kinase. *Am J Clin Nutr.* 2011;93:891S–896S.
336. Stewart LK, Soileau JL, Ribnicky D, et al. Quercetin transiently increases energy expenditure but persistently decreases circulating markers of inflammation in C57BL/6J mice fed a high-fat diet. *Metabolism.* 2008;57(7 Suppl 1):S39–S46.
337. Davis JM, Murphy EA, Carmichael MD, Davis B. Quercetin increases brain and muscle mitochondrial biogenesis and exercise tolerance. *Am J Physiol Regul Integr Comp Physiol.* 2009;2(96):R1071–R1077.
338. Zhou M, Wang S, Zhao A, et al. Transcriptomic and metabonomic profiling reveal synergistic effects of quercetin and resveratrol supplementation in high-fat diet fed mice. *J Proteome Res.* 2012;11:4961–4971.
339. Sharma RD, Raghuram TC, Rao NS. Effect of fenugreek seeds on blood glucose and serum lipids in type I diabetes. *Eur J Clin Nutr.* 1990;44:301–306.
340. Mohammad S, Taha A, Akhtar K, Bamezai RN, Baquer NZ. In vivo effect of Trigonella foenum graecum on the expression of pyruvate kinase, phosphoenolpyruvate carboxykinase, and distribution of glucose transporter (GLUT4) in alloxan-diabetic rats. *Can J Physiol Pharmacol.* 2006;84:647–654.
341. Saxena A, Vikram NK. Role of selected Indian plants in management of type 2 diabetes: a review. *J Altern Complement Med.* 2004;10:369–378.
342. Haeri MR, Izaddoost M, Ardekani MR, Nobar MR, White KN. The effect of fenugreek 4-hydroxyisoleucine on liver function biomarkers and glucose in diabetic and fructose-fed rats. *Phytother Res.* 2009;23:61–64.
343. Punkt K, Psinia I, Welt K, Barth W, Asmussen G. Effects on skeletal muscle fibres of diabetes and Ginkgo biloba extract treatment. *Acta Histochem.* 1999;101:53–69.
344. Cheng D, Liang B, Li Y. Antihyperglycemic effect of Ginkgo biloba extract in streptozotocin-induced diabetes in rats. *Biomed Res Int.* 2013;2013:162724.
345. Jung CH, Ahn J, Jeon TI, Kim TW, Ha TY. Syzygium aromaticum ethanol extract reduces high-fat diet-induced obesity in mice through downregulation of adipogenic and lipogenic gene expression. *Exp Ther Med.* 2012;4:409–414.

346. Fan S, Zhang Y, Hu N, et al. Extract of Kuding tea prevents high-fat diet-induced metabolic disorders in C57BL/6 mice via liver X receptor (LXR) beta antagonism. *PLoS One*. 2012;7:e51007.

347. Juan YC, Tsai WJ, Lin YL, et al. The novel anti-hyperglycemic effect of Paeoniae radix via the transcriptional suppression of phosphoenopyruvate carboxykinase (PEPCK). *Phytomedicine*. 2010;17:626–634.

348. Limaye PB, Bowen WC, Orr AV, Luo J, Tseng GC, Michalopoulos GK. Mechanisms of hepatocyte growth factor-mediated and epidermal growth factor-mediated signaling in transdifferentiation of rat hepatocytes to biliary epithelium. *Hepatology*. 2008;47:1702–1713.

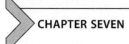

CHAPTER SEVEN

The Roles of Peroxisome Proliferator-Activated Receptors in the Metabolic Syndrome

Mahmoud Mansour
Department of Anatomy, Physiology and Pharmacology, College of Veterinary Medicine, Auburn University, Auburn, Alabama, USA

Contents

Abstract

The epidemic of obesity and its association with insulin resistance, glucose intolerance, hypertension, and dyslipidemia, collectively known as the metabolic syndrome or syndrome X, is one of the most challenging health problems facing industrialized countries. The nuclear receptors, peroxisome proliferator-activated receptors (PPARs alpha (α), beta (β) also known as delta (δ), and gamma (γ)), have well-documented roles in lipid and glucose metabolism. Pharmacologically, PPARα is activated by fibrate hypolipidemic drugs, whereas PPARγ is activated by insulin sensitizers thiazolidinediones (TZDs). No marketed drug is yet available for PPARβ(δ). The identification of fibrates and TZDs as respective ligands for PPARα and PPARγ was a groundbreaking finding

Progress in Molecular Biology and Translational Science, Volume 121
ISSN 1877-1173
http://dx.doi.org/10.1016/B978-0-12-800101-1.00007-7

217

that sparked notable pharmaceutical interest in PPARs as potential drug targets for treatment of the metabolic syndrome. Limiting side effects associated with clinical use of TZDs have emerged in recent years. New and novel PPAR drugs with broad safety margins and therapeutic potentials for the metabolic syndrome are in development. These include partial, dual, or pan PPAR agonists; PPAR antagonists; and selective PPAR modulators. The objective of this chapter is to highlight the therapeutic benefits of targeting more than one PPAR subtype in the treatment of the metabolic syndrome. The pros and cons observed during clinical use of TZDs and the strategies and progress made in the production of new generations of safe and effective PPAR ligands are discussed.

1. INTRODUCTION

The propensity for overconsumption of "unhealthy" food high in calories from saturated fatty acids (FAs) and sugars increases the risk of obesity and metabolic syndrome. The World Health Organization (WHO) has ranked obesity as one of the top global health problems in Western societies. In 2010, more than one-third of adults (35.7%) and almost 17% of children and adolescents in the United States were either overweight or obese.[1,2] Obesity results from "caloric imbalances" between energy intake and energy expenditure for the amount of calories consumed. Excess adiposity is influenced by several factors including genetics, sedentary lifestyle, and environmental factors such as smoking. A "healthy weight" is deduced from calculating the body mass index (BMI), which is measured by dividing an adult's body weight in kilograms by the height squared in meters (kg/m^2). A normal adult BMI generally ranges from 18.5 to 24.9. BMI numbers above this range are considered a reflection of being either overweight (25–29.5) or obese (30 and above).

The metabolic syndrome is defined as a constellation of health disorders that include increased blood pressure, hyperglycemia (increased blood glucose associated with type II diabetes), and excess visceral body fat accompanied with abnormal blood lipid levels (dyslipidemia).[3] In men, the metabolic syndrome is defined as the coexistence of at least three of the following: (1) waist circumference over 102 cm, (2) serum triglyceride level over 150 mg/dl, (3) serum high-density lipoprotein (HDL) cholesterol (HDL-c) level below 40 mg/dl, (4) systolic blood pressure over 130 mmHg and/or diastolic blood pressure over 85 mmHg, and (5) fasting plasma glucose level over 110 mg/dl.[4]

Basic understanding of lipids and carbohydrates metabolism is an essential prelude to discussion of the critical roles of PPARs in the metabolic syndrome and their exploitation as targets for therapeutic intervention. Lipids from the diet are composed of 98–99% triglycerides (TGs) with the remainder consisting of cholesterol, phospholipids, mono- and diglycerides, fat-soluble vitamins, terpenes, and steroids.[5] Energy overload in the form of excess TGs and excessive consumption of carbohydrates (over 60% of the total energy intake) are risk factors for the development of dyslipidemia. Dyslipidemia is considered a major predisposing factor for type II diabetes mellitus (T2DM) and/or for the development and progression of atherogenic dyslipidemia (atherosclerotic plaque buildup in the arteries) that may lead to coronary heart disease (CHD). Generally, dyslipidemia is classified into at least five types (depending on elevated lipid particle) characterized by abnormal concentrations of one or more lipids (hypercholesterolemia and/or TGs), apolipoproteins (i.e., major apolipoproteins A, B, C, and E with additional subtypes), or lipoproteins. Lipoproteins are complex transport vehicles for FAs, cholesterol (LDL and HDL cholesterol), and their esters.[6] The lipoproteins are classified according to their density (relative amounts of lipids to proteins) into chylomicrons, low-density lipoproteins (LDLs), very-low-density lipoproteins (VLDLs), and HDLs. The protein components of lipoproteins are called apolipoproteins. The major apolipoproteins and their subtypes (e.g., A-I and A-II; B-26, B-48, B-74, and B-100; and C-I, C-II, and C-III, among others) are linked by disulfide bonds to specific lipoproteins (e.g., B-100 with LDL and apo-A-I and apo-A-II with HDL). As such, they function as lipid transport vehicles, enzymatic cofactors (such as C-II for lipid hydrolysis by lipoprotein lipase (LPL) enzyme and A-I for lecithin cholesterol acyltransferase enzyme), and more importantly as ligands for lipoprotein receptor binding (e.g., B-100 and apo-E for LDL receptor, apo-E for the remnant receptor, and apo-A-I for HDL receptor).[7]

Measurements of total cholesterol, LDL cholesterol (LDL-c), HDL-c, and TGs are recommended by the National Cholesterol Education Program (NCEP) as important risk factors for the development of CHD.[4] High total TGs, high LDL-c, and high total cholesterol in association with low HDL-c are typical findings in patients with the metabolic syndrome. According to NCEP guidelines, the total cholesterol of less than 200 mg/dl is desirable, while levels between 200 and 239 mg/dl are at borderline risk for CHD. Levels greater than 240 mg/dl are considered high risk. LDL carries the bulk of cholesterol particles; LDL levels less than 100 mg/dl are considered optimal, levels 130–159 mg/dl are borderline high, and levels between 160 and 189 mg/dl are considered high. For HDL-c, levels less than 40 mg/dl are considered low while levels greater than or equal to 60 mg/dl are considered high. Total serum TG levels less than 200 mg/dl are considered normal, ranges between 150 and 199 mg/dl are borderline high, levels 200–400 mg/dl are high, and values greater than or equal to 500 mg/dl are considered very high.

Lipid and glucose metabolism are intimately linked together. Diabetes is defined as a group of metabolic diseases characterized by hyperglycemia (high blood glucose) resulting from the lack of insulin (type 1) or defects in insulin action, secretion, or both (type 2).[8] According to the American Diabetes Association's 2011 National Diabetes Fact Sheet, 25.8 million children and adults in the United States (8.3% of the population) have diabetes.[9] Globally, the total number of people with diabetes is projected to rise from 171 million in 2000 to 366 million in 2030.[10] Adult T2DM is closely associated with obesity, cardiovascular disease (CVD), and dyslipidemia with inflammation in white adipose tissue as a common denominator[11] (Fig. 7.1). Increased mortality in T2DM is strongly associated with microvascular (nephropathy, retinopathy, and neuropathy) and macrovascular (coronary

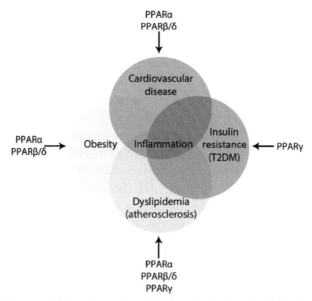

Figure 7.1 The metabolic syndrome is a collection of obesity-related disorders. The lipid-sensing PPARs (α, β(δ), and γ) are critical therapeutic targets for treatment of inflammation in fat[11] and for treatment of overlapping maladies of the metabolic disorder. Arrows indicate beneficial effects from activation of PPAR subtypes.[12] (See color plate.)

artery, cerebrovascular, and peripheral) complications.[13] The mechanism responsible for cardiovascular complications is not fully understood. The formation of advanced glycosylated products from long-term hyperglycemia in diabetic patients was postulated as a plausible hypothesis for the development of atherosclerosis.[14]

Six classes of antidiabetic drugs are currently used for T2DM management that include (1) sulfonylureas (e.g., glimepiride (Amaryl)), (2) biguanides (e.g., metformin), (3) meglitinides (e.g., nateglinide), (4) dipeptidyl (e.g., DPP-4) inhibitors (e.g., alogliptin), (5) alpha-glucosidase inhibitors (e.g., acarbose), and (6) thiazolidinediones (TZDs, pioglitazone, and rosiglitazone). With the exception of alpha-glucosidase inhibitors, which inhibit the digestion of starches, all of these agents lower blood glucose concentrations in part by either stimulating insulin secretion (meglitinides and sulfonylureas) and enhancing glucose-dependent insulin secretion (DPP-4 inhibitors) or increasing insulin sensitivity (biguanides and TZDs). Other drugs include glucagon-like peptide-1 (GLP-1) agonists (exenatide) and amylin analog (pramlintide). Given the lack of impact on macrovascular endpoints by intensive reduction of hyperglycemia alone, increased attention has shifted towards simultaneous reduction of both lipids and blood glucose using PPAR agonists.[15]

Nuclear receptor peroxisome proliferator-activated receptors (PPARs) are involved in the regulation of lipid and glucose metabolism and atherogenic macrovascular complications associated with dyslipidemia (fatty streaks that eventually develop into fibrous plaques).[15] Drugs that bind PPARs, fibrates for PPARα (Table 7.4) and TZDs for PPARγ (Table 7.6), have direct positive impact on the components of the metabolic syndrome.[12,15,29–32] TZD agents with beneficial impact on lipid profiles, such as pioglitazone or the newly developed aleglitazar glitazar (a dual PPARα/PPARγ agonist), are effective in reducing hyperglycemia and TGs while simultaneously increasing HDL-c or the so-called good cholesterol.[15]

Two TZDs (rosiglitazone and pioglitazone) are clinically approved as oral hypoglycemic drugs for treatment of T2DM. Both rosiglitazone and pioglitazone are PPARγ full agonists. In human clinical trials and meta-analysis, pioglitazone increased HDL-c and reduced several parameters associated with T2DM including blood glucose, macrovascular complications, and TGs.[33,34] Relatively, pioglitazone has fewer side effects compared with rosiglitazone.[35]

Because of the side effects of TZDs, such as skeletal muscle myopathy by fibrates[36] and weight gain and cardiovascular risks by rosiglitazone,[37,38]

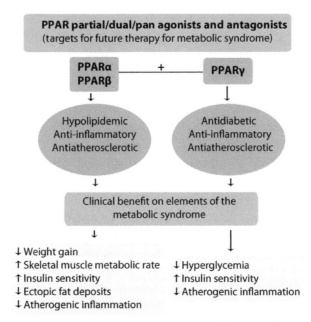

Figure 7.2 Therapeutic potential of PPAR subtypes. Development of partial, dual, or pan PPAR agonists or antagonists is an area of active research that holds hope for the development of single compounds that activate more than one PPAR subtype resulting in broad therapeutic benefits without the side effects. (See color plate.)

development of dual antilipidemic/antiglycemic drugs (e.g., glitazars) is underway (Fig. 7.2). Given the favorable effects of fibrates on lipid profile and that of TZDs on hyperglycemia, a single compound that delivers both effects (good glycemic control and improved lipid profile) would be ideal for treatment of the metabolic syndrome. Among other strategies, novel dual PPAR agonists that activate more than one PPAR subtype are sought as one way to produce a "balanced" therapeutic response with increased efficacy and minimal side effects. For example, aleglitazar (a dual PPARα/γ agonist) captures the ability of PPARα to lower lipids and that of PPARγ to increase insulin sensitivity without the side effects seen with TZD monotherapy.[34,39–42]

PPARs are involved in multiple functions that include regulation of tissue metabolism, cell differentiation (proliferation/apoptosis), and host immunity.[19,43] The name (peroxisome proliferator-activated receptor) was inferred from the fact that fibrate antilipidemic drugs (PPARα agonists) and other chemicals cause proliferation of cytoplasmic cell organelles known as peroxisomes in the liver of rodents.[44,45] PPARs are composed of three members, encoded by separate genes, that include PPARα (NR1C1), PPARβ (also known as δ, NR1C2, or NUC-1), and PPARγ (NR1C3) isotypes.[46] They are nonsteroidal receptors that form part of the class I family in nuclear hormone receptor superfamily. Other members of this family include retinoic acid, thyroid hormone, and vitamin D receptors. In contrast to subfamily III nuclear receptors (e.g., estrogen and androgen receptors), PPARs are located in the cell nucleus and bound by corepressor proteins in the unliganded state. When bound to their specific ligands, each PPAR functions as a transcriptional factor that heterodimerizes with retinoid X receptor α (RXRα).[47] The RXR is independently activated by 9-*cis*-retinoic acid. Subsequent to ligand binding and recruitment of coactivators, the heterodimers bind to specific DNA sequences called peroxisome proliferator response elements (PPREs) located in the 5'-end flanking regions of target genes. Genes activated by PPARs include enzymes involved in the β-oxidation of long-chain FAs, such as acyl-CoA oxidase (ACOX), peroxisomal enoyl-CoA (hydratase-3-hydroxyacyl-CoA dehydrogenase bifunctional enzyme), and 3-ketoacyl-CoA thiolase, and genes of the cytochrome P450 4A family involved in ω-oxidation systems.[48] In addition to lipid homeostasis, PPARs are involved in a broad range of functions that include glucose homeostasis, inflammation, tissue remodeling, angiogenesis, and prostaglandin production.[49]

2. TISSUE DISTRIBUTION OF PPARs SUGGESTS KEY ROLE IN LIPID AND GLUCOSE METABOLISM

In humans and rodents, PPAR isoforms exist in different tissues; nevertheless, each PPAR subtype exhibits a discrete tissue pattern.[19,50–52] For instance, PPARα is highly expressed in cells with high catabolic rates and cells that perform β-oxidation of FAs in tissues such as the liver,[53] heart,[54,55] skeletal muscles, brown fat, and kidneys.[19] PPARα regulates FA oxidation and is the primary target for drugs that lower blood TGs such as fibrates.[56]

PPARγ is highly expressed in the white adipose tissue,[57–59] gastrointestinal tract[60], and macrophages.[61] Two splice isoforms of PPARγ, PPARγ1 and PPARγ2, have been identified. In addition, two mRNA splice variants, PPARγ3 and PPARγ4, produce PPARγ1. Of the three known PPARs, PPARα and PPARγ are the most widely studied. PPARγ is essential for adipocyte differentiation[62–64] and lipid accumulation[65] while also activating several genes involved in lipid uptake and lipid metabolism.[20] The link between PPARγ, adipocyte function, obesity, and T2DM stems from the

1995 discovery that PPARγ is the bona fide receptor for the antidiabetic drugs TZDs.[65–67] Consistent with this finding, genetic mutations (gain or loss of function) in PPARγ, though rare, are either associated with obesity,[68] lipodystrophy, or insulin sensitivity.[69–71]

In contrast to PPARα and PPARγ, PPARβ(δ) exhibits a ubiquitous expression and is abundant in most tissues, including metabolically active tissues, with the highest expression in the gastrointestinal tract, kidneys, and heart.[19,51] PPARβ(δ) has received less scientific attention, but recent studies have shown promising broad therapeutic potentials for treatment of the metabolic syndrome with PPARβ(δ).[72] PPARβ(δ) regulates genes involved in FA uptake, β-oxidation, and energy uncoupling, all of which result in protection against lipotoxicity, ectopic fat buildup, and the improvement of insulin sensitivity.[31] Although animal studies have demonstrated the beneficial roles of PPARβ(δ) in the metabolic syndrome, no pharmaceutical drugs that target this receptor are currently on the market.[31]

3. PROTOTYPIC NUCLEAR RECEPTOR DOMAINS OF PPARs

PPARs have a quasistructural and evolutionary kinship. Like other nuclear receptors, PPARs retain the typical or classical hormone receptor modular structure with six functional domains named A/B, C, D, and E/F from the N- to the C-terminus, respectively[47,73] (Fig. 7.3). These domains function as ligand-independent activation factor-1 (AF-1) site in the variable A/B N-terminal region, DNA-binding domain (DBD, C region), nuclear localization hinge (D region), ligand-binding domain (LBD, E region), and ligand-independent AF-2 (F, C-terminal region). AF-1 in the A/B region contributes to the constitutive ligand-independent activation and is the site for receptor phosphorylation by kinases and for the binding of some coactivators. The most highly conserved DBD binds to cognate PPRE DNA elements located in the promoter regions of target genes. Typically, the PPREs recognized by the obligate heterodimers of PPAR and RXR are direct repeats of hexanucleotides separated by one spacer nucleotide (n) in-between the two half sites (DR1: **AGGTCAnAGGTCA**). Located upstream of the first half site is a short sequence (AAACT) that confers polarity on the PPREs. In the PPAR–RXR heterodimer, the PPAR moiety binds the 5′ to the RXR half site. Like other nuclear receptors, two zinc atoms are present in the LBD; each zinc atom binds to four cysteine residues forming what are called zinc

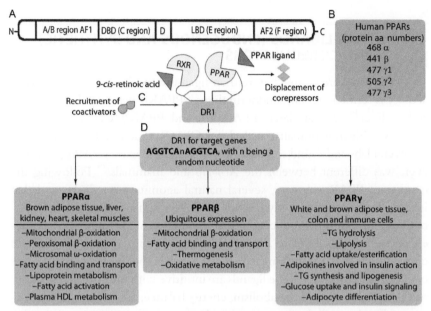

Figure 7.3 (A) Modular domains of a PPAR. (B) Protein size (in amino acids (aa)) for human PPARs. The PPAR proteins γ2 contain 28 more aa (30 aa in rodents) than γ1, while γ1 and γ3 are identical in size. (C) Transactivation of PPAR by obligate heterodimerizing with RXRα and specific ligand binding, leading to preferential recruitment of coactivators (decondensers of chromatin) and dismissal of corepressors. (D) PPRE (DR1) for PPAR target genes, tissue distribution, and metabolic functions for each PPAR subtype. (See color plate.)

fingers. These fingers contain two sets of amino acids known as the P and D boxes responsible for PPRE binding and dimerization, respectively. The hinge region contains the short carboxy (C)-terminal extensions and is responsible for nuclear localization. The hinge region of PPAR forms extensive interactions with the upstream AAACT element.[47] The LBD has an extensive secondary structure of multiple alpha helices and a beta sheet. The binding pocket contains the ligand–dependent AF-2 site responsible for receptor transactivation by agonists and coactivator proteins. The PPARγ protein has a relatively large-sized ligand-binding pocket, but the amino acid sequence of the LBD for the three main PPAR subtypes is closely similar. For example, PPARα exhibits 68% homology with PPARγ and 70% with PPARβ(δ).[74] Isoform-specific gene targets are generally regulated by binding of selective coactivators and subtle cis-sequence differences that flank the core PPREs.

4. PPAR LIGANDS (ENDOGENOUS AND SYNTHETIC) AND COREGULATORS

PPARs were cloned before their ligands were discovered; such work activity is termed "reverse endocrinology." PPARα was first cloned from a rodent liver[45]; 2 years later, PPARβ(δ) and PPARγ were identified in *Xenopus*.[46] Next, mammalian orthologs of PPARα and PPARγ were cloned by several labs and found to be conserved among species. PPARβ(δ), however, was different between the *Xenopus* and mammals.[75] Following the cloning of PPAR subtypes, several natural agonists were discovered that include free mono- or polyunsaturated FAs, oxidized lipids, and eicosanoids.[76,77] Micromolar concentrations (indication of low affinity) of a variety of essential FAs (linoleic, linolenic, and arachidonic acids) and FA analogs derived from diet metabolism activate PPAR subtypes with varying degrees of selectivity. Although these ligands are intuitive with the critical regulatory roles of PPARs in lipid metabolism, energy balance, and inflammatory processes, their low cellular concentrations leave room for more potent agonists to be discovered. The exact role of natural ligands on the biology of the metabolic syndrome is yet to be elucidated.

Eicosanoids (prostaglandins and leukotrienes) are lipid mediators derived from phospholipase-released arachidonic acid metabolism.[78] Eicosanoids, 8 (S)-hydroxyeicosatetraenoic acid (8-HETE) and 15-deoxy-D12,14-prostaglandin J2 (15d-PGJ$_2$), function as selective ligands for PPARα[79,80] and PPARγ,[81,82] respectively.[77] Other eicosanoids include leukotriene B4 (LKB4) for PPARα[83] and prostaglandin I$_2$ or prostacyclin I$_2$ (PGI$_2$) for PPARβ(δ).[84] An unidentified ligand, produced during early stages of preadipocyte differentiation in 3T3-L1 preadipocyte cells, is implicated as PPARγ adipogenic ligand.[85]

Long-chain FAs, especially polyunsaturated FAs, are more selective for PPARα[86] followed by PPARβ(δ), with the lowest selectivity for PPARγ.[87] Although some of the natural PPAR ligands show some form of subtype specificity, the vast majority are promiscuous and activate more than one PPAR subtype. Likewise, each PPAR subtype can be activated by more than one ligand.

Synthetic ligands include fibrates for PPARα (Table 7.4)[76,83] and glitazones or TZDs for PPARγ (Table 7.6).[67] Examples of fibrates include gemfibrozil, bezafibrate, and ciprofibrate, which are currently used as antilipidemic drugs and for the protection of the heart and kidney from ischemia/reperfusion injury.[96] The PPARγ activation by TZDs causes insulin sensitization via alterations in the transcription of genes involved in glucose and FA metabolism (Fig. 7.3). The currently marketed TZDs include pioglitazone (brand name Actos, Takeda Pharmaceuticals, and Eli Lilly and Company) and rosiglitazone (brand name Avandia, GlaxoSmithKline). There is no commercially available drug for PPARβ(δ). A new series of promising novel selective partial PPARβ(δ) agonists, phenoxyacetic and anthranilic (GW9371) acids, are currently being developed.[97,98]

Translocation and protection of PPAR hydrophobic ligands during their passage from extracellular fluid to intracellular sites are mediated by binding to intracellular lipid-binding proteins (iLBPs), which are members of FA-binding proteins (FABPs). The complex formed by the ligand and the iLBPs then translocates to the nucleus and binds to different PPAR isotypes for a short period of time.[99–101] PPARs without ligand bound are associated with protein corepressors when located in the cell nucleus; upon ligand binding, PPARs change conformation and bind to RXR. Depending on the cell type and availability, 9-cis-retinoid acid then binds the RXR. The PPAR–RXR complex is considered a permissive heterodimer. It can thus be activated by either PPAR or its partner receptor and the presence of both ligands adds synergism to transcription.

Like other nuclear receptors, interaction of PPAR–RXR heterodimers with the basal transcriptional machinery and a variety of proteins, known as cofactors or coregulators made up of coactivators and corepressors,[102,103] is required for modulation of target genes. The newly assembled or preassembled coactivators facilitate dismissal of corepressors, heterodimerization of PPAR–RXR, chromatin decompaction (by intrinsic histone acetyltransferase (HAT) or methyltransferase activities), and recruitment of the basal transcriptional machinery to the promoter region of target genes in a tissue-/cell-specific manner. Several families of PPAR coactivators and corepressors are known, with most families common to other nuclear receptors.[103] The two most conserved families include p160/SRC family that possesses HAT activity (with molecular mass of 160 kDa) and cyclic AMP-responsive element-binding protein (CREB) and its homologue p300 (CBP/p300, a cointegrator with HAT activity). The p160 family has three distinct related members that include steroid receptor coactivator-1 (SRC-1 or NCoA-1), SRC-2 (TIF2 or GRIP1), and SRC-3 (pCIP, TRAM-1, or AIB1).[104] In rodents, some of these coactivators have different names; for example, SRC-1 is termed NCoA-2 in mice.

The second important family of PPAR coactivators has no known enzymatic activity. It is a multiprotein complex, termed the TRAP/DRIP/ARC/mediator complex, consisting of 15–30 proteins involved in the interaction of RNA polymerase II holoenzyme of the basal transcriptional machinery.[103] In the screening for novel antidiabetic selective PPARγ modulators, a ligand-dependent transcriptional activity that specifically recruited DRIP205/TRAP220 coactivator complex promoted undesirable adipogenic effects while recruitment of SRC-1 coactivator enhanced insulin sensitivity.[105] The ability of coactivators to differentially modulate PPAR function is supported by genetic studies. For example, DRIP205/TRAP220-deficient embryonic fibroblasts lack the ability to induce adipogenesis, while TIF2 knockout mice are resistant to diet-induced obesity and are more insulin-sensitive.[106,107] Further, SRC-1-deficient mice have reduced energy expenditure and are prone to obesity. As such, this differential cofactor recruitment is successfully exploited to predict the adipogenic and antidiabetic abilities of compounds screened in cell- and non-cell-based assays.[108]

A specific coactivator, PPARγ coactivator-1 (PGC-1), was demonstrated to interact with PPARγ[109] and PPARα.[110] PGC-1 plays an important role in adaptive thermogenesis, energy homeostasis in brown fat and skeletal muscles, and increased hyperglycemia and insulin resistance in T2DM.[111,112]

Coactivators possess conserved LXXLL motifs (L is leucine and X is any amino acid) that interact with the coactivator-binding groove in the AF-2 domain. Different combinations of LXXLL motifs are required for PPAR activation (e.g., in SRC-1, four consecutive LXXLL motifs make identical contacts with both subunits of a PPAR–RXR heterodimer[113]). In contrast to SRC-1, fragments of both p300 and CBP that interact with PPARα contain one LXXLL motif.[114] The conserved glutamate and lysine residues in the LBD form a "charge clamp" with the backbone atoms of LXXLL helices.[115]

In the absence of ligand, the PPAR–RXR heterodimer remains in a repressed state by the nuclear receptor corepressor (NCoR) and the silencing mediator of retinoid and thyroid hormone receptor (SMRT).[116] Both NCoR and SMRT interact with the Sin3 complex (global regulator of transcription) to form a multisubunit repressor complex.[117] SMRT facilitates the recruitment of histone deacetylases (HDACs) to the DNA promoters bound by specific interacting transcription factors. Another corepressor receptor-interacting protein 140 (RIP140), also known as

NRIP1 (nuclear receptor-interacting protein 1), recruits HDACs and represses the activity of PPARs.[118,119]

PPAR coactivators are used in *in vitro* transactivation assays using cell-based and/or cell-free assays to screen for PPARγ modulators.[30] Knowledge of selective PPAR/ligand/coactivator activity in the context of target tissues is a critical first step for the successful development of new drugs for the treatment of the metabolic syndrome.[120–122]

5. KEY TARGET GENES OF PPARs RELATED TO THE METABOLIC SYNDROME

Fat and glucose metabolisms are inextricably linked together by a complex transcriptional network.[123] Elevated TGs in nonadipose tissue such as the liver and skeletal muscles induce insulin resistance by what is called "lipotoxicity effect".[124] Although the exact mechanism by which fat induces insulin resistance is unknown, inhibitions of genes related to insulin signaling (e.g., insulin receptor substrates, IRS-1 in the muscle and IRS-2 in the liver, among others) are suggested as possible causes.[125,126] In rodents and humans, PPARs are genetic sensors for fat and modulate genes that play pivotal roles in regulation of lipid homoeostasis, including promotion of reverse choles-terol transport, reduction of TGs, and regulation of apolipoproteins, ther-mogenesis, and glucose metabolism in various metabolic tissues such as the liver, heart, fat, and skeletal muscles. As such, PPARs are considered important targets for treatment of the metabolic syndrome and as the cho-reographers of metabolic gene transcription.[12,20,29,72,127–129]

PPARα has a crucial role in FA oxidation[48] and is highly upregulated during fasting when FAs are mobilized from the adipose tissue to the liver for energy production.[130] This role is consistent with genetic animal studies where fasting PPARα-null mice are hyperlipidemic, hypoglycemic, and hypoketonemic with fatty liver.[130]

Drugs that activate PPARα (fibrates) lower serum TGs and increase HDL-c in patients with metabolic syndrome via modulation of genes involved in peroxisomal and mitochondrial FA oxidation, microsomal FA hydroxylation, and production of apolipoproteins (components of HDL-c) in tissues that obtain their energy preferentially from fat[16,131] (Fig. 7.4). These processes involve a considerable number of genes in the liver, heart, kidney, and skeletal muscles and monocytes/macrophages[16] (Table 7.1). Hepatic genes regulated by PPARα positively promote FA activation into acyl-CoA derivatives, FA binding and FA transport (across cell, peroxisomal,

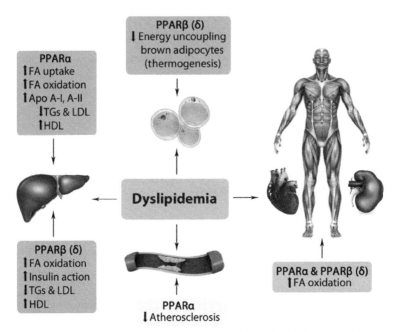

Figure 7.4 PPARα and PPARβ(δ) improve lipid profile in dyslipidemia with secondary antiatherogenic effect.[132] PPARα[133] and PPARβ(δ)[134] agonists increase expression of the reverse cholesterol transporter ATP-binding cassette A1 (ABCA1) and induce apo-lipoprotein A-I (a transporter that controls apo-A-I-mediated cholesterol efflux from macrophages). (See color plate.)

and mitochondrial membranes), β-oxidation in cell mitochondria (mostly for long-chain FAs) or cell peroxisomes (mostly for very long-chain FAs), lipoprotein metabolism, and mitochondrial ω-oxidation. The end result is increased FA uptake by the liver, FA oxidation, and reduced free FA in the blood. Fibrates also mediate TG lipolysis by upregulation of LPL and inhibition of apolipoprotein (apo)-C-III (LPL inhibitor).[135] Depending on the genes involved, differences in regulation between rodent and humans are noted.[18]

Similar to PPARα, PPARβ(δ) regulates gene targets in the adipose tissue (white and brown), skeletal and cardiac muscles, and macrophages.[72] Adipose differentiation-related protein, a VLDL protein that transports lipid droplets and cholesterol, was identified as a PPARβ(δ) target that regulates gene expression in atherosclerotic lesions.[136] PPARβ(δ) knockout mice are meta-bolically less active and glucose-intolerant, whereas PPARβ(δ) activation in *db/db* (diabetic and leptin receptor-deficient) mice improves insulin

Table 7.1 PPARα-regulated genes in the liver[16,17]

Function	Gene symbol	Gene name	Function
Lipoprotein metabolism			
	apo-A-l	Apolipoprotein-A-I	Plasma HDL metabolism
	apo-A-II	Apolipoprotein-A-II	Plasma HDL metabolism
	apo-C-III	Apolipoprotein-C-III	Plasma HDL metabolism
	apo-AV	Apolipoprotein-AV	Plasma TG metabolism
	PLTP	Phospholipid transfer protein	Plasma HDL metabolism
Mitochondrial β-oxidation			
	CPT-1	Carnitine palmitoyltransferase-1	Mitochondrial fatty acid β-oxidation
	CPT-2	Carnitine palmitoyltransferase-2	Mitochondrial fatty acid β-oxidation
	MCAD	Medium-chain acyl-CoA dehydrogenase	Mitochondrial fatty acid β-oxidation
Mitochondrial ω-oxidation			
	CYP450 4A1	Cytochrome P450 4A1	Mitochondrial fatty acid ω-hydroxylation
	CYP450 4A6-Z	Cytochrome P450 4A6-Z	Mitochondrial fatty acid ω-hydroxylation
Peroxisomal β-oxidation			
	ACO	Acyl-CoA oxidase	Peroxisomal β-oxidation
	Bien	Bifunctional enzyme	Fatty acid β-oxidation
	Thiolase B	Thiolase B	Fatty acid oxidation
	SCP-X	Sterol carrier protein-X	Fatty acid oxidation
Fatty acid-binding and transport			
	ACBP	Acyl-CoA-binding protein	Fatty acyl-CoA ester transport
	L-FABP	Cytosolic fatty acid-binding protein	Fatty acid-binding transport
	FATP-1	Fatty acid transport protein-1	Fatty acid transport
Fatty acid activation			
	LCAS	Long-chain acyl-CoA synthetase	Fatty acid activation

PPARα plays an important role in metabolic response to fasting and is the selective receptor for fibrates used as second choice after statins for treatment of dyslipidemia. PPARα mediates oxidation of fatty acids in the liver by activating the expression of genes involved in multiple pathways. Likewise, it mediates xenobiotic-induced peroxisome proliferation in rodent liver but not in humans.[18]

sensitivity.[137] In brown fat, PPARβ(δ) increases expression of hormone-sensitive lipase (HSL, involved in TG hydrolysis), long-chain acyl-CoA dehydrogenase (LCAD), long-chain acyl-CoA synthetase (LCAS) and very long-chain acyl-CoA oxidase (VLCAS), carnitine palmitoyltransferase-1 (CPT-1, involved in FAs oxidation), and uncoupling proteins 1 and 3 (UCP1 and UCP3, involved in uncoupling of oxidative phosphorylation during the process of thermogenesis). In the white adipose tissue, PPARβ(δ) induces expression of genes required for FA oxidation and energy dissipation, which in turn leads to improved lipid profiles and reduced adiposity.[138] More significantly, PPARβ(δ) senses and catabolizes VLDL particles in macrophages (major contributor to the pathogenesis of atherosclerosis).[132]

These PPARβ(δ)-regulated genes are consistent with the beneficial role for PPARβ(δ) in the treatment of obesity and amelioration of the metabolic syndrome.[138] In the skeletal muscles, PPARβ(δ) regulated FA-binding protein (FABP) and FA transport protein (FATP/CD36) involved in FA transport and oxidation.[139,140] PPARβ(δ) also increased the numbers of type I muscle fibers (associated with increased endurance),[141] succinate dehydrogenase and citrate synthase (oxidative metabolism), cytochrome c and cytochrome oxidase IV (involved in mitochondrial respiration),[139] and UCP2 and UCP3 (involved in thermogenesis)[139] (Table 7.2).

PPARγ is the target for TZD–insulin sensitizers and regulates genes involved in adipocyte differentiation and fat metabolism.[143] The protein product of genes regulated by PPARγ was categorized by Savage[20] into four groups. The first group is composed of proteins involved in hydrolysis and/or resynthesis of TGs, FA uptake and esterification, and lipogenesis. The second group contains proteins that regulate lipolysis. Proteins in the third group regulate adipokines, while the fourth group consists of proteins implicated in glucose homeostasis (insulin signaling and glucose uptake) (Table 7.3). Examples of human PPARγ target genes with documented PPRE in the first group include LPL (mediates TG hydrolysis),[19] CD36, FA transport protein-1 (FATP-1), FA-binding protein 4 (FABP4 and/or aP2), phosphoenolpyruvate carboxykinase (PEPCK) (involved in lipogenesis and TG production destined for storage in the fat cells),[21,22] and acyl-CoA synthase (involved in conversion of nonesterified FAs into acyl-CoA and TGs).[19] Perilipin (involved in lipolysis) is an example of a protein in the second group.[23] Finally, CbI-associated protein,[24] insulin receptor substrate-2 (IRS-2),[25] and glucose transporter 4 (GLUT4)[26] represent genes involved in insulin signaling and glucose metabolism in the fourth group.

Table 7.2 PPARβ(δ) gene targets in the liver, fat, and skeletal muscles[31,72,142]

Function	Gene	Tissue
Lipogenesis and increase glucose consumption	Pentose phosphate pathway, *FAS*, *ACC2*	Liver
Fatty acid transport and oxidation	*LCAS* (long-chain acyl-CoA synthetase) *VLCAS*, *ACOX1*, *m-CPT-1*, *LCAD*, *VLCAD*, *HSL*	Brown fat
	Heart fatty acid-binding protein, fatty acid transport protein/CD36, *m-CPT-1*, *PDK4*, *HMGCS2*, thiolase, *LCAD*	Skeletal muscle
	CPT-1, *m-CTP1*, thiolase, *UCP3*, *LCAD*, *MCAD* (medium-chain acyl-CoA dehydrogenase), *PDK4*, *VLCAD*, *ACOX1*	Cardiac muscle
Thermogenesis	*UCP1*, *UCP3*	Brown fat
	UCP1	White fat
Oxidative metabolism	Succinate dehydrogenase, citrate synthase	Skeletal muscles
Mitochondrial respiration and thermogenesis	Cytochrome c, cytochrome oxidase II, cytochrome oxidase IV, *UCP2*, *UCP3*	Skeletal muscles
Slow-twitch fibers	Myoglobin, troponin I slow	Skeletal muscles

In the liver, PPARβ(δ) increases lipogenesis and reduces glucose output. In adipocytes, it increases fatty acid oxidation and energy expenditure and reduces adiposity. Likewise, it increases energy expenditure, reduces adiposity, and increases oxidative metabolism in the skeletal muscles. In addition, PPARβ(δ) controls switch in the type of muscle fibers. In macrophages, it controls metabolism of VLDL particles and suppresses inflammation.

In addition to genes involved in buildup of lipids and glucose metabolism, PPARγ is also induced during adipocyte differentiation and regulates multiple cascades and arrays of genes that force conversion of preadipocytes into mature fat cells.[144–146] In adipogenesis, a number of signaling pathways with multiple genes (Wnt pathways, C/EBPβ(δ), ADD1/SREBP1, EBFs, and Pref-1, among others) were identified as acting upstream of PPARγ, with C/EBPα and PPARγ mediating the terminal adipocyte differentiation.[147,148] Indeed, one of the side effects of PPARγ activation is undesirable weight gain that results from the increased subcutaneous adipose tissue deposits by TZDs-induced PPARγ activation.[149]

Table 7.3 PPARγ target genes[19-28]

Function	Gene involved	Gene symbol
TGs hydrolysis	Lipoprotein lipase	*LPL*
Fatty acid uptake/ esterification	CD36, fatty acid transport protein 1 Fatty acid-binding protein 4 (aP2) acyl-CoA synthetase (ACS)	*CD36, FATP-1, FATP-4*
Lipogenesis and TG synthesis	Phosphoenolpyruvate carboxykinase	*PEPCK*
	Glycerol kinase	*GK*
Lipolysis regulation	Perilipin	*PLIN*
Adipokines	Adiponectin (adipocyte complement-related protein)	*Acrp30*
	Resistin	*ADSF*
Insulin signaling and glucose uptake	Cbl-associated protein	*CAP*
	Insulin receptor substrate 2	*IRS-2*
	Glucose transporter 4	*GLUT4*

PPARγ regulates fat cell formation and whole-body glucose hemostasis by regulation of genes involved in several pathways in the adipose tissue, liver, and skeletal muscles. PPARγ mediates the action of TZDs and other antidiabetic drugs (rosiglitazone and pioglitazone) and is target for selective PPARγ modulators, a novel class of TZD and non-TZD new antidiabetic drugs (see Table 7.11).

Homozygous deletion of PPARγ (−/−) is lethal, but using cre/loxP strategy for selective deletion of PPARγ in the liver, fat, and muscles, key genetic regulatory roles for PPARγ in glucose and lipid homeostasis were revealed.[150] Hyperlipidemia, elevated plasma glucose, and insulin with increased adiposity were reported in mice with liver-specific PPARγ knockout.[151] Likewise, deletion of PPARγ in fat results in hyperlipidemia and progressive loss of fat and fatty liver. Despite hyperphagia, adipose-specific PPARγ knockout mice exhibited diminished weight gain when fed a high-fat diet. In addition, these mice had diminished serum concentrations of both leptin and adiponectin.[152] TZD treatments of fat-PPARγ-deficient mice reversed liver insulin resistance but failed to lower plasma free FAs.[153] Although PPARγ is expressed at low level in the skeletal muscles, targeted disruption of PPARγ in the muscle causes insulin resistance and increased adiposity.[154,155] Collectively, these studies suggest that PPARγ regulates tissue–tissue cross talk in glucose and lipid homeostasis.

6. LESSONS LEARNED FROM CLINICAL USE OF TZD CLASS OF ANTIDIABETIC DRUGS

The PPARγ full agonists, rosiglitazone (Avandia, GlaxoSmithKline, Research Triangle, NC) and pioglitazone (Actos, Pharmaceuticals North America, IL), were marketed in 1999 as oral hypoglycemic drugs. The first drug in this class of insulin sensitizers, troglitazone (Rezulin), was approved in 1997 but was later withdrawn from the market due to idiosyncratic hepatic toxicity.

Both Avandia and Actos lower blood glucose and glycated hemoglobin A1c. In T2DM patients, 1–1.5% reduction of HbA1c is achieved when TZDs are used as monotherapy.[29] As a 2.5–4.5% HbA1c reduction is considered optimal, the therapeutic action of TZDs is boosted by combination with insulin, sulfonylurea insulin secretagogues, or metformin to achieve a higher percentage of HbA1c reduction. Metformin is a biguanide family antidiabetic drug that acutely decreases hepatic glucose production by inhibition of the mitochondrial respiratory chain complex and activation of the AMPK (AMP-activated protein kinase). As an added benefit, TZD PPARγ agonists were also reported to preserve β-cell function in preclinical studies.[156]

Mechanistically, activation of PPARγ by TZDs causes insulin sensitization in the skeletal muscles and the liver, production of new insulin–sensitive adipocytes, and mobilization of ectopic lipid away from the liver and skeletal muscles to the adipose tissue[143,157] (Fig. 7.5). Activation of PPARγ by TZDs also modulates the expression of several adipokines such as the induction of adiponectin (insulin sensitizer)[162,163] and inhibition of proinflammatory adipokines that include resistin, tumor necrosis factor alpha (TNFα), and interleukin 6 (IL6).[164] Likewise, PPARγ treatment suppresses the expression of the adipocyte 11β-HSD1 (an enzyme responsible for conversion of cortisone into active cortisol) and plasminogen activator inhibitor-1 (PAI-1). Both proteins are involved in the exacerbation of insulin resistance.[165,166] Collectively, these PPARγ-mediated actions lead to improvement of whole-body glucose metabolism (Fig. 7.5).

Despite early success of TZDs in the treatment of T2DM, their clinical use was dampened by the development of significant adverse effects.[38,167–169] These effects include pulmonary and macular edema in 10–15% of treated patients,[170] plasma volume expansion (shown by decreased hematocrit used as a marker for increased plasma volume),

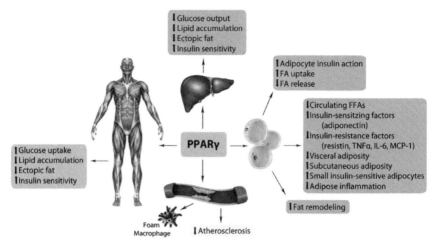

Figure 7.5 Activation of PPARγ by TZDs improves insulin sensitivity by modulating several processes in the white adipose tissue as a primary target for PPARγ action. Significant improvement in insulin sensitivity is achieved in the liver and skeletal muscles (predominant sites for insulin-mediated glucose uptake in the postprandial state). The antidiabetic mechanism of PPARγ involves regulation of fat metabolism, remodeling, and adipokine secretion. PPARγ activation may have an antiatherogenic effect independent of insulin sensitivity[158] but this role is not shown by TZDs.[159–161] Other PPARγ activators induce cholesterol removal from human macrophage foam cells through stimulation of the ABCA1 pathway (a transporter for apo-A-I-mediated cholesterol efflux from macrophages).[133] TNFα, tumor necrosis factor alpha; IL6, interleukin 6; MCP-1, monocyte chemotactic protein-1. (See color plate.)

decreased bone density (especially in women),[171] hemodilution, weight gain (2–3 kg for every percent decrease in HbA1c values),[29,172] and 40% increase in the risk of myocardial infarction among patients receiving the TZD rosiglitazone.[167] TZD monotherapy or TZD–insulin combination therapy is problematic in patients with congestive heart due to TZD-induced extra fluid volume.[172,173] Recent meta-analysis of rosiglitazone clinical data showed significant tendency towards an increase in macrovascular events in diabetic patients.[95,174] Because of these safety issues, the American Diabetes Associations has now advised against the use of TZDs in diabetic patients with advanced heart failure.[175]

The unfortunate side effects of TZDs prompted the FDA to implement new risk evaluation and mitigation strategies for rosiglitazone (Avandia) and its combinations (Avandamet (rosiglitazone plus metformin) and Avandaryl (rosiglitazone plus glimepiride))[172,176] (Table 7.3). The adverse event of edema is likely caused by increased peripheral vascular permeability and TZDs' induced expression of epithelial Na^+ channel gamma subunit, resulting in increased fluid retention by the kidneys.[177,178] TZD–PPARγ-induced weight gain is attributed to a combination of several factors that include generation of new adipocytes, increased edema, and increased feed intake as the result of PPARγ-induced reduction of adipose tissue-derived hormone leptin (involved in the regulation of energy balance and neuroendocrine function).[176]

According to the FDA, the nationally projected number of patients filling a rosiglitazone-containing product declined by 50% from 235,500 patients in January 2010 to approximately 119,000 patients in October 2010.[179] With these setbacks, new generations of TZDs and non-TZD PPAR ligands, with desirable efficacy and improved safety, are working their way through the different stages of development.[180] These new novel PPAR ligands can be categorized into (a) dual PPARα/γ; (b) selective PPARβ(δ); (c) dual PPARα/β(δ); (d) dual PPARβ(δ)/γ; (e) selective PPARγ modulators (SPPARγMs); (f) PPARα, β(δ), and γ pan agonists; and (g) PPAR antagonists (Tables 7.7–7.12). The strategy for the development of dual PPAR agonists is based on the premise of a "one size fits all" approach. As such, dual agonists are designed as single molecules with an intrinsic potency for more than one receptor subtype with broad spectrum metabolic effects. The goal is to combine the antilipidemic effects of PPARα and/or PPARβ(δ) with the antiglycemic and insulin-sensitizing properties of PPARγ. Added to the mix is the cancelation of the adipogenic and other side effects associated with the use of TZD PPARγ full agonists. For example, a dual PPARα/γ agonist is predicted to lower dyslipidemia and improve insulin sensitivity while simultaneously impeding or minimizing the side effects associated with PPARγ activation (Tables 7.4–7.6).

7. NEW, NOVEL PPAR LIGANDS FOR TREATMENT OF THE METABOLIC SYNDROME

The quest for new novel PPAR drugs is driven by two goals in mind: (1) the desire to treat all components of the metabolic syndrome that include hyperlipidemia, obesity, insulin resistance, and the CVD with a single drug and (2) to mitigate the documented side effects now evident from the clinical use of TZDs.[42,196] Highly potent lead compounds that are dual PPAR

Table 7.4 PPARα synthetic agonists (marketed, clinical, or preclinical) and natural agonists

Selective PPARα agonists	Status/indication/comments
Clofibrate	Marketed (generic)—fibric acid derivative. Lowers TGs and very low-density lipoprotein (VLDL). Several brand names and analogs (e.g., Atromid-S)
Fenofibrate	Marketed (generic)—fibric acid derivative. Lowers TGs and total cholesterol and increases HDL-c. Several brand names (e.g., Fenoglide, Lipofen, Lofibra, Antara, Fibricor, Tricor, Triglide, and Trilipix)
Bezafibrate	Marketed (generic)—fibric acid derivative. Lowers LDL-c and TGs and increases HDL-c. Brand names: Beza-XL, Bezalip, Fenolip, Globez
Gemfibrozil	Marketed (generic)—fibric acid derivatives. Lowers TGs and total cholesterol. Brand names: Lopid, Jezil, and Gen-Fibro
LY518674	Phase II (Eli Lilly and Company)[88]
GW590735	Phase II (GlaxoSmithKline)[89,90]
AVE8134	Phase II (Sano-Aventis)[91,92]
DRF-10945	Phase II (Dr. Reddy's Laboratories)
WY 14643 (Wyeth Pharmaceutical), GW 7647 and GW 9578 (GlaxoSmithKline), PLX974 (Plexxikon)	Preclinical stage—these agents improve insulin sensitivity and reduce adiposity[a]
Unsaturated and saturated (weak) fatty acids, LKB4, 8-HETE	Endogenous ligands

Marketed drugs (fibrates) are used for treatment of dyslipidemia.

[a]The U.S. Food and Drug Administration (FDA) now requires that human clinical trials longer than 6 months with PPAR agonists be preceded by a 2-year mouse carcinogenicity study to rule out tumor formation. Mechanistically, PPARα agonists reduce dyslipidemia by reduction of TGs (via reduction of VLDL), mediation of lipolysis, and stimulation of human HDL-c via increased production of apo-A-1 and apo-A-2 (major apolipoproteins associated with HDL-c receptor activation that confer protection against the development of atherosclerosis).[16,93,94]

Table 7.5 PPARβ(δ) endogenous[193] and synthetic agonists[194] (preclinical stage)

Selective PPARβ(δ)	Status/indication/comments
GW 501516 and GW 0742 (analog of GW 501516) (GlaxoSmithKline)	Phase II clinical trials—candidates for metabolic syndrome (reduce weight and TGs and induce reverse cholesterol transport with moderate improvement in glucose tolerance and insulin sensitivity in animal models).[134] Skeletal muscle myopathy and tumor formation as potential side effects.
L-165041 (Merck)	Preclinical—increased HDL-c in *db/db* mice[195]
PLX447	Preclinical
Prostacyclin, 13-s-hydroxy-octadecadienoic acid (13-HODE)	Endogenous ligands

Synthetic PPARβ(δ) agonists showed promising results in amelioration of obesity, dyslipidemia, and insulin resistance.[134,138,139]

agonists (α/γ, α/β(δ), and β(δ)/γ) (collectively known as glitazars), PPAR pan agonists (α, β(δ), and γ), selective PPAR modulators (SPPARMs), and PPAR antagonists are aggressively sought after by pharmaceutical companies with several promising drugs in development.[12,39,42,197–199]

7.1. Dual PPARα/γ

The role of PPARγ in increased insulin sensitivity is well established.[29,200] Likewise, PPARα activation corrects dyslipidemia and indirectly decreases diet-induced insulin resistance.[201] Thus, single compounds that combine the antilipidemic effects of PPARα with the antidiabetic effects of PPARγ should have broad therapeutic benefits in the clinical management of the metabolic syndrome (Fig. 7.2). Based on preclinical studies and early-phase clinical trials, the most challenging problem facing this endeavor is drug toxicity. Furthermore, deciphering the biological pathways that lead to balanced responses by two or three PPAR subtypes is arduous. However, substantial evidence suggests that the antilipidemic effects of PPARα and PPARβ(δ) are likely to offset the propensity of PPARγ activation to induce weight gain.[42]

To date, several TZDs (glitazars) and non-TZD drug candidates were described in this category with potential benefits in reversing

Table 7.6 TZDs list of PPARγ synthetic agonists (marketed, clinical, or preclinical) and natural endogenous ligands

Selective PPARγ agonists	Status/indication/comments
Pioglitazone (Actos, Pharmaceuticals North America)	Marketed—glitazone or TZD. Used as adjunct to diet and exercise to improve glycemic control in adults with T2DM (Actos, Pharmaceuticals North America, IL). Dual pioglitazone/glimepiride (Duetact) and pioglitazone/metformin (Actoplus Met), generic or extended release (Actoplus Met XR, Takeda), are indicated in diabetic adult patients who require both drugs for effective glycemic control
Rosiglitazone[a] (Avandia, GlaxoSmithKline)	Marketed—glitazone or TZD. Used as adjunct with diet and exercise to improve glycemic control in adults with T2DM (Avandia, GlaxoSmithKline). Dual rosiglitazone/glimepiride (Avandaryl) and rosiglitazone/metformin (Avandamet, GlaxoSmithKline) are indicated in diabetic adult patients who require both drugs for effective glycemic control
Balaglitazone	Phase III—Rheoscience/Dr. Reddy's Laboratories (India)
Rivoglitazone (CS-011)	Phase III—Daiichi Sankyo, Inc.
KRP-297 (Merck)	Phase I
GW 1929 (GlaxoSmithKline)	Preclinical
GW 7845	Preclinical
L-165041 (Eli Lilly and Company)	Preclinical
PLX589	Preclinical
RWJ-348260	Preclinical
RWJ-348260	Preclinical
Indone derivatives	Preclinical
Compound 14c (S26948)	Preclinical
Non-TZD PPARγ agonists	Preclinical

Table 7.6 TZDs list of PPARγ synthetic agonists (marketed, clinical, or preclinical) and natural endogenous ligands—cont'd

Selective PPARγ agonists	Status/indication/comments
Troglitazone, ciglitazone, and TT-501 (Japan Tobacco Inc.)	Discontinued
Ciglitazone	Experimental
Unsaturated fatty acids, 15-HETE, 9- and 13-HODE, and 15-D PGJ2	Endogenous ligands from diet and arachidonic acid metabolism

[a]Following the recent advent of the adverse increased cardiovascular risk associated with rosiglitazone,[95] the FDA requires that healthcare providers and patients to enroll in the Avandia-Rosiglitazone Medicines Access Program. Rosiglitazone or rosiglitazone combinations are no longer available through retail pharmacies. Prescriptions for rosiglitazone can only be obtained by mail order through certified pharmacies.

insulin resistance, hyperglycemia, and atherogenic dyslipidemia.[180,181,186] Drugs described in this category include the following: muraglitazar, tesaglitazar, ragaglitazar, chiglitazar, netoglitazone, BVT-142, O-arylmandelic acid derivatives, azaindole-α-alkyloxyphenylpropionic acid, amide substituted with α-substituted-β-phenylpropionic acid derivatives, 2-alkoxydihydrocinnamate derivatives, alpha-aryloxy-alpha-methylhydrocinnamic acids (LY51029), TZD18, MK0767, α-aryloxyphenyl acetic acid derivatives, tricyclic-α-alkyloxyphenyl propionic acids, TAK559, KRP-297, and compound 3q JTT-501 (Table 7.7). Based on positive preclinical data, several compounds have advanced into clinical trials. However, an equally large number of clinical trials on promising drug compounds (muraglitazar,[182] tesaglitazar,[183,202] ragaglitazar,[184,203] TAK559,[204] KRP 297,[205] naveglitazar,[206] and MK0767[207]) were discontinued due to adverse events.[208] Other drugs (aleglitazar,[179] chiglitazar,[209] and TZD18[210], among others) await further advancement (Table 7.7).

TZD18 suppressed hyperglycemia in rodent model of T2DM and lowered both cholesterol and TGs in rodents, hamsters, and dogs.[42] Likewise, alpha-aryloxy-alpha-methylhydrocinnamic acids (e.g., LY51029) improved insulin sensitivity and reversed diabetic hyperglycemia with significant improvement of whole-body lipid profile.[211] Aleglitazar (Hoffmann-La Roche) is one of the most promising drug candidates in this category. Phase I and II clinical trials showed that therapy with aleglitazar reduced hyperglycemia and significantly improved lipid profiles with less side effects compared with pioglitazone.[212] Aleglitazar is now in phase III clinical trials.

Table 7.7 Dual PPARα/γ single molecule agonists[39,181-185] currently in development to achieve safer and more effective therapy for the metabolic syndrome

Dual PPARα/γ	Status/indication/comments
Chiglitazar (CS038) (Hoffmann-La Roche)	Phase II clinical trial—type II diabetes and metabolic disorders
AVE0847 (Shenzhen Chipscreen Biosciences)	Phase II clinical trials
Aleglitazar (R1439) (GlaxoSmithKline)	Phase III clinical trials—reduces TGs and CVD and increases HDL-c
5-Substituted 2-benzoylamino-benzoic acid derivatives (BVT-142)	Preclinical—type II diabetes
O-Arylmandelic acid derivatives	Preclinical
Azaindole-α-alkyloxyphenylpropionic acid	Preclinical
Amide substituted with α-substituted-β-phenylpropionic acid derivatives	Preclinical
2-Alkoxydihydro cinnamate derivatives	Preclinical—type II diabetes and atherosclerosis
Alpha-aryloxy-alpha-methylhydrocinnamic acids (LY51029)	Preclinical—antidiabetic and antilipidemic
TZD18	Preclinical
α-Aryloxyphenyl acetic acid derivatives	Preclinical
Tricyclic-α-alkyloxyphenyl propionic acids	Preclinical—type II diabetes atherosclerosis
PLX429	Preclinical
Muraglitazar, tesaglitazar, naveglitazar, ragaglitazar, farglitazar, imiglitazar, netoglitazone, compound 3q JTT-501, MK0767, KRP-297, AZD 6610	Clinical trials discontinued for safety issues (edema, weight gain, cardiovascular risks, impairment of glomerular filtration, and carcinogenic effects in mice[186,187])

The aim is to combine the antilipidemic effects of PPARα with the antiglycemic effect of PPARγ for maximum therapeutic potentials without the side effects.

The hope for quick progress, however, is slowed by a recent FDA request for 2-year rodent carcinogenicity studies on any PPAR compound intended for clinical human trials. This requirement was based on preclinical data that suggested potential tumor formation by several of the newly developed dual PPAR agonists.[213]

7.2. Selective PPARβ(δ)

A limited number of compounds with selective PPARβ(δ) activity are in preclinical development for treatment of dyslipidemia (Table 7.8). Initial data suggest amelioration of Western diet-induced hepatic lipid accumulation and inflammation in LDL receptor-deficient (LDLR −/−) mice by L-165041 (Merck).[195]

A novel and selective PPARβ(δ) agonist, GW501516, induced reverse cholesterol transport and rectifies lipoprotein and TGs in obese rhesus monkeys.[134,188] GW501516 reached phase II clinical trials but was unfortunately discontinued in 2007 due to induction of intestinal adenoma in mice.[214]

Discovery of a novel class of PPARβ(δ) agonists was recently reported with both antilipidemic and antidiabetic effects. Among several partial PPARβ(δ) agonists, compounds 46 and 47 decreased body weight and significantly improved insulin sensitivity in insulin-resistant *ob/ob* mice when used at 30 and 100 mg/kg, respectively.[97] Likewise, two compounds (GSK1115 and GSK7227) were shown to induce partial activation of PPARβ(δ)-regulated target genes in human skeletal muscles.[98] The biology of GSK1115 and GSK7227, however, awaits further clarification.

Table 7.8 Selective PPARβ(δ) agonists

Selective PPARβ(δ)	Status/indication/comments
L-165041	Preclinical–experimental
GW501516	Reached phase II but discontinued 2007—antilipidemic. Reduces TGs and increases HDL-c
Compounds 46 and 47	Preclinical (antidiabetic)
GSK1115 and GSK7227	Preclinical

7.3. Dual PPARα/β(δ)

In preclinical studies, dual PPARα/β(δ) agonist T913659 (T659) increased HDL-c by approximately 43% from the baseline measurement.[188] T913659 and another dual PPARα/β(δ) agonist, PLX682, are also classified as dual PPARα/β(δ) agonists in early preclinical development.

GFT505 is a promising dual PPARα/β(δ) agonist currently in phase II clinical trials. GFT505 is effective in treatment of lipid and glucose disorders associated with the metabolic syndrome[189] (Table 7.9).

7.4. Dual PPARβ(δ)/γ

Two synthetic ligands, compound 20 and compound 23, were characterized in this category. In preclinical studies, both compounds are promising candidates for the treatment of the metabolic syndrome. Dual PPARβ(δ)/γ compound 23 (GlaxoSmithKline) increased HDL-c by 24%, decreased plasma glucose by 47%, and lowered serum TGs by 51% when administered to male Zucker diabetic fatty (ZDF) rats at 30 mg/kg for 7 days.[190] Similarly, compound 20 (Eli Lilly and Company) improved insulin sensitivity and reversed diabetic hyperglycemia with approximately 50% less weight gain relative to the PPARγ full agonist rosiglitazone in ZDF rats[190–192] (Table 7.10). Compound PLX350 is reported as a dual PPARβ(δ)/γ agonist with an EC50 of 0.01 μM for PPARβ(δ) and 0.052 μM for PPARγ; however, no other preclinical data are available.[215]

7.5. Selective PPARγ modulators

The concept of SPPARγMs is based on identification of chemical compounds that induce partial activation of PPARγ with broad separation of

Table 7.9 Novel dual PPARα/β(δ) agonists with antilipidemic effects

Dual PPARα/β(δ)	Status/indication/comments
T913659 (T659)(Amgen)	Preclinical—beneficial effects on HDL-c
GW2433 (GlaxoSmithKline)	Preclinical
GFT505 (Genfit SA, France)	Phase II clinical trials—primates[188,189]
PLX682 (Plexxikon, Daiichi Sankyo Company)	Preclinical

GFT505 is in phase II clinical trials.[188,189]

Table 7.10 Dual PPARβ(δ)/γ agonists in development with beneficial effects on diabetes and dyslipidemia[190–192]

Dual PPARβ(δ)/γ	Status/indication/comments
Compound 23 (GlaxoSmithKline)	Preclinical—reduces plasma glucose and TGs and elevates HDL-c in rodents. No data on body weight
Compound 20 (Eli Lilly and Company)	Preclinical—activates PPARγ/β(δ) by 17-fold greater than PPARγ
PLX350	Preclinical
Propionic acid derivative	Preclinical

efficacy from side effects.[216] The clinical proof of this concept is shown for some PPARγM compounds (e.g., telmisartan-marketed) with others in phase II clinical trials (halofenate/metaglidasen and INT131).[105,216] Partial activation of PPARγ by several compounds increased insulin sensitivity without the side effects (adipogenesis and weight gain). Several SPPARγMs, halofenate, metaglidasen, PA-082, telmisartan, DRF-2593 (balaglitazone), MCC555 (netoglitazone), YM440, PA-082, KR-62980, compound 5 (aryl indole-2-carboxylic acid), compound 12 (N-benzyl-indole), L-764406, GW 0072, FMOC-L-leucine, nTZDpa, FK614, AMG-131, compound 24 (benzoyl-2-methyl indole), and T-131, were pursued by various pharmaceutical firms as potential candidates for improved safety and for treatment of multiple features of the metabolic syndrome[217] (Table 7.11).

Halofenate is rapidly modified *in vivo* to halofenic acid (HA). In dyslipidemic subjects, HA lowered TGs and significantly decreased fasting plasma glucose in T2DM patients.[218] *In vitro*, HA caused partial activation of PPARγ, with activity approximately 10–15% of the maximal activity of the full PPARγ agonist rosiglitazone. In receptor reporter assays and as PPARγ partial agonist, HA efficaciously displaced NCoR and SMRT corepressors but failed to recruit coactivators p300, CBP, and DRIP205/TRAP220.[218] In 3T3-L1 adipocyte cell model, HA failed to fully activate genes involved in FA storage (e.g., GK and PEPCK) and FA transport (FABP4 and CD36). In both diabetic *ob/ob* mice and Zucker fatty (*fa/fa*) rats, HA causes insulin sensitization without weight gain.[105] Metaglidasen (formerly MBX-102) is the (-) enantiomer of halofenate with both antidiabetic and antilipidemic effects in rodent models of T2DM. In animal studies (*db/db* mice and ZDF rats), metaglidasen has both antidiabetic and antilipidemic benefits.[219,220] Like HA, metaglidasen has a weak ability to

Table 7.11 Selective PPARγ modulators (SPPARγMs)

Selective PPARγ modulators (SPPARγMs)	Status/indication/comments
Telmisartan (ARB) (formerly MBX-102)	Marketed
Halofenate	Phase II clinical trials
Metaglidasen	Phase II clinical trials
DRF-2593 (balaglitazone)	Phase II clinical trials
MCC555 (netoglitazone)	Phase II clinical trials
INT131 (T0903131, T131, AMG-131)	Phase II clinical trials
MBX-2044	Phase II clinical trials
YM440	Phase II clinical trials discontinued
PA-082	Preclinical
KR-62980	Preclinical
Compound 5 (aryl indole-2-carboxylic acid)	Preclinical
Compound 12 (N-benzyl-indole)	Preclinical
L-764406	Preclinical
GW 0072	Preclinical
FMOC-L-leucine	Preclinical
nTZDpa	Preclinical
Compounds 14, 15, 16 (Merck)	Preclinical
compound 24 (benzoyl-2-methyl indole)	Preclinical
NHRI compound (Taiwan)	Preclinical
PAT-5A	Preclinical
CLX-0921	Preclinical discontinued
T-131, MK-0533, DRF-2593, FK614	Discontinued

recruit CBP, DRIP205/TRAP220, and P300 coactivators; it is currently in phase II clinical trials.

PA-082 (Roche) is a unique PPARγ partial agonist with antilipidemic and insulin-sensitizing effect *in vitro*.[108] No *in vivo* data are currently available for PA-082.

Telmisartan is currently marketed as an angiotensin receptor blocker (ARB) for management of hypertension. ARBs were reported to have selective PPARγ agonism.[221-225] Clinical studies ranging from 6 to 12 months showed that telmisartan treatment is associated with increased insulin sensitivity and decreased lipids including TGs and total cholesterol.[226-228]

GW0072 was reported to lower insulin and TGs with reduced weight gain (antagonist of adipocyte differentiation) compared with the PPARγ full agonist rosiglitazone.[42,229] Likewise, FMOC-L-leucine was shown to possess low adipogenic activity but high antiglycemic effects in diabetic *ob/ob* mice and in mice with normal blood glucose.[230] These characteristics were attributed to differential cofactor recruitment by FMOC-L-leucine.

The non-TZD SPPARγM, nTZDpa (Merck), was identified as a partial agonist for PPARγ in cell-based transactivation assays (using adipocyte 3T3-L1 cell model) with a different gene expression pattern compared with PPARγ full agonists.[231] Furthermore, chronic treatment of fat-fed C57BL/6 J mice with nTZDpa reduced both hyperglycemia and hyperinsulinemia.[231]

FK614, another non-TZD PPARγ partial agonist, was shown to improve insulin sensitivity in animal models of T2DM by a mechanism that involved lower recruitment of coactivators compared with PPARγ full agonists rosiglitazone and pioglitazone.[232,233] In ZDF rats, FK614 ameliorated peripheral and hepatic insulin resistance in a dose-dependent manner (0.32, 1, and 3.2 mg/kg for 14 days) with a larger safety window than rosiglitazone.[234]

INT131 (formerly T0903131, T131, and AMG-131) (Amgen) is a highly selective PPARγM compound with a high affinity for PPARγ (displaces PPARγ full agonists with a Ki of ~10 nM and approximately 20-fold higher affinity for PPARγ than rosiglitazone or pioglitazone[235]). In addition, INT131 has 1000× selectivity for PPARγ than for PPARα, PPARβ(δ), or other nuclear receptors.[235-237] In cell-based reporter assays, INT131 activated PPARγ with an efficacy of about 10% compared to rosiglitazone.[216] Similarly, in fluorescence resonance energy transfer assays (FRETs), INT131 produced only 20–25% recruitment of the coactivator DRIP205 (important for adipocyte differentiation) when compared with rosiglitazone, pioglitazone, or troglitazone.[216] Consistent with the lower recruitment of DRIP205 coactivator, INT131 caused little adipocyte differentiation and lipid accumulation. In ZDF rats, INT131 was effective in reducing TGs, blood glucose, and insulin with no adverse side effects such as weight gain, edema, or cardiac hypertrophy.[236] Robust toxicological studies of INT131 in rodents showed lack of side effects inherently associated with the PPARγ full agonists.[216] INT131 increased the secretion of insulin sensitizer adiponectin in healthy volunteers in phase I studies.[238] INT131 is currently in phase II clinical trials with ongoing multicenter, double-blind placebo-controlled studies.

Compound 24 is another novel selective PPARγM molecule.[239] In animal studies, it possesses strong antidiabetic activity with a good therapeutic index in prevention of weight gain, adiposity, and cardiac hypertrophy that are associated with the metabolic syndrome.

The antidiabetic and antilipidemic effects of the earlier-mentioned selective modulators of PPARγ with no or little weight gain provide proof of concept for therapeutic value of these agents. The fact that some PPARγMs have advanced to clinical trials (see Table 7.11) is encouraging news for the development of new therapy with tangible value in management of the metabolic syndrome.

7.6. PPARα, β(δ), and γ pan agonists

Following the success of development of dual PPAR and SPPARMs, the pharmaceutical industry and academia took this concept further by developing single drug molecules that target the three PPARs (α, β(δ), and γ). PPAR pan agonist GW677954 (GlaxoSmithKline), compounds PLX204 and PLX134 (Plexxikon), and LY-465608 are currently in various stages of development (Table 7.12).

Table 7.12 PPAR (α, β(δ), γ) pan agonists

PPAR (α, β(δ), γ) pan agonists	Status/indication/comments
GW677954	Phase II (GlaxoSmithKline)
PLX204	Phase II (Plexxikon)
GW4148	Preclinical (GlaxoSmithKline)
GW9135	Preclinical (GlaxoSmithKline)
LY-465608	Preclinical (Eli Lilly and Company)
PLX134	Preclinical (Plexxikon)
DRL 11605	Preclinical (Perlecan Pharma)
Sipoglitazar, indeglitazar, sodeglitazar, GW-625019	Discontinued

PPAR pan agonist GW677954 was developed to treat all of the clinical disorders of the metabolic syndrome including insulin resistance, hyperglycemia, and dyslipidemia[240]; this drug is currently in phase II of clinical trials. Furthermore, a randomized, double-blind, and parallel placebo-controlled study is currently underway (GlaxoSmithKline) to evaluate the safety and tolerability of oral GW677954 capsules (15 mg) in combination with insulin. In preclinical studies, GW677954 achieved 30% reduction in glucose, insulin, and TGs. Furthermore, a 20% reduction in LDL-c while simultaneously increasing HDL-c by 20% was achieved without increased weight gain or fluid retention.[42]

In rodents, PLX204 lowered both glucose and HbA1c by approximately 50%. HbA1c is a blood marker of the mean blood glucose concentration over the earlier period of approximately 2–3 months. HbA1c is formed by nonenzymatic attachment of an N-terminal amino acid of hemoglobin to hexose. In addition to HbA1C reduction, PLX204 achieved a 4-fold reduction in TGs and elevation of HDL-c by approximately 25%.[42] Like GW677954, PLX204 is in phase II clinical trials for the treatment of T2DM.

LY465608 (Eli Lilly and Company) was reported to ameliorate insulin resistance and diabetic hyperglycemia while improving cardiovascular risk factors in preclinical ZDF rat animal models.[241] It was reclassified as a PPAR pan agonist after it was first reported as a dual PPARα/γ. LY465608 was described as effective in reducing plasma glucose coupled with improved insulin sensitivity, increased HDL-c, and reduced TGs.[42]

DRL 11605 (Perlecan Pharma) is at an early stage of development as an antiobesity drug.[242] Two PPAR pan agonists (GW4148 and GW9135) with different reactivity to PPAR subtypes were recently described.[243] In obese AKR/J mice, GW4148 decreased fat mass and produced insulin sensitization while producing weight loss. GW9135 on the other hand had no effect on body weight.

Several of the PPAR pan agonists belonging to the glitazar class were unfortunately discontinued with some at advanced clinical phases (e.g., sipoglitazar was discontinued at phase III).[244]

7.7. PPAR antagonists

All of the chemical compounds described under this category are currently being used in preclinical studies. PPAR partial or full antagonists were pursued as a therapy for T2DM based on genetic studies that suggest reduced PPARγ activity is associated with lowered risk for development of

T2DM. One missense mutation that alters a Pro (CCA)-to-Ala (GCA) at codon 12 of the PPARγ2 isoform is associated with reduced risk and increased resistance to T2DM.[245–249] Bisphenol A diglycidyl ether (BADGE), PD068235, LG100641, GW9662, SR-202, and GW6741 are PPARγ antagonists that inhibit TZD-induced adipocyte differentiation, suggesting a pharmacological role in the suppression of high-fat diet-induced obesity[250] (Table 7.13). BADGE is considered a low affinity PPARγ (Kd 100 µM) blocker of adipocyte differentiation.[251] In 3T3-L1 and 3T3-F442A preadipocyte cells, BADGE inhibited rosiglitazone-induced adipocyte differentiation.[251] Like BADGE, PD068235 blocked several of the biological activities of the PPARγ full agonist rosiglitazone.[252]

LG100641 is described as an interesting PPARγ blocker that inhibited TZD-induced adipogenesis but stimulated glucose uptake in 3T3-L1 adipocytes.[253]

GW9662 is a potent selective PPARγ antagonist with higher affinity for PPARγ (Ki of 13 nM) than rosiglitazone (110 nM).[250] In in vitro GST fusion protein pull-down assay, GW9662 did not recruit SRC-1 and p300/CBP coactivators compared with the PPARγ full agonist rosiglitazone.[250]

Animal studies with GW9662 and SR-202 PPARγ blockers suggest their utility as antiobesity drugs. For example, GW9662 blocked PPARγ in several cell types.[254] In animal studies, GW9662 blocked high-fat diet-induced obesity without affecting food intake.[250] Likewise, SR-202 blocked high-fat diet-induced adiposity and improved insulin sensitivity in ob/ob mice.[255] The antidiabetic and antiobesity properties of SR-202 point to potential therapeutic value in treatment of the metabolic syndrome.

Table 7.13 List of several PPARγ and one PPARα antagonists with therapeutic potential for the metabolic syndrome[42]

PPARα/γ antagonists	Status/receptor blocked
Bisphenol A diglycidyl ether (BADGE)	Preclinical—PPARγ
PD068235	Preclinical—PPARγ
LG100641	Preclinical—PPARγ
GW9662	Preclinical—PPARγ
SR-202 (phosphonophosphate)	Preclinical—PPARγ
GW6741	Preclinical—PPARγ
N-acylsulfonamides A and B	Preclinical—PPARα

The previously mentioned data suggest the therapeutic utility of PPARγ antagonists in treatment of the metabolic syndrome, but their detailed molecular mechanisms and clinical utility require further studies.

Structural characterization (receptor conformational changes and reactivity with coregulators) of GW6471[256] and a series of triazolone-based PPARα antagonists (N-acylsulfonamides A and B)[257] were reported. Their pharmacological role in the context of the metabolic syndrome, however, awaits further studies.

8. SUMMARY

The metabolic syndrome is a collection of obesity-related disorders that include T2DM, dyslipidemia, and CVD. The future for deciphering safe and effective therapy for the metabolic syndrome is promising despite arduous challenges. The PPARs are key conduits for insulin signaling (PPARγ) and fat burning (PPARα and PPARβ(δ)) that positively reduce hyperglycemia, dyslipidemia, and atherosclerosis. Replacement of the current antidiabetic TZDs with PPAR drug targets that have broad spectrum of metabolic effects is underway.

As shown by several large clinical human trials (UKPDS 33, ACCORD, and ADVANCE), aggressive reduction of hyperglycemia by several antidiabetic drugs (e.g., sulfonylureas and insulin) reduces microvascular complications but disappointingly does not ameliorate the progressive macrovascular complications of diabetes and other symptoms of the metabolic syndrome.[258–261] As by far the greatest cause of death in people with diabetes and metabolic syndrome is CVD,[5] there is an urgent need to reduce the major cardiovascular events (cardiovascular mortality, heart attack, or myocardial infarction), especially in patients who have experienced an acute coronary syndrome and have T2DM or patients with stable CVD and T2DM or even prediabetics. One compound that fulfills the hope for a drug with broad metabolic effects is aleglitazar (Hoffmann-La Roche). In preclinical studies, aleglitazar decreased nonfasted blood glucose levels, increased glucose clearance, and improved insulin resistance while simultaneously increasing HDL-c and decreasing LDL-c.[262] In phase I and II clinical trials in patients with T2DM, aleglitazar demonstrated beneficial antidiabetic activities and had a higher glucose-lowering effect than pioglitazone.[263] In addition, there were either little or no significant side effects in terms of kidney function or weight gain.[179] The drug is currently in phase III human clinical trial (acronym AleNephro) for evaluation of kidney function.

Based on the pre- and clinical studies delineated in the preceding text, the strategy of selective modulation of PPARs to increase efficacy and safety is doable. In theory, compounds that induce proportionate agonism of PPARs are likely to have increased safety margins and enhanced therapeutic efficacy. In practice, the risk of skeletal muscle myopathy from PPARα activation and kidney, lipid, and cardiac complications from PPARγ activation remains real. This was apparent by the higher incidence of adverse effects in several preclinical studies and in advanced clinical trials of initially very promising dual PPARα/γ agonists muraglitazar and tesaglitazar that led to their unfortunate demise in May 2006. Based on this and other preclinical data that suggest possible receptor-mediated tumor growth, the FDA now requires 2-year carcinogenicity studies in rodents for any future PPAR clinical trials. This FDA preclinical requirement can only lead to clinical advancement of rigorously tested compounds.

In closing, one effective approach to reduce the metabolic syndrome that remains at our immediate reach for now is to eat healthy and stay active.

ACKNOWLEDGMENT

The author thanks Caitlin Trebelhorn for technical assistance.

REFERENCES

1. Ogden CL, Carroll MD, Kit BK, Flegal KM. Prevalence of obesity and trends in body mass index among US children and adolescents, 1999–2010. *JAMA*. 2012;307:483–490.
2. Fakhouri TH, Ogden CL, Carroll MD, Kit BK, Flegal KM. Prevalence of obesity among older adults in the United States, 2007–2010. *NCHS Data Brief*. 2012;106:1–8.
3. Cornier M-A, Dabelea D, Hernandez TL, et al. The metabolic syndrome. *Endocr Rev*. 2008;29:777–822.
4. National Cholesterol Education Program (NCEP) Expert Panel on Detection, Evaluation, and Treatment of High Blood Cholesterol in Adults (Adult Treatment Panel III) . Third Report of the National Cholesterol Education Program (NCEP) Expert Panel on Detection, Evaluation, and Treatment of High Blood Cholesterol in Adults (Adult Treatment Panel III) Final Report. *Circulation*. 2002;106:3143–3421.
5. Ogedegbe HO, Brown DW. The ubiquitous lipids and related diseases: a laboratory perspective. *MLO Med Lab Obs*. 2001;33(18–24):26, quiz 28–29.
6. Katsiki N, Nikolic D, Montalto G, Banach M, Mikhailidis DP, Rizzo M. The role of fibrate treatment in dyslipidemia: an overview. *Curr Pharm Des*. 2013;19:3124–3131.
7. Hegele RA. Plasma lipoproteins: genetic influences and clinical implications. *Nat Rev Genet*. 2009;10:109–121.
8. American Diabetes Association. Diagnosis and classification of diabetes mellitus. *Diabetes Care*. 2010;33:S62–S69.
9. American Diabetes Association. National Diabetes Fact Sheet, 2011. http://www.diabetesorg/diabetes-basics/diabetes-statistics/; 2011.
10. Wild S, Roglic G, Green A, Sicree R, King H. Global prevalence of diabetes: estimates for the year 2000 and projections for 2030. *Diabetes Care*. 2004;27:1047–1053.

11. Xu H, Barnes GT, Yang Q, et al. Chronic inflammation in fat plays a crucial role in the development of obesity-related insulin resistance. *J Clin Invest.* 2003;112: 1821–1830.

12. Berger JP, Akiyama TE, Meinke PT. PPARs: therapeutic targets for metabolic disease. *Trends Pharmacol Sci.* 2005;26:244–251.

13. Fowler MJ. Microvascular and macrovascular complications of diabetes. *Clin Diabetes.* 2008;26:77–82.

14. The relationship of glycemic exposure (HbA1c) to the risk of development and progression of retinopathy in the diabetes control and complications trial. *Diabetes.* 1995;44:968–983.

15. Cavender MA, Lincoff AM. Therapeutic potential of aleglitazar, a new dual PPAR-alpha/gamma agonist: implications for cardiovascular disease in patients with diabetes mellitus. *Am J Cardiovasc Drugs.* 2010;10:209–216.

16. Mandard S, Muller M, Kersten S. Peroxisome proliferator-activated receptor alpha target genes. *Cell Mol Life Sci.* 2004;61:393–416.

17. Patsouris D, Mandard S, Voshol PJ, et al. PPARalpha governs glycerol metabolism. *J Clin Invest.* 2004;114:94–103.

18. Peters JM, Hennuyer N, Staels B, et al. Alterations in lipoprotein metabolism in peroxisome proliferator-activated receptor alpha-deficient mice. *J Biol Chem.* 1997;272:27307–27312.

19. Desvergne B, Wahli W. Peroxisome proliferator-activated receptors: nuclear control of metabolism. *Endocr Rev.* 1999;20:649–688.

20. Savage DB. PPAR gamma as a metabolic regulator: insights from genomics and pharmacology. *Expert Rev Mol Med.* 2005;7:1–16.

21. Tontonoz P, Hu E, Devine J, Beale EG, Spiegelman BM. PPAR gamma 2 regulates adipose expression of the phosphoenolpyruvate carboxykinase gene. *Mol Cell Biol.* 1995;15:351–357.

22. Guan HP, Li Y, Jensen MV, Newgard CB, Steppan CM, Lazar MA. A futile metabolic cycle activated in adipocytes by antidiabetic agents. *Nat Med.* 2002;8:1122–1128.

23. Nagai S, Shimizu C, Umetsu M, et al. Identification of a functional peroxisome proliferator-activated receptor responsive element within the murine perilipin gene. *Endocrinology.* 2004;145:2346–2356.

24. Baumann CA, Chokshi N, Saltiel AR, Ribon V. Cloning and characterization of a functional peroxisome proliferator activator receptor-gamma-responsive element in the promoter of the CAP gene. *J Biol Chem.* 2000;275:9131–9135.

25. Smith U, Gogg S, Johansson A, Olausson T, Rotter V, Svalstedt B. Thiazolidinediones (PPARgamma agonists) but not PPARalpha agonists increase IRS-2 gene expression in 3T3-L1 and human adipocytes. *FASEB J.* 2001;15:215–220.

26. Dana SL, Hoener PA, Bilakovics JM, et al. Peroxisome proliferator-activated receptor subtype-specific regulation of hepatic and peripheral gene expression in the Zucker diabetic fatty rat. *Metabolism.* 2001;50:963–971.

27. Maeda N, Takahashi M, Funahashi T, et al. PPARgamma ligands increase expression and plasma concentrations of adiponectin, an adipose-derived protein. *Diabetes.* 2001;50:2094–2099.

28. Steppan CM, Bailey ST, Bhat S, et al. The hormone resistin links obesity to diabetes. *Nature.* 2001;409:307–312.

29. Moller DE, Greene DA. Peroxisome proliferator-activated receptor (PPAR) gamma agonists for diabetes. *Adv Protein Chem.* 2001;56:181–212.

30. Cho MC, Lee K, Paik SG, Yoon DY. Peroxisome proliferators-activated receptor (PPAR) modulators and metabolic disorders. *PPAR Res.* 2008;2008:679137.

31. Reilly SM, Lee CH. PPAR delta as a therapeutic target in metabolic disease. *FEBS Lett.* 2008;582:26–31.

32. Burkart EM, Sambandam N, Han X, et al. Nuclear receptors PPARbeta/delta and PPARalpha direct distinct metabolic regulatory programs in the mouse heart. *J Clin Invest.* 2007;117:3930–3939.
33. Dormandy JA, Charbonnel B, Eckland DJ, et al. Secondary prevention of macrovascular events in patients with type 2 diabetes in the PROactive Study (PROspective pioglitAzone Clinical Trial In macroVascular Events): a randomised controlled trial. *Lancet.* 2005;366:1279–1289.
34. Lincoff AM, Wolski K, Nicholls SJ, Nissen SE. Pioglitazone and risk of cardiovascular events in patients with type 2 diabetes mellitus: a meta-analysis of randomized trials. *JAMA.* 2007;298:1180–1188.
35. Chappuis B, Braun M, Stettler C, et al. Differential effect of pioglitazone (PGZ) and rosiglitazone (RGZ) on postprandial glucose and lipid metabolism in patients with type 2 diabetes mellitus: a prospective, randomized crossover study. *Diabetes Metab Res Rev.* 2007;23:392–399.
36. Hodel C. Myopathy and rhabdomyolysis with lipid-lowering drugs. *Toxicol Lett.* 2002;128:159–168.
37. Nesto RW, Bell D, Bonow RO, et al. Thiazolidinedione use, fluid retention, and congestive heart failure: a consensus statement from the American Heart Association and American Diabetes Association. *Diabetes Care.* 2004;27:256–263.
38. Nissen SE. Perspective: effect of rosiglitazone on cardiovascular outcomes. *Curr Cardiol Rep.* 2007;9:343–344.
39. Balakumar P, Rose M, Ganti SS, Krishan P, Singh M. PPAR dual agonists: are they opening Pandora's Box? *Pharmacol Res.* 2007;56:91–98.
40. Benardeau A, Verry P, Atzpodien EA, et al. Effects of the dual PPAR-α/γ agonist aleglitazar on glycaemic control and organ protection in the Zucker diabetic fatty rat. *Diabetes Obes Metab.* 2013;15:164–174.
41. Benardeau A, Benz J, Binggeli A, et al. Aleglitazar, a new, potent, and balanced dual PPARalpha/gamma agonist for the treatment of type II diabetes. *Bioorg Med Chem Lett.* 2009;19:2468–2473.
42. Shearer BG, Billin AN. The next generation of PPAR drugs: do we have the tools to find them? *Biochim Biophys Acta.* 2007;1771:1082–1093.
43. Wahli W. Peroxisome proliferator-activated receptors (PPARs): from metabolic control to epidermal wound healing. *Swiss Med Wkly.* 2002;132:83–91.
44. Holden PR, Tugwood JD. Peroxisome proliferator-activated receptor alpha: role in rodent liver cancer and species differences. *J Mol Endocrinol.* 1999;22:1–8.
45. Issemann I, Green S. Activation of a member of the steroid hormone receptor superfamily by peroxisome proliferators. *Nature.* 1990;347:645–650.
46. Dreyer C, Krey G, Keller H, Givel F, Helftenbein G, Wahli W. Control of the peroxisomal beta-oxidation pathway by a novel family of nuclear hormone receptors. *Cell.* 1992;68:879–887.
47. Chandra V, Huang P, Hamuro Y, et al. Structure of the intact PPAR-gamma-RXR–nuclear receptor complex on DNA. *Nature.* 2008;456:350–356.
48. Reddy JK, Hashimoto T. Peroxisomal beta-oxidation and peroxisome proliferator-activated receptor alpha: an adaptive metabolic system. *Annu Rev Nutr.* 2001;21:193–230.
49. Berger J, Moller DE. The mechanisms of action of PPARs. *Annu Rev Med.* 2002;53:409–435.
50. Ferré P. The biology of peroxisome proliferator-activated receptors: relationship with lipid metabolism and insulin sensitivity. *Diabetes.* 2004;53:S43–S50.
51. Braissant O, Foufelle F, Scotto C, Dauca M, Wahli W. Differential expression of peroxisome proliferator-activated receptors (PPARs): tissue distribution of PPAR-alpha, -beta, and -gamma in the adult rat. *Endocrinology.* 1996;137:354–366.

52. Auboeuf D, Rieusset J, Fajas L, et al. Tissue distribution and quantification of the expression of mRNAs of peroxisome proliferator-activated receptors and liver X receptor-alpha in humans: no alteration in adipose tissue of obese and NIDDM patients. *Diabetes.* 1997;46:1319–1327.

53. Jia Y, Qi C, Zhang Z, et al. Overexpression of peroxisome proliferator-activated receptor-alpha (PPARalpha)-regulated genes in liver in the absence of peroxisome proliferation in mice deficient in both L- and D-forms of enoyl-CoA hydratase/dehydrogenase enzymes of peroxisomal beta-oxidation system. *J Biol Chem.* 2003;278: 47232–47239.

54. Gilde AJ, Van Bilsen M. Peroxisome proliferator-activated receptors (PPARS): regulators of gene expression in heart and skeletal muscle. *Acta Physiol Scand.* 2003;178:425–434.

55. Barger PM, Kelly DP. PPAR signaling in the control of cardiac energy metabolism. *Trends Cardiovas Med.* 2000;10:238–245.

56. Duval C, Fruchart JC, Staels B. PPAR alpha, fibrates, lipid metabolism and inflammation. *Arch Mal Coeur Vaiss.* 2004;97:665–672.

57. Tontonoz P, Graves RA, Budavari AI, et al. Adipocyte-specific transcription factor ARF6 is a heterodimeric complex of two nuclear hormone receptors, PPAR gamma and RXR alpha. *Nucleic Acids Res.* 1994;22:5628–5634.

58. Tontonoz P, Hu E, Graves RA, Budavari AI, Spiegelman BM. mPPAR gamma 2: tissue-specific regulator of an adipocyte enhancer. *Genes Dev.* 1994;8: 1224–1234.

59. Escher P, Braissant O, Basu-Modak S, Michalik L, Wahli W, Desvergne B. Rat PPARs: quantitative analysis in adult rat tissues and regulation in fasting and refeeding. *Endocrinology.* 2001;142:4195–4202.

60. Thompson EA. PPARgamma physiology and pathology in gastrointestinal epithelial cells. *Mol Cells.* 2007;24:167–176.

61. Hevener AL, Olefsky JM, Reichart D, et al. Macrophage PPAR gamma is required for normal skeletal muscle and hepatic insulin sensitivity and full antidiabetic effects of thiazolidinediones. *J Clin Invest.* 2007;117:1658–1669.

62. Tontonoz P, Hu E, Spiegelman BM. Stimulation of adipogenesis in fibroblasts by PPAR gamma 2, a lipid-activated transcription factor. *Cell.* 1994;79:1147–1156.

63. Saladin R, Fajas L, Dana S, Halvorsen YD, Auwerx J, Briggs M. Differential regulation of peroxisome proliferator activated receptor gamma1 (PPARgamma1) and PPARgamma2 messenger RNA expression in the early stages of adipogenesis. *Cell Growth Differ.* 1999;10:43–48.

64. Spiegelman BM, Hu E, Kim JB, Brun R. PPAR gamma and the control of adipogenesis. *Biochimie.* 1997;79:111–112.

65. Spiegelman BM. PPAR-gamma: adipogenic regulator and thiazolidinedione receptor. *Diabetes.* 1998;47:507–514.

66. Day C. Thiazolidinediones: a new class of antidiabetic drugs. *Diabet Med.* 1999; 16:179–192.

67. Lehmann JM, Moore LB, Smith-Oliver TA, Wilkison WO, Willson TM, Kliewer SA. An antidiabetic thiazolidinedione is a high affinity ligand for peroxisome proliferator-activated receptor gamma (PPAR gamma). *J Biol Chem.* 1995;270: 12953–12956.

68. Ristow M, Muller-Wieland D, Pfeiffer A, Krone W, Kahn CR. Obesity associated with a mutation in a genetic regulator of adipocyte differentiation. *N Engl J Med.* 1998;339:953–959.

69. Barroso I, Gurnell M, Crowley VE, et al. Dominant negative mutations in human PPARgamma associated with severe insulin resistance, diabetes mellitus and hypertension. *Nature.* 1999;402:880–883.

70. Hegele RA, Cao H, Frankowski C, Mathews ST, Leff T. PPARG F388L, a transactivation-deficient mutant, in familial partial lipodystrophy. *Diabetes.* 2002;51:3586–3590.

71. Agarwal AK, Garg A. A novel heterozygous mutation in peroxisome proliferator-activated receptor-gamma gene in a patient with familial partial lipodystrophy. *J Clin Endocrinol Metab.* 2002;87:408–411.

72. Barish GD, Narkar VA, Evans RM. PPAR delta: a dagger in the heart of the metabolic syndrome. *J Clin Invest.* 2006;116:590–597.

73. Schoonjans K, Staels B, Auwerx J. Role of the peroxisome proliferator-activated receptor (PPAR) in mediating the effects of fibrates and fatty acids on gene expression. *J Lipid Res.* 1996;37:907–925.

74. Shearer BG, Hoekstra WJ. Peroxisome proliferator-activated receptors (PPARs): choreographers of metabolic gene transcription. *DNA.* 2002;83:68.

75. Takada I, Yu RT, Xu HE, et al. Alteration of a single amino acid in peroxisome proliferator-activated receptor-α (PPARα) generates a PPARδ phenotype. *Mol Endocrinol.* 2000;14:733–740.

76. Krey G, Braissant O, L'Horset F, et al. Fatty acids, eicosanoids, and hypolipidemic agents identified as ligands of peroxisome proliferator-activated receptors by coactivator-dependent receptor ligand assay. *Mol Endocrinol.* 1997;11:779–791.

77. Kliewer SA, Sundseth SS, Jones SA, et al. Fatty acids and eicosanoids regulate gene expression through direct interactions with peroxisome proliferator-activated receptors alpha and gamma. *Proc Natl Acad Sci U S A.* 1997;94:4318–4323.

78. Funk CD. Prostaglandins and leukotrienes: advances in eicosanoid biology. *Science.* 2001;294:1871–1875.

79. Caijo F, Mosset P, Gree R, et al. Synthesis of new carbo- and heterocyclic analogues of 8-HETE and evaluation of their activity towards the PPARs. *Bioorg Med Chem Lett.* 2005;15:4421–4426.

80. Yu K, Bayona W, Kallen CB, et al. Differential activation of peroxisome proliferator-activated receptors by eicosanoids. *J Biol Chem.* 1995;270:23975–23983.

81. Forman BM, Tontonoz P, Chen J, Brun RP, Spiegelman BM, Evans RM. 15-Deoxy-delta 12, 14-prostaglandin J2 is a ligand for the adipocyte determination factor PPAR gamma. *Cell.* 1995;83:803–812.

82. Kliewer SA, Lenhard JM, Willson TM, Patel I, Morris DC, Lehmann JM. A prostaglandin J2 metabolite binds peroxisome proliferator-activated receptor gamma and promotes adipocyte differentiation. *Cell.* 1995;83:813–819.

83. Forman BM, Chen J, Evans RM. Hypolipidemic drugs, polyunsaturated fatty acids, and eicosanoids are ligands for peroxisome proliferator-activated receptors alpha and delta. *Proc Natl Acad Sci U S A.* 1997;94:4312–4317.

84. Lim H, Dey SK. A novel pathway of prostacyclin signaling-hanging out with nuclear receptors. *Endocrinology.* 2002;143:3207–3210.

85. Tzameli I, Fang H, Ollero M, et al. Regulated production of a peroxisome proliferator-activated receptor-gamma ligand during an early phase of adipocyte differentiation in 3T3-L1 adipocytes. *J Biol Chem.* 2004;279:36093–36102.

86. Hostetler HA, Petrescu AD, Kier AB, Schroeder F. Peroxisome proliferator-activated receptor alpha interacts with high affinity and is conformationally responsive to endogenous ligands. *J Biol Chem.* 2005;280:18667–18682.

87. Xu HE, Lambert MH, Montana VG, et al. Molecular recognition of fatty acids by peroxisome proliferator-activated receptors. *Mol Cell.* 1999;3:397–403.

88. Millar JS, Duffy D, Gadi R, et al. Potent and selective PPAR-α agonist LY518674 upregulates both ApoA-I production and catabolism in human subjects with the metabolic syndrome. *Arterioscler Thromb VascBiol.* 2009;29:140–146.

89. Sierra ML, Beneton V, Boullay AB, et al. Substituted 2-[(4-aminomethyl)phenoxy]-2-methylpropionic acid PPARalpha agonists. 1. Discovery of a novel series of potent HDLc raising agents. *J Med Chem.* 2007;50:685–695.
90. Hansen MK, McVey MJ, White RF, et al. Selective CETP inhibition and PPARα agonism increase HDL cholesterol and reduce LDL cholesterol in human ApoB100/human CETP transgenic mice. *J Cardiovas Pharmacol Ther.* 2010;15:196–202.
91. Schafer HL, Linz W, Falk E, et al. AVE8134, a novel potent PPAR[alpha] agonist, improves lipid profile and glucose metabolism in dyslipidemic mice and type 2 diabetic rats. *Acta Pharmacol Sin.* 2012;33:82–90.
92. Linz W, Wohlfart P, Baader M, et al. The peroxisome proliferator-activated receptor-[alpha] (PPAR-[alpha]) agonist, AVE8134, attenuates the progression of heart failure and increases survival in rats. *Acta Pharmacol Sin.* 2009;30:935–946.
93. Duval C, Müller M, Kersten S. PPARα and dyslipidemia. *Biochim Biophys Acta.* 2007;1771:961–971.
94. Joy T, Hegele RA. Is raising HDL a futile strategy for atheroprotection? *Nat Rev Drug Discov.* 2008;7:143–155.
95. Nissen SE, Wolski K. Effect of rosiglitazone on the risk of myocardial infarction and death from cardiovascular causes. *N Engl J Med.* 2007;356:2457–2471.
96. Balakumar P, Rohilla A, Mahadevan N. Pleiotropic actions of fenofibrate on the heart. *Pharmacol Res.* 2011;63:8–12.
97. Evans KA, Shearer BG, Wisnoski DD, et al. Phenoxyacetic acids as PPARdelta partial agonists: synthesis, optimization, and *in vivo* efficacy. *Bioorg Med Chem Lett.* 2011;21:2345–2350.
98. Shearer BG, Patel HS, Billin AN, et al. Discovery of a novel class of PPARdelta partial agonists. *Bioorg Med Chem Lett.* 2008;18:5018–5022.
99. Zimmerman AW, Veerkamp JH. New insights into the structure and function of fatty acid-binding proteins. *Cell Mol Life Sci.* 2002;59:1096–1116.
100. Tan NS, Shaw NS, Vinckenbosch N, et al. Selective cooperation between fatty acid binding proteins and peroxisome proliferator-activated receptors in regulating transcription. *Mol Cell Biol.* 2002;22:5114–5127.
101. Adida A, Spener F. Adipocyte-type fatty acid-binding protein as inter-compartmental shuttle for peroxisome proliferator activated receptor gamma agonists in cultured cell. *Biochim Biophys Acta.* 2006;1761:172–181.
102. Aranda A, Pascual A. Nuclear hormone receptors and gene expression. *Physiol Rev.* 2001;81:1269–1304.
103. Viswakarma N, Jia Y, Bai L, et al. Coactivators in PPAR-regulated gene expression. *PPAR Res.* 2010;2010:250126.
104. Li S, Paulsson KM, Chen S, Sjogren HO, Wang P. Tapasin is required for efficient peptide binding to transporter associated with antigen processing. *J Biol Chem.* 2000;275:1581–1586.
105. Zhang F, Lavan BE, Gregoire FM. Selective modulators of PPAR-gamma activity: molecular aspects related to obesity and side-effects. *PPAR Res.* 2007;2007:32696.
106. Picard F, Gehin M, Annicotte J, et al. SRC-1 and TIF2 control energy balance between white and brown adipose tissues. *Cell.* 2002;111:931–941.
107. Knouff C, Auwerx J. Peroxisome proliferator-activated receptor-gamma calls for activation in moderation: lessons from genetics and pharmacology. *Endocr Rev.* 2004;25:899–918.
108. Burgermeister E, Schnoebelen A, Flament A, et al. A novel partial agonist of peroxisome proliferator-activated receptor-gamma (PPARgamma) recruits PPARgamma-coactivator-1alpha, prevents triglyceride accumulation, and potentiates insulin signaling *in vitro*. *Mol Endocrinol.* 2006;20:809–830.

109. Pyper SR, Viswakarma N, Jia Y, Zhu YJ, Fondell JD, Reddy JK. PRIC295, a nuclear receptor coactivator, identified from PPARalpha-interacting cofactor complex. *PPAR Res.* 2010;2010:173907.
110. Vega RB, Huss JM, Kelly DP. The coactivator PGC-1 cooperates with peroxisome proliferator-activated receptor alpha in transcriptional control of nuclear genes encoding mitochondrial fatty acid oxidation enzymes. *Mol Cell Biol.* 2000;20:1868–1876.
111. Koo SH, Satoh H, Herzig S, et al. PGC-1 promotes insulin resistance in liver through PPAR-alpha-dependent induction of TRB-3. *Nat Med.* 2004;10:530–534.
112. Yoon JC, Puigserver P, Chen G, et al. Control of hepatic gluconeogenesis through the transcriptional coactivator PGC-1. *Nature.* 2001;413:131–138.
113. Chen S, Johnson BA, Li Y, et al. Both coactivator LXXLL motif-dependent and -independent interactions are required for peroxisome proliferator-activated receptor gamma (PPARgamma) function. *J Biol Chem.* 2000;275:3733–3736.
114. Mark M, Yoshida-Komiya H, Gehin M, et al. Partially redundant functions of SRC-1 and TIF2 in postnatal survival and male reproduction. *Proc Natl Acad Sci U S A.* 2004;101:4453–4458.
115. Nolte RT, Wisely GB, Westin S, et al. Ligand binding and co-activator assembly of the peroxisome proliferator-activated receptor-gamma. *Nature.* 1998;395:137–143.
116. Karagianni P, Wong J. HDAC3: taking the SMRT-N-CoRrect road to repression. *Oncogene.* 2007;26:5439–5449.
117. Nagy L, Kao HY, Chakravarti D, et al. Nuclear receptor repression mediated by a complex containing SMRT, mSin3A, and histone deacetylase. *Cell.* 1997;89:373–380.
118. Wei LN, Hu X, Chandra D, Seto E, Farooqui M. Receptor-interacting protein 140 directly recruits histone deacetylases for gene silencing. *J Biol Chem.* 2000;275:40782–40787.
119. Miyata KS, McCaw SE, Meertens LM, Patel HV, Rachubinski RA, Capone JP. Receptor-interacting protein 140 interacts with and inhibits transactivation by, peroxisome proliferator-activated receptor alpha and liver-X-receptor alpha. *Mol Cell Endocrinol.* 1998;146:69–76.
120. Miles PD, Barak Y, He W, Evans RM, Olefsky JM. Improved insulin-sensitivity in mice heterozygous for PPAR-gamma deficiency. *J Clin Invest.* 2000;105:287–292.
121. Olefsky JM, Saltiel AR. PPAR gamma and the treatment of insulin resistance. *Trends Endocrinol Metab.* 2000;11:362–368.
122. Zhang J, Fu M, Cui T, et al. Selective disruption of PPARgamma 2 impairs the development of adipose tissue and insulin sensitivity. *Proc Natl Acad Sci U S A.* 2004;101:10703–10708.
123. Savage DB, Semple RK. Recent insights into fatty liver, metabolic dyslipidaemia and their links to insulin resistance. *Curr Opin Lipidol.* 2010;21:329–336.
124. Kim JK, Fillmore JJ, Chen Y, et al. Tissue-specific overexpression of lipoprotein lipase causes tissue-specific insulin resistance. *Proc Natl Acad Sci U S A.* 2001;98:7522–7527.
125. Dresner A, Laurent D, Marcucci M, et al. Effects of free fatty acids on glucose transport and IRS-1-associated phosphatidylinositol 3-kinase activity. *J Clin Invest.* 1999;103:253–259.
126. Yu C, Chen Y, Cline GW, et al. Mechanism by which fatty acids inhibit insulin activation of insulin receptor substrate-1 (IRS-1)-associated phosphatidylinositol 3-kinase activity in muscle. *J Biol Chem.* 2002;277:50230–50236.
127. Fei J. PPAR: a pivotal regulator in metabolic syndromes. *Endocrinol Metab Syndr.* 2012;1.
128. Semple RK, Chatterjee VK, O'Rahilly S. PPAR gamma and human metabolic disease. *J Clin Invest.* 2006;116:581–589.

129. Shulman AI, Mangelsdorf DJ. Retinoid X receptor heterodimers in the metabolic syndrome. *N Engl J Med*. 2005;353:604–615.
130. Kersten S, Seydoux J, Peters JM, Gonzalez FJ, Desvergne B, Wahli W. Peroxisome proliferator-activated receptor alpha mediates the adaptive response to fasting. *J Clin Invest*. 1999;103:1489–1498.
131. Fruchart JC. Peroxisome proliferator-activated receptor-alpha activation and high-density lipoprotein metabolism. *Am J Cardiol*. 2001;88:24N–29N.
132. Lee CH, Kang K, Mehl IR, et al. Peroxisome proliferator-activated receptor delta promotes very low-density lipoprotein-derived fatty acid catabolism in the macrophage. *Proc Natl Acad Sci U S A*. 2006;103:2434–2439.
133. Chinetti G, Lestavel S, Bocher V, et al. PPAR-alpha and PPAR-gamma activators induce cholesterol removal from human macrophage foam cells through stimulation of the ABCA1 pathway. *Nat Med*. 2001;7:53–58.
134. Oliver Jr WR, Shenk JL, Snaith MR, et al. A selective peroxisome proliferator-activated receptor delta agonist promotes reverse cholesterol transport. *Proc Natl Acad Sci U S A*. 2001;98:5306–5311.
135. Staels B, Schoonjans K, Fruchart JC, Auwerx J. The effects of fibrates and thiazolidinediones on plasma triglyceride metabolism are mediated by distinct peroxisome proliferator activated receptors (PPARs). *Biochimie*. 1997;79:95–99.
136. Chawla A, Lee CH, Barak Y, et al. PPARdelta is a very low-density lipoprotein sensor in macrophages. *Proc Natl Acad Sci U S A*. 2003;100:1268–1273.
137. Lee CH, Olson P, Hevener A, et al. PPARdelta regulates glucose metabolism and insulin sensitivity. *Proc Natl Acad Sci U S A*. 2006;103:3444–3449.
138. Wang YX, Lee CH, Tiep S, et al. Peroxisome-proliferator-activated receptor delta activates fat metabolism to prevent obesity. *Cell*. 2003;113:159–170.
139. Tanaka T, Yamamoto J, Iwasaki S, et al. Activation of peroxisome proliferator-activated receptor delta induces fatty acid beta-oxidation in skeletal muscle and attenuates metabolic syndrome. *Proc Natl Acad Sci U S A*. 2003;100:15924–15929.
140. Holst D, Luquet S, Nogueira V, Kristiansen K, Leverve X, Grimaldi PA. Nutritional regulation and role of peroxisome proliferator-activated receptor delta in fatty acid catabolism in skeletal muscle. *Biochim Biophys Acta*. 2003;1633:43–50.
141. Wang YX, Zhang CL, Yu RT, et al. Regulation of muscle fiber type and running endurance by PPARdelta. *PLoS Biol*. 2004;2:e294.
142. Barish GD. Peroxisome proliferator-activated receptors and liver X receptors in atherosclerosis and immunity. *J Nutr*. 2006;136:690–694.
143. Rangwala SM, Lazar MA. Peroxisome proliferator-activated receptor gamma in diabetes and metabolism. *Trends Pharmacol Sci*. 2004;25:331–336.
144. Spiegelman BM, Flier JS. Adipogenesis and obesity: rounding out the big picture. *Cell*. 1996;87:377–389.
145. Rosen ED, Walkey CJ, Puigserver P, Spiegelman BM. Transcriptional regulation of adipogenesis. *Genes Dev*. 2000;14:1293–1307.
146. Lowell BB. PPARgamma: an essential regulator of adipogenesis and modulator of fat cell function. *Cell*. 1999;99:239–242.
147. Tontonoz P, Spiegelman BM. Fat and beyond: the diverse biology of PPARgamma. *Annu Rev Biochem*. 2008;77:289–312.
148. Kim JB, Spiegelman BM. ADD1/SREBP1 promotes adipocyte differentiation and gene expression linked to fatty acid metabolism. *Genes Dev*. 1996;10:1096–1107.
149. Miyazaki Y, Mahankali A, Matsuda M, et al. Effect of pioglitazone on abdominal fat distribution and insulin sensitivity in type 2 diabetic patients. *J Clin Endocrinol Metab*. 2002;87:2784–2791.
150. Kintscher U, Law RE. PPARγ-mediated insulin sensitization: the importance of fat versus muscle. *Am J Physiol Endocrinol Metab*. 2005;288:E287–E291.

151. Gavrilova O, Haluzik M, Matsusue K, et al. Liver peroxisome proliferator-activated receptor gamma contributes to hepatic steatosis, triglyceride clearance, and regulation of body fat mass. *J Biol Chem*. 2003;278:34268–34276.
152. Jones JR, Barrick C, Kim K-A, et al. Deletion of PPARγ in adipose tissues of mice protects against high fat diet-induced obesity and insulin resistance. *Proc Natl Acad Sci U S A*. 2005;102:6207–6212.
153. He W, Barak Y, Hevener A, et al. Adipose-specific peroxisome proliferator-activated receptor gamma knockout causes insulin resistance in fat and liver but not in muscle. *Proc Natl Acad Sci U S A*. 2003;100:15712–15717.
154. Hevener AL, He W, Barak Y, et al. Muscle-specific Pparg deletion causes insulin resistance. *Nat Med*. 2003;9:1491–1497.
155. Norris AW, Chen L, Fisher SJ, et al. Muscle-specific PPARgamma-deficient mice develop increased adiposity and insulin resistance but respond to thiazolidinediones. *J Clin Invest*. 2003;112:608–618.
156. Diani AR, Sawada G, Wyse B, Murray FT, Khan M. Pioglitazone preserves pancreatic islet structure and insulin secretory function in three murine models of type 2 diabetes. *Am J Physiol Endocrinol Metab*. 2004;286:E116–E122.
157. Okuno A, Tamemoto H, Tobe K, et al. Troglitazone increases the number of small adipocytes without the change of white adipose tissue mass in obese Zucker rats. *J Clin Invest*. 1998;101:1354–1361.
158. Duan SZ, Usher MG, Mortensen RM. Peroxisome proliferator-activated receptor-γ-mediated effects in the vasculature. *Circ Res*. 2008;102:283–294.
159. Moore KJ, Rosen ED, Fitzgerald ML, et al. The role of PPAR-gamma in macrophage differentiation and cholesterol uptake. *Nat Med*. 2001;7:41–47.
160. Rosen ED, Spiegelman BM. Peroxisome proliferator-activated receptor gamma ligands and atherosclerosis: ending the heartache. *J Clin Invest*. 2000;106:629–631.
161. Chawla A. Control of macrophage activation and function by PPARs. *Circ Res*. 2010;106:1559–1569.
162. Bajaj M, Suraamornkul S, Piper P, et al. Decreased plasma adiponectin concentrations are closely related to hepatic fat content and hepatic insulin resistance in pioglitazone-treated type 2 diabetic patients. *J Clin Endocrinol Metab*. 2004;89:200–206.
163. Yamauchi T, Kamon J, Waki H, et al. The fat-derived hormone adiponectin reverses insulin resistance associated with both lipoatrophy and obesity. *Nat Med*. 2001;7:941–946.
164. Sigrist S, Bedoucha M, Boelsterli UA. Down-regulation by troglitazone of hepatic tumor necrosis factor-alpha and interleukin-6 mRNA expression in a murine model of non-insulin-dependent diabetes. *Biochem Pharmacol*. 2000;60:67–75.
165. Gottschling-Zeller H, Rohrig K, Hauner H. Troglitazone reduces plasminogen activator inhibitor-1 expression and secretion in cultured human adipocytes. *Diabetologia*. 2000;43:377–383.
166. Harte AL, McTernan PG, McTernan CL, Smith SA, Barnett AH, Kumar S. Rosiglitazone inhibits the insulin-mediated increase in PAI-1 secretion in human abdominal subcutaneous adipocytes. *Diabetes Obes Metab*. 2003;5:302–310.
167. Drazen JM, Morrissey S, Curfman GD. Rosiglitazone—continued uncertainty about safety. *N Engl J Med*. 2007;357:63–64.
168. Shah P, Mudaliar S. Pioglitazone: side effect and safety profile. *Expert Opin Drug Saf*. 2010;9:347–354.
169. Yki-Jarvinen H. Thiazolidinediones. *N Engl J Med*. 2004;351:1106–1118.
170. Nesto RW, Bell D, Bonow RO, et al. Thiazolidinedione use, fluid retention, and congestive heart failure: a consensus statement from the American Heart Association and American Diabetes Association. *Circulation*. 2003;108:2941–2948.

171. Murphy CE, Rodgers PT. Effects of thiazolidinediones on bone loss and fracture. *Ann Pharmacother.* 2007;41:2014–2018.
172. Scheen AJ. Combined thiazolidinedione-insulin therapy: should we be concerned about safety? *Drug Saf.* 2004;27:841–856.
173. Mudaliar S, Chang AR, Henry RR. Thiazolidinediones, peripheral edema, and type 2 diabetes: incidence, pathophysiology, and clinical implications. *Endocr Pract.* 2003;9:406–416.
174. Lago RM, Singh PP, Nesto RW. Congestive heart failure and cardiovascular death in patients with prediabetes and type 2 diabetes given thiazolidinediones: a meta-analysis of randomised clinical trials. *Lancet.* 2007;370:1129–1136.
175. Nathan DM, Buse JB, Davidson MB, et al. Management of hyperglycemia in type 2 diabetes: a consensus algorithm for the initiation and adjustment of therapy: update regarding thiazolidinediones: a consensus statement from the American Diabetes Association and the European Association for the Study of Diabetes. *Diabetes Care.* 2008;31:173–175.
176. Wilding J. Thiazolidinediones, insulin resistance and obesity: finding a balance. *Int J Clin Pract.* 2006;60:1272–1280.
177. Guan Y, Hao C, Cha DR, et al. Thiazolidinediones expand body fluid volume through PPARgamma stimulation of ENaC-mediated renal salt absorption. *Nat Med.* 2005;11:861–866.
178. Zhang H, Zhang A, Kohan DE, Nelson RD, Gonzalez FJ, Yang T. Collecting duct-specific deletion of peroxisome proliferator-activated receptor gamma blocks thiazolidinedione-induced fluid retention. *Proc Natl Acad Sci U S A.* 2005;102: 9406–9411.
179. Younk LM, Uhl L, Davis SN. Pharmacokinetics, efficacy and safety of aleglitazar for the treatment of type 2 diabetes with high cardiovascular risk. *Expert Opin Drug Metab Toxicol.* 2011;7:753–763.
180. Chu NN, Li XN, Chen WL, Xu HR. Determination of chiglitazar, a dual alpha/gamma peroxisome proliferator-activated receptor (PPAR) agonist, in human plasma by liquid chromatography-tandem mass spectrometry. *Pharmazie.* 2007;62: 825–829.
181. Pourcet B, Fruchart JC, Staels B, Glineur C. Selective PPAR modulators, dual and pan PPAR agonists: multimodal drugs for the treatment of type 2 diabetes and atherosclerosis. *Expert Opin Emerg Drugs.* 2006;11:379–401.
182. Buse JB, Rubin CJ, Frederich R, et al. Muraglitazar, a dual (alpha/gamma) PPAR activator: a randomized, double-blind, placebo-controlled, 24-week monotherapy trial in adult patients with type 2 diabetes. *Clin Ther.* 2005;27:1181–1195.
183. Fagerberg B, Edwards S, Halmos T, et al. Tesaglitazar, a novel dual peroxisome proliferator-activated receptor alpha/gamma agonist, dose-dependently improves the metabolic abnormalities associated with insulin resistance in a non-diabetic population. *Diabetologia.* 2005;48:1716–1725.
184. Pickavance LC, Brand CL, Wassermann K, Wilding JP. The dual PPARalpha/gamma agonist, ragaglitazar, improves insulin sensitivity and metabolic profile equally with pioglitazone in diabetic and dietary obese ZDF rats. *Br J Pharmacol.* 2005;144:308–316.
185. Calkin AC, Thomas MC, Cooper ME. MK-767. Kyorin/Banyu/Merck. *Curr Opin Investig Drugs.* 2003;4:444–448.
186. Fievet C, Fruchart JC, Staels B. PPARalpha and PPARgamma dual agonists for the treatment of type 2 diabetes and the metabolic syndrome. *Curr Opin Pharmacol.* 2006;6:606–614.
187. Long GG, Reynolds VL, Lopez-Martinez A, Ryan TE, White SL, Eldridge SR. Urothelial carcinogenesis in the urinary bladder of rats treated with naveglitazar, a

gamma-dominant PPAR alpha/gamma agonist: lack of evidence for urolithiasis as an inciting event. *Toxicol Pathol.* 2008;36:218–231.

188. Wallace JM, Schwarz M, Coward P, et al. Effects of peroxisome proliferator-activated receptor alpha/delta agonists on HDL-cholesterol in vervet monkeys. *J Lipid Res.* 2005;46:1009–1016.

189. Cariou B, Zaïr Y, Staels B, Bruckert E. Effects of the new dual PPARα/δ agonist GFT505 on lipid and glucose homeostasis in abdominally obese patients with combined dyslipidemia or impaired glucose metabolism. *Diabetes Care.* 2011;34:2008–2014.

190. Liu KG, Lambert MH, Leesnitzer LM, et al. Identification of a series of PPAR gamma/delta dual agonists via solid-phase parallel synthesis. *Bioorg Med Chem Lett.* 2001;11:2959–2962.

191. Gonzalez IC, Lamar J, Iradier F, et al. Design and synthesis of a novel class of dual PPARgamma/delta agonists. *Bioorg Med Chem Lett.* 2007;17:1052–1055.

192. Xu Y, Etgen GJ, Broderick CL, et al. Design and synthesis of dual peroxisome proliferator-activated receptors gamma and delta agonists as novel euglycemic agents with a reduced weight gain profile. *J Med Chem.* 2006;49:5649–5652.

193. Shureiqi I, Jiang W, Zuo X, et al. The 15-lipoxygenase-1 product 13-S-hydroxyoctadecadienoic acid down-regulates PPAR-delta to induce apoptosis in colorectal cancer cells. *Proc Natl Acad Sci U S A.* 2003;100:9968–9973.

194. Brown PJ, Smith-Oliver TA, Charifson PS, et al. Identification of peroxisome proliferator-activated receptor ligands from a biased chemical library. *Chem Biol.* 1997;4:909–918.

195. Lim HJ, Park JH, Lee S, Choi HE, Lee KS, Park HY. PPARdelta ligand L-165041 ameliorates Western diet-induced hepatic lipid accumulation and inflammation in LDLR-/- mice. *Eur J Pharmacol.* 2009;622:45–51.

196. Faiola B, Falls JG, Peterson RA, et al. PPAR alpha, more than PPAR delta, mediates the hepatic and skeletal muscle alterations induced by the PPAR agonist GW0742. *Toxicol Sci.* 2008;105:384–394.

197. Akiyama TE, Meinke PT, Berger JP. PPAR ligands: potential therapies for metabolic syndrome. *Curr Diab Rep.* 2005;5:45–52.

198. Balakumar P, Rose M, Singh M. PPAR ligands: are they potential agents for cardiovascular disorders? *Pharmacology.* 2007;80:1–10.

199. Higgins LS, Depaoli AM. Selective peroxisome proliferator-activated receptor gamma (PPARgamma) modulation as a strategy for safer therapeutic PPARgamma activation. *Am J Clin Nutr.* 2010;91:267S–272S.

200. Moller DE. New drug targets for type 2 diabetes and the metabolic syndrome. *Nature.* 2001;414:821–827.

201. Guerre-Millo M, Gervois P, Raspe E, et al. Peroxisome proliferator-activated receptor alpha activators improve insulin sensitivity and reduce adiposity. *J Biol Chem.* 2000;275:16638–16642.

202. Hegarty BD, Furler SM, Oakes ND, Kraegen EW, Cooney GJ. Peroxisome proliferator-activated receptor (PPAR) activation induces tissue-specific effects on fatty acid uptake and metabolism *in vivo*—a study using the novel PPARalpha/gamma agonist tesaglitazar. *Endocrinology.* 2004;145:3158–3164.

203. Chakrabarti R, Vikramadithyan RK, Misra P, et al. Ragaglitazar: a novel PPAR alpha PPAR gamma agonist with potent lipid-lowering and insulin-sensitizing efficacy in animal models. *Br J Pharmacol.* 2003;140:527–537.

204. Sakamoto J, Kimura H, Moriyama S, et al. A novel oxyiminoalkanoic acid derivative, TAK-559, activates human peroxisome proliferator-activated receptor subtypes. *Eur J Pharmacol.* 2004;495:17–26.

205. Murakami K, Tsunoda M, Ide T, Ohashi M, Mochizuki T. Amelioration by KRP-297, a new thiazolidinedione, of impaired glucose uptake in skeletal muscle from obese insulin-resistant animals. *Metabolism.* 1999;48:1450–1454.

206. Yi P, Hadden CE, Annes WF, et al. The disposition and metabolism of naveglitazar, a peroxisome proliferator-activated receptor α-γ dual, γ-dominant agonist in mice, rats, and monkeys. *Drug Metab Dispos.* 2007;35:51–61.

207. Doebber TW, Kelly LJ, Zhou G, et al. MK-0767, a novel dual PPARalpha/gamma agonist, displays robust antihyperglycemic and hypolipidemic activities. *Biochem Biophys Res Commun.* 2004;318:323–328.

208. Chang F, Jaber LA, Berlie HD, O'Connell MB. Evolution of peroxisome proliferator-activated receptor agonists. *Ann Pharmacother.* 2007;41:973–983.

209. Li PP, Shan S, Chen YT, et al. The PPARalpha/gamma dual agonist chiglitazar improves insulin resistance and dyslipidemia in MSG obese rats. *Br J Pharmacol.* 2006;148:610–618.

210. Guo Q, Sahoo SP, Wang PR, et al. A novel peroxisome proliferator-activated receptor alpha/gamma dual agonist demonstrates favorable effects on lipid homeostasis. *Endocrinology.* 2004;145:1640–1648.

211. Xu Y, Rito CJ, Etgen GJ, et al. Design and synthesis of alpha-aryloxy-alpha-methylhydrocinnamic acids: a novel class of dual peroxisome proliferator-activated receptor alpha/gamma agonists. *J Med Chem.* 2004;47:2422–2425.

212. Lecka-Czernik B. Aleglitazar, a dual PPARalpha and PPARgamma agonist for the potential oral treatment of type 2 diabetes mellitus. *IDrugs.* 2010;13:793–801.

213. El Hage J. *Preclinical and clinical safety assessments for PPAR agonists;* 2004. http://wwwfdagov/CDER/present/DIA2004/Elhageppt.

214. Gupta RA, Wang D, Katkuri S, Wang H, Dey SK, DuBois RN. Activation of nuclear hormone receptor peroxisome proliferator-activated receptor-delta accelerates intestinal adenoma growth. *Nat Med.* 2004;10:245–247.

215. Lin J. Scaffold based discovery of indeglitazar (PLX204), a pan-PPAR agent for NIDDM. In: *Power Point Presentation, AAPS Conference;* 2009.

216. Higgins LS, Mantzoros CS. The development of INT131 as a selective PPARgamma modulator: approach to a safer insulin sensitizer. *PPAR Res.* 2008;2008:936906.

217. Rangwala SM, Lazar MA. The dawn of the SPPARMs? *Sci STKE.* 2002;2002:pe9.

218. Allen T, Zhang F, Moodie SA, et al. Halofenate is a selective peroxisome proliferator-activated receptor gamma modulator with antidiabetic activity. *Diabetes.* 2006;55:2523–2533.

219. Zhang F, Clemens E, Gregoire M, et al. Metaglidasen, a novel selective peroxisome proliferator-activated receptor-gamma modulator, preserves pancreatic islet structure and function in db/db mice. *Diabetes.* 2006;55:1396-P.

220. Zhang F, Clemens E, Gregoire M, et al. Metaglidasen, a selective PPARgamma modulator (SPPARgammaM) with anti-diabetic and hypo-lipidemic activity in multiple diabetic and insulin resistant rat models. In: *Conference on Diabetes Mellitus and the Control of Cellular Energy Metabolism,* Vancouver, BC, Canada: Keystone Symposia; 2006.

221. Benson SC, Pershadsingh HA, Ho CI, et al. Identification of telmisartan as a unique angiotensin II receptor antagonist with selective PPARgamma-modulating activity. *Hypertension.* 2004;43:993–1002.

222. Schupp M, Clemenz M, Gineste R, et al. Molecular characterization of new selective peroxisome proliferator-activated receptor gamma modulators with angiotensin receptor blocking activity. *Diabetes.* 2005;54:3442–3452.

223. Schupp M, Lee LD, Frost N, et al. Regulation of peroxisome proliferator-activated receptor gamma activity by losartan metabolites. *Hypertension.* 2006;47:586–589.

224. Pershadsingh HA. Treating the metabolic syndrome using angiotensin receptor antagonists that selectively modulate peroxisome proliferator-activated receptor-gamma. *Int J Biochem Cell Biol.* 2006;38:766–781.
225. Di Filippo C, Lampa E, Tufariello E, et al. Effects of irbesartan on the growth and differentiation of adipocytes in obese zucker rats. *Obes Res.* 2005;13:1909–1914.
226. Derosa G, Ragonesi PD, Mugellini A, Ciccarelli L, Fogari R. Effects of telmisartan compared with eprosartan on blood pressure control, glucose metabolism and lipid profile in hypertensive, type 2 diabetic patients: a randomized, double-blind, placebo-controlled 12-month study. *Hypertens Res.* 2004;27:457–464.
227. Vitale C, Mercuro G, Castiglioni C, et al. Metabolic effect of telmisartan and losartan in hypertensive patients with metabolic syndrome. *Cardiovasc Diabetol.* 2005;4:6.
228. Michel MC, Bohner H, Koster J, Schafers R, Heemann U. Safety of telmisartan in patients with arterial hypertension: an open-label observational study. *Drug Saf.* 2004;27:335–344.
229. Oberfield JL, Collins JL, Holmes CP, et al. A peroxisome proliferator-activated receptor gamma ligand inhibits adipocyte differentiation. *Proc Natl Acad Sci U S A.* 1999;96:6102–6106.
230. Rocchi S, Picard F, Vamecq J, et al. A unique PPARgamma ligand with potent insulin-sensitizing yet weak adipogenic activity. *Mol Cell.* 2001;8:737–747.
231. Berger JP, Petro AE, Macnaul KL, et al. Distinct properties and advantages of a novel peroxisome proliferator-activated protein [gamma] selective modulator. *Mol Endocrinol.* 2003;17:662–676.
232. Fujimura T, Sakuma H, Konishi S, et al. FK614, a novel peroxisome proliferator-activated receptor gamma modulator, induces differential transactivation through a unique ligand-specific interaction with transcriptional coactivators. *J Pharmacol Sci.* 2005;99:342–352.
233. Fujimura T, Kimura C, Oe T, et al. A selective peroxisome proliferator-activated receptor gamma modulator with distinct fat cell regulation properties. *J Pharmacol Exp Ther.* 2006;318:863–871.
234. Minoura H, Takeshita S, Yamamoto T, et al. Ameliorating effect of FK614, a novel nonthiazolidinedione peroxisome proliferator-activated receptor gamma agonist, on insulin resistance in Zucker fatty rat. *Eur J Pharmacol.* 2005;519:182–190.
235. Willson TM, Brown PJ, Sternbach DD, Henke BR. The PPARs: from orphan receptors to drug discovery. *J Med Chem.* 2000;43:527–550.
236. Li Y, Wang Z, Motani A, et al. T090313 (T131): a selective modulator of PPARgamma. In: *64th Annual Scientific Sessions of the American Diabetes Association (ADA), Orlando, Florida, USA*; 2004.
237. McGee LR, Rubenstein SM, Houze JB, et al. Discovery of AMG131: a selective modulator of PPARgamma. In: *231st American Chemical Society National Meeting, Atlanta, Georgia, USA*; 2006.
238. Kersey K, Floren LC, Pendleton B, Stempien MJ, Buchanan J, Dunn F. T0903131, a selective modulator of PPAR-gamma activity, increases adiponectin levels in healthy subjects. In: *64th Annual Scientific Sessions of the American Diabetes Association (ADA), Orlando, Florida, USA*; 2004.
239. Acton 3rd JJ, Black RM, Jones AB, et al. Benzoyl 2-methyl indoles as selective PPARgamma modulators. *Bioorg Med Chem Lett.* 2005;15:357–362.
240. Ramachandran U, Kumar R, Mittal A. Fine tuning of PPAR ligands for type 2 diabetes and metabolic syndrome. *Mini Rev Med Chem.* 2006;6:563–573.
241. Etgen GJ, Oldham BA, Johnson WT, et al. A tailored therapy for the metabolic syndrome: the dual peroxisome proliferator-activated receptor-alpha/gamma agonist LY465608 ameliorates insulin resistance and diabetic hyperglycemia while improving cardiovascular risk factors in preclinical models. *Diabetes.* 2002;51:1083–1087.

242. Cho N, Momose Y. Peroxisome proliferator-activated receptor gamma agonists as insulin sensitizers: from the discovery to recent progress. *Curr Top Med Chem*. 2008;8:1483–1507.
243. Harrington WW, Britt CS, Wilson JG, et al. The effect of PPARα, PPARδ, PPARγ, and PPARpan agonists on body weight, body mass, and serum lipid profiles in diet-Induced obese AKR/J mice. *PPAR Res*. 2007;2007:13. Article ID 97125, http://dx.doi.org/10.1155/2007/97125.
244. Azhar S. Peroxisome proliferator-activated receptors, metabolic syndrome and cardiovascular disease. *Future Cardiol*. 2010;6:657–691.
245. Yen CJ, Beamer BA, Negri C, et al. Molecular scanning of the human peroxisome proliferator activated receptor gamma (hPPAR gamma) gene in diabetic Caucasians: identification of a Pro12Ala PPAR gamma 2 missense mutation. *Biochem Biophys Res Commun*. 1997;241:270–274.
246. Deeb SS, Fajas L, Nemoto M, et al. A Pro12Ala substitution in PPARgamma2 associated with decreased receptor activity, lower body mass index and improved insulin sensitivity. *Nat Genet*. 1998;20:284–287.
247. Altshuler D, Hirschhorn JN, Klannemark M, et al. The common PPARgamma Pro12Ala polymorphism is associated with decreased risk of type 2 diabetes. *Nat Genet*. 2000;26:76–80.
248. Hara K, Okada T, Tobe K, et al. The Pro12Ala polymorphism in PPAR gamma2 may confer resistance to type 2 diabetes. *Biochem Biophys Res Commun*. 2000;271:212–216.
249. Yamauchi T, Kamon J, Waki H, et al. The mechanisms by which both heterozygous peroxisome proliferator-activated receptor gamma (PPARgamma) deficiency and PPARgamma agonist improve insulin resistance. *J Biol Chem*. 2001;276:41245–41254.
250. Nakano R, Kurosaki E, Yoshida S, et al. Antagonism of peroxisome proliferator-activated receptor gamma prevents high-fat diet-induced obesity *in vivo*. *Biochem Pharmacol*. 2006;72:42–52.
251. Wright HM, Clish CB, Mikami T, et al. A synthetic antagonist for the peroxisome proliferator-activated receptor gamma inhibits adipocyte differentiation. *J Biol Chem*. 2000;275:1873–1877.
252. Camp HS, Chaudhry A, Leff T. A novel potent antagonist of peroxisome proliferator-activated receptor gamma blocks adipocyte differentiation but does not revert the phenotype of terminally differentiated adipocytes. *Endocrinology*. 2001;142:3207–3213.
253. Mukherjee R, Hoener PA, Jow L, et al. A selective peroxisome proliferator-activated receptor-gamma (PPARgamma) modulator blocks adipocyte differentiation but stimulates glucose uptake in 3T3-L1 adipocytes. *Mol Endocrinol*. 2000;14:1425–1433.
254. Miyahara T, Schrum L, Rippe R, et al. Peroxisome proliferator-activated receptors and hepatic stellate cell activation. *J Biol Chem*. 2000;275:35715–35722.
255. Rieusset J, Touri F, Michalik L, et al. A new selective peroxisome proliferator-activated receptor gamma antagonist with antiobesity and antidiabetic activity. *Mol Endocrinol*. 2002;16:2628–2644.
256. Xu HE, Stanley TB, Montana VG, et al. Structural basis for antagonist-mediated recruitment of nuclear co-repressors by PPARalpha. *Nature*. 2002;415:813–817.
257. Etgen GJ, Mantlo N. PPAR ligands for metabolic disorders. *Curr Top Med Chem*. 2003;3:1649–1661.
258. UK Prospective Diabetes Study (UKPDS) Group. Intensive blood-glucose control with sulphonylureas or insulin compared with conventional treatment and risk of complications in patients with type 2 diabetes (UKPDS 33). *Lancet*. 1998;352:837–853.
259. Launer LJ, Miller ME, Williamson JD, et al. Effects of intensive glucose lowering on brain structure and function in people with type 2 diabetes (ACCORD MIND): a randomised open-label substudy. *Lancet Neurol*. 2011;10:969–977.
260. Mancia G. Effects of intensive blood pressure control in the management of patients with type 2 diabetes mellitus in the Action to Control Cardiovascular Risk in Diabetes (ACCORD) trial. *Circulation*. 2010;122:847–849.

261. Skyler JS, Bergenstal R, Bonow RO, et al. Intensive glycemic control and the prevention of cardiovascular events: implications of the ACCORD, ADVANCE, and VA diabetes trials: a position statement of the American Diabetes Association and a scientific statement of the American College of Cardiology Foundation and the American Heart Association. *Diabetes Care*. 2009;32:187–192.

262. Hansen BC, Tigno XT, Benardeau A, Meyer M, Sebokova E, Mizrahi J. Effects of aleglitazar, a balanced dual peroxisome proliferator-activated receptor alpha/gamma agonist on glycemic and lipid parameters in a primate model of the metabolic syndrome. *Cardiovas Diabetol*. 2011;10:7.

263. Balasubramanian J, Narayanan N. Role of Aleglitazar in T2DM: a bench-to-bedside. *Drug Discov*. 2013;3:3–4.

CHAPTER EIGHT

Free Fatty Acids and Skeletal Muscle Insulin Resistance

Lyudmila I. Rachek
Department of Cell Biology and Neuroscience, College of Medicine, University of South Alabama, Mobile, Alabama, USA

Contents

Abstract

Insulin resistance plays a key role in the development of type 2 diabetes mellitus and is also associated with several other diseases, such as obesity, hypertension, and cardiovascular diseases. Type 2 diabetes and obesity have become epidemic worldwide in the past few decades, and epidemiological and metabolic evidence indicates that the two conditions are linked closely through insulin resistance. The perturbation of free fatty acid (FFA) metabolism is now accepted to be a major factor contributing to whole-body insulin resistance, including that in skeletal muscle. Acute exposure to FFAs and excess dietary lipid intake are strongly associated with the pathogenesis of muscle insulin resistance. Despite an enormous amount of published research and the proposal of numerous hypotheses, however, the mechanisms underlying FFA-induced skeletal muscle insulin resistance have not been fully elucidated. This chapter describes existing hypotheses, recent findings, and debates about the role of FFAs in the development of muscle insulin resistance. Therapeutic options for this condition are also discussed.

Progress in Molecular Biology and Translational Science, Volume 121
ISSN 1877-1173
http://dx.doi.org/10.1016/B978-0-12-800101-1.00008-9

1. HISTORIC OVERVIEW OF THE SCIENTIFIC UNDERSTANDING OF INSULIN RESISTANCE

Insulin resistance can be broadly defined as an impaired response to the physiological effects of insulin, including glucose, lipid, and protein metabolism. This condition also affects vascular endothelial function. Insulin action is impaired in the liver, skeletal muscle, and adipose tissue in insulin-resistant subjects. This chapter focuses on insulin resistance in skeletal muscle.

Himsworth first described insulin resistance in 1930,[1–3] although this physician was not convinced that a primary deficiency of insulin was the only (or even a major) cause of diabetes. He observed that patients had variable responses to insulin treatment; some were "readily responsive to insulin," whereas others were "surprisingly insusceptible."[4] He believed that "the question of variation in insulin activity required deeper analysis"[4] and he invented the insulin–glucose test, in which oral glucose and intravenous insulin are administered simultaneously. Using this test, he distinguished "insulin-sensitive" and "insulin-insensitive" patterns in patients with diabetes mellitus based on the ability of subcutaneously administered insulin to dispose of an oral glucose load.[5]

With the development of clinical methods for the measurement of insulin action, researchers came to understand that almost all individuals with glucose intolerance or type 2 diabetes were insulin-resistant. Insulin resistance was further associated with hypertriglyceridemia in the 1970s and with essential hypertension thereafter. Currently, insulin resistance is recognized as a major metabolic risk factor. This condition is thus a target of interventions; for example, insulin resistance and associated disorders can be minimized or prevented by weight loss and exercise programs, which improve metabolic flexibility in skeletal muscle.

2. LIPIDS AND SKELETAL MUSCLE INSULIN RESISTANCE

Skeletal muscle is a major site of glucose uptake, storage, and disposal. Only 30% of glucose uptake is insulin-dependent in the basal state, whereas insulin-mediated glucose disposal increases to 85% in the postprandial state (after a meal).[6] Limb catheterization studies have shown that 80–90% of this increased disposal enters skeletal muscle.[6] This increased glucose flux into skeletal muscle, together with the activation of key enzymes in glucose metabolism by insulin, leads to a marked increase in muscle glucose oxidation.[7]

Free fatty acids (FFAs) are important oxidative fuels for many tissues, such as the heart, skeletal muscles, and liver.[8] Their roles become particularly important during starvation, prolonged exercise, and pregnancy.[9,10] Muscle energy metabolism, characterized predominantly by the oxidation of fat during fasting, switches to predominantly oxidation of glucose under post-prandial conditions.[11] This ability of skeletal muscle to change oxidation patterns is termed metabolic flexibility.[12] In the postprandial state, insulin promotes carbohydrate uptake at key storage sites (skeletal muscle and liver) and prompts the conversion of carbohydrates and protein to lipids, which store calories more efficiently. Conversely, metabolic inflexibility is defined as the inability to efficiently take up and store fuel and to transition from fat to glucose as the primary fuel source during times of plenty (increased insulin).

Insulin resistance is now accepted to be closely associated with lipid accumulation in muscle cells.[13] Lipotoxicity, characterized by the accumulation of ectopic lipids in skeletal muscle, is a major factor in the etiologies of insulin resistance and type 2 diabetes. Under this condition, lipid metabolites interfere with insulin signaling and action. Skeletal muscle insulin resistance can precedes overt manifestations and complications of diabetes, such as hyperglycemia and beta-cell failure, by decades.[14] Moreover, insulin-stimulated whole-body and muscle glucose disposal are compromised in healthy, normal weight individuals with diabetic parents, to a similar extent as in the parents.[15–17] However, the development of skeletal muscle insulin resistance can be independent of a family history of type 2 diabetes.

Skeletal muscle and the liver normally contain small amounts of triglycerides (TGs). In contrast, ectopic fat storage is characterized by the accumulation of large TG droplets in nonadipose tissues. The consequences of ectopic fat accumulation depend on the organ involved, but similar mechanisms leading to the disruption of organ function are present at the cellular level. Lipids can be dispersed intercellularly or accumulate intracellularly. Intercellular lipid accumulation may impair organ function via the paracrine effects of released adipokines, whereas intracellular lipid accumulation is associated with decreased insulin sensitivity in skeletal muscle.

In obese individuals and those with type 2 diabetes, FFA plasma concentrations are usually elevated due to increased adipose tissue mass and lipolysis caused by insulin resistance.[11,18] Because of its higher lipolytic activity in comparison with other adipose tissue, visceral fat was long thought to be a primary source of increased FFA flux responsible for whole-body insulin resistance.[19] However, recent studies have found that subcutaneous adipose tissue in the upper abdomen, which has greater absolute adiposity than visceral fat, is a major site of systemic FFA flux.[20] The development of obesity has also been associated with abundant and large adipocytes,[21] and these cells appear to contribute to insulin resistance through higher lipolysis rates and altered adipokine secretion patterns. Abdominal adipocyte size thus plays a major role in the etiologies of insulin resistance and type 2 diabetes, constituting a risk factor independent of overall and central adiposity.[22]

Several mechanisms lead to the accumulation of ectopic fat. First, continuous FFA oversupply due to enhanced lipolysis resulting from adipocyte dysfunction causes muscle TG accumulation under the condition of insulin resistance. Second, FFA oversupply results from lipoprotein lipase in liver from very-low-density lipoprotein TG or directly from gut-derived chylomicrons in the postprandial state. Third, enhanced FFA transport into cells under the condition of insulin resistance and reduced muscle FFA oxidation in mitochondria have the combined effect of excess lipid accumulation in the cytosol. Free FFAs are taken up by cells via passive diffusion and specific transport proteins in the cell membrane (CD36, FFA transport protein).[23] Within cells, FFA binding protein is the most important cytosolic protein guiding long-chain FFAs to oxidation or etherification sites. Long-chain fatty acyl-coenzyme A (LCACoA) is taken up by mitochondria via carnitine-palmitoyltransferase 1 (CPT1). Within mitochondria, β-oxidation and further degradation in the tricarboxylic acid cycle occur. Taken together, intramuscular lipid accumulation is thus a consequence of continuous FFA oversupply (caused by enhanced lipolysis, adipocyte, and liver dysfunction) combined with impaired FFA oxidation in the mitochondria.

2.1. The Randle hypothesis: Glucose–FFA competition

Several hypotheses have been proposed to explain how FFAs interfere with glucose metabolism and insulin sensitivity. In the early 1960s, Randle et al.[24] proposed the "glucose–FFA cycle." Based on studies in isolated rat cardiac muscle, these researchers argued that insulin sensitivity and other metabolic disturbances associated with type 2 diabetes were associated with preferential oxidation of FFAs rather than glucose. The glucose–FFA cycle consists of two major processes occurring under contrasting conditions: (1) Under increased glucose availability, FFA oxidation is inhibited while glucose oxidation and storage, as well as lipid storage, are enhanced, and (2) under increased FFA availability, glucose oxidation is inhibited while FFA oxidation is enhanced. Randle *et al.* suggested that mitochondrial acetyl-CoA production increased with FFA oxidation, inhibiting pyruvate dehydrogenase activity and elevating citrate in the tricarboxylic acid cycle. In turn, this process inhibited phosphofructokinase, causing the accumulation of glucose-6 phosphate, a hexokinase inhibitor, and thereby increasing intracellular glucose content while reducing glucose uptake. The Boden group confirmed Randle *et al.*'s hypothesis, demonstrating a relationship between increased fat oxidization and decreased carbohydrate oxidation in healthy human skeletal muscle after 1 h of acute lipid infusion.[25]

However, the findings of other studies contradicted Randle's hypothesis. In the glucose–FFA cycle, impaired glucose uptake follows intracellular glucose accumulation; in contrast, other researchers demonstrated that such inhibited uptake was associated primarily with FFA–induced insulin resistance, rather than altered glucose metabolism.[26,27] Studies combining lipid infusion with methods such as nuclear magnetic resonance (NMR) imaging and glucose and insulin clamping revealed a rapid (within 2 h) reduction in glycolysis, followed (at 4–6 h) by impaired glucose disposal and glycogen synthesis.[28,29] Roden *et al.*[29] found that a reduction in intramuscular glucose 6-phosphate preceded reduced glycogen synthesis in muscle, suggesting that insulin resistance is induced by increased plasma FFA concentration via the inhibition of glucose transport or phosphorylation. Recent studies have also indicated that FFA-induced insulin resistance involves mechanisms other than the glucose–FFA cycle, which cannot fully explain FFAs' effects on glucose metabolism.[30] Using ^{13}C NMR spectroscopy and hyperinsulinemic–euglycemic clamping after a 5 h infusion of heparin with TGs or glycerol, the Shulman group found that increased FFA availability reduced, rather than increased (as expected in the Randle hypothesis), glucose-6 phosphate levels.[31] Thus, in summary, several lipid infusion studies have demonstrated that glycogen synthesis is inhibited in the absence of an increase in glucose-6 phosphate, in contradiction to Randle's proposed glucose–FFA cycle. Research findings have suggested that FFA-induced insulin resistance is limited primarily by glucose transport and revealed molecular mechanisms beyond the glucose–FFA cycle underlying insulin resistance.

2.2. Current prospects: The revised Randle hypothesis: FFA intermediation with insulin signaling

As described earlier, insulin resistance is closely associated with TG accumulation in muscle. However, studies of athletes with endurance training have suggested that insulin sensitivity is associated with increased intramyocellular lipids in the form of TGs.[32,33] This "athlete paradox" has been replicated in transgenic mice with overexpression of the enzyme diacylglycerol (DAG) acyltransferase 1, which catalyzes the final step of TG synthesis, in skeletal muscle. These insulin–sensitive mice reflect the situation observed in athletes; their intramuscular lipid content is similar to that observed in murine models of fat-induced insulin resistance, but DAG and ceramide levels are reduced and mitochondrial FFA oxidation is increased.[34] This model suggests that TG accumulation is a neutral factor, whereas accumulated intermediate lipid metabolites may have a lipotoxic effect, disrupting insulin action.[35]

Dysfunctional adipose tissue may be created by visceral adiposity or limited adipose tissue expandability under positive energy balance conditions and may ultimately lead to ectopic FFA accumulation through the uptake of excess FFA in the circulation by the muscle, the liver, and other organs. In all tissues, FFA metabolism begins with the activation of acyl-CoA synthase, which transforms FFA to its LCACoA derivative. LCACoA metabolism then follows three pathways: (1) Acylcarnitine transferase transports LCACoA to the mitochondria, where it enters the β-oxidation pathway; (2) LCACoA is esterified to monoacylglycerol, DAG, and TGs in lipid droplets; and (3) LCACoA is integrated into membrane phospholipids (reviewed in Ref. 36). The rate of LCACoA β-oxidation is limited by its transport to the mitochondria by CPT1. In TG synthesis from FFAs, LCACoA moieties are added sequentially to a glycerol backbone and intermediary FFA metabolites, including lysophosphatidic acid (LPA), phosphatidic acid (PA), and DAG, are formed. Ceramide is also created when saturated FFAs are abundant.[37]

To understand how FFAs intermediate with insulin signaling, we now briefly describe the mechanisms of insulin signaling in skeletal muscle under normal physiological conditions. Insulin action in target cells (muscle, for our purpose here) begins when insulin binds to the insulin receptor on cell membranes. This activates the insulin receptor tyrosine kinase, which phosphorylates and activates a protein called insulin receptor substrate 1 (IRS-1). In turn, IRS-1 activates two parallel cell signaling pathways, the (1) PI3K/Akt pathway, which activates glucose transport, protein synthesis, and glycogen synthesis (Fig. 8.2) and (2) ERK/MAPK pathway, which activates downstream genes responsible for the growth-promoting effects of insulin. Insulin by itself is just the messenger; the real action happens at the protein glucose transporter 4 (GLUT4), which gets activated downstream of PI3K/Akt. Upon activation, GLUT4 is sent from the cytosol to the plasma membrane, where it acts as a gateway for glucose to enter the cell (Fig. 8.1). The more GLUT4 molecules that are sent to the plasma membrane, the more glucose will be allowed to enter and utilized inside the cell, thus, the less glucose will be left outside the cell, and overall, the less hyperglycemia is. The exact details of how insulin activates GLUT4 translocation and glucose uptake still need to be elucidated, but the basics here are the key to understanding insulin sensitivity: the less insulin needed

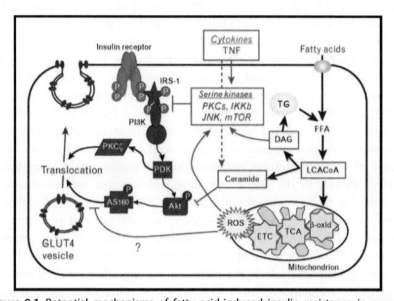

Figure 8.1 Potential mechanisms of fatty acid-induced insulin resistance in muscle. Exposure to excess fatty acids results in the accumulation of intramyocellular lipid species such as diacylglycerol (DAG) and ceramides. DAG is thought to activate serine kinases that can serine phosphorylate and reduce the signal transduction capacity of insulin receptor substrate 1 (IRS-1). Ceramides can interfere with insulin signalling at the level of Akt and are produced *de novo* from fatty acids or by release from sphingolipids in response to stress cytokines such as tumour necrosis factor (TNF). TNF and other cytokines associated with fat accumulation can also activate serine kinases directly via inflammatory signalling pathways. Excess fatty acid oxidation in the mitochondria, via the tricarboxylic acid (TCA) cycle and the electron transport chain (ETC), can lead to an increase in fatty acid metabolites and reactive oxygen species (ROS) which can activate intracellular stress kinases but may possibly have effects on insulin signalling to GLUT4 translocation at unknown points downstream of Akt and AS160 function. TG, triglyceride. *Reprinted with permission from Copyright Clearance Center*[38].

to send GLUT4 to the cell membrane for glucose uptake, the more insulin-sensitive the skeletal muscle, and overall the whole body, is.

What happens with insulin signaling under pathophysiological condition of increased FFAs in the muscle cell? Studies on high-fat diet (HFD) or palmitate-induced insulin resistance showed the accumulation of FFA metabolites (i.e., LPA, PA, and DAG) in muscle cytosol occurs when FFA flux in muscle cells exceeds LCACoA and β-oxidation capability and insulin-dependent FFA storage sites for TG synthesis are saturated. These FFA metabolites can activate serine kinases (e.g., c-Jun N-terminal kinase (JNK), I kappa-β kinase (IKKβ), and novel protein kinase Cs (PKCs)), which inhibit insulin activity and promote degradation by proteasome through serine phosphorylation of IRS-1 protein (Fig. 8.1).[39–41]

Among the novel PKCs, DAG directly activates PKCθ, that catalyzes the phosphorylation of the serine-307 residue at IRS-1, reducing its tyrosine phosphorylation and activation by insulin. In line with this study, Schmitz-Peiffer et al. demonstrated increased concentration of DAG in rodent muscle and activation of PKCs induced by HFD.[42] Similarly, infusion of lipid and heparin caused insulin resistance in muscles that was associated with accumulation of intracellular DAG and specific activation of PKCθ.[26] Insulin resistance in this model was due to lipid-induced defects in the insulin signaling pathway that was caused by a reduction in tyrosine phosphorylation of IRS1, increasing its phosphorylation in serine-307 residue.[43] However, as indicated and reviewed by Samuel et al.[44], there is still no evidence to explain how the activation of novel PKCs might relate to serine phosphorylation of IRS1 and which kinases might have a role in the pathway. Among other serine/threonine kinases that are activated in HFD-induced or palmitate-induced insulin resistance is a mammalian target of rapamycin[45] but the mechanisms involved are unknown.

Ceramides can also inhibit insulin activity through protein kinase B (Akt) inhibition (Fig. 8.1).[41] There are several proposed mechanisms of how ceramides may impair Akt activation (reviewed in Ref. 44): (1) Ceramides may lead to activation of protein phosphatase 2A,[46] which can dephosphorylate Akt, inhibiting its translocation to the plasma membrane and thus effectively impairing insulin signaling, and (2) ceramides may impair insulin signaling via the atypical PKC isoform, PKCζ.[47] PKCζ and Akt interact intracellularly but dissociate upon insulin stimulation; ceramides impair this disassociation, and furthermore, via PKCζ phosphorylation of Akt, prevent Akt activation.

3. SATURATED VERSUS UNSATURATED FFAs AND INSULIN RESISTANCE

It has been postulated that the type of FFAs, saturated or unsaturated, is critical in the development of insulin resistance, as the degree of saturation correlates with insulin resistance. Previously, it has been shown that in diabetic patients, saturated FFAs cause insulin resistance[48,49] whereas unsaturated FFAs improve insulin sensitivity.[50,51] Also, studies *in vitro* have demonstrated that the saturated FFA palmitate-induced insulin resistance in skeletal muscle cells,[52–54] whereas the unsaturated FFA oleate improved insulin sensitivity.[53,55,56] A recent study showed that palmitate-induced insulin resistance and apoptosis were positively related, while oleate did not induce apoptosis and improved insulin sensitivity in L6 skeletal muscle cells.[54] However, the mechanisms by which these two common dietary FFAs cause different effects on insulin resistance and apoptosis have not been fully elucidated. Potential candidates mediating the effects of the saturated FFA palmitate (which is the most common saturated FFA in human diet) on insulin resistance include: (1) an increase in production of ceramide[52,57] and accumulation of DAG, leading to activation of PKCθ;[58] (2) mitochondrial dysfunction[59,60] and increased oxidative stress;[60–62] (3) activation of proinflammatory nuclear factor-κB (NF-κB) and mitogen-activated protein kinases;[63] and (4) decreased peroxisome proliferator-activated receptor (PPAR) γ coactivator-1α/β (PGC-1α/β) activation[64] and reduced mitochondrial gene expression and reduced cellular oxygen consumption rates.[64]

As it has been mentioned earlier, the role of ceramide in mediating insulin resistance may be limited to saturated FFAs, because the rate of ceramide synthesis depends largely on the availability of long-chain saturated FFAs, which participate in the initial, rate-limiting step in *de novo* ceramide synthesis.[65] As it has been reviewed by Samuels,[44] treating fat-fed mice with myriocin, an inhibitor of serine palmitoyltransferase 1 (the first enzyme in the *de novo* synthesis of ceramide from palmitate), specifically attenuates the increase in muscle ceramide content without any change in LCACoA, DAGs, or TG and improves glucose tolerance. Myriocin prevents acute skeletal muscle insulin resistance following infusion of palmitate, but not oleate.[66] Mice fed 12 weeks of a lard-based diet (enriched with saturated fat) are protected from glucose intolerance when treated with myriocin.[67]

Compelling evidence in the literature has suggested that all of the aforementioned factors linking palmitate-induced insulin resistance are somehow connected. Indeed, a very recent study has shown that mitochondrial super-oxide production is a common feature of many different models of insulin resistance, including saturated fat-induced insulin resistance in skeletal muscle cells.[62] The consumption of a Western diet that is high in saturated fats significantly worsens insulin resistance,[68,69] whereas diets rich in mono- and polyunsaturated FFA have a less pronounced effect or even improve insulin sensitivity.[70,71] Therefore, supplementing diets with unsaturated fat may have a favorable affect on the prevention of insulin resistance and development of type 2 diabetes. Although several different mechanisms for the beneficial effect of oleate on insulin signaling have been proposed,[63,72] the exact mechanisms remain to be elucidated.

4. ADDITIONAL INSIGHTS INTO MECHANISMS INVOLVED INTO FFA-INDUCED MUSCLE INSULIN RESISTANCE

In addition to pathways described earlier, it has been shown that other mechanisms are involved in FFA-induced insulin resistance in skeletal muscle, such as the following: (1) FFAs activate inflammatory signals that lead to insulin resistance, (2) FFAs induce endoplasmic reticulum (ER) stress, (3) FFAs induce mitochondrial dysfunction and increased mitochondrial oxidative stress, (4) FFAs have an effect on gene regulation that contributes to impaired glucose metabolism; and (5) FFA composition of membrane phospholipids influences insulin action.

Recently, a growing amount of evidence has been reported to support the role for cytokines and activation of inflammatory signaling pathways to insulin resistance. As reviewed in Ref. 73, saturated FFAs activate inflammation by two ways: (1) indirectly through the secretion of cytokines including TNF-α, IL-1β, and IL-6[74,75] and (2) directly through interaction with members of Toll-like receptor (TLR) family. Saturated FFAs activate TLR-4 in skeletal muscle promoting JNK and IKKβ complex activation, which results in degradation of the inhibitor of IKKβ and NF-κB activation. Activation of JNK and IKKβ by saturated FFAs is associated with a marked inhibition of insulin action due to the phosphorylation of serine residues on the insulin IRS-1, and inhibition of its stimulatory phosphorylation of tyrosine residues by the insulin receptor.[76,77] The importance of TLR4 in insulin resistance has been proven by studies using mice containing a loss of function mutation in this receptor. These mice are partially protected from fat-induced inflammation and insulin resistance.[78] In addition, diabetic and obese mice have increased skeletal muscle IKK and JNK activities, whose pharmacological and genetic inhibition leads to an improvement in insulin sensitivity and glucose tolerance[79,80] (reviewed in Ref. 73).

In addition, inflammatory signals may link saturated fatty acids to ceramide synthesis. Palmitate infusion increases plasma cytokine concentrations, and mice lacking TLR4 are protected from ceramide accumulation and insulin resistance following lard, but not soy oil, infusions.[81] Some have also suggested that saturated FFAs may also be the ligands for TLR4.[82] Thus, as reviewed in Ref. 44, intracellular ceramides may also act as "second messengers" that coordinate a cell's response to circulating cytokines or possibly nutrient (e.g., saturated FFAs) signals.

Regarding the link between FFA-induced ER stress and development of insulin resistance, activation of ER stress has been shown in liver of leptin-deficient mice.[83] A significant amount of publications has shown that ER stress plays an important role in hepatic insulin resistance (reviewed in Ref. 44). Concerning skeletal muscle, there are conflicting data as to whether a HFD-induced ER stress, which possibly can be explained by the difference in study duration and diet composition.[83,84] Recently, consistent with previous data,[84] our group has shown that HFD increased markers of ER stress in both skeletal muscle and liver.[85] Also, there is a link between ER stress and mitochondrial dysfunction: it is widely accepted that ER stress induces mitochondrial dysfunction.[86] Furthermore, it has been shown that mitochondrial dysfunction increased the level of ER stress markers in adipocytes.[87] Mechanisms describing the link between mitochondrial function and development of insulin resistance are discussed separately (Section 5). In summary, models describing skeletal muscle insulin resistance due to inflammation, ER stress and mitochondrial stress, and redox imbalance state that each of cellular insults is thought to engage stress-sensitive serine kinases disrupting insulin signaling as discussed in Section 2.2 and shown in Fig. 8.1.[88]

Increased lipid alters expression of specific genes, mostly related to pathways of lipid metabolism.[89] FFAs can bind and thus activate nuclear receptors, the most important of which are PPARs. Use of pharmacological agonists of these receptors to ameliorate insulin resistance in skeletal muscle will be discussed in Section 7. There is a distinct pattern of change in skeletal muscle gene expression from insulin-resistant subjects.[89–91] Also, it has been shown that plasma FFA is negatively correlated with the expression of PGC-1α, and nuclear encoded mitochondrial genes and also increases the expression of extracellular matrix genes in a manner reminiscent of inflammation.[92] In animal and human studies, a significant correlation has been demonstrated between the FFA composition of muscle membranes and insulin action at a tissue and whole-body level.[93,94] The mechanisms by which membrane phospholipid composition influences insulin action are not clear but may involve (1) changing membrane fluidity that could affect insulin and other membrane receptor action,[95] (2) release of different DAG molecules after phospholipase activity that altered modulatory effects on cellular processes such as PKC activity, and (3) altered cellular energy expenditure influencing accumulation of intracellular TG.[96] In summary, the current opinion now is that HFD-induced insulin resistance reverses too quickly to implicate significant changes in membrane structural lipids; nevertheless, these issues remain to be elucidated.

5. FFAs, MITOCHONDRIAL FUNCTION, AND INSULIN RESISTANCE

Increasing evidence accumulated over the last decade indicates that mitochondrial dysfunction and oxidative stress have been implicated in the skeletal muscle insulin resistance[97,98] but the underlying mechanisms are still unknown. Mitochondria are the primary site of skeletal muscle fuel metabolism and ATP production. In addition, they are responsible for β-oxidation of FFA. Therefore, when mitochondria are less capable of burning FFAs, for example, because the mitochondria are damaged, accumulation of lipid intermediates can occur inside a cell. As reviewed in Ref. 99, the widespread hypothesis that a "defect" in mitochondrial β-oxidation of FFA is a major contributing factor to the etiology of HFD-induced insulin resistance raised the prospect that any intervention that promote FFA oxidation should relieve the toxicity caused by accumulation of lipid metabolites and thus improve insulin sensitivity. According to this hypothesis, LCACoA derived from circulating lipids or intramuscular TGs are diverted away from CPT1, the mitochondrial enzyme that catalyzes the first and essential step of β-oxidation of FFA, the transport of FFA into mitochondria, and are instead preferentially partitioned forward the synthesis of DAG and ceramide.[100] However, recent studies have challenged this theory by showing that genetic manipulations that increase skeletal muscle levels of these metabolites do not necessarily lead to insulin resistance.[101] Moreover, it has been shown that obesity-associated glucose intolerance might arise from excessive, rather than reduced β-oxidation, and suggested that FFAs must penetrate mitochondria to exert their insulin-desensitizing actions in skeletal muscle.[102] Also, a more recent study performed by the same group has shown that mitochondrial overload and incomplete FFA oxidation improved insulin sensitivity in skeletal muscle.[103] Taken together, the role of β-oxidation per se as an underlying cause of obesity-induced insulin resistance is still controversial.

Another hypothesis that explains mechanisms leading to FFA-induced muscle insulin resistance is that mitochondria derived oxidative stress impairs insulin signaling. Apart from producing energy, mitochondria also are a major source of reactive oxygen/nitrogen species (ROS/RNS).[104] Oxidative stress results from an increased content of ROS and/or RNS. ROS and RNS directly oxidize and damage DNA, proteins, and lipids and are believed to play a direct key role in the pathogenesis of insulin resistance.[62,97,98] Skeletal muscle is particularly vulnerable to oxidative stress because it is composed of postmitotic cells that are capable of accumulating oxidative damage over time and because they consume a large amount of oxygen for their action. A study by Houstis *et al.* confirmed that insulin resistance can be prevented by blocking the increase in ROS levels, including mitochondrial ROS (mtROS).[105] Additionally, a more recent study showed that mitochondrial superoxide production is a unifying element of insulin resistance.[62] Also, a very recent study by the Shulman group showed that targeted expression of catalase to mitochondria prevented age-associated reductions in mitochondrial function and insulin resistance.[60] Elevated saturated FFA levels have numerous deleterious effects on mitochondria, including increased production of ROS.[60,61,105,106] Moreover, it has been postulated that impaired mitochondrial function could directly contribute to insulin resistance by impairing the production of ATP, which is essential for the support of all the reactions in the insulin signaling pathway that require phosphorylation.[107] It is still arguable whether mitochondrial dysfunction precedes or results from insulin resistance in the skeletal muscle of type 2 diabetes patients or whether these are parallel processes. A recent study by Bonnard *et al.*[98] showed that mitochondrial dysfunction results from oxidative stress in the skeletal muscle of diet-induced insulin-resistant mice. Also, data obtained in this study[98] and in the more recent work describing the prevention of age-associated insulin resistance in mice with targeted overexpression of human catalase in mitochondria,[108] suggested that mitochondrial dysfunction per se is not the initial event that triggers the inhibition of insulin action, as observed in type 2 diabetes, but it is rather the increased oxidative stress that promotes mitochondrial alterations, lipid accumulation, and insulin resistance.

Previous studies have shown that multiple stress-sensitive kinases such as JNK, NF-κB, and p38 MAP kinase, and some novel and atypical PKC isoforms, are activated by oxidative stress, which lead to the pathological condition of insulin resistance[109–114] and protective strategies including using ROS scavengers and antioxidants have improved insulin signaling both *in vitro* and *in vivo*.[115,116] One such major kinase target for oxidative stress is JNK, activation of which by oxidative stress has been previously shown to interfere with insulin action both *in vivo* (fat-fed animals) and in skeletal muscle cells treated with hydrogen peroxide by decreasing insulin-stimulated tyrosine phosphorylation of IRS-1.[115] Furthermore, antioxidants preserved redox balance, inhibited JNK activation, and thus improved insulin signaling in both fat-fed animals and in skeletal muscle cells treated with hydrogen peroxide.[115] Also, α-lipoic acid has been shown to improve insulin sensitivity in skeletal muscle both *in vivo* and *in vitro* through inhibition of JNK activation.[116] Additionally, it has been found that selective inhibition of JNK in adipose tissue protected against diet-induced obesity and improved insulin sensitivity in both liver and skeletal muscle in mice.[117]

Previously, our group and others have shown that saturated FFA palmitate induced the generation of ROS in skeletal muscle cells.[60,105] In addition, it has been shown that ROS production occurred through activation of the NAPDH oxidase system, and, also by the alteration of mitochondrial electron transport chains.[61] Moreover, we found that palmitate-induced mtROS generation through the *de novo* synthesis of ceramide.[105] In addition, we demonstrated that ceramide treatment led to the increase of mtROS generation.[105] Ceramide has been directly connected with mitochondrial oxidative stress: ceramide triggers mitochondrial oxidative stress, and oxidative stress in turn promotes further ceramide synthesis.[118] In addition, we showed that palmitate decreased the expression of two major mitochondrial transcription factors, PGC-1α and mitochondrial transcription factor A (TFAM), which regulate mitochondrial biogenesis.[105] Also, palmitate radically decreased the promoter activity of PGC-1α. Moreover, we identified that palmitate-induced downregulation of those transcription factors, as well as the promoter activity of PGC-1α, is mediated by oxidative stress, since the ROS scavenger significantly restored expression of both TFAM and PGC-1α and the promoter activity of PGC-1α.[105] In summary, we proposed a model for the palmitate-induced mitochondrial dysfunction and ROS production, and consequent insulin resistance (Fig. 8.2). We suggested that palmitate-mediated ROS-induced downregulation of mitochondrial transcription factors (TFAM and PGC-1α)

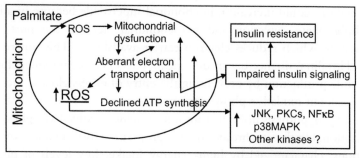

Figure 8.2 A schematic model of the proposed links between palmitate-induced mitochondrial dysfunction and ROS production, and consequent insulin resistance in skeletal muscle cell.

may further contribute to palmitate-induced mitochondrial dysfunction, oxidative stress, and consequent insulin resistance by decreasing mitochondrial biogenesis and thus establishing a vicious cycle of events in which palmitate-induced oxidative stress causes mitochondrial dysfunction, which causes a concomitant increase in ROS production and impairment of insulin signaling (Fig. 8.2).[105]

6. IS FFA-INDUCED INSULIN RESISTANCE A CELLULAR ADAPTATION PROCESS?

In current theoretical frameworks, pathophysiological processes such as insulin resistance are regarded as mechanisms contributing to disease development. However, another perspective suggests that insulin resistance is "a highly regulated adaptation that protects, or at least partially protects, the very organ systems (muscle, adipocytes, and liver) that are pivotal in generating insulin resistance."[119] At the cellular level, insulin resistance can be viewed as a defense against overnutrition and further excess of the substrate. Hoehn et al.[62] further developed this alternative perspective, providing convincing evidence that many different models of insulin resistance in adipocytes, myotubes, and mice were linked by a single specific radical superoxide generated in the mitochondria. They placed this mitochondrial superoxide at the nexus between intracellular metabolism and the control of insulin action, potentially defining it as a metabolic sensor of excess energy.

How can insulin resistance act as an antioxidant cellular defense? The provision of a greater flux of nutrients through mitochondrial oxidative phosphorylation with insulin, with no consequent increase in ATP consumption, could increase mitochondrial superoxide because of adenosine diphosphate depletion and mitochondrial electron transport chain dysfunction, which reduces electron carrier availability. This superoxide-induced insulin resistance has an antioxidant effect because it reduces further superoxide production by preventing the entry of excess glucose into cells and subsequent oxidative phosphorylation.

Another hypothesis holds that insulin resistance evolved as a physiological adaption, but that its chronic and inappropriate activation under modern conditions, characterized by high- energy intake, low levels of physical activity, and chronic stress, ultimately causes metabolic syndrome (reviewed in Ref. 120). Insulin's anabolic activity regulates the storage of energy, mainly as glycogen (in the liver) and TGs (in adipose tissue). When the body is stressed (e.g., by infection or trauma), insulin resistance in target tissues serves to mobilize this stored energy via proinflammatory cytokines and stress hormones. Under modern conditions, large amounts of excess fat are stored in adipose tissue, overloading storage sites and liporegulation capacity; in central fat tissue, such overloading activates an inflammatory response and causes adipocyte dysfunction, resulting in low-grade systemic inflammation and lipid overflow to peripheral tissues. Consequently, insulin resistance is activated (as under stressful conditions) by proinflammatory cytokines and the accumulation of nonoxidized lipid metabolites in the liver and muscle cells. Thus, the negative regulation of insulin signaling may be viewed as an "adaptive" physiological mechanism that is activated when the body needs to switch from an anabolic to a catabolic or "insulin resistance" state and mobilize energy, primarily in the form of glucose released from the liver and free FFAs released from adipocytes, to support vital metabolic processes.

7. CLINICAL PERSPECTIVES AND THERAPEUTIC IMPLICATIONS

It is clear now that insulin resistance is not simply a defect in glucose disposal but rather it is a much more widespread dysregulation of metabolism that significantly contributes to the development of obesity-induced disorders. An understanding of the pathogenic mechanisms behind insulin resistance is of great interest as it potentially allows design of appropriate therapeutic interventions. Substantial progress has been achieved in recent years in our understanding of the intracellular signaling pathways mediating insulin's varied biological effects. A further area of progress is the understanding of "cross talk" that exists between metabolically active tissues as exemplified by the adipocytokines, which can influence glucose homeostasis by acting on nonadipose tissues. Also of profound importance have been studies of the role played by PPARs in insulin signaling and the effect of their agonists, the thiazolidinediones (TZDs). These agents represent a major advance in this field, as they are the first direct pharmacological means available to treat insulin resistance. While much remains to be learned about their exact mode of action, a more complete understanding of this could permit development of more efficacious treatments for insulin resistance.

The PPARs are nuclear receptors that regulate transcription in response to FFAs and, as such, are potential therapeutic targets for obesity-related diseases. Three different human PPAR subtypes have been identified so far, designated as PPARs, α, β/δ, and γ; they are activated by lipids and are targets for current and prospective drug therapies for components of the metabolic syndrome.[121] PPARα, a target for the fibrate class of lipid-lowering drugs,[122,123] is primarily expressed in the liver, where it upregulates genes involved in lipid oxidation in the fasted state.[124] PPARγ is highly expressed in adipose tissue and regulates adipogenesis and insulin sensitivity.[125] TZDs are a class of drugs that increase insulin sensitivity through activating PPARγ.[126] PPARδ is expressed in many tissues, including metabolically active sites such as liver, muscle, and fat, and its role in the metabolic syndrome is only now being elucidated.[127] Treatment with a high-affinity PPARδ agonist GW501516 has been shown to increase high-density lipoprotein cholesterol,[127] affect lesion progression in a mouse model of atherosclerosis,[128] cause weight loss, and regulate muscle fiber-type switching when constitutively activated.[129] In the latter two cases, the phenotypes appeared to be mediated through upregulation of lipid catabolism and oxidative phosphorylation in fat and muscle. In addition, ligands for PPARδ have been proposed to be potential insulin sensitizers, based on improvements in standard glucose tolerance tests.[130] These studies, however, used long-term ligand treatment regimens that resulted in significant weight loss and a decrease in fat mass. These effects alone enhance insulin sensitivity. Therefore, it remains unclear whether PPARδ can directly regulate insulin sensitivity and, if so, through which tissue and what mechanism.

The overall improvement of insulin sensitivity observed upon glitazone treatment may potentially result from PPARγ activation in skeletal muscle. Even though PPARγ is expressed at a low level in myofibers of humans and rodents, the net result of skeletal muscle PPARγ activation is potentially relevant, because skeletal muscle is the largest glucose-utilizing organ in the body. Mice with genetic deletion of PPARγ in skeletal muscle showed significantly increased whole-body insulin resistance, demonstrated either by insulin/glucose-tolerance tests or by hyperinsulinemic–euglycemic clamp studies, and developed dyslipidemia, enlarged fat pads, and obesity on HFD.[131] It appears that the pharmacological response to TZDs is preserved, at least under some experimental conditions, in mice lacking PPARγ selectively in muscle.[131] Thus, it is unlikely that a direct action on muscle is the primary basis for the clinical effects of PPARγ agonists, again underscoring the importance of adipose tissue as the main mediator of TZD actions.

Another key regulator of both glucose and lipid metabolism, which is associated with improved insulin signaling and enhanced insulin sensitivity in skeletal muscle, is 5′-AMP kinase (AMPK). AMPK activation increases fatty acid oxidation in skeletal muscle by decreasing malonyl-CoA concentrations. Both TZDs (i.e., pioglitazone)[132] and metformin[133] have been shown to improve glucose tolerance via AMPK. Activation of AMPK by metformin decreased the level of plasma glucose and plasma triglycerides by promoting muscle glucose uptake and inhibiting hepatic glucose output.[134] Recently, Coletta et al. have demonstrated that pioglitazone activates AMPK and acetyl-CoA carboxylase (ACC) in human muscle biopsies from patients with type 2 diabetes, leading to increased expression of genes involved in mitochondrial function and fat oxidation and reduced toxic burden of intracellular lipid metabolites (LCoCoA, DAG, and ceramides).[132]

One of the most commonly used drugs for the treatment of type 2 diabetes is metformin. It is an effective hypoglycemic drug that also improves lipid profiles[135] and reduces cardiovascular risk.[136] Despite years of research, the effects of metformin on glucose uptake in skeletal muscle remain controversial. It has been shown that metformin induced increases in AMPK activity that was associated with higher rates of glucose disposal and muscle glycogen concentrations.[133]

Another opportunity to alleviate the excess of FFA is both fasting or undernourishment, but both happen rarely in modern life. Indeed, with a schedule of three meals daily, a majority of the 24 h is spent in postprandial metabolism. The other physiological circumstance for promoting fat oxidation is low to moderate intensity physical activity, especially a sustained duration of physical activity. For many people in modern society, careful scheduling and a commitment to undertake daily exercise are required, because work and daily living no longer require manual labor. Thus, the structure of modern life impedes the physiology of fat oxidation and disposes instead to fat storage. Physical activity has a beneficial effect on insulin sensitivity in normal as well as insulin-resistant populations (reviewed in Ref. 137). A distinction should be made between the acute effects of exercise and genuine training effects. Up to two hours after exercise, glucose uptake is in part elevated due to insulin independent mechanisms that bypass the typical insulin signaling defects associated with these conditions (reviewed in Ref. 138). However, this "insulin sensitizing" effect is short-lived and disappears after 48 h. In contrary, repeated physical activity (i.e., exercise training) results in a persistent increase in insulin action in skeletal muscle from obese and insulin-resistant individuals. Physical training potentiates the effect of exercise on insulin sensitivity through multiple adaptations in glucose transport and metabolism (reviewed in Ref. 138). The molecular mechanism(s) for the enhanced glucose uptake with exercise training has been attributed to the increased expression and/or activity of key signaling proteins involved in insulin signal transduction and regulation of glucose uptake and metabolism in muscle such as the AMPK and the PKB (Akt) substrate AS160 (reviewed in Ref. 138). In addition, increased lipid oxidation and/or turnover is likely to be another mechanism by which exercise improves insulin sensitivity: exercise training results in an increase in the oxidative capacity of skeletal muscle by upregulating lipid oxidation and the expression of proteins involved in mitochondrial biogenesis (reviewed in Ref. 138). In conclusion, physical training plays an important, if not essential role, in the treatment and prevention of insulin insensitivity.

8. CONCLUSIONS

In summary, increased FFA flux, resulting from increased lipolysis secondary to adipose tissue insulin resistance, induces or aggravates insulin resistance in skeletal muscle leading to intracellular accumulation of FFA intermediates and/or induction of ER or mitochondrial stress, all of which can activate the mechanism of insulin resistance. Alleviating the excess of FFAs is a target for the treatment of insulin resistance in skeletal muscle.

REFERENCES

1. Himsworth H, Kerr R. Insulin-sensitive and insulin-insensitive types of diabetes mellitus. *Clin Sci.* 1939;4:119–152.
2. Himsworth H. The mechanism of diabetes mellitus. I. *Lancet.* 1939;2:1–6.
3. Himsworth H. The mechanism of diabetes mellitus. II. The control of the blood sugar level. *Lancet.* 1939;2:65–68.
4. Himsworth HP. The syndrome of diabetes mellitus and its causes. *Lancet.* 1949;1:465–473.

5. Himsworth HP. Diabetes mellitus: its differentiation into insulin-sensitive and insulin-insensitive types. *Lancet.* 1936;i:127–130.
6. DeFronzo RA, Gunnarson R, Bjorkman O, Olsson M, Wahren J. Effects of insulin on peripheral and splanchnic glucose metabolism in noninsulin-dependent (type II) diabetes mellitus. *J Clin Invest.* 1985;76:149–155.
7. DeFronzo RA. Pathogenesis of type 2 diabetes: metabolic and molecular implications for identifying diabetes genes. *Diabetes Rev.* 1997;5:177–269.
8. Coppack SW, Jensen MD, Miles JM. In vivo regulation of lipolysis in humans. *J Lipid Res.* 1994;35:177–193.
9. Felber JP, Magneanat G, Casthelaz M, et al. Carbohydrate and lipid oxidation in normal and diabetic subjects. *Diabetes.* 1977;26:693–699.
10. Boden G. Fuel metabolism in pregnancy and in gestational diabetes mellitus. *Obstet Gynecol Clin N Am.* 1996;23:1–10.
11. Blaak EE. Metabolic fluxes in skeletal muscle in relation to obesity and insulin resistance. *Best Pract Res Clin Endocrinol Metab.* 2005;19:391–403.
12. Kiens B. Skeletal muscle lipid metabolism in exercise and insulin resistance. *Physiol Rev.* 2006;86:205–243.
13. Shulman GI. Cellular mechanisms of insulin resistance. *J Clin Invest.* 2000;106:171–176.
14. DeFronzo RA. Lilly lecture 1987. The triumvirate: beta-cell, muscle, liver. A collusion responsible for NIDDM. *Diabetes.* 1988;37:667–687.
15. Kashyap SR, Belfort R, Berria R, et al. Discordant effects of a chronic physiological increase in plasma FFA on insulin signaling in healthy subjects with or without a family history of type 2 diabetes. *Am J Physiol Endocrinol Metab.* 2004;287:E537–E546.
16. Danadian K, Balasekaran G, Lewy V, et al. Insulin sensitivity in African-American children with and without family history of type 2 diabetes. *Diabetes Care.* 1999;22:1325–1329.
17. Petersen KF, Dufour S, Befroy D, Garcia R, Shulman GI. Impaired mitochondrial activity in the insulin-resistant offspring of patients with type 2 diabetes. *N Engl J Med.* 2004;350:664–671.
18. Kovacs P, Stumvoll M. Fatty acids and insulin resistance in muscle and liver. *Best Pract Res Clin Endocrinol Metab.* 2005;19:625–635.
19. Wajchenberg BL. Subcutaneous and visceral adipose tissue: their relation to the metabolic syndrome. *Endocr Rev.* 2000;21:697–738.
20. Basu A, Basu R, Shah P, et al. Systemic and regional free fatty acid metabolism in type 2 diabetes. *Am J Physiol Endocrinol Metab.* 2001;280:E1000–E1006.
21. Salans LB, Cushman SW, Weismann RE. Studies of human adipose tissue. Adipose cell size and number in nonobese and obese patients. *J Clin Invest.* 1973;52:929–941.
22. Weyer C, Wolford JK, Hanson RL, et al. Subcutaneous abdominal adipocyte size, a predictor of type 2 diabetes, is linked to chromosome 1q21–q23 and is associated with a common polymorphism in LMNA in Pima Indians. *Mol Genet Metab.* 2001;72:231–238.
23. Glatz JF, Bonen A, Luiken JJ. Exercise and insulin increase muscle fatty acid uptake by recruiting putative fatty acid transporters to the sarcolemma. *Curr Opin Clin Nutr Metab Care.* 2002;5:365–370.
24. Randle PJ, Garland PB, Hales CN, Newsholme EA. The glucose fatty-acid cycle. Its role in insulin sensitivity and the metabolic disturbances of diabetes mellitus. *Lancet.* 1963;1:785–789.
25. Boden G, Jadali F, White J, et al. Effects of fat on insulin-stimulated carbohydrate metabolism in normal men. *J Clin Invest.* 1991;88:960–966.
26. Griffin ME, Marcucci MJ, Cline GW, et al. Free fatty acid-induced insulin resistance is associated with activation of protein kinase C theta and alterations in the insulin signaling cascade. *Diabetes.* 1999;48:1270–1274.

27. Dresner A, Laurent D, Marcucci M, et al. Effects of free fatty acids on glucose transport and IRS-1-associated phosphatidylinositol 3-kinase activity. *J Clin Invest.* 1999;103:253–259.

28. Roden M. How free fatty acids inhibit glucose utilization in human skeletal muscle. *News Physiol Sci.* 2004;19:92–96.

29. Roden M, Price TB, Perseghin G, et al. Mechanism of free fatty acid-induced insulin resistance in humans. *J Clin Invest.* 1996;97:2859–2865.

30. Rothman DL, Shulman RG, Shulman GI. 31P nuclear magnetic resonance measurements of muscle glucose-6-phosphate. Evidence for reduced insulin-dependent muscle glucose transport or phosphorylation activity in non-insulin-dependent diabetes mellitus. *J Clin Invest.* 1992;89:1069–1075.

31. Shulman GI. Unraveling the cellular mechanism of insulin resistance in humans: new insights from magnetic resonance spectroscopy. *Physiology.* 2004;19:183–190.

32. Machann J, Haring H, Schick F, Stumvoll M. Intramyocellular lipids and insulin resistance. *Diabetes Obes Metab.* 2004;6:239–248.

33. Russell AP. Lipotoxicity: the obese and endurance-trained paradox. *Int J Obes Relat Metab Disord.* 2004;28(Suppl 4):S66–S71.

34. Liu L, Zhang Y, Chen N, et al. Upregulation of myocellular DGAT1 augments triglyceride synthesis in skeletal muscle and protects against fat-induced insulin resistance. *J Clin Invest.* 2007;117:1679–1689.

35. Muoio DM. Revisiting the connection between intramyocellular lipids and insulin resistance: a long and winding road. *Diabetologia.* 2012;55:2551–2554.

36. Faergeman NJ, Knudsen J. Role of long-chain fatty acyl-CoA esters in the regulation of metabolism and in cell signalling. *Biochem J.* 1997;323:1–12.

37. Chavez JA, Summers SA. Lipid oversupply, selective insulin resistance, and lipotoxicity: molecular mechanisms. *Biochim Biophys Acta.* 1801;2010:252–265.

38. Kraegen E, Cooney G. Free fatty acids and skeletal muscle insulin resistance. *Curr Opin Lipidol.* 2008;19(3):235–241. http://dx.doi.org/10.1097/01.mol.0000319118.44995.9a.

39. Schenk S, Saberi M, Olefsky JM. Insulin sensitivity: modulation by nutrients and inflammation. *J Clin Invest.* 2008;118:2992–3002.

40. Itani SI, Ruderman NB, Schmieder F, Boden G. Lipid-induced insulin resistance in human muscle is associated with changes in diacylglycerol, protein kinase C, and IkappaB-alpha. *Diabetes.* 2002;51:2005–2011.

41. Savage DB, Petersen KF, Shulman GI. Disordered lipid metabolism and the pathogenesis of insulin resistance. *Physiol Rev.* 2007;87:507–520.

42. Schmitz-Peiffer C, Browne CL, Oakes ND, et al. Alterations in the expression and cellular localization of protein kinase C isozymes epsilon and theta are associated with insulin resistance in skeletal muscle of the high fat-fed rat. *Diabetes.* 1997;46:169–178.

43. Yu C, Chen Y, Cline GW, et al. Mechanism by which fatty acids inhibit insulin activation of insulin receptor substrate-1 (IRS-1)-associated phosphatidylinositol 3-kinase activity in muscle. *J Biol Chem.* 2002;277:50230–50236.

44. Samuel VT, Shulman GI. Mechanisms for insulin resistance: common threads and missing links. *Cell.* 2012;148:852–871.

45. Khamzina L, Veilleux A, Bergeron S, Marette A. Increased activation of the mammalian target of rapamycin pathway in liver and skeletal muscle of obese rats: possible involvement in obesity-linked insulin resistance. *Endocrinology.* 2005;146:1473–1481.

46. Teruel T, Hernandez R, Lorenzo M. Ceramide mediates insulin resistance by tumor necrosis factor-alpha in brown adipocytes by maintaining Akt in an inactive dephosphorylated state. *Diabetes.* 2001;50:2563–2571.

47. Powell DJ, Hajduch E, Kular G, Hundal HS. Ceramide disables 3-phosphoinositide binding to the pleckstrin homology domain of protein kinase B (PKB)/Akt by a PKCzeta-dependent mechanism. *Mol Cell Biol.* 2003;23:7794–7808.

48. Hunnicutt JW, Hardy RW, Williford J, McDonald JM. Saturated fatty acid-induced insulin resistance in rat adipocytes. *Diabetes.* 1994;43:540–545.
49. Vessby B, Unsitupa M, Hermansen K, et al. Substituting dietary saturated for mono-unsaturated fat impairs insulin sensitivity in healthy men and women: the KANWU Study. *Diabetologia.* 2001;44:312–319.
50. Ryan M, McInerney D, Owens D, et al. Diabetes and the Mediterranean diet: a beneficial effect of oleic acid on insulin sensitivity, adipocyte glucose transport and endothelium-dependent vasoreactivity. *QJM.* 2000;93:85–91.
51. Schrauwen P, Hesselink MK. Oxidative capacity, lipotoxicity, and mitochondrial damage in type 2 diabetes. *Diabetes.* 2004;53:1412–1417.
52. Chavez JA, Summers SA. Characterizing the effects of saturated fatty acids on insulin signaling and ceramide and diacylglycerol accumulation in 3T3-L1 adipocytes and C2C12 myotubes. *Arch Biochem Biophys.* 2003;419:101–109.
53. Dimopoulos N, Watson M, Sakamoto K, Hundal HS. Differential effects of palmitate and palmitoleate on insulin action and glucose utilization in rat L6 skeletal muscle cells. *Biochem J.* 2006;399:473–481.
54. Turpin SM, Lancaster GI, Darby I, Febbraio MA, Watt MJ. Apoptosis in skeletal muscle myotubes is induced by ceramides and is positively related to insulin resistance. *Am J Physiol Endocrinol Metab.* 2006;291:E1341–E1350.
55. Coll T, Jove M, Rodriguez-Calvo R, et al. Palmitate-mediated downregulation of peroxisome proliferator-activated receptor-gamma coactivator 1alpha in skeletal muscle cells involves MEK1/2 and nuclear factor-kappaB activation. *Diabetes.* 2006;55: 2779–2787.
56. Sabin MA, Stewart CE, Crowne EC, et al. Fatty acid-induced defects in insulin signalling, in myotubes derived from children, are related to ceramide production from palmitate rather than the accumulation of intramyocellular lipid. *J Cell Physiol.* 2007;211:244–252.
57. Powell DJ, Turban S, Gray A, Hajduch E, Hundal HS. Intracellular ceramide synthesis and protein kinase Czeta activation play an essential role in palmitate-induced insulin resistance in rat L6 skeletal muscle cells. *Biochem J.* 2004;382:619–629.
58. Itani SI, Zhou Q, Pories WJ, MacDonald KG, Dohm GL. Involvement of protein kinase C in human skeletal muscle insulin resistance and obesity. *Diabetes.* 2000;49:1353–1358.
59. Hirabara SM, Curi R, Maechler P. Saturated fatty acid-induced insulin resistance is associated with mitochondrial dysfunction in skeletal muscle cells. *J Cell Physiol.* 2010;222:187–194.
60. Rachek LI, Musiyenko SI, LeDoux SP, Wilson GL. Palmitate induced mitochondrial deoxyribonucleic acid damage and apoptosis in l6 rat skeletal muscle cells. *Endocrinology.* 2007;148:293–299.
61. Lambertucci RH, Hirabara SM, Silveira Ldos R, et al. Palmitate increases superoxide production through mitochondrial electron transport chain and NADPH oxidase activity in skeletal muscle cells. *J Cell Physiol.* 2008;216:796–804.
62. Hoehn KL, Salmon AB, Hohnen-Behrens C, et al. Insulin resistance is a cellular antioxidant defense mechanism. *Proc Natl Acad Sci U S A.* 2009;106:17787–17792.
63. Coll T, Eyre E, Rodriguez-Calvo R, et al. Oleate reverses palmitate-induced insulin resistance and inflammation in skeletal muscle cells. *J Biol Chem.* 2008;283: 11107–11116.
64. Crunkhorn S, Dearie F, Mantzoros C, et al. Peroxisome proliferator activator receptor gamma coactivator-1 expression is reduced in obesity: potential pathogenic role of saturated fatty acids and p38 mitogen-activated protein kinase activation. *J Biol Chem.* 2007;282:15439–15450.

65. Merrill Jr AH. De novo sphingolipid biosynthesis: a necessary, but dangerous, pathway. *J Biol Chem.* 2002;277:25843–25846.
66. Holland WL, Brozinick JT, Wang LP, et al. Inhibition of ceramide synthesis ameliorates glucocorticoid-, saturated-fat-, and obesity-induced insulin resistance. *Cell Metab.* 2007;5:167–179.
67. Ussher JR, Koves TR, Cadete VJ, et al. Inhibition of de novo ceramide synthesis reverses diet-induced insulin resistance and enhances whole-body oxygen consumption. *Diabetes.* 2010;59:2453–2464.
68. Maron DJ, Fair JM, Haskell WL. Saturated fat intake and insulin resistance in men with coronary artery disease. The Stanford coronary risk intervention project investigators and staff. *Circulation.* 1991;84:2020–2027.
69. Parker DR, Weiss ST, Troisi R, et al. Relationship of dietary saturated fatty acids and body habitus to serum insulin concentrations: the normative aging study. *Am J Clin Nutr.* 1993;58:129–136.
70. Parillo M, Rivellese AA, Ciardullo AV, et al. A high-monounsaturated-fat/low-carbohydrate diet improves peripheral insulin sensitivity in non-insulin-dependent diabetic patients. *Metabolism.* 1992;41:1373–1378.
71. Riccardi G, Giacco R, Rivellese AA. Dietary fat, insulin sensitivity and the metabolic syndrome. *Clin Nutr.* 2004;23:447–456.
72. Sabin MA, Stewart CE, Crowne EC, et al. Fatty acid-induced defects in insulin signalling, in myotubes derived from children, are related to ceramide production from palmitate rather than rather than the accumulation of intramyocellular lipid. *J Cell Physiol.* 2007;211:244–252.
73. Martins AR, Nachbar RT, Gorjao R, et al. Mechanisms underlying skeletal muscle insulin resistance induced by fatty acids: importance of the mitochondrial function. *Lipids Health Dis.* 2012;11:30.
74. Hotamisligil GS. Inflammation and metabolic disorders. *Nature.* 2006;444:860–867.
75. Wen H, Gris D, Lei Y, et al. Fatty acid-induced NLRP3-ASC inflammasome activation interferes with insulin signaling. *Nat Immunol.* 2011;12:408–415.
76. Hotamisligil GS, Shargill NS, Spiegelman BM. Adipose expression of tumor necrosis factor-alpha: direct role in obesity-linked insulin resistance. *Science.* 1993;259:87–91.
77. Hotamisligil GS. Inflammation and endoplasmic reticulum stress in obesity and diabetes. *Int J Obes (Lond).* 2008;32:52–54.
78. Tsukumo DM, Carvalho-Filho MA, Carvalheira JB, et al. Loss-of function mutation in Toll-like receptor 4 prevents diet-induced obesity and insulin resistance. *Diabetes.* 2007;56:1986–1998.
79. Hirosumi J, Tuncman G, Chang L, et al. A central role for JNK in obesity and insulin resistance. *Nature.* 2002;420:333–336.
80. Kaneto H, Nakatani Y, Kawamori D, et al. Role of oxidative stress, endoplasmic reticulum stress, and c-Jun N-terminal kinase in pancreatic beta-cell dysfunction and insulin resistance. *Int J Biochem Cell Biol.* 2006;38:782–793.
81. Holland WL, Bikman BT, Wang LP, et al. Lipid-induced insulin resistance mediated by the proinflammatory receptor TLR4 requires saturated fatty acid-induced ceramide biosynthesis in mice. *J Clin Invest.* 2011;121:1858–1870.
82. Shi H, Kokoeva MV, Inouye K, et al. TLR4 links innate immunity and fatty acid-induced insulin resistance. *J Clin Invest.* 2006;116:3015–3025.
83. Ozcan U, Cao Q, Yilmaz E, et al. Endoplasmic reticulum stress links obesity, insulin action, and type 2 diabetes. *Science.* 2004;306:457–461.
84. Deldicque L, Cani PD, Philp A, et al. The unfolded protein response is activated in skeletal muscle by high-fat feeding: potential role in the downregulation of protein synthesis. *Am J Physiol Endocrinol Metab.* 2010;299:E695–E705.

85. Yuzefovych LV, Musiyenko SI, Wilson GL, Rachek LI. Mitochondrial DNA damage and dysfunction, and oxidative stress are associated with endoplasmic reticulum stress, protein degradation and apoptosis in high fat diet-induced insulin resistance mice. *PLoS One*. 2013;8:e54059.

86. Csordas G, Hajnoczky G. SR/ER-mitochondrial local communication: calcium and ROS. *Biochim Biophys Acta*. 2009;1787:1352–1362.

87. Kim PK, Hailey DW, Mullen RT, Lippincott-Schwartz J. Ubiquitin signals autophagic degradation of cytosolic proteins and peroxisomes. *Proc Natl Acad Sci U S A*. 2008;105:20567–20574.

88. Muoio DM, Newgard CB. Mechanisms of disease: molecular and metabolic mechanisms of insulin resistance and beta-cell failure in type 2 diabetes. *Nat Rev Mol Cell Biol*. 2008;9:193–205.

89. Patti ME, Butte AJ, Crunkhorn S, et al. Coordinated reduction of genes of oxidative metabolism in humans with insulin resistance and diabetes: potential role of PGC1 and NRF1. *Proc Natl Acad Sci U S A*. 2003;100:8466–8471.

90. Mootha VK, Lindgren CM, Eriksson KF, et al. PGC-1alpha-responsive genes involved in oxidative phosphorylation are coordinately downregulated in human diabetes. *Nat Genet*. 2003;34:267–273.

91. Yang X, Pratley RE, Tokraks S, Bogardus C, Permana PA. Microarray profiling of skeletal muscle tissues from equally obese, non-diabetic insulin-sensitive and insulin-resistant Pima Indians. *Diabetologia*. 2002;45:1584–1593.

92. Richardson DK, Kashyap S, Bajaj M, et al. Lipid infusion decreases the expression of nuclear encoded mitochondrial genes and increases the expression of extracellular matrix genes in human skeletal muscle. *J Biol Chem*. 2005;280:10290–10297.

93. Borkman M, Storlien LH, Pan DA, et al. The relation between insulin sensitivity and the fatty-acid composition of skeletal-muscle phospholipids. *N Engl J Med*. 1993;328:238–244.

94. Pan DA, Hulbert AJ, Storlien LH. Dietary fats, membrane phospholipids and obesity. *J Nutr*. 1994;124:1555–1565.

95. Storlien LH, Baur LA, Kriketos AD, et al. Dietary fats and insulin action. *Diabetologia*. 1996;39:621–631.

96. Storlien LH, Kriketos AD, Calvert GD, Baur LA, Jenkins AB. Fatty acids, triglycerides and syndromes of insulin resistance. *Prostaglandins Leukot Essent Fatty Acids*. 1997;57:379–385.

97. Anderson EJ, Lustig ME, Boyle KE, et al. Mitochondrial H2O2 emission and cellular redox state link excess fat intake to insulin resistance in both rodents and humans. *J Clin Invest*. 2009;119:573–581.

98. Bonnard C, Durand A, Peyrol S, et al. Mitochondrial dysfunction results from oxidative stress in the skeletal muscle of diet-induced insulin-resistant mice. *J Clin Invest*. 2008;118:789–800.

99. Muoio DM, Neufer PD. Lipid-induced mitochondrial stress and insulin action in muscle. *Cell Metab*. 2012;15:595–605.

100. Morino K, Petersen KF, Shulman GI. Molecular mechanisms of insulin resistance in humans and their potential links with mitochondrial dysfunction. *Diabetes*. 2006;55(Suppl 2):S9–S15.

101. An J, Muoio DM, Shiota M, et al. Hepatic expression of malonyl-CoA decarboxylase reverses muscle, liver and whole-animal insulin resistance. *Nat Med*. 2004;10:268–274.

102. Koves TR, Li P, An J, et al. Peroxisome proliferator-activated receptor-gamma co-activator 1alpha-mediated metabolic remodeling of skeletal myocytes mimics exercise training and reverses lipid-induced mitochondrial inefficiency. *J Biol Chem*. 2005;280:33588–33598.

103. Koves TR, Ussher JR, Noland RC, et al. Mitochondrial overload and incomplete fatty acid oxidation contribute to skeletal muscle insulin resistance. *Cell Metab*. 2008;7:45–56.

104. Chance B, Sies H, Boveris A. Hydroperoxide metabolism in mammalian organs. *Physiol Rev.* 1979;59:527–605.
105. Houstis N, Rosen ED, Lander ES. Reactive oxygen species have a causal role in multiple forms of insulin resistance. *Nature.* 2006;440:944–948.
106. Yuzefovych L, Wilson G, Rachek L. Different effects of oleate vs. palmitate on mitochondrial function, apoptosis, and insulin signaling in L6 skeletal muscle cells: role of oxidative stress. *Am J Physiol Endocrinol Metab.* 2010;299:E1096–E1105.
107. Gerbitz KD, Gempel K, Brdiczka D. Mitochondria and diabetes. Genetic, biochemical, and clinical implications of the cellular energy circuit. *Diabetes.* 1996;45:113–126.
108. Lee HY, Choi CS, Birkenfeld AL, et al. Targeted expression of catalase to mitochondria prevents age-associated reductions in mitochondrial function and insulin resistance. *Cell Metab.* 2010;12:668–674.
109. Adler V, Yin Z, Tew KD, Ronai Z. 1999 Role of redox potential and reactive oxygen species in stress signaling. *Oncogene.* 1999;18:6104–6111.
110. Dey D, Mukherjee M, Basu D, et al. Inhibition of insulin receptor gene expression and insulin signaling by fatty acid: interplay of PKC isoforms therein. *Cell Physiol Biochem.* 2005;16:217–228.
111. Tirosh A, Potashnik R, Bashan N, Rudich A. Oxidative stress disrupts insulin-induced cellular redistribution of insulin receptor substrate-1 and phosphatidylinositol 3-kinase in 3T3-L1 adipocytes. A putative cellular mechanism for impaired protein kinase B activation and GLUT4 translocation. *J Biol Chem.* 1999;274:10595–10602.
112. Jove M, Laguna JC, Vázquez-Carrera M. Agonist-induced activation releases peroxisome proliferator-activated receptor beta/delta from its inhibition by palmitate-induced nuclear factor-kappaB in skeletal muscle cells. *Biochim Biophys Acta.* 2005;1734:52–61.
113. Bloch-Damti A, Potashnik R, Gual P, et al. Differential effects of IRS1 phosphorylated on Ser307 or Ser632 in the induction of insulin resistance by oxidative stress. *Diabetologia.* 2006;49:2463–2473.
114. Watt MJ, Steinberg GR, Chen ZP, Kemp BE, Febbraio MA. Fatty acids stimulate AMP-activated protein kinase and enhance fatty acid oxidation in L6 myotubes. *J Physiol.* 2006;574:139–147.
115. Vinayagamoorthi R, Bobby Z, Sridhar MG. Antioxidants preserve redox balance and inhibit c-Jun-N-terminal kinase pathway while improving insulin signaling in fat-fed rats: evidence for the role of oxidative stress on IRS-1 serine phosphorylation and insulin resistance. *J Endocrinol.* 2008;197:287–296.
116. Gupte AA, Bomhoff GL, Morris JK, Gorres BK, Geiger PC. Lipoic acid increases heat shock protein expression and inhibits stress kinase activation to improve insulin signaling in skeletal muscle from high-fat-fed rats. *J Appl Physiol.* 2009;106:1425–1434.
117. Zhang X, Xu A, Chung SK, et al. Selective inactivation of c-Jun NH2-terminal kinase in adipose tissue protects against diet-induced obesity and improves insulin sensitivity in both liver and skeletal muscle in mice. *Diabetes.* 2011;60:486–495.
118. Andrieu-Abadie N, Gouaze V, Salvayre R, Levade T. Ceramide in apoptosis signaling: relationship with oxidative stress. *Free Radic Biol Med.* 2001;31:717–728.
119. Kelley DE. Free fatty acids, insulin resistance, and ectopic fat. In: Kushner RF, Bessesen DH, eds. *Contemporary Endocrinology: Treatment of the Obese Patient.* Totowa, NJ: Humana Press Inc.; 2007:87–97
120. Tsatsoulis A, Mantzaris MD, Bellou S, Andrikoula M. Insulin resistance: an adaptive mechanism becomes maladaptive in the current environment—an evolutionary perspective. *Metabolism.* 2013;62:622–633.
121. Lee CH, Olson P, Evans RM. Minireview: lipid metabolism, metabolic diseases, and peroxisome proliferator-activated receptors. *Endocrinology.* 2003;144:2201–2207.
122. Forman BM, Chen J, Evans RM. Hypolipidemic drugs, polyunsaturated fatty acids, and eicosanoids are ligands for peroxisome proliferator-activated receptors alpha and delta. *Proc Natl Acad Sci U S A.* 1997;94:4312–4317.

123. Krey G, Braissant O, L'Horset F, et al. Fatty acids, eicosanoids, and hypolipidemic agents identified as ligands of peroxisome proliferator-activated receptors by coactivator-dependent receptor ligand assay. *Mol Endocrinol*. 1997;11:779–791.

124. Leone TC, Weinheimer CJ, Kelly DP. A critical role for the peroxisome proliferator-activated receptor alpha (PPARalpha) in the cellular fasting response: the PPARalpha-null mouse as a model of fatty acid oxidation disorders. *Proc Natl Acad Sci U S A*. 1999;96:7473–7478.

125. He W, Barak Y, Hevener A, et al. Adipose-specific peroxisome proliferator-activated receptor gamma knockout causes insulin resistance in fat and liver but not in muscle. *Proc Natl Acad Sci U S A*. 2003;100:15712–15717.

126. Lehmann JM, Moore LB, Smith-Oliver TA, et al. An antidiabetic thiazolidinedione is a high affinity ligand for peroxisome proliferator-activated receptor gamma (PPAR gamma). *J Biol Chem*. 1995;270:12953–12956.

127. Barak Y, Liao D, He W, et al. Effects of peroxisome proliferator-activated receptor delta on placentation, adiposity, and colorectal cancer. *Proc Natl Acad Sci U S A*. 2002;99:303–308.

128. Oliver Jr WR, Shenk JL, Snaith MR, et al. A selective peroxisome proliferator-activated receptor delta agonist promotes reverse cholesterol transport. *Proc Natl Acad Sci U S A*. 2001;98:5306–5311.

129. Lee CH, Chawla A, Urbiztondo N, et al. Transcriptional repression of atherogenic inflammation: modulation by PPARdelta. *Science*. 2003;302:453–457.

130. Wang YX, Lee CH, Tiep S, et al. Peroxisome-proliferator-activated receptor delta activates fat metabolism to prevent obesity. *Cell*. 2003;113:159–170.

131. Norris AW, Chen L, Fisher SJ, et al. Muscle-specific PPARgamma-deficient mice develop increased adiposity and insulin resistance but respond to thiazolidinediones. *J Clin Invest*. 2003;112:608–618.

132. Coletta DK, Sriwijitkamol A, Wajcberg E, et al. Pioglitazone stimulates AMP-activated protein kinase signalling and increases the expression of genes involved in adiponectin signalling, mitochondrial function and fat oxidation in human skeletal muscle in vivo: a randomised trial. *Diabetologia*. 2009;52:723–732.

133. Musi N, Hirshman MF, Nygren J, et al. Metformin increases AMP-activated protein kinase activity in skeletal muscle of subjects with type 2 diabetes. *Diabetes*. 2002;51:2074–2081.

134. Zhou G, Myers R, Li Y, et al. Role of AMP-activated protein kinase in mechanism of metformin action. *J Clin Invest*. 2001;108:1167–1174.

135. Cusi K, DeFronzo RA. Metformin: a review of its metabolic effects. *Diabetes Rev*. 1998;6:89–131.

136. UKPDS Group. Effects of intensive blood-glucose control with metformin on complications in overweight patients with type 2 diabetes. *Lancet*. 1998;352:854–855.

137. Goodyear LJ, Kahn BB. Exercise, glucose transport, and insulin sensitivity. *Annu Rev Med*. 1998;49:235–261.

138. Hawley JA, Lessard SJ. Exercise training-induced improvements in insulin action. *Acta Physiol (Oxf)*. 2008;192:127–135.

CHAPTER NINE

Adiponectin Signaling and Metabolic Syndrome

Yuchang Fu
Department of Nutrition Sciences, School of Health Professions, University of Alabama at Birmingham, Birmingham, Alabama, USA

Contents

Abstract

Metabolic syndrome is a combination of several serious metabolic disorders, including obesity, insulin resistance, type II diabetes, and cardiovascular disease. A class of drugs called thiazolidinediones (TZDs) has been used for treatment of metabolic syndrome; however, TZDs also show side effects. Therefore, additional alternative medications that are both effective and safe for the prevention and treatment of metabolic syndrome are a big challenge for us. Adiponectin is exclusively expressed and secreted from adipocyte, and it has been proved as one thiazolidinediones with antidiabetic, anti-inflammatory, and antiatherogenic properties for metabolic syndrome. Studies conducted in human and animal models of metabolic diseases have clearly demonstrated that adiponectin and adiponectin receptors as well as the signaling pathways involved can indeed have beneficial effects on these metabolic disorders. The use of macrophage cells as carriers for adiponectin and its receptors will provide a novel and unique strategy for studying the actions of adiponectin *in vivo*, and it also serves as a potential innovative therapeutic approach for treatment of metabolic syndrome in the future.

Progress in Molecular Biology and Translational Science, Volume 121
ISSN 1877-1173
http://dx.doi.org/10.1016/B978-0-12-800101-1.00009-0

1. INTRODUCTION

Metabolic syndrome, defined by multiple metabolic disorders including obesity, insulin resistance, type II diabetes mellitus, and cardiovascular disease, has become increasingly common in the United States. It is estimated that over 50 million Americans have these diseases.

Adipocyte has been gradually recognized not only as an inert storage site for fat but also as an active endocrine organ, secreting numerous hormones known as adipokines. Adiponectin (also known as apM1, AdipoQ, Gbp28, and Acrp30) is one such adipokine, expressed and secreted only from adipocyte, which circulates in high- (HMW), middle-, and low-molecular-weight (LMW) multimeric forms. Epidemiological evidence has indicated that plasma adiponectin levels are reduced in patients with the metabolic syndrome. In the circulation, adiponectin exerts biological effects on multiple cell types/tissues and has insulin-sensitizing, anti-inflammatory, and anti-atherosclerotic properties. Accordingly, administration of adiponectin to intact rodents improved the metabolic syndrome. Transgenic and knockout mouse models of adiponectin and its receptors have confirmed the importance of adiponectin in the metabolic syndrome.

2. METABOLIC SYNDROME AND THE LIMITATION OF CURRENT TREATMENT

The metabolic syndrome (also known as syndrome X or the dysmetabolic syndrome) is a combination of medical disorders that exist in a single individual, including central obesity, dyslipidemia characterized by high blood triglycerides and low high-density lipoprotein (HDL) cholesterol levels, hypertension, and hyperglycemia (high blood glucose levels). Without treatments of these medical disorders, the metabolic syndrome would develop into serious diseases, such as type II diabetes mellitus and cardiovascular disease. The prevalence of metabolic syndrome among adults in the United States is estimated to be ~25%, and it is considered one of the main causes of mortality in the United States and in developing countries.[1–3]

Since the metabolic syndrome is a cluster of conditions that occur together, except for aggressive lifestyle changes and treatment of each component, such as hyperlipidemia, high blood pressure, and high blood glucose, of the metabolic syndrome separately, taking on all of them together might seem overwhelming and impossible; lifestyle modification is currently the preferred treatment of metabolic syndrome. Thus, finding an effective way to clinically treat the multiple risk factors linked to the metabolic syndrome is a big challenge for biomedical scientists and clinicians.

In addition to changes in lifestyle, medications to control each component of the metabolic syndrome, such as cholesterol levels, lipid levels, and high blood pressure, have been used as the complementally treating methods with the lifestyle modification. However, since some medications used in the metabolic syndrome often have more than one effect, either beneficial or harmful, to several components in metabolic syndrome, correct selection of the medications for treating metabolic syndrome is an important consideration when discussing the medication(s) for treating the metabolic syndrome. For example, a class of drugs called thiazolidinediones (TZDs) including pioglitazone (Actos), rosiglitazone (Avandia), and troglitazone (Rezulin) had been used to lower blood glucose levels to achieve glycemic goals in metabolic syndrome or diabetes. However, due to reported increase in heart attack (rosiglitazone) and serious liver toxicity (troglitazone) in patients taking Avandia and Rezulin, the US FDA (Food and Drug Administration) has removed them from the market based on their harmful side effects on physiological metabolism.[4,5] Recently, even the sale of Actos (pioglitazone) has been suspended in European markets based on its potential increased risk of bladder cancer in patients using pioglitazone for treatment of diabetes.[6] Although at this time, the FDA has not concluded that Actos increases the risk of bladder cancer; in anticipation, an updated formal recommendation from the FDA about Actos will be coming in the very near future. Thus, the limitation of clinical utilization of TZDs highlights the need for additional alternative medications that are both effective and safe for the prevention and treatment of metabolic syndrome.

3. ADIPONECTIN AND ADIPONECTIN RECEPTORS

Adiponectin[7] is a 244-amino acid protein, which has four distinct regions, and it shows similarity to the complement factor C1q family of proteins. The first region of adiponectin protein is a short signal peptide sequence that targets the hormone for secretion outside the cell; the second region is a short sequence that varies between species with no homology to other proteins; the third region is a 65-amino acid region involved in collagen-triple helix formation with similarity to collagenous proteins; the last region of adiponectin protein is a globular domain at the carboxyl terminus, which shows a very high degree of homology with subunits of C1q and the globular domains of type VIII and type X collagens.

Adiponectin, also known as apM1 (adipose most abundant gene transcript 1),[8] AdipoQ (adipose gene Q),[9] Gbp28 (gelatin-binding protein of 28 kDa),[10] and Acrp30 (adipocyte complement-related protein of 30 kDa),[11] is an adipokine identified by screening adipose-specific genes in the human cDNA projects.[10] Adiponectin is exclusively expressed and secreted from adipocytes and circulates in bloodstream as a full-length protein of 30 kDa (fAd) in high (larger than hexamers)-, middle (hexamers)-, and low (trimers or smaller)-molecular-weight multimeric forms or as a fragment in trimer containing the globular adiponectin domain (gAd) generated by proteolysis of the full-length adiponectin.[12–14] Studies of these adiponectin multimeric forms have mainly demonstrated that the HMW form is the most metabolically active compared with the LMW form.[15,16] Adiponectin is very abundant in human plasma (2–30 μg/mL) relative to many other adipokines.

Two cell-surface transmembrane receptors, named as AdipoR1 and AdipoR2, for adiponectin binding, have been identified by expression cloning.[17] Experiments for the expression of AdipoR1 and AdipoR2 or for the suppression of AdipoR1 and AdipoR2 expression by small-interfering RNA (siRNA) have indicated that AdipoR1 is a high-affinity receptor for globular adiponectin and a low-affinity receptor for full-length adiponectin, whereas AdipoR2 has an intermediate affinity for both the full-length and globular species. Both adiponectin receptors are ubiquitously expressed; but AdipoR1 is abundantly expressed in skeletal muscle, whereas AdipoR2 is predominantly expressed in the liver.[17] These two adiponectin receptors are predicted to contain seven transmembrane domains with an extracellular carboxy terminus and an intracellular amino terminus, a structure that is opposite to that of all G-protein-coupled receptors. Adiponectin also interacts with a putative receptor, T-cadherin[18]; however, since T-cadherin lacks the transmembrane and cytoplasmic domains, the biological significance of this nontransmembrane receptor is not clear yet;[19] T-cadherin receptor probably acts as a coreceptor to exert its functions with the integrin pathway or AdipoR1 and AdipoR2.[20]

Epidemiological evidence has indicated that circulating adiponectin levels are reduced in metabolic syndrome patients with insulin resistance, type II diabetes, obesity, or cardiovascular disease.[21–23] When examined, low plasma adiponectin levels in these metabolic disease states are accompanied by reduced adiponectin gene expression in adipose tissue.[24,25] There is also evidence that adiponectin gene polymorphisms may be associated with hypoadiponectinemia together with insulin resistance and type II diabetes.[26] Conversely, in the circulation, adiponectin exerts biological effects on multiple cell types and has insulin-sensitizing, anti-inflammatory, and anti-atherosclerotic properties; therefore, increased circulating levels of adiponectin are associated with improvement in the metabolic syndrome. The metabolic effects and mechanisms of action of adiponectin and adiponectin receptors on metabolic syndrome are interesting research topics, and recent studies have established a better understanding of the functions of adiponectin and adiponectin receptors in metabolic syndrome and diseases.

It has been demonstrated that adiponectin knockout mice develop insulin resistance, glucose intolerance, hyperglycemia, and hypertension, all characteristics of metabolic syndrome.[27–29] These adiponectin knockout mice have shown moderate insulin resistance with glucose intolerance despite body weight gain similar to that of wild-type mice and twofold more neointimal formation in response to external vascular cuff injury than wild-type mice, suggesting that adiponectin plays a protective role against insulin resistance and atherosclerosis *in vivo*. When examining these adiponectin knockout mice in details, there is also evidence to show delayed clearance of free fatty acid in plasma, low levels of fatty acid transport protein 1 (FATP-1) mRNA in muscle, high levels of tumor necrosis factor-alpha (TNF-α) mRNA in adipose tissue, and high plasma TNF-α concentrations in these knockout mice. These knockout mice also exhibit severe diet-induced insulin resistance with reduced insulin receptor substrate 1 (IRS-1)-associated phosphatidylinositol-3 kinase (PI3K) activity in muscle. More importantly, viral-mediated adiponectin expression in these knockout mice can reverse some phenotypes in these knockout mice. These studies have further revealed important roles of adiponectin in regulating inflammatory response and glucose and lipid metabolism related to the metabolic syndrome.

In terms of adiponectin receptor functions, it has been demonstrated that knockout of AdipoR1 gene in mice causes the abrogation of adiponectin-induced 5′ adenosine monophosphate-activated protein kinase (AMPK) activation, whereas disruption of AdipoR2 gene in mice results in decreased activity of peroxisome proliferator-activated receptor-α (PPARα) signaling pathways.[30] Furthermore, simultaneous disruption of both AdipoR1 and AdipoR2 has abolished adiponectin binding and actions, resulting in increased tissue triglyceride content, inflammation, and oxidative stress in mice and thus leading to insulin resistance and marked glucose intolerance. Therefore, AdipoR1 and AdipoR2 serve as the predominant receptors for adiponectin *in vivo* and play important roles in the regulation of glucose and lipid metabolism, inflammation, and oxidative stress in metabolic syndrome.

Impaired glucose uptake and insulin sensitivity are two typical hallmarks for obesity and type II diabetes, and there are three main insulin target cells/tissues, adipocyte/adipose tissue, myocyte/skeletal muscle, and hepatocyte/liver tissue, involved in the insulin–mediated metabolic activity. For vascular diseases such as hypertension and atherosclerosis, inflammatory response and lipid metabolism are two important hallmarks in endothelial cell and macrophage/foam cell in vascular wall. Adiponectin signaling plays significant roles in these active metabolic cells/tissues for the metabolic activity related to the metabolic syndrome as detailed later (Fig. 9.1).

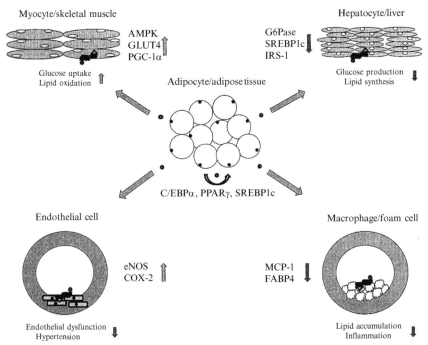

Figure 9.1 Adiponectin signaling in major metabolically active tissues. Adiponectin is expressed and secreted from adipocyte/adipose tissue (red-colored solid small circles) and binding to adiponectin receptors (red-colored receptor domain structures) and APPL1/2 adaptor protein complexes (blue triangles) in other cells/tissues. Green arrows indicate positive or upregulated effects from adiponectin and red arrows represent negative or downregulated regulations from adiponectin. Only the major regulated molecules and involved metabolism are listed in these metabolic tissues. (See color plate.)

4. ADIPONECTIN SIGNALING IN KEY METABOLIC CELLS/ TISSUES RELATED TO METABOLIC SYNDROME

4.1. Adipocyte/adipose tissue

Adipocyte/adipose tissue is a highly active metabolic and endocrine cell/ tissue, which secrets numerous regulatory factors that circulate in the blood-stream and act on distal tissues, and these metabolic factors also mediate local autocrine/paracrine effects, to influence many physiological metabolic activities in humans and animals, including food intake, energy expenditure, and carbohydrate and lipid metabolism.[31] The expression of adiponectin is almost exclusively expressed and secreted from adipocyte/adipose tissue, including white and brown adipose tissues,[32,33] but with trace levels in other types of cells/tissues.[34,35]

During differentiations of adipocyte/adipose tissue, several early expressed key transcription factors, including CCAAT/enhancer-binding protein α (C/EBPα),[36,37] peroxisome proliferator-activated receptor γ (PPARγ),[38,39] and sterol regulatory element-binding protein 1c (SREBP1c),[40,41] which bind to the response elements of adiponectin gene promoter region, have been demonstrated to regulate adiponectin gene expression in adipocyte/adipose tissue. As a cell model for adipocyte/ adipose tissue, 3T3–L1 fibroblasts, which exclusively express and secrete abundant adiponectin after these cells differentiate under appropriate conditions, have been broadly used in biological research for studying the adipocyte-like phenotypes.[42] When adiponectin gene is overexpressed in the 3T3–L1 adipocytes, it functions as an autocrine factor in these adipocytes promoting cell differentiation from preadipocytes to adipocytes, augmenting programmed gene expression, such as C/EBPα, PPARγ, and SREBP1c, which are responsible for adipogenesis, and increasing lipid content and insu-lin responsiveness of the glucose transport system in adipocytes.[43]

Similar results have been observed in adipocyte-specific adiponectin transgenic (Ad-TG) mice, in which adiponectin gene is specifically over-expressed in adipose tissue with an adipocyte protein 2 (aP2, also known as fatty acid-binding protein 4, FABP4) enhancer/promoter sequence.[44] In these Ad-TG mice, there is a significant increase in percent total body fat, which reached approximately 30% of their total body weight, in contrast to 18% in their wild-type siblings. These adipocyte-specific Ad-TG mice with two- to threefold elevated adiponectin levels are protected against high-fat diet (HFD)-induced insulin resistance. When these Ad-TG mice have been crossed with the diabetic *ob/ob* mice, which lack leptin and display hyperglycemia, hyperinsulinemia, and dyslipidemia, overexpressing adiponectin in these mice has completely rescued the diabetic phenotype in *ob/ob* mice with dramatically improved glucose as well as positively affected serum triglyceride levels even though these crossed transgenic mice are morbidly obese with significantly higher levels of adipose tissue than their *ob/ob* littermates.[45] Recent studies have further shown overexpressed adiponectin in adipose tissue as a factor to increase the metabolic flexibility of the adipose tissue, enhancing its ability to maintain proper function under metabolically challenging conditions.[46]

Adiponectin receptor signaling in adipocyte/adipose tissue is also regulated by adiponectin levels in the bloodstream and reflects metabolic conditions in these cells/tissues. There was evidence showing that low adiponectin receptors were expressed in visceral adipocytes in adipose tissues of humans and rats, and decreased adiponectin receptor expression was detected in adipose tissues of insulin-resistant animals; these results have also indicated disturbed adiponectin bioactivity in the insulin-resistant animals by low adiponectin receptor activity.[47] Moreover, studies have also shown that PPARα and PPARγ stimulations by their agonists in obese diabetic KKAy mice can enhance the action of adiponectin by increasing the expression of both adiponectin and adiponectin receptors in adipocyte/adipose tissue, which can result in the amelioration of obesity-induced insulin resistance in these animals.[48]

4.2. Myocyte/skeletal muscle

In human, about 78% of glucose uptake occurs in skeletal muscle; therefore, myocyte/skeletal muscle is the primary cell/tissue responsible for insulin-dependent glucose uptake for metabolism, and glucose uptake by this cell/tissue plays an important role in determining the concentration of glucose in the blood. Glucose uptake in myocyte/muscle occurs by a system of facilitated diffusion involving at least two distinct glucose transporters, GLUT-1 and GLUT-4.[49] Glucose traverses the myocyte cell membrane by these glucose transporter proteins, primarily by GLUT4, for oxidation to provide energy or stored as glycogen in skeletal muscle. For the glucose uptake process, insulin stimulation initiates translocation of GLUT4 from the cell interior to the cell-surface membrane. In myocyte/skeletal muscle from patients with insulin resistance or diabetes, this insulin-dependent glucose uptake process is impaired because the insulin signaling pathways are defective for glucose uptake and thus blocking other cellular glucose metabolic processes.

One of the downstream key molecules in the insulin signaling pathways, the serine/threonine-specific protein kinase Akt (also known as protein kinase B, PKB), plays an important role in the insulin-mediated glucose uptake in myocyte/skeletal muscle, and since adiponectin is well known to sensitize this insulin-mediated glucose uptake process in skeletal muscle cell/tissue, recent studies have even shown that globular adiponectin itself can activate Akt phosphorylation independently from insulin in rat skeletal muscle L6 cells and mouse C2C12 myocytes in dose- and time-dependent manners and, more importantly, that these globular adiponectin-induced effects on the Akt phosphorylation are not additive to those of insulin.[50] These results suggest that adiponectin and insulin may have a similar mechanism of actions on Akt activation.

Myocytes/skeletal muscle cells are densely populated with complex organelles, mitochondria. Mitochondrial dysfunction has been implicated in the etiology of skeletal muscle insulin resistance and diabetes, and interventions that stimulate skeletal muscle mitochondrial biogenesis can improve insulin sensitivity in insulin resistance and diabetes. Mitochondrial biogenesis is activated by numerous different signals during times of cellular stress or in response to environmental stimuli. Recently, studies have shown that adiponectin, as one of the signals, can enhance mitochondrial biogenesis and oxidative metabolism in the skeletal muscle.[51] In these studies, several mouse models, including adiponectin gene knockout, adiponectin-reconstituted, and control mice, are used to investigate the mitochondrial contents, expression, and activation status of p38 mitogen-activated protein kinase (MAPK) and PPARγ coactivator 1α (PGC-1α) in skeletal muscle cells. An inhibitory effect of adiponectin on MAPK phosphatase 1 (MKP1) gene expression has been also observed in mouse skeletal muscle cells and cultured mouse C2C12 myotubes. In contrast, the overexpression of MKP1 attenuated adiponectin-enhanced mitochondrial biogenesis with significantly decreased PGC-1α expression and p38 MAPK phosphorylation in these cells. These results demonstrate that adiponectin enhances p38 MAPK/PGC-1α signaling and mitochondrial biogenesis in the skeletal muscle by suppressing the MKP1 expression for improving insulin sensitivity in these cells/tissues.

Since enhanced levels of nuclear factor (NF)-κB-inducing kinase (NIK), which is an upstream kinase in the NF-κB pathway, have been implicated in the pathogenesis of chronic inflammation in insulin resistance and diabetes, recent studies demonstrated that adiponectin can inhibit this inflammatory response of NIK in the skeletal muscle.[52] These studies showed that skeletal muscle NIK protein is significantly reduced in association with increased plasma adiponectin and enhanced AMPK phosphorylation and insulin sensitivity in obese human subjects, and enhanced NIK expression in cultured rat L6 myotubes by using wild-type NIK plasmid transfection showed a dose-dependent decrease in insulin-stimulated glucose uptake, which was associated with a significant decrease in PI3K activity and PKB/Akt phosphorylation. In contrast, adiponectin treatment of these cells inhibited NIK-induced NF-κB activation and restored insulin sensitivity by restoring PI3K activation and subsequent Akt phosphorylation. These results indicate that NIK induces insulin resistance and that adiponectin exerts its insulin-sensitizing effect by suppressing NIK-induced skeletal muscle inflammation in obesity and diabetes.

As mentioned previously, the effects of adiponectin on enhancement of glucose uptake and oxidation in myocyte/skeletal muscle are mediated through its two membrane receptors, AdipoR1 and AdipoR2. To investigate the downstream signaling of the two adiponectin receptors, studies have recently demonstrated that the C-terminal extracellular domain of AdipoR1 interacts with adiponectin, whereas the N-terminal cytoplasmic domain of AdipoR1 interacts with an adaptor protein containing pleckstrin homology domain, phosphotyrosine-binding domain, and leucine zipper motif (APPL). This interaction of APPL with AdipoR1 in mouse skeletal muscle C2C12 cells is stimulated by adiponectin binding, and this adiponectin signaling enhances the adiponectin-mediated downstream events such as AMPK activation, lipid oxidation, GLUT4 membrane translocation, and glucose uptake in these cells.[53] Therefore, APPL1 is the first identified protein that interacts directly with both adiponectin receptors, and APPL1 acts as a critical regulator of the cross talk between adiponectin signaling and insulin signaling pathways in skeletal muscle cells, providing a molecular mechanism for the insulin-sensitizing function of adiponectin.[54] In contrast to the role of APPL1 for positively mediating adiponectin signaling in muscle cells, recent studies have further shown that APPL2, an isoform of APPL1 that forms a dimer with APPL1, can interact with both AdipoR1 and AdipoR2 and acts as a negative regulator of adiponectin signaling in skeletal muscle cells.[55] These studies demonstrated that overexpression of APPL2 inhibits the interaction between APPL1 and AdipoR1, leading to downregulation of adiponectin signaling in mouse C2C12 myotubes, and suppressing APPL2 expression significantly enhances adiponectin-stimulated glucose uptake and fatty acid oxidation. Furthermore, adiponectin can promote dissociation of the APPL1/APPL2 complex and translocation of cytosolic APPL1 to the plasma membrane, and adiponectin can also stimulate disruption of the AdipoR1/APPL2 complex, thus enabling binding of APPL1 to AdipoR1 in muscle cells. These results reveal that APPL isoforms function as an integrated Yin–Yang regulator of adiponectin signaling and mediate the cross talk between adiponectin and insulin signaling pathways in skeletal muscle cells.

4.3. Hepatocyte/liver

Hepatocyte/liver plays a central role in glucose production process by balancing the uptake and storage of glucose via glycogenesis (process of glycogen synthesis) and the release of glucose via glycogenolysis (breakdown of glycogen to glucose-1-phosphate and glucose) and gluconeogenesis (generation of glucose from noncarbohydrate carbon substrates such as pyruvate, lactate, glycerol, and glucogenic amino acids).[56]

To directly gain insight into the mechanism(s) of the hypoglycemic effect of adiponectin in hepatocyte/liver, studies have been performed and demonstrated that intraperitoneal injection of purified recombinant adiponectin can lower glucose levels in mice.[57] In these studies, purified recombinant adiponectin was infused in conscious mice during a pancreatic euglycemic hyperinsulinemic clamp (a method for quantifying insulin secretion and resistance). In the presence of physiological hyperinsulinemia, this treatment increased circulating adiponectin levels by approximately twofold and stimulated glucose metabolism. The effect of adiponectin on *in vivo* insulin action was completely accounted for by a 65% reduction in the rate of glucose production in liver. Similarly, glucose flux through glucose-6-phosphatase (G6Pase) decreased with adiponectin injection, whereas the activity of the direct pathway of glucose-6-phosphate biosynthesis, an index of hepatic glucose phosphorylation, increased significantly. Hepatic expression of the gluconeogenic enzymes such as phosphoenolpyruvate carboxykinase and G6Pase mRNAs was reduced by more than 50% following adiponectin infusion compared with vehicle infusion. Thus, a moderate rise in circulating levels of the adipose-derived protein adiponectin inhibits both the expression of hepatic gluconeogenic enzymes and the rate of endogenous glucose production. These results indicate that an acute increase in circulating adiponectin levels can lower hepatic glucose production *in vivo*.

Similar studies to investigate adiponectin-mediated mechanisms for glucose metabolism have also shown that adiponectin lowers mouse serum glucose levels through suppression of hepatic glucose production by activating AMPK activity in the liver.[58] In these studies, a mouse model that can be induced to have a liver-specific knockout of liver kinase B1 (LKB1, also known as serine/threonine kinase 11, STK11), an upstream regulator of AMPK, was used to investigate the adiponectin signaling for glucose production in mouse hepatocytes. These results have shown that loss of LKB1 in the mouse liver partially impaired the ability of adiponectin to lower serum glucose, including reduction of gluconeogenic gene expression and hepatic glucose production as assessed by euglycemic hyperinsulinemic clamp. Furthermore, in mouse hepatocytes, the absence of LKB1, AMPK, or the cAMP response element-binding protein (CREB)-regulated transcription coactivator 2 (CRTC2) did not prevent adiponectin from inhibiting glucose output or reducing gluconeogenic gene expression. Thus, these results have revealed that whereas some adiponectin actions *in vivo* may be LKB1 dependent, substantial LKB1-, AMPK-, and CRTC2-independent signaling pathways also mediate effects of adiponectin on glucose production in the liver.

Metabolic pathways in liver are very complex and well orchestrated with other signaling molecules, and adiponectin receptors are widely distributed in many tissues, including the liver. Adiponectin has direct actions in the liver with prominent roles to improve hepatic insulin sensitivity, increase fatty acid oxidation, and decrease inflammation. Nonalcoholic steatohepatitis (NASH) is one of the most frequent causes of abnormal liver dysfunction associated with accumulated lipid, inflammation, and cell/tissue damage in the liver. Adiponectin receptors (AdipoR1 and AdipoR2) and insulin receptor substrates (IRS-1/-2) are known as modulators of these fatty acid metabolisms in the liver. Studies using an animal model, obese *fa/fa* Zucker rat, which is fed with a high-fat and high–cholesterol diet and developed fatty liver spontaneously with inflammation and fibrosis that are characteristic of NASH, have shown that expression levels of AdipoR1, AdipoR2, and IRS-2 were significantly decreased, whereas IRS-1 was significantly increased, in these NASH animals.[59] As a result of the decreased AdipoR1 and AdipoR2 expression, the messenger RNA expression levels of genes located downstream of AdipoR1 and AdipoR2, including AMPK $\alpha 1/\alpha 2$, which inhibits fatty acid synthesis, and PPARα, which activates fatty acid oxidation, were also decreased in these animals. Expression level of SREBP1c has been found to be elevated, suggesting upregulation of IRS-1 that resulted in increased fatty acid synthesis. Furthermore, increased forkhead box protein A2 (FOXA2) expression has been observed, which might be associated with the downregulation of IRS-2, facilitating fatty acid oxidation. These results indicate increased synthesis and oxidation of fatty acids by up- or downregulation of AdipoR1 and AdipoR2 or IRS may contribute to the progression of NASH. Thus, AdipoR1, AdipoR2, and IRS might be crucially important regulators in the signaling pathway for the synthesis and oxidation of fatty acids in the liver of NASH.

4.4. Endothelial cell

Adiponectin exerts the beneficial effects on vascular disorders by directly affecting components of vascular tissue, including endothelial cells in endothelium. Vascular insulin resistance has been generally considered to contribute to elevated peripheral vascular resistance and subsequent hypertension, and some clinical observations have already shown that lower plasma adiponectin concentration is significantly associated with hypertension.[60–63] Therefore, endothelial dysfunction in vascular system is an important feature predisposing to vascular disease and is closely associated with obesity-linked complications including hypertension and insulin resistance.[64] Currently, numerous studies have confirmed that adiponectin is beneficial for endothelial function in vascular system and plasma adiponectin level is closely correlated with the vasodilator response to reactive hyperemia in hypertensive patients.[29]

Nitric oxide synthases (NOSs) are a family of enzymes catalyzing the production of nitric oxide (NO) from L-arginine, and the NO is an important cellular signaling molecule that modulates vascular tone, with insulin secretion as one of its multiple functions. Endothelial nitric oxide synthase (eNOS) and NO are crucial regulators of vascular homeostasis, in particular endothelial function.[65,66] Current accumulating evidences indicate that one of the adiponectin functions in endothelial cell/endothelium is acting as an endogenous modulator of endothelial-derived NO production.[67,68] It has been reported that adiponectin knockout mice have lower levels of eNOS transcripts in the aorta and NO metabolites in plasma as well as higher blood pressure when compared with wild-type mice after mice were fed with a high-salt diet.[69] In contrast, systemic administration of adiponectin to the adiponectin knockout mice fed with high salt lowers the elevated blood pressure and restores the reduced eNOS transcripts in the aorta. Moreover, the inhibition of eNOS by adiponectin administration completely reverses the reduction of blood pressure in adiponectin knockout mice. Therefore, these studies have proved that adiponectin deficiency participates in salt-sensitive hypertension through modulation of eNOS function in endothelial cells. The stimulatory effects of adiponectin on eNOS activity and NO production have also been demonstrated to be mediated through adiponectin receptors, AdipoR1 and AdipoR2, as well as their intracellular adaptor molecule APPL1.[70]

A number of clinical studies have already demonstrated the existence of an inverse relationship between serum adiponectin concentration and human blood pressure.[60,62] To determine that hypoadiponectinemia (low concentration of adiponectin) indeed induces vascular insulin resistance before the systemic hypertension and the involved regulatory mechanisms, animal models, including 4-week-old young spontaneously hypertensive rats (ySHRs, normotensive) and adiponectin knockout mice, are used to evaluate the role of hypoadiponectinemia in insulin-induced vasodilation of resistance vessels.[71] These studies have shown that ySHRs present significant vascular insulin resistance as evidenced by the blunted vasorelaxation response to insulin in mesenteric arterioles compared with that of age-matched control rats and serum adiponectin concentration and mesenteric arteriolar APPL1 expression of ySHRs are also significantly reduced. In addition, Akt and eNOS phosphorylation and NO production in arterioles are markedly reduced, whereas extracellular signal-regulated protein kinases 1/2 (ERK1/2) phosphorylation and endothelin-1 secretion are augmented in ySHRs. Consistent with the results from ySHRs, adiponectin knockout mice also show significantly decreased APPL1 expression and vasodilation evoked by insulin. Furthermore, treatment of ySHRs *in vivo* with adiponectin for 1 week increases APPL1 expression and insulin-induced vasodilation and restores the balance between insulin-stimulated endothelial vasodilator NO and vasoconstrictor endothelin-1. When using adiponectin to treat cultured human umbilical vein endothelial cells (HUVEC), APPL1 expression in these cells has been found to be upregulated. In contrast, suppression of APPL1 expression with siRNA markedly blunts the effects of adiponectin-induced insulin sensitization as evidenced by reduced Akt/eNOS and potentiated ERK1/2 phosphorylations. These results indicate that hypoadiponectinemia induces APPL1 downregulation in the resistance vessels, contributing to the development of vascular insulin resistance by differentially modulating the Akt/eNOS/NO and ERK1/2/endothelin-1 pathways in vascular endothelium in normotensive ySHRs.

Since coexisting hyperglycemia and systemic inflammation predisposes to dysregulated angiogenesis and vascular disease, how adiponectin modulates these processes have been investigated by using human microvascular endothelial cells (HMEC-1).[72] These studies have found that endothelial cell proliferation, *in vitro* migration, and angiogenesis are significantly increased by adiponectin mediated by AdipoR1 via the AMPK–Akt pathways; and adiponectin also significantly increases matrix metalloproteinase 2 and 9 (MMP-2 and MMP-9) and vascular endothelial growth factor (VEGF) expression levels in these endothelial cells. The effect of adiponectin on VEGF appears to be mediated by AdipoR1, while the effect of adiponectin on MMP-2 and MMP-9 appears to be mediated by both AdipoR1 and AdipoR2. Importantly, adiponectin decreases glucose and C-reactive protein-induced angiogenesis with a concomitant reduction in MMP-2, MMP-9, and VEGF in the HMEC-1 cells. These results have provided novel insights into the regulatory mechanisms of adiponectin on angiogenesis in endothelial cells.

Similar studies using HUVEC have also been performed to investigate apoptosis that may lead to endothelial dysfunction and contribute to vascular complications. These studies have shown that apoptosis induced by an intermittent high-glucose media is modulated by adiponectin receptors and AMPK signaling pathways.[73] HUVEC apoptosis is increased more significantly in an intermittent high-glucose medium than in a constant high-glucose medium, and the HUVEC apoptosis induced by an intermittent high-glucose medium is inhibited when the cells are pretreated with adiponectin, which rapidly activated AMPK and AdipoR1 in HUVECs. However, AdipoR2 is not activated by adiponectin for apoptosis inhibition in these cells. These results have indicated that AdipoR1, but not AdipoR2, is involved in mediating intermittent high-concentration glucose-evoked apoptosis in endothelial cells and adiponectin can activate AMPK through AdipoR1, leading to the partial inhibition of HUVEC apoptosis.

Recent clinical studies also indicated that adiponectin can reduce the risk of hypertension in humans based on its anti-inflammatory, antiatherogenic, and insulin-sensitizing properties.[74] These studies have evaluated the interrelationships of adiponectin, blood pressure, obesity, body fat distribution, puberty, and insulin resistance in a selected group of children. In these studies, the subject's blood glucose, insulin, and adiponectin concentrations were assayed, and their homeostatic model assessment index was calculated as an estimate of insulin resistance. These data have shown that in childhood, serum levels of adiponectin are inversely related to hypertension and this relationship is partly independent of obesity, fat distribution, and insulin resistance. These results also indicate that low values of adiponectin in both obese and normal weight children are associated with a higher probability of hypertension.

In addition to the eNOS/NO signaling pathway, adiponectin also promotes endothelial cell function through another signaling pathway, cyclooxygenese-2 (COX-2). Endothelial COX-2 and its metabolites (prostaglandins) play an important role in the control of vascular functions such as vascular tone and endothelial function.[75-78] Studies have demonstrated that in cultured human or mouse endothelial cells, adiponectin increases COX-2 expression, and ablation of COX-2 abrogates the adiponectin-stimulated increases in endothelial cell migration, differentiation, and survival.[79] These studies have also shown that ablation of calreticulin (CRT) or its adaptor protein CD91 can diminish the adiponectin-stimulated COX-2 expression and endothelial cell responses. These data have provided evidence that adiponectin promotes endothelial cell function through CRT/CD91-mediated increases in COX-2 signaling. Therefore, these results indicate that disruption of the adiponectin-COX-2 regulatory axis in endothelial cells could participate in the pathogenesis of obesity-related vascular diseases. More importantly, the adiponectin-mediated endothelial cell protection is through an Akt–COX-2 regulatory axis that is largely dependent on the ability of adiponectin to associate with the adiponectin-binding protein, CRT, and its adaptor protein CD91 on the surface of endothelial cells.[80]

Previous studies have shown that adiponectin knockout mice display reduced levels of prostaglandin I synthase mRNA in the aorta and circulating prostaglandin I_2 metabolite following high-salt feeding in these mice[69] and adiponectin supplementation using adenovirus system can restore the reduced levels of prostaglandin I synthase transcript in the aorta and prostaglandin I_2 metabolite in plasma in these high-salt-fed adiponectin knockout mice. These results indicate that adiponectin improves endothelial cell function, at least in part, through COX-2-prostaglandin I_2-dependent pathway.[80] Taken together, adiponectin protects against endothelial dysfunction via at least two regulatory pathways involving AMPK–eNOS signaling and COX-2-prostaglandin I_2 signaling within endothelial cells. Therefore, adiponectin directly acts on vascular endothelial cells and exerts salutary effects on endothelial function through eNOS-dependent and COX-2-dependent regulatory mechanisms.

4.5. Macrophage/foam cell

Macrophages are a heterogeneous population of phagocytic cells found throughout the body that originate from the mononuclear phagocytic system.[81] These are highly plastic cells that arise from circulating myeloid-derived blood monocytes that have entered target tissues and gained the phenotypic and functional attributes of their tissue of residence. Recent attention has focused on the potential role of macrophages in the process of metabolic diseases.[82,83] It has been shown that in obesity, adipose tissue contains an increased number of resident macrophages and that in some circumstances, macrophages can constitute up to 40% of the cell population within an adipose tissue depot[84,85]; and macrophages are obviously a potential source of secreted proinflammatory factors to other tissues for insulin resistance. This correlative evidence has led to the concept that macrophages can directly influence other insulin target tissues, such as the adipose tissue, skeletal muscle, and liver, and cause whole body systemic insulin resistance. Furthermore, animal models have also been reported to demonstrate the causal role of the macrophage in leading to insulin resistance.[86] When these animals were fed with an HFD, the macrophage-specific inflammatory pathway knockout mice were relatively protected from glucose intolerance and hyperinsulinemia, and these results showed a global improvement in insulin sensitivity in all insulin target tissues. These studies are consistent with the interpretation that the macrophage is an important, and potentially initiating, cell type in the process of inflammation-induced insulin resistance.[87] Preventing macrophage infiltration into these insulin target tissues will have beneficial effects on the inflammatory response and the abnormal metabolic state.

Except for the effects on insulin sensitivity, macrophages also have critical impacts on cardiovascular disease in metabolic syndrome, such as atherosclerosis. Atherosclerosis is considered a chronic inflammatory disease and a disorder of lipid metabolism.[88] The complex physiopathologic process is initiated by the formation of cholesterol-rich lesions in the arterial wall. The accumulation of cholesterol-rich lipoproteins in the vascular artery wall results in the recruitment of circulating monocytes, their adhesion to the endothelium, and their differentiation into macrophages. Macrophages play a crucial role in this process because they accumulate large amounts of lipid to form the foam cells that initiate the formation of the lesion and participate actively in the development of the atherosclerotic lesion. Because the transformation of macrophage into foam cell is a critical component of atherosclerotic lesion formation,[89] the prevention or reversal of cholesterol accumulation or the production of inflammatory mediators in macrophage foam cells could result in protection from multiple pathological effects of atherosclerosis and other abnormal metabolic disorders in metabolic syndrome.

As mentioned earlier, adiponectin is almost exclusively expressed and secreted from adipocyte/adipose tissue. Adiponectin expression has not been detected in macrophage cells. Since two adiponectin receptors, the

AdipoR1 and AdipoR2, have been found to be expressed in human macrophages[90] with AdipoR1 predominating in these macrophage cells,[91,92] therefore, it is generally considered that the binding of adiponectin to its receptors, and subsequent signaling events initiated by this binding, is probably responsible for the physiological effects in macrophage cells.

Human THP-1 monocytic cell line[93] is a well-characterized cell model for studying processes and mechanisms of monocyte differentiation into macrophage and macrophage transformation into foam cell. These THP-1 monocytic cells can be induced to differentiate into macrophages following treatment with phorbol myristate acetate, and the resulting macrophages can then be induced to form foam cells following treatment with modified low-density lipoprotein (LDL).[94] Adiponectin has been reported to inhibit lipid accumulation by downregulating scavenger receptor A (SR-A) expression and acyl-CoA: cholesterol-acyltransferase 1 (ACAT1) expression, two key players in foam cell formation, in cultured human monocyte-derived macrophages when the adipocyte-derived adiponectin protein was added to the THP-1 cell cultures.[95,96] These results suggest that adiponectin has inhibitory role in macrophage foam cell formation from human monocyte-derived macrophages. Interestingly, when adiponectin gene is directly expressed in THP-1 cells, these monocyte-derived macrophage foam cells can significantly decrease triglyceride and cholesterol accumulation in these cells by reducing oxidized low-density lipoprotein (oxLDL) uptake into the cells while enhancing HDL-mediated cholesterol efflux.[97] There are two potential mechanisms for the reduced lipid accumulation in these adiponectin-transduced macrophage foam cells: the first mechanism involves the PPARγ and liver X receptor signaling pathways, which upregulate the expression of ATP-binding cassette A1 gene and promote lipid efflux from these cells; the second mechanism involves decreased lipid uptake and increased lipid hydrolysis, which may result from decreased scavenger receptor A1 (SR-A1) and increased scavenger receptor BI (SR-BI) and hormone-sensitive lipase gene activities in the adiponectin-transduced macrophage foam cells. Furthermore, two inflammatory cytokines, monocyte chemoattractant protein 1 (MCP-1) and TNF-α, are also decreased in the adiponectin-transduced macrophage foam cells. These studies suggest that expression of adiponectin in human THP-1 macrophage cells can modulate multiple signaling pathways for both lipid metabolism and inflammatory response to reduce macrophage foam cell formation during atherosclerosis.

Macrophages can be functionally divided into two subsets with two different phenotypes, classically activated M1 and alternatively activated M2 macrophages, based on these cells in response to exposure to microenvironmental stimuli and signals.[98,99] M1 and M2 macrophages in adipose tissue can be distinguished by the presence or the absence of CD11c (a type I transmembrane protein), an M1 macrophage marker.[100] M1 macrophages produce high levels of proinflammatory cytokines, such as TNF-α, MCP-1, and interleukin 6 (IL-6), and play a role in tissue destruction. In contrast, M2 are more prominent producers of anti-inflammatory cytokines, IL-1 receptor antagonist (IL-1Ra), IL-10, and transforming growth factor β (TGF-β), which are involved in tissue repair or remodeling. It has been reported that adiponectin can modulate macrophage polarization from a classically activated M1 phenotype to an alternatively activated M2 phenotype in mouse adipose tissue.[101] These studies have shown that peritoneal macrophages and the stromal vascular fraction (SVF) cells of adipose tissue isolated from adiponectin knockout mice display increased M1 markers and decreased M2 markers and the systemic delivery of adenovirus expressing adiponectin significantly augmented M1 markers in peritoneal macrophages and SVF cells in both wild-type and adiponectin knockout mice. Adiponectin also stimulates the expression of M2 markers and attenuated the expression of M1 markers in human monocyte-derived macrophages and SVF cells isolated from human adipose tissue. These studies have demonstrated that adiponectin functions as a regulator of macrophage polarization, and they indicate that conditions of high adiponectin expression may deter metabolic and cardiovascular disease progression by favoring an anti-inflammatory phenotype in macrophages.

Recent studies using Ad-TG mice that specifically express adiponectin in mouse macrophage cells have further demonstrated that adiponectin can physiologically modulate metabolic activities *in vivo* by improving metabolism in distal tissues for metabolic syndrome.[102,103] In these studies, a macrophage-specific Ad-TG mouse model was generated using an SR-A1 gene enhancer/promoter; data from these Ad-TG mice have shown that expression of adiponectin in mouse macrophages can significantly reduce cholesterol and triglyceride accumulation in these mouse macrophage cells and the bloodstream with significantly lower levels of total cholesterol due to significantly decreased LDL and VLDL (very low-density lipoprotein) cholesterol concentrations together with a significant increase in HDL cholesterol, which contribute to the reduction of macrophage foam cell formation in the atherosclerotic lesions when these macrophage-modified transgenic mice are crossed with a LDL receptor (Ldlr)-deficient mouse model under the HFD condition. In addition, Ad-TG mice also exhibit higher total and HMW adiponectin levels in plasma. Furthermore, the adiponectin-modified mouse macrophages also alter systemic metabolism in other metabolic active tissues, such as the adipose tissue, skeletal muscle, and liver, with cell interactions through either infiltration or circulation by bloodstream, which influences the levels of inflammatory cytokines, MCP-1, TNF-α, and IL-6 in these metabolically active tissues and thus physiologically improves whole body glucose tolerance and insulin sensitivity. These studies indicate that macrophages engineered to produce adiponectin can significantly improve intercellular microenvironment and alter the important metabolic activities in other tissues through macrophage infiltration and circulation to reduce inflammation response and lipid accumulation in these tissues. Although adiponectin is a specific adipokine expressed and secreted physiologically mainly from adipocytes, these studies have demonstrated that adiponectin from the modified macrophages can also exert powerful anti-inflammatory and antiatherogenic effects on whole body metabolism.

However, since the Ad-TG mice described earlier are also exhibiting increased circulating adiponectin concentrations in the bloodstream, the elevated adiponectin itself could be influencing multiple tissues directly, not necessarily as a result of a specific interaction with the adiponectin-modified macrophage cells. To exclusively point out the mechanisms for those antidiabetic and anti-atherosclerotic roles of adiponectin and adiponectin receptors in macrophage cells, further studies have recently developed a novel approach to manipulate adiponectin action at the level of the macrophage in order to examine systemic effects related to metabolic diseases by genetic manipulation of the major receptor for adiponectin in macrophages, AdipoR1.[104] These macrophage-specific AdipoR1 transgenic (AdR1-TG) mice have presented reduced whole body weight, fat accumulation, and liver steatosis when these mice were fed with an HFD. Moreover, these AdR1-TG mice exhibited enhanced whole body glucose tolerance and insulin sensitivity by activating Akt and AMPK phosphorylation, with reduced proinflammatory cytokines, MCP-1 and TNF-α, both in the serum and in the insulin target metabolic tissues. Additional studies have also demonstrated that these AdR1-TG animals exhibited decreased lipid accumulation in macrophages and other active metabolic tissues as indicated by reduced expression of molecular markers, such as F4/80 (a marker specific for mature macrophages associated with "crown-like structure"), chemokine receptor 2 (CCR2, a marker for M1 macrophage phenotype), and fatty acid-binding protein 4 (FABP4/aP2, which promotes lipid loading into cells), and reduced macrophage foam cell formation in the arterial wall when these transgenic mice were crossed with a low-density lipoprotein receptor (Ldlr)-deficient mouse model. These new studies, for the first time, suggest that overexpression of AdipoR1 can alter macrophage biology and impact systemic metabolism *in vivo*. These data point to the central role of macrophage cells and the singular ability of adiponectin to regulate macrophage function, in metabolic diseases.

Interestingly, the transgenic mouse model with macrophages over-expressing AdipoR1 (AdR1-TG) has shown not only improved insulin sensitivity and reduced inflammation detected in the adipose tissue, skeletal muscle, and liver, but also favorably reduced fat mass; macrophage and plasma lipid accumulation; and foam cell formation resulting from the enhanced adiponectin actions in both macrophage cells and other metabolically active tissues.[104] When the AdR1-TG mouse is crossed with a TALLYHO (TH) diabetic mouse that is an inbred polygenic mouse model for type II diabetes, to generate an adiponectin receptor 1 transgenic/TALLYHO (AdR1-TG/TH) mouse,[105] these transgenic crossed mice have significantly improved lipid accumulation and insulin sensitivity *in vivo*. In addition to the improved lipid and glucose metabolism, increases of insulin secretion and Sirtuin 1 (Sirt1) gene expression for insulin secretion in pancreas are also detected in the AdR1-TG/TH mice, and there are reduced pancreatic hypotrophy and decreased apoptotic gene expression observed in these animals. These results further suggest that enhanced adiponectin actions by overexpressing AdipoR1 in macrophage cells can provide unique interactions with other metabolic cells/tissues, improving symptoms of the metabolic syndrome. Moreover, these studies not only provide new insights for investigating the mechanisms of metabolic syndrome *in vivo*, using macrophages as the carriers for *in vivo* enhanced actions of adiponectin, which is an antidiabetic, anti-inflammatory, and antiatherogenic cytokine, but also may provide a novel, unique strategy (Fig. 9.2) to develop a new therapeutic application for the treatment of metabolic syndrome or metabolic disorders in the future.

5. SUMMARY

Metabolic syndrome is a combination of several serious metabolic disorders, including obesity, insulin resistance, type II diabetes, and cardiovascular disease. Currently, there is no effective way to treat all the components in metabolic syndrome except by lifestyle changes. Adiponectin has been proved as one adipokine with antidiabetic, anti-inflammatory, and

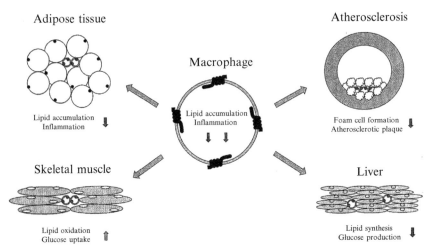

Figure 9.2 AdipoR1-modified macrophages to influence other metabolically active tissues. AdipoR1 molecules (four red-colored receptor domain structures) are overexpressed in macrophages and the modified macrophages infiltrate/circulate or reside in metabolically active tissues such as the adipose tissue, vascular artery, liver, and skeletal muscle through the blood vessels and bloodstream (green arrows). The modified macrophages interact with these metabolic cells/tissues to improve intercellular microenvironment for whole body systemic metabolism. Red-colored arrows indicate downregulation effects and a small green-colored arrow represents upregulation effects in the interacting skeletal muscle. (See color plate.)

antiatherogenic properties mainly expressed and secreted from the adipocyte/adipose tissue. Studies conducted in human and animal models of obesity, diabetes, and atherosclerosis have clearly demonstrated that adiponectin and adiponectin receptors as well as the involved signaling pathways can indeed have beneficial effects on these metabolic diseases. The use of macrophage cells as carriers for overexpression of AdipoR1 will provide a novel and unique strategy for studying the mechanisms of adiponectin–mediated alterations in body metabolism *in vivo*. This unique strategy can probably serve as a potential novel and innovative therapeutic approach for treatment of the metabolic syndrome in the future.

REFERENCES

1. Reaven GM. Role of insulin resistance in human disease. *Diabetes*. 1988;37:1595–1607.
2. Reaven GM. Insulin resistance, hyperinsulinemia, hypertriglyceridemia, and hypertension: parallels between human disease and rodent models. *Diabetes Care*. 1991;14:195–202.

3. DeFronzo RA, Ferrannini E. Insulin resistance: a multifaceted syndrome responsible for NIDDM, obesity, hypertension, dyslipidemia, and atherosclerotic cardiovascular disease. *Diabetes Care*. 1991;14:173–194.
4. Nesto RW, Bell D, Bonow RO. Thiazolidinedione use, fluid retention, and congestive heart failure. *Circulation*. 2003;108:2941–2948.
5. Nissen SE, Wolski K. Effect of rosiglitazone on the risk of myocardial infarction and death from cardiovascular causes. *N Engl J Med*. 2007;356:2457–2471.
6. Lewis JD, Ferrara A, Peng T, et al. Risk of bladder cancer among diabetic patients treated with pioglitazone. *Diabetes Care*. 2011;34:916–922.
7. Arita Y, Kihara S, Ouchi N, et al. Paradoxical decrease of an adipose-specific protein, adiponectin, in obesity. *Biochem Biophys Res Commun*. 1999;257:79–83.
8. Maeda K, Okubo K, Shimomura I, Funahashi T, Matsuzawa Y, Matsubara K. cDNA cloning and expression of a novel adipose specific collagen-like factor, apM1 (AdiPose Most abundant Gene transcript 1). *Biochem Biophys Res Commun*. 1996;221:286–289.
9. Hu E, Liang P, Spiegelman BM. AdipoQ is a novel adipose-specific gene dysregulated in obesity. *J Biol Chem*. 1996;271:10697–10703.
10. Nakano Y, Tobe T, Choi-Miura NH, Mazda T, Tomita M. Isolation and characterization of GBP28, a novel gelatin-binding protein purified from human plasma. *J Biochem*. 1996;120:803–812.
11. Scherer PE, Williams S, Fogliano M, Baldini G, Lodish HF. A novel serum protein similar to C1q, produced exclusively in adipocytes. *J Biol Chem*. 1995;270: 26746–26749.
12. Berg AH, Combs TP, Scherer PE. ACRP30/adiponectin: an adipokine regulating glucose and lipid metabolism. *Trends Endocrinol Metab*. 2002;13:84–89.
13. Pajvani UB, Hawkins M, Combs TP, et al. Complex distribution, not absolute amount of adiponectin, correlates with thiazolidinedione-mediated improvement in insulin sensitivity. *J Biol Chem*. 2004;279:12152–12162.
14. Lara-Castro C, Fu Y, Chung BH, Garvey WT. Adiponectin and the metabolic syndrome: mechanisms mediating risk for metabolic and cardiovascular disease. *Curr Opin Lipidol*. 2007;18:263–270.
15. Bobbert T, Rochlitz H, Wegewitz U, et al. Changes of adiponectin oligomer composition by moderate weight reduction. *Diabetes*. 2005;54:2712–2719.
16. Lara-Castro C, Luo N, Wallace P, Klein RL, Garvey WT. Adiponectin multimeric complexes and the metabolic syndrome trait cluster. *Diabetes*. 2006;55:249–259.
17. Yamauchi T, Kamon J, Ito Y, et al. Cloning of adiponectin receptors that mediate antidiabetic metabolic effects. *Nature*. 2003;423:762–769.
18. Hug C, Wang J, Ahmad NS, Bogan JS, Tsao TS, Lodish HF. T-cadherin is a receptor for hexameric and high-molecular-weight forms of Acrp30/adiponectin. *Proc Natl Acad Sci U S A*. 2004;101:10308–10313.
19. Kadowaki T, Yamauchi T, Kubota N, Hara K, Ueki K, Tobe K. Adiponectin and adiponectin receptors in insulin resistance, diabetes, and the metabolic syndrome. *J Clin Invest*. 2006;116:1784–1792.
20. Denzel MS, Scimia MC, Zumstein PM, Walsh K, Ruiz-Lozano P, Ranscht B. T-cadherin is critical for adiponectin-mediated cardioprotection in mice. *J Clin Invest*. 2010;120:4342–4352.
21. Hotta K, Funahashi T, Arita Y, et al. Plasma concentrations of a novel, adipose-specific protein, adiponectin, in type 2 diabetic patients. *Arterioscler Thromb Vasc Biol*. 2000;20:1595–1599.
22. Lindsay RS, Funahashi T, Hanson RL, et al. Adiponectin and development of type 2 diabetes in the Pima Indian population. *Lancet*. 2002;360:57–58.
23. Lindsay RS, Funahashi T, Krakoff J, et al. Genome-wide linkage analysis of serum adiponectin in the Pima Indian population. *Diabetes*. 2003;52:2419–2425.

24. Hotta K, Funahashi T, Bodkin NL, et al. Circulating concentrations of the adipocyte protein adiponectin are decreased in parallel with reduced insulin sensitivity during the progression to type 2 diabetes in rhesus monkeys. *Diabetes.* 2001;50:1126–1133.

25. Weyer C, Funahashi T, Tanaka S, et al. Hypoadiponectinemia in obesity and type 2 diabetes: close association with insulin resistance and hyperinsulinemia. *J Clin Endocrinol Metab.* 2001;86:1930–1935.

26. Menzaghi C, Trischitta V, Doria A. Genetic influences of adiponectin on insulin resistance, type 2 diabetes, and cardiovascular disease. *Diabetes.* 2007;56:1198–1209.

27. Kubota N, Terauchi Y, Yamauchi T, et al. Disruption of adiponectin causes insulin resistance and neointimal formation. *J Biol Chem.* 2002;277:25863–25866.

28. Maeda N, Shimomura I, Kishida K, et al. Diet-induced insulin resistance in mice lacking adiponectin/ACRP30. *Nat Med.* 2002;8:731–737.

29. Ouchi N, Ohishi M, Kihara S, et al. Association of hypoadiponectinemia with impaired vasoreactivity. *Hypertension.* 2003;42:231–234.

30. Yamauchi T, Nio Y, Maki T, et al. Targeted disruption of AdipoR1 and AdipoR2 causes abrogation of adiponectin binding and metabolic actions. *Nat Med.* 2007;13:332–339.

31. Havel P. Control of energy homeostasis and insulin action by adipocyte hormones: leptin, acylation stimulating protein, and adiponectin. *Curr Opin Lipidol.* 2002;13:51–59.

32. Fujimoto N, Matsuo N, Sumiyoshi H, et al. Adiponectin is expressed in the brown adipose tissue and surrounding immature tissues in mouse embryos. *Biochim Biophys Acta.* 2005;1731:1–12.

33. Viengchareun S, Zennaro MC, Pascual-Le Tallec L, Lombes M. Brown adipocytes are novel sites of expression and regulation of adiponectin and resistin. *FEBS Lett.* 2002;532:345–350.

34. Maddineni S, Metzger S, Ocón O, Hendricks 3rd G, Ramachandran R. Adiponectin gene is expressed in multiple tissues in the chicken: food deprivation influences adiponectin messenger ribonucleic acid expression. *Endocrinology.* 2005;146:4250–4256.

35. Nishida M, Funahashi T, Shimomura I. Pathophysiological significance of adiponectin. *Med Mol Morphol.* 2007;40:55–67.

36. Saito K, Tobe T, Yoda M, Nakano Y, Choi-Miura NH, Tomita M. Regulation of gelatin-binding protein 28 (GBP28) gene expression by C/EBP. *Biol Pharm Bull.* 1999;22:1158–1162.

37. Qiao L, Maclean PS, Schaack J, et al. C/EBPα regulates human adiponectin gene transcription through an intronic enhancer. *Diabetes.* 2005;54:1744–1754.

38. Iwaki M, Matsuda M, Maeda N, et al. Induction of adiponectin, a fat-derived antidiabetic and antiatherogenic factor, by nuclear receptors. *Diabetes.* 2003;52:1655–1663.

39. Maeda N, Takahashi M, Funahashi T, et al. PPARgamma ligands increase expression and plasma concentrations of adiponectin, an adipose-derived protein. *Diabetes.* 2001;50:2094–2099.

40. Seo JB, Moon HM, Noh MJ, et al. Adipocyte determination- and differentiation dependent factor 1/sterol regulatory element-binding protein 1c regulates mouse adiponectin expression. *J Biol Chem.* 2004;279:22108–22117.

41. Doran AC, Meller N, Cutchins A, et al. The helix–loop–helix factors Id3 and E47 are novel regulators of adiponectin. *Circ Res.* 2008;103:624–634.

42. Green H, Kehinde O. An established preadipose cell line and its differentiation in culture II. Factors affecting the adipose conversion. *Cell.* 1975;5:19–27.

43. Fu Y, Luo N, Klein RL, Garvey WT. Adiponectin promotes adipocyte differentiation, insulin sensitivity, and lipid accumulation. *J Lipid Res.* 2005;46:1369–1379.

44. Combs TP, Pajvani UB, Berg AH, et al. A transgenic mouse with a deletion in the collagenous domain of adiponectin displays elevated circulating adiponectin and improved insulin sensitivity. *Endocrinology.* 2004;145:367–383.

45. Kim JY, van de Wall E, Laplante M, et al. Obesity-associated improvements in metabolic profile through expansion of adipose tissue. *J Clin Invest.* 2007;117:2621–2637.
46. Asterholm IW, Scherer PE. Enhanced metabolic flexibility associated with elevated adiponectin levels. *Am J Pathol.* 2010;176:1364–1376.
47. Bauer S, Weigert J, Neumeier M, et al. Low-abundant adiponectin receptors in visceral adipose tissue of humans and rats are further reduced in diabetic animals. *Arch Med Res.* 2010;41:75–82.
48. Tsuchida A, Yamauchi T, Takekawa S, et al. Peroxisome proliferator-activated receptor (PPAR) alpha activation increases adiponectin receptors and reduces obesity-related inflammation in adipose tissue: comparison of activation of PPARalpha, PPARgamma, and their combination. *Diabetes.* 2005;54:3358–3370.
49. Klip A, Paquet MR. Glucose transport and glucose transporters in muscle and their metabolic regulation. *Diabetes Care.* 1990;13:228–243.
50. Sattar AA, Sattar R. Globular adiponectin activates Akt in cultured myocytes. *Biochem Biophys Res Commun.* 2012;424:753–757.
51. Qiao L, Kinney B, Yoo HS, Lee B, Schaack J, Shao J. Adiponectin increases skeletal muscle mitochondrial biogenesis by suppressing mitogen-activated protein kinase phosphatase-1. *Diabetes.* 2012;61:1463–1470.
52. Choudhary S, Sinha S, Zhao Y, et al. NF-kappaB-inducing kinase (NIK) mediates skeletal muscle insulin resistance: blockade by adiponectin. *Endocrinology.* 2011;152:3622–3627.
53. Mao X, Kikani CK, Riojas RA, et al. APPL1 binds to adiponectin receptors and mediates adiponectin signalling and function. *Nat Cell Biol.* 2006;8:516–523.
54. Deepa SS, Dong LQ. APPL1: role in adiponectin signaling and beyond. *Am J Physiol Endocrinol Metab.* 2009;296:E22–E36.
55. Wang C, Xin X, Xiang R, et al. Yin-Yang regulation of adiponectin signaling by APPL isoforms in muscle cells. *J Biol Chem.* 2009;284:31608–31615.
56. Nordlie RC, Foster JD, Lange AJ. Regulation of glucose production by the liver. *Annu Rev Nutr.* 1999;19:379–406.
57. Combs TP, Bergm AH, Obici S, Scherer PE, Rossetti L. Endogenous glucose production is inhibited by the adipose-derived protein Acrp30. *J Clin Invest.* 2001;108:1875–1881.
58. Miller RA, Chu Q, Le Lay J, et al. Adiponectin suppresses gluconeogenic gene expression in mouse hepatocytes independent of LKB1-AMPK signaling. *J Clin Invest.* 2011;121:2518–2528.
59. Matsunami T, Sato Y, Ariga S, et al. Regulation of synthesis and oxidation of fatty acids by adiponectin receptors (AdipoR1/R2) and insulin receptor substrate isoforms (IRS-1/-2) of the liver in a nonalcoholic steatohepatitis animal model. *Metabolism.* 2011;60:805–814.
60. Adamczak M, Wiecek A, Funahashi T, Chudek J, Kokot F, Matsuzawa Y. Decreased plasma adiponectin concentration in patients with essential hypertension. *Am J Hypertens.* 2003;16:72–75.
61. Iwashima Y, Katsuya T, Ishikawa K, et al. Hypoadiponectinemia is an independent risk factor for hypertension. *Hypertension.* 2004;43:1318–1323.
62. Kazumi T, Kawaguchi A, Sakai K, Hirano T, Yoshino G. Young men with high normal blood pressure have lower serum adiponectin, smaller LDL size, and higher elevated heart rate than those with optimal blood pressure. *Diabetes Care.* 2002;25:971–976.
63. Murakami H, Ura N, Furuhashi M, Higashiura K, Miura T, Shimamoto K. Role of adiponectin in insulin-resistant hypertension and atherosclerosis. *Hypertens Res.* 2003;26:705–710.
64. Lüscher TF. The endothelium and cardiovascular disease—a complex relation. *N Engl J Med.* 1994;330:1081–1083.
65. Huang PL, Huang Z, Mashimo H, et al. Hypertension in mice lacking the gene for endothelial nitric oxide synthase. *Nature.* 1995;377:239–242.

66. Guzik TJ, Black E, West NE, et al. Relationship between the G894T polymorphism (Glu298Asp variant) in endothelial nitric oxide synthase and nitric oxide-mediated endothelial function in human atherosclerosis. *Am J Med Genet*. 2001;100:130–137.

67. Chen H, Montagnani M, Funahashi T, Shimomura I, Quon MJ. Adiponectin stimulates production of nitric oxide in vascular endothelial cells. *J Biol Chem*. 2003;278:45021–45026.

68. Motoshima H, Wu X, Mahadev K, Goldstein BJ. Adiponectin suppresses proliferation and superoxide generation and enhances eNOS activity in endothelial cells treated with oxidized LDL. *Biochem Biophys Res Commun*. 2004;315:264–271.

69. Ohashi K, Kihara S, Ouchi N, et al. Adiponectin replenishment ameliorates obesity-related hypertension. *Hypertension*. 2006;47:1108–1116.

70. Cheng KK, Lam KS, Wang Y, et al. Adiponectin induced endothelial nitric oxide synthase activation and nitric oxide production are mediated by APPL1 in endothelial cells. *Diabetes*. 2007;56:1387–1394.

71. Xing W, Yan W, Liu P, et al. A novel mechanism for vascular insulin resistance in normotensive young SHRs: hypoadiponectinemia and resultant APPL1 downregulation. *Hypertension*. 2013;61:1028–1035.

72. Adya R, Tan BK, Chen J, Randeva HS. Protective actions of globular and full-length adiponectin on human endothelial cells: novel insights into adiponectin-induced angiogenesis. *J Vasc Res*. 2012;49:534–543.

73. Zhao HY, Zhao M, Yi TN, Zhang J. Globular adiponectin protects human umbilical vein endothelial cells against apoptosis through adiponectin receptor 1/adenosine monophosphate-activated protein kinase pathway. *Chin Med J*. 2011;124:2540–2547.

74. Brambilla P, Antolini L, Street ME, et al. Adiponectin and hypertension in normal-weight and obese children. *Am J Hypertens*. 2013;26:257–264.

75. Bulut D, Liaghat S, Hanefeld C, Koll R, Miebach T, Mügge A. Selective cyclooxygenase-2 inhibition with parecoxib acutely impairs endothelium-dependent vasodilatation in patients with essential hypertension. *J Hypertens*. 2003;21:1663–1667.

76. Oshima M, Oshima H, Taketo MM. Hypergravity induces expression of cyclooxygenase-2 in the heart vessels. *Biochem Biophys Res Commun*. 2005;330:928–933.

77. Sun D, Liu H, Yan C, et al. COX-2 contributes to the maintenance of flow-induced dilation in arterioles of eNOS-knockout mice. *Am J Physiol Heart Circ Physiol*. 2006;291:H1429–H1435.

78. Hennan JK, Huang J, Barrett TD, et al. Effects of selective cyclooxygenase-2 inhibition on vascular responses and thrombosis in canine coronary arteries. *Circulation*. 2001;104:820–825.

79. Ohashi K, Ouchi N, Sato K, et al. Adiponectin promotes revascularization of ischemic muscle through a cyclooxygenase 2-dependent mechanism. *Mol Cell Biol*. 2009;29:3487–3499.

80. Ohashi K, Ouchi N, Matsuzawa Y. Adiponectin and hypertension. *Am J Hypertens*. 2011;24:263–269.

81. Mantovani A, Sica A, Locati M. Macrophage polarization comes of age. *Immunity*. 2005;23:344–346.

82. Bouloumié A, Curat CA, Sengenès C, Lolmède K, Miranville A, Busse R. Role of macrophage tissue infiltration in metabolic diseases. *Curr Opin Clin Nutr Metab Care*. 2005;8:347–354.

83. Kanda H, Tateya S, Tamori Y, et al. MCP-1 contributes to macrophage infiltration into adipose tissue, insulin resistance, and hepatic steatosis in obesity. *J Clin Invest*. 2006;116:1494–1505.

84. Weisberg SP, McCann D, Desai M, Rosenbaum M, Leibel RL, Ferrante Jr AW. Obesity is associated with macrophage accumulation in adipose tissue. *J Clin Invest*. 2003;112:1796–1808.

85. Xu H, Barnes GT, Yang Q, et al. Chronic inflammation in fat plays a crucial role in the development of obesity-related insulin resistance. *J Clin Invest.* 2003;112:1821–1830.

86. Arkan MC, Hevener AL, Greten FR, et al. IKK-beta links inflammation to obesity-induced insulin resistance. *Nat Med.* 2005;11:191–198.

87. Nguyen MT, Favelyukis S, Nguyen AK, et al. A subpopulation of macrophages infiltrates hypertrophic adipose tissue and is activated by free fatty acids via Toll-like receptors 2 and 4 and JNK-dependent pathways. *J Biol Chem.* 2007;282:35279–35292.

88. Lusis AJ. Atherosclerosis. *Nature.* 2000;407:233–241.

89. Ross R. The pathogenesis of atherosclerosis: a perspective for the 1990s. *Nature.* 1993;362:801–809.

90. Chinetti G, Zawadski C, Fruchart JC, Staels B. Expression of adiponectin receptors in human macrophages and regulation by agonists of the nuclear receptors PPARalpha, PPARgamma, and LXR. *Biochem Biophys Res Commun.* 2004;314:151–158.

91. Yamaguchi N, Argueta JG, Masuhiro Y, et al. Adiponectin inhibits toll-like receptor family-induced signaling. *FEBS Lett.* 2005;579:6821–6826.

92. Tian L, Luo N, Zhu X, Chung BH, Garvey WT, Fu Y. Adiponectin-AdipoR1/2-APPL1 signaling axis suppresses human macrophage foam cell transformation; differential ability of AdipoR1 and AdipoR2 to regulate inflammatory cytokine responses. *Atherosclerosis.* 2012;221:66–75.

93. Auwerx J. The human leukemia cell line, THP-1: a multifaceted model for the study of monocyte-macrophage differentiation. *Experientia.* 1991;47:22–31.

94. Steinberg D. Low density lipoprotein oxidation and its pathobiological significance. *J Biol Chem.* 1997;272:20963–20966.

95. Ouchi N, Kihara S, Arita Y, et al. Adipocyte-derived plasma protein, adiponectin, suppresses lipid accumulation and class A scavenger receptor expression in human monocyte-derived macrophages. *Circulation.* 2001;103:1057–1063.

96. Furukawa K, Hori M, Ouchi N, et al. Adiponectin down-regulates acyl-coenzyme A: cholesterol acyltransferase-1 in cultured human monocyte-derived macrophages. *Biochem Biophys Res Commun.* 2004;317:831–836.

97. Tian L, Luo N, Klein RL, Chung BH, Garvey WT, Fu Y. Adiponectin reduces lipid accumulation in macrophage foam cells. *Atherosclerosis.* 2009;202:152–161.

98. Gordon S. Alternative activation of macrophages. *Nat Rev Immunol.* 2003;3:23–35.

99. Mantovani A, Sica A, Sozzani S, Allavena P, Vecchi A, Locati M. The chemokine system in diverse forms of macrophage activation and polarization. *Trends Immunol.* 2004;25:677–686.

100. Lumeng CN, Bodzin JL, Saltiel AR. Obesity induces a phenotypic switch in adipose tissue macrophage polarization. *J Clin Invest.* 2007;117:175–184.

101. Ohashi K, Parker JL, Ouchi N, et al. Adiponectin promotes macrophage polarization toward an anti-inflammatory phenotype. *J Biol Chem.* 2010;285:6153–6160.

102. Luo N, Liu J, Chung BH, et al. Macrophage adiponectin expression improves insulin sensitivity and protects against inflammation and atherosclerosis. *Diabetes.* 2010;59:791–799.

103. Luo N, Wang X, Chung BH, et al. Effects of macrophage-specific adiponectin expression on lipid metabolism in vivo. *Am J Physiol Endocrinol Metab.* 2011;301:E180–E186.

104. Luo N, Chung BH, Wang X, et al. Enhanced adiponectin actions by overexpression of adiponectin receptor 1 in macrophages. *Atherosclerosis.* 2013;228:124–135.

105. Luo N, Wang X, Zhang W, Garvey WT, Fu Y. AdR1-TG/TALLYHO mice have improved lipid accumulation and insulin sensitivity. *Biochem Biophys Res Commun.* 2013;433:567–572.

CHAPTER TEN

Regulation of Pancreatic Islet Beta-Cell Mass by Growth Factor and Hormone Signaling

Yao Huang*, Yongchang Chang†

*Department of Obstetrics and Gynecology, St. Joseph's Hospital and Medical Center, Phoenix, Arizona, USA
†Barrow Neurological Institute, St. Joseph's Hospital and Medical Center, Phoenix, Arizona, USA

Contents

Abstract

Dysfunction and destruction of pancreatic islet beta cells is a hallmark of diabetes. Better understanding of cellular signals in beta cells will allow development of therapeutic strategies for diabetes, such as preservation and expansion of beta-cell mass and improvement of beta-cell function. During the past several decades, the number of studies analyzing the molecular mechanisms, including growth factor/hormone signaling pathways that impact islet beta-cell mass and function, has increased exponentially. Notably, somatolactogenic hormones including growth hormone (GH), prolactin (PRL), and insulin-like growth factor-1 (IGF-1) and their receptors (GHR, PRLR, and IGF-1R) are critically involved in beta-cell growth, survival, differentiation, and insulin secretion. In this chapter, we focus more narrowly on GH, PRL, and IGF-1 signaling, and GH–IGF-1

Progress in Molecular Biology and Translational Science, Volume 121
ISSN 1877-1173
http://dx.doi.org/10.1016/B978-0-12-800101-1.00010-7
321

cross talk. We also discuss how these signaling aspects contribute to the regulation of beta-cell proliferation and apoptosis. In particular, our novel findings of GH-induced formation of GHR–JAK2–IGF-1R protein complex and synergistic effects of GH and IGF-1 on beta-cell signaling, proliferation, and antiapoptosis lead to a new concept that IGF-1R may serve as a proximal component of GH/GHR signaling.

ABBREVIATIONS

BrdU 5-bromo-2′-deoxyuridine
ERK extracellular signal-regulated kinase
GH growth hormone
GHR GH receptor
GSIS glucose-stimulated insulin secretion
IGF-1 insulin-like growth factor-1
IGF-1R IGF-1 receptor
IR insulin receptor
IRS insulin receptor substrate
JAK Janus kinase
PI3K phosphatidylinositol 3-kinase
PRL prolactin
PRLR PRL receptor
pTyr phosphotyrosine
RNAi RNA interference
STAT signal transducer and activator of transcription

1. INTRODUCTION

Diabetes mellitus is a disease in which the body does not produce or properly respond to insulin. There are two main types of diabetes. Type 1 diabetes is due to destruction of the insulin-producing beta cells in the pancreas and can be treated with insulin injections. In type 2 diabetes, the body is resistant to insulin and the insulin secretion by pancreatic islet beta cells is also defective. Type 2 diabetes accounts for 90% of all diabetes and is characterized by peripheral insulin resistance and/or abnormal insulin secretion associated with pancreatic beta-cell dysfunction or failure.[1–3] Obesity-linked type 2 diabetes is approaching epidemic proportions and becoming a serious problem, especially in developed countries like the United States.[4–8] Under normal conditions, the number of beta cells has a positive correlation with the body mass, that is, nondiabetic obese individuals have more beta cells than lean individuals or obese type 2 diabetic patients.[9–11] Obesity often correlates with an increased beta-cell mass that initially compensates for the insulin resistance; however, beta-cell compensation may eventually fail and result in type 2 diabetes.[4,7,12]

It is widely believed that the balance of beta-cell growth and death (reflected in beta-cell mass) plays a pivotal role in the pathogenesis of type 2 diabetes.[4,12–15] At least four factors contribute to adult pancreatic beta-cell mass: replication from preexisting beta cells,[16,17] neogenesis from precursor cells,[18] beta-cell size,[14] and apoptosis.[12,15] In fact, beta-cell mass changes in response to metabolic status and insulin demand.[19,20] Certain nutrients (e.g., glucose and free fatty acids) and growth factors, especially somatolactogenic hormones including growth hormone (GH), insulin-like growth factor-1 (IGF-1), and prolactin (PRL) (during pregnancy), can influence beta-cell growth, survival, differentiation, and insulin secretion.[4,9,10,21,22] The best example is that during pregnancy, there is a marked hyperplasia of the beta cells, which is reversed to normal by apoptosis after delivery.[23] On the other hand, the high incidence of gestational diabetes (approximately 4% of all pregnancies) could be due to the lack of compensatory increase in beta-cell mass,[11] which may precede the development of type 2 diabetes.[24] Thus, a better understanding of growth factor/hormone-mediated signal transduction pathways that contribute to the modulation of beta-cell mass and function will be essential to allow the development of therapeutic strategies for type 2 diabetes. In the current review, we mainly focus on GH (including PRL that shares important structural and functional features with GH) and IGF-1 signaling and their cross talk. We also discuss how these signaling aspects are involved in the regulation of beta-cell proliferation and apoptosis given their growing importance in the field.

2. GH, PRL, AND THEIR RECEPTOR SIGNALING AND ACTIONS IN BETA CELLS

2.1. Overview of GH and PRL signaling

GH (also known as somatotropin) is a 22 kDa peptide hormone mainly derived from the anterior pituitary gland. It exerts powerful somatogenic and metabolic regulatory effects in various tissues by interacting with its high-affinity receptor, the GH receptor (GHR).[25,26] The GHR is a cell-surface glycoprotein member of the cytokine receptor superfamily.[26] Like other cytokine receptors,[26] the GHR lacks intrinsic tyrosine kinase activity but is physically and functionally coupled to Janus kinase 2 (JAK2), a non-receptor cytoplasmic tyrosine kinase member of the JAK family.[27,28] GH-triggered JAK2 activation results in tyrosine phosphorylation of

GHR and engagement of several intracellular signaling pathways, including signal transducers and activators of transcription (STATs) (most notably STAT5b), Ras/extracellular signal-regulated kinases (ERKs), and phosphatidylinositol 3-kinase (PI3K)/Akt[29–32] (Fig. 10.1). GH-induced STAT5b activation has been extensively studied and is critical for regulation of several GH-responsive gene transcription, such as IGF-1, serine protease inhibitor 2.1 (Spi2.1), and suppressors of cytokine signaling (SOCS) genes.[33–38] The Ras/ERK and PI3K/Akt pathways are more common targets of a variety of growth factors and cytokines. In particular, we have shown that activation of ERKs fosters the cross talk between the GHR and epidermal growth factor receptor (EGFR) (and ErbB-2) signaling.[39,40]

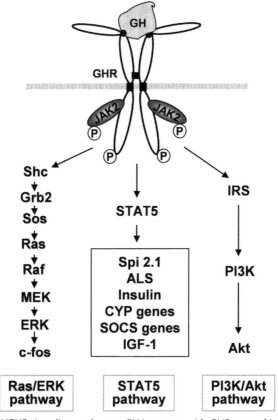

Figure 10.1 *GH/GHR signaling pathways.* GH interacts with GHR, a cytokine receptor lacking a kinase activity. GH engagement enhances the association of GHR with tyrosine kinase JAK2 and causes JAK2 activation and GHR phosphorylation. Three major GH signaling pathways (STAT5, PI3K/Akt, and Ras/ERK) are then activated.

PRL is a peptide hormone that has similar size and structure to GH.[41] Like GH, PRL is produced mainly by the anterior pituitary gland. To a lesser extent, it is also produced by multiple extrapituitary sites, such as the mammary epithelium, placenta, uterus, bone, brain, and immune system, and is involved in diverse physiological processes including reproduction and lactation, growth and development, metabolism, brain function, immunomodulation, and osmoregulation.[42,43] Similar to the GH/GHR signaling, PRL signals via the PRL receptor (PRLR), a cytokine receptor family member that possesses no intrinsic tyrosine kinase activity and couples to JAK2.[26,44,45] PRL binding leads to autophosphorylation and activation of PRLR-associated JAK2, followed by phosphorylation of PRLR and activation of STATs (predominantly STAT5a), PI3K/Akt, Ras/ERK, and other signaling cascades that control mitogenic, apoptotic, and cell differentiation responses.[46,47] Cross talk between ERKs and other PRL/PRLR pathways has been investigated recently, which suggests that the PI3K/Akt pathway is required for activation of ERK signaling upon PRL stimulation.[48] In addition, we and others have demonstrated that in some cases, PRL synergizes with other growth factors in activating ERKs (the so-called signaling synergy).[46,49]

Given their structural similarity, the existence of interactions between ligands (GH and PRL) and their receptors (GHR and PRLR) has been long appreciated. For example, human GH binds not only the GHR but also the PRLR. Although incompletely known, it is believed that the physiological consequences of human GH and PRLR interaction may diversify the role of GH in humans.[50–52] In contrast, human PRL only binds the PRLR but not GHR.

2.2. GH and PRL signaling in beta cells

GH was originally considered a diabetogenic hormone because of its effect in counteracting insulin on glucose metabolism. However, many studies have provided strong evidence that supports a direct stimulatory role of both GH and PRL on beta-cell mass and function.[11] A marked beta-cell hyperplasia occurs in both pregnant women and rodents.[53] A number of early studies showed that both GH and PRL induce insulin production, glucose-stimulated insulin secretion (GSIS), DNA synthesis, and beta-cell replication in human fetal pancreas, isolated human islets, and islets of newborn and adult rats and mice.[54–63] It has been found that the circulating levels of PRL and placental GH (variant GH) are elevated during pregnancy.

In the meantime, the expression of both GHR and PRLR is also markedly increased in the pancreas from pregnant rats.[64,65] In fact, the relative potencies of GH and PRL on beta-cell proliferation depend on the respective receptor expression levels, which are prone to differential hormonal regulation. PRLR expression is stimulated by GH, PRL, and estrogens, whereas GHR expression is stimulated by glucocorticoids.[64]

Although the beta cells of the pancreatic islets of Langerhans express both GHR and PRLR,[64,66,67] how GH/GHR and PRL/PRLR signaling regulates beta-cell mass and function remains poorly understood. It has been previously reported that, under certain conditions, GH can stimulate IGF-1 production in fetal rat islets and human fetal pancreas in tissue culture.[60,68] However, the role of IGF-1 in GH-induced beta-cell or islet cell growth remains controversial (see the succeeding text for more details). Rather, GH (and its related peptide hormone PRL) is more likely to exert a direct effect on beta-cell proliferation, predominantly via the JAK2/STAT5 signaling pathway.[69–72] A number of studies have shown that activated STAT5a and STAT5b translocate to the nuclei and regulate the transcription of genes involved in beta-cell proliferation.[9,73–77] A previous study in the rat insulinoma INS-1 cells has provided more insights into how GH signaling regulates beta-cell function.[70] GH stimulates INS-1 cell proliferation in a glucose-dependent manner. The combination of GH and glucose synergistically increases beta-cell growth. It was further concluded that GH promotes INS-1 cell growth directly via the JAK2/STAT5 pathway. However, only limited mitogenic signaling elements were examined in that study.[70] Similar to GH, PRL promotes INS-1 cell growth in a glucose-dependent manner.[78] In addition to mitogenic effects, STATs are implicated in regulation of cell death and survival. It has been shown that GH (and PRL) protects beta cells against the effects of cytotoxic cytokines via STAT5-mediated regulation of Bcl-2- and Bcl-xL-related genes.[72,79–83]

A family of proteins called SOCS is capable of inhibiting the JAK/STAT signaling.[84–86] So far, eight members of the SOCS family have been identified, including SOCS-1-7 and cytokine-inducible SH2-containing protein (CIS). Although the interaction of SOCS-2 or CIS with GHR was reported,[87] GH has been found to preferentially promote the expression of SOCS-3, which inhibits GH-induced gene transcription and STAT5 activation.[88,89] Interestingly, SOCS-2 influences GH's effects, and gigantism is seen in mice lacking SOCS-2.[90] SOCS-1 knockout mice have a low blood glucose level and increased insulin signaling.[91] SOCS-3 inhibits tumor necrosis factor-alpha (TNF-α)-induced apoptosis and signaling in the pancreatic insulin-producing beta cells.[92] However, the roles of the SOCS proteins in GH- and PRL-mediated beta-cell mass and function remain to be further investigated.

The roles of GH and PRL signaling in beta cells are also supported by knockout mouse models. The first physiological evidence for a role of GHR and PRLR in beta cells was from *GHR-null*[93] and *PRLR-null*[94] mice, which exhibited reduced beta-cell mass, impaired glucose tolerance, and increased insulin sensitivity. Thus, GH signaling (and PRL signaling especially during pregnancy) is essential for maintaining islet size, normal insulin sensitivity, and glucose homeostasis. However, these knockout mice displayed compromised growth and significant changes in body adiposity; thus, a direct causal effect could not be established. Later, Lee *et al.* used the *Cre* transgene to specifically ablate STAT5a/b (the downstream mediators of GH/GHR and PRL/PRLR) in beta cells and showed that these mice developed functional islets and were glucose-tolerant.[95] They also reported that mild glucose intolerance occurred with aging. Thus, they concluded that STAT5 is not essential for islet development but may modulate beta-cell function.[95] Very recently, Wu *et al.* have successfully generated pancreatic islet beta-cell-specific GHR knockout (*βGHRKO*) mice.[96] When fed with a standard chow diet, the mice exhibited impaired GSIS but had no changes in beta-cell mass. In contrast, when challenged with a high-fat diet (HFD), the obese *βGHRKO* mice were impaired in beta-cell hyperplasia with decreased beta-cell proliferation and overall reduced beta-cell mass. Thus, they have concluded that GHR plays critical roles in GSIS and beta-cell compensation in response to HFD-induced obesity.[96] Further, while the exact molecular mechanism(s) involved are largely unknown, data from this elegant study suggest that the GHR signaling for these important beta-cell functions may not solely depend on STAT5.

3. IGF-1 AND ITS RECEPTOR SIGNALING AND ACTIONS IN BETA CELLS

3.1. Overview of IGF-1 signaling

IGF-1 (also known as somatomedin-C) is synthesized largely in the liver and also produced extrahepatically to promote postnatal growth in bone, muscle, fat, and other tissues.[97,98] Classically, IGF-1 is considered a major physiological effector of GH. As originally articulated, the somatomedin hypothesis of GH action postulated that GH stimulated the hepatic secretion of IGF-1, which then functioned in an endocrine manner to interact with IGF-1 receptors (IGF-1R) in tissues responded with growth.[99,100] Studies in mice with unrestricted targeted deletion of the IGF-1 gene not only validated the importance of IGF-1 in growth mediation, but also suggested that GH may promote growth in some tissues independent of IGF-1.[101,102] Liver-specific knockout of IGF-1, although significantly lowering serum IGF-1 levels, does not prevent normal growth, suggesting that autocrine/paracrine IGF-1 (rather than liver-derived IGF-1) may predominate for normal post-natal growth.[97,98]

IGF-1 signals through IGF-1R, which is a cell-surface heterotetramer consisting of two α- and two β-subunits with intrinsic kinase activity embedded in the cytoplasmic domain of the β-subunit.[103] Unlike GHR and PRLR, IGF-1R is a member of receptor tyrosine kinase superfamily.[104] IGF-1 binding causes IGF-1R intrinsic kinase activation and phosphorylation of the insulin receptor (IR) substrate (IRS) and Shc proteins, which in turn activates downstream signaling pathways, mainly including the PI3K/Akt and the Ras/ERK pathways[104] (Fig. 10.2). In general, these pathways are thought to be critical for antiapoptosis and proliferation.[105,106] In addition, activation of the JAK-related pathway by IGF-1 and insulin has been reported in some cases[107–110] (see Section 4.1 for more details). The specific roles of IGF-1/IGF-1R signaling in pancreatic islets and islet beta cells will be discussed later. While we focus on IGF-1 acting through its cognate receptor (IGF-1R), the highly homologous IR also plays a role in islet biology, which will not be discussed here.

3.2. IGF-1 signaling in beta cells

Compared with GH and PRL signaling, relatively more is known about how IGF-1 signaling regulates pancreatic islet beta-cell function. IGF-1R and key IGF-1 signaling components are present and functional in islets.[4,111–114] Regarding IGF-1, it was previously reported that its expression in the pancreas *in vivo* mainly occurs in endothelial cells and proliferating duct cells.[115] However, several studies have shown that IGF-1 appears to be specifically localized to alpha cells in the islets.[116–118] This somewhat supports direct effect of GH (independent of IGF-1) on beta-cell proliferation[71] (see Section 2.2 earlier).

Islet beta-cell mass is a key element in the development of autoimmunity-induced type 1 diabetes and in compensating insulin resistance in type 2 diabetes.[119,120] A number of animal models have suggested that insulin resistance and beta-cell dysfunction are tightly linked to the

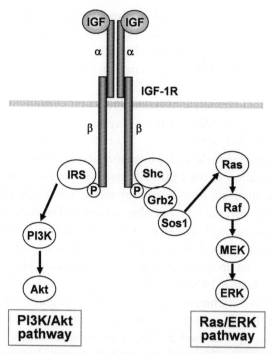

Figure 10.2 *IGF-1/IGF-1R signaling pathways.* IGF-1 exerts its biological effects by binding to the heterotetrameric IGF-1R (a receptor tyrosine kinase consisting of two α- and two β-subunits). IGF-1 engagement leads to autophosphorylation and activation of IGF-1R and phosphorylation of IRS and Shc. The IRS and Shc proteins, in turn, activate downstream signaling of primarily the PI3K/Akt and the Ras/ERK pathways.

pathogenesis of type 2 diabetes.[121] Both IRS-1$^{-/-}$ and IRS-2$^{-/-}$ mice displayed insulin resistance. However, diabetes did not develop in IRS-1$^{-/-}$ mice because the beta-cell mass expanded to compensate for the insulin resistance. In contrast, IRS-2$^{-/-}$ mice became diabetic due to the lack of beta cells in compensation for insulin resistance.[112,122,123] The markedly reduced beta-cell mass in IRS-2$^{-/-}$ mice was caused by the increased beta-cell apoptosis, as found in type 2 diabetes.[113] These studies have demonstrated that IRS-2 is a key player in the signal transduction pathway(s) controlling the beta-cell mass to compensate for peripheral insulin resistance. Interestingly, IGF-1R allelic insufficiency reduced the life span of IRS2$^{-/-}$ mice to only a month due to the near absence of pancreatic β cells and extreme hyperglycemia,[113] suggesting that IRS-2 coordinates the IGF/IGF-1R-mediated beta-cell development and maintenance.[121] In addition, beta-cell-specific IGF-1R knockout (β*Igf1r*$^{-/-}$) mice exhibited defective GSIS,[124,125] and mice lacking functional receptors for both insulin and IGF-1 only in beta cells (βDKO) developed overt diabetes.[126] Taken together, all these findings suggest that IGF-1/IGF-1R/IRS-2 signaling pathways are critical for beta-cell function and thereby for the development of type 2 diabetes.

In vitro studies in the rat insulinoma INS-1 cells reaffirmed that IGF-1-mediated IRS-2/PI3K/Akt pathway plays a pivotal role in maintaining beta-cell growth and survival.[70,127–132] IGF-1 augmented the glucose-induced beta-cell growth and phosphorylation of Shc and IRS proteins.[70,127] Activation of the PI3K/Akt pathway was not only essential for IGF-1/glucose-induced beta-cell proliferation (although the Ras/ERK pathway is likely involved)[127] but also important for promoting beta-cell survival and maintaining beta-cell mass.[133,134] Regulation of IRS-2 expression appeared to be critical. Increased IRS-2 expression prevented free fatty acid (FFA)-induced apoptosis in beta cells, most likely via Akt activation. Conversely, reducing IRS-2 resulted in apoptosis, which was further enhanced by FFA.[132] Chronic exposure to IGF-1 or glucose-induced phosphorylation of IRS-2, correlated with decreased IRS-2 protein levels. Such IRS-2 degradation was mediated by a mammalian target of rapamycin (mTOR), associated with inhibition of IGF-1 and glucose-induced Akt activation.[131] Furthermore, depletion of IGF-1R in the mouse insulinoma MIN6 cells by RNA silencing inhibited GSIS.[135] Collectively, these *in vitro* experimental data provide new insights into the molecular mechanisms of how IGF-1/IGF-1R-directed signaling pathways contribute to the regulation of beta-cell mass.

4. GH–IGF-1 SIGNALING CROSS TALK AND COLLABORATIVE ROLES IN REGULATION OF BETA-CELL MASS

With increasing knowledge of the complexity of signaling pathways and utilization of common pathways by diverse receptors, cross talk or interplay among various signaling systems, which is less understood but critical for signal integration and diversification, has emerged as an important and interesting area. As described earlier, GH and IGF-1 can independently promote beta-cell growth, inhibit apoptosis, and are potentially involved in normal islet growth and maintenance.[59,127,136,137] However, it remains poorly understood whether GH and IGF-1 signaling pathways interact with each other to regulate pancreatic beta-cell function and how they are involved, as collaborators, in islet growth, insulin biosynthesis, and secretion. GH and IGF-1 share a close and complex physiological relationship.[97,98,100,102,138] GHR$^{-/-}$IGF-1$^{-/-}$ mice exhibit more severe growth retardation (>80%) when compared to either GHR$^{-/-}$ or IGF-1$^{-/-}$ mice,[139] suggesting that the GH and IGF-1 signaling may serve both independent and overlapping functions. We have recently uncovered the physical and functional interactions between the GH/GHR and IGF-1/IGF-1R signaling systems and evaluated the collaborative roles of GH and IGF-1 signaling in regulation of beta-cell growth and survival.[140] These previously understudied aspects are further discussed in detail in the succeeding text.

4.1. GH-induced GHR–JAK2–IGF-1R protein complex formation

In the past several years, we have been investigating the signaling cross talk between cytokine receptors (e.g., GHR and PRLR) and receptor tyrosine kinases (e.g., IGF-1R and EGFR) in several cell model systems,[38–40,49,140] a situation more like *in vivo*. In mouse preadipocytes, we demonstrated that GH (but not IGF-1) promoted IGF-1R association with the GHR–JAK2 complex.[38] Recently, we have investigated such protein complex formation in rodent beta-cell lines, including rat INS-1, mouse MIN6, and mouse BTC6.[140] As mentioned earlier (see Section 2.1), because human GH interacts with both GHR and PRLR,[9] we used bovine GH that only binds to GHR[41] in our study. We showed that all three beta-cell lines endogenously express GHR, JAK2, and IGF-1R, although the expression levels of these proteins differ among the cell lines.[140] Using reciprocal coimmunoprecipitation (co-IP) assays, we detected a GHR–JAK2–IGF-1R protein complex in these beta-cell lines upon GH stimulation (Fig. 10.3). We also confirmed that rat GH had a similar effect, promoting the GHR–JAK2–IGF-1R complex formation in rodent beta cells.[140]

It is well known that the active signaling GHR upon GH engagement is a dimer bound to a single GH molecule (GHR/GH = 2:1).[141] Accumulating evidence suggests that GHR is more likely a predimer in the absence of ligand (Fig. 10.4A). Active assembly arises by the GH-induced GHR conformational change, which could allow more productive interaction between GHR and JAK2 and resultant activation of the GH/GHR signaling pathways (Fig. 10.4B).[142–144] G120K is one of the human GH antagonists harboring mutations at residues known to be critical for the active GHR conformation[145] (Fig. 10.4C). We found that G120K can antagonize the ability of bovine GH to induce GHR phosphorylation and JAK2 activation

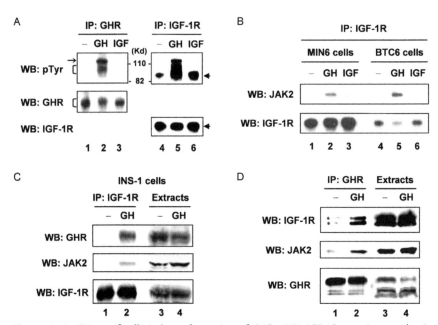

Figure 10.3 *GH specifically induces formation of GHR–JAK2–IGF-1R protein complex in beta cells.* (A) GH-induced phosphoprotein complex in the anti-IGF-1R precipitates. Serum-starved MIN6 cells were stimulated with vehicle (−), bovine GH (500 ng/ml), or IGF-1 (20 ng/ml) for 15 min. Cell extracts were immunoprecipitated with anti-GHR (lanes 1–3) or anti-IGF-1R (lanes 4–6). The immunoprecipitates were analyzed by immunoblotting with antiphosphotyrosine (pTyr) (lanes 1–6), anti-GHR (lanes 1–3), or anti-IGF-1R (lanes 4–6). The GHR, JAK2, and IGF-1R bands are indicated by bracket, arrow, and arrowhead, respectively. (B) JAK2 is a component of the GH-induced protein complex from the anti-IGF-1R precipitates. MIN6 and BTC6 cells treated as in (A) were subjected to immunoprecipitation with anti-IGF-1R antibody followed by immunoblotting with anti-JAK2 and anti-IGF-1R, respectively. (C) GHR is also a component of the GH-induced protein complex from the anti-IGF-1R precipitates. Serum-starved INS-1 cells were stimulated with vehicle (−) or bovine GH (500 ng/ml) for 15 min. Cell extracts were subjected to immunoprecipitation with anti-IGF-1R followed by immunoblotting (lanes 1 and 2) or direct immunoblotting (lanes 3 and 4) with anti-GHR, anti-JAK2, and anti-IGF-1R, respectively. (D) Reverse coimmunoprecipitation experiment. INS-1 cells as treated in (C) were subjected to immunoprecipitation with anti-GHR followed by immunoblotting (lanes 1 and 2) or direct immunoblotting (lanes 3 and 4) with anti-IGF-1R, anti-JAK2, and anti-GHR, respectively. For details, see Ref. 140. *Reprinted with permission from Molecular Endocrinology.*

in beta cells[140] (Fig. 10.4D–E). More importantly, in the same experimental setting, G120K prevented the bovine GH-induced formation of GHR–JAK2–IGF-1R complex formation[140] (Fig. 10.4F). These results from rodent beta cells together with our previous findings in mouse

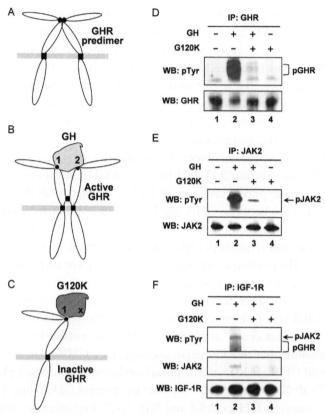

Figure 10.4 *GH antagonist, G120K, prevents the GH-induced GHR–JAK2–IGF-1R complex formation in beta cells.* (A) GHR is a predimer in the absence of ligand (inactive form). (B) GH binding to GHR results in GHR conformational change (active form). (C) The GH antagonist, G120K, prevents the ligand-induced active GHR conformation. (D–F) Serum-starved BTC6 cells were stimulated with vehicle (−) or bovine GH (500 ng/ml) in the absence (lanes 1 and 2) or presence (lanes 3 and 4) of G120K (2500 ng/ml) for 15 min. Cell extracts were subjected to immunoprecipitation with anti-GHR (D), anti-JAK2 (E), and anti-IGF-1R (F), respectively, followed by immunoblotting with anti-pTyr, anti-GHR, anti-JAK2, or anti-IGF-1R, as indicated. The phosphorylated forms of GHR and JAK2 are indicated by bracket and arrow, respectively. For details, see Ref. 140. *Reprinted with permission from Molecular Endocrinology.*

preadipocytes[38] suggest that the GHR conformational change upon GH engagement may be a prerequisite for the GH-induced GHR–JAK2–IGF-1R protein complex formation.

Several previous studies have suggested that, in addition to causing activation of their own receptor kinases, IGF-1 and insluin may activate members of the JAK family, including JAK1 and JAK2.[107–110] The impact of JAK activation by these growth factors is unclear, and in most cases it has been observed in the setting of overexpression of at least one of the components. In particular, Gual *et al.* reported physical associations between IGF-1R and IR with JAKs, which required tyrosine phosphorylation of the receptors.[146] The associations detected in that study were under the conditions of *in vitro* overexpression of both binding partners but not in a normal cellular context (e.g., endogenous proteins). In our cases, the physical interaction of IGF-1R with the GHR–JAK2 complex upon GH stimulation was studied in the cells where all these components were endogenously expressed.[38,140] As mentioned earlier (see Section 2.1), GH binding to GHR leads to activation of JAK2 and tyrosine phosphorylation of GHR, coupled with the enhanced physical association between GHR and JAK2. Thus, a fundamentally important question is whether the GH-induced GHR–JAK2–IGF-1R protein complex formation requires phosphorylation/activation of any or all of the components. The existing experimental data do not favor such a requirement although we cannot completely rule out the possibility. Firstly, the complex formation is strictly dependent upon GH engagement to GHR but not IGF-1 binding to IGF-1R (subsequent activation/phosphorylation of IGF-1R) in both mouse preadipocytes[38] and islet beta cells.[140] Secondly, the kinase inhibitor staurosporine, which is known to uncouple GH/GHR engagement from GH-induced GHR–JAK2 tyrosine phosphorylation by inhibiting phosphorylation of both GHR and JAK2,[38,147] does not prevent GH-induced physical association between GHR–JAK2 and IGF-1R.[38] Furthermore, deletion of IGF-1R in primary osteoblasts reduces GH-mediated STAT5 signaling, which can be partially rescued by reexpression of a truncated IGF-1R lacking the intracellular domain of its beta chain that contains the kinase domain.[148] In contrast, IGF-1R tyrosine kinase inhibitor has no effect on GH-induced STAT5 signaling.[148] Finally, IGF-1R knockdown in beta cells diminishes GH-dependent STAT5 activation[140] (see below). Collectively, these results suggest that an IGF-1R component(s) other than the kinase domain impacts GH action and GH-induced GHR–JAK2–IGF-1R complex formation. Thus, it is reasonable to speculate that the extracellular domain of IGF-1R may affect GHR signaling conformation and could be essential for GHR–JAK2 and IGF-1R physical interaction.

4.2. Synergy of GH and IGF-1 signaling in beta cells

A previous study in INS-1 cells showed that GH promoted cell proliferation in a glucose-dependent manner, which was enhanced further by IGF-1.[70]

Given that either glucose or the combination of glucose and IGF-1 activated JAK2 and STAT5 in these cells, it was concluded that although dependent on glucose, GH promoted INS-1 cell growth directly via the JAK2/STAT5 pathway with no cross talk to IGF-1 signaling.[70] On the other hand, it has been reported that GH and IGF-1 act synergistically in promoting gene activation, cell signaling, and proliferation in several cell systems.[38,149,150] One of the possible molecular mechanisms underlying the GH–IGF-1 collaborative signaling effects is the GHR–JAK2–IGF-1R protein complex, which provides a platform allowing the cross talk between GH/GHR and IGF-1/IGF-1R signaling systems.

To this end, we have recently probed the potential signaling cross talk between GH and IGF-1 in rodent beta cells.[140] By comparing the acute signaling in response to GH or IGF-1 alone with that in response to their combination (GH–IGF-1 cotreatment), we revealed that GH and IGF-1 synergized in activating STAT5 and Akt pathways (Fig. 10.5). This strongly suggests that signaling cross talk between GH and IGF-1 is present in beta cells. Our findings of Akt signaling[140] also differ somewhat from those reported by Cousin et al.[70] In their study, using the physical association of the 85 kDa subunit of PI3K (p85 PI3K) with IRS-1 or IRS-2 (revealed by co-IP) as a readout, they concluded that glucose and IGF-1, but not rat GH, activated the IRS/PI3K pathway in INS-1 cells, although the effect of GH and IGF-1 cotreatment was not examined.[70] In contrast, we found that IGF-1 and GH each activated Akt (the downstream effector of PI3K), and more importantly, they synergized in Akt signaling (revealed by direct immunoblotting).[140] The discrepancy in these results could be due to different experimental conditions used. In the study of Cousin et al., cells were quiescent by serum and glucose deprivation and then exposed to GH or IGF-1 with the addition of glucose (0–15 mM) for 10 min.[70] However, in our case, cells were only serum-starved in normal culture media containing glucose (11.2 mM for INS-1 and 25 mM for both MIN6 and BTC6) and then stimulated with GH and IGF-1 for 15 min.[140] Whether, or to what extent glucose exposure dictates GH and IGF-1 responsiveness and impacts the GH–IGF-1 cross talk, deserves further investigation.

4.3. Collaborative effects of GH and IGF-1 signaling on beta-cell mass

It is known that beta-cell mass is determined and maintained by a dynamic balance of cell proliferation and cell death. Insufficient understanding of the signals regulating the growth and survival of adult beta cells remains one of

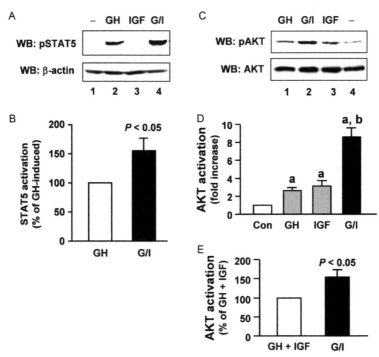

Figure 10.5 *GH and IGF-1 synergize in activating STAT5 and Akt in beta cells.* (A) Serum-starved MIN6 cells were treated with vehicle (−), bovine (500 ng/ml), or IGF-1 (20 ng/ml) or cotreated with bovine GH plus IGF-1 (G/I) for 15 min. Cell extracts were analyzed by immunoblotting with anti-pSTAT5 and anti-β-actin, respectively. (B) Densitometric analysis of pooled data, as in (A), from eight independent experiments. Comparison of the level of STAT5 activation induced by bovine GH and IGF-1 cotreatment (referred to as G/I) with that induced by bovine GH alone (set as 100%) is shown. Data are mean ± SEM ($n = 8$). $P < 0.05$ (G/I vs. GH). (C) Serum-starved MIN6 cells were treated with vehicle (−), bovine GH (500 ng/ml), or IGF-1 (20 ng/ml) or cotreated with bovine GH plus IGF-1 (G/I) for 15 min. Cell extracts were analyzed by immunoblotting with anti-pAkt and anti-total Akt, respectively. (D) Densitometric analysis of pooled data, as in (C), from five independent experiments. Fold increase in Akt activation is shown for each condition with the basal (vehicle control) set as 1. Data are mean ± SEM ($n = 5$). a, $P < 0.01$ (compared to control); b, $P < 0.01$ (compared to either GH or IGF). (E) Reanalyzed results of pooled data of Akt activation. In this display, the sum of pAkt levels induced by bovine GH alone plus IGF-1 alone (referred to as GH + IGF) is set as 100% and compared to the pAkt level induced by bovine GH and IGF-1 cotreatment (referred to as G/I). Data are mean ± SEM ($n = 5$). $P < 0.05$ (G/I vs. GH + IGF). For details, see Ref. 140. *Reprinted with permission from Molecular Endocrinology.*

the main challenges in diabetes research. Previous studies have shown that GH and IGF-1 each can independently promote beta-cell mitogenesis and prevent apoptosis *in vitro*.[59,70,127,136,137,151] However, little is known about the collaborative role (signaling cross talk) of GH and IGF-1 in the regulation of beta-cell mass.

STAT5 and Akt pathways have been implicated in cell proliferation and antiapoptosis (see Sections 2 and 3). Existing data suggest that GH-mediated JAK2/STAT5 and IGF-1-mediated (and perhaps GH-mediated) IRS/PI3K/Akt signaling can directly contribute to beta-cell proliferation and apoptosis.[70,127] Our findings of synergy in STAT5 and Akt activation by GH and IGF-1 cotreatment in beta cells prompted us to evaluate these essential cellular processes. Using multiple experimental approaches, we have recently demonstrated that beta cells proliferate more robustly and are better protected from apoptosis when exposed to GH and IGF-1 in combination versus GH or IGF-1 alone.[140]

Although cultured rodent pancreatic insulinoma cell lines retain many key functional features of normal islets and have been proven to be a valuable tool for studying islet beta-cell signaling,[152,153] whether some cellular processes in these immortalized cell lines, especially cell proliferation, represent the actual scenario in pancreatic islets remains a concern. To address this, we isolated and purified islets from mice and confirmed the synergistic effects of GH and IGF-1 in promoting cell proliferation in isolated islets[140] (Fig. 10.6). Taken together, we are intrigued by our recent exciting findings that GH and IGF-1 cotreatment augments beta-cell growth and antiapoptosis, which could be ultimate manifestations of synergistic activation of STAT5 and Akt observed in these cells.

4.4. IGF-1R serves as a proximal component of GH/GHR signaling

GH and IGF-1 are key somatotropic and metabolic hormones in humans and animals. Classically, IGF-1 is a GH effector. However, not all effects of GH are exerted through stimulation of IGF-1 from liver and peripheral tissues. For example, GH can promote skeletal muscle cell fusion, an essential process of muscle growth, independent of IGF-1.[154] In pancreatic islets, IGF-1 expression appears to be specifically in alpha cells.[116–118] Thus, it is now believed that GH most likely mediates a direct effect on beta-cell proliferation via the JAK2/STAT5 signaling cascade.[69,70] GH and IGF-1 bind to their respective receptors, GHR and IGF-1R, to engage diverse signaling

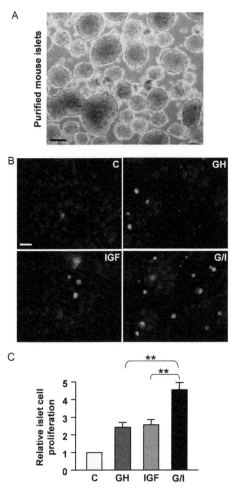

Figure 10.6 *GH and IGF-1 cotreatment augments islet cell proliferation.* (A) Purified mouse islets. Scale bar, 50 μm. (B) Representative images from BrdU (5-bromo-2'-deoxyuridine) incorporation assays. Single islet cells were treated with vehicle (control), bovine GH (500 ng/ml), or IGF-1 (20 ng/ml) or cotreated with bovine GH plus IGF-1 (G/I) in serum-free medium for 96 h in which BrdU was added in the last 24 h treatment (chase phase). Cells with incorporated BrdU (proliferating cells) were detected by immunostaining with an anti-BrdU mAb (red). Total cell numbers were measured by DAPI staining (blue). Scale bar, 10 μm. (C) Cell proliferation rates determined by BrdU incorporation assays. A total of approximately 1000 cells from 6 random imaging fields under each condition, as in (B), were counted and the ratios of BrdU-positive cells (red) to the total cells (blue) (representing cell proliferation rates) were plotted with the control set as 1. Data are mean ± SEM (*n* = 6). **P < 0.01. For details, see Ref. 140. *Reprinted with permission from Molecular Endocrinology. (See color plate.)*

pathways. While studying the cross talk between GH and IGF-1, we uncovered that GH specifically induces formation of a protein complex including GHR, JAK2, and IGF-1R in mouse preadipocytes,[38] rodent pancreatic islet beta cells,[140] and some human cancer cells (Y. Huang et al., unpublished observations). Our data from preadipocytes and islet beta cells indicated that the GHR–JAK2–IGF-1R complex assembly does not seem to require phosphorylation/activation of the components, but is rather governed by a proper GHR conformation upon GH engagement. Furthermore, IGF-1 cotreatment with GH augments GH-induced assembly of conformationally active GHRs and subsequently GH-induced STAT5 activation. Finally, GH and IGF-1 synergize in activating ERKs in preadipocytes and in activating Akt in beta cells. These intriguing findings raise a possibility that IGF-1R serves as a proximal component of GH/GHR signaling. To test this, we knocked down the endogenous IGF-1R in MIN6 cells by RNA interference (RNAi) using a lentiviral-based shRNA delivery system.[140] We found that the GH-induced STAT5 activation was substantially diminished in these IGF-1R knockdown cells.[140] Evidence supporting this hypothesis also came from two recent studies in which genetic deletion of IGF-1R in mouse primary osteoblasts impaired some of the GH-directed signaling.[148,155]

The classical somatomedin hypothesis of GH action implies that GHR acts in series with IGF-1R.[156] Based on recent new data regarding GH and IGF-1 cross talk in several cell types,[38,140,148,155] we propose that there also exists a parallel signaling relationship between their receptors (Fig. 10.7).[140] Under resting conditions, GHR (most likely predimerized) and IGF-1R are present on the cell plasma membrane (Fig. 10.7A). GH engagement causes GHR conformational changes, allowing achievement of an active signaling status (with enhanced GHR–JAK2 interaction). This, in turn, allows the association of GHR–JAK2 with IGF-1R (Fig. 10.7B). Further IGF-1 binding to IGF-1R within such a complex facilitates, stabilizes, and enhances the GHR signaling (augmenting STAT5 activation) and simultaneously activates IGF-1R signaling (yielding synergistic Akt activation), which in part contributes to increased beta-cell proliferation and survival under the condition of GH and IGF-1 cotreatment (Fig. 10.7C).[140]

5. CONCLUDING REMARKS

Dysfunction and destruction of pancreatic islet beta cells is a hallmark of diabetes. Better understanding of cell signals controlling beta-cell growth,

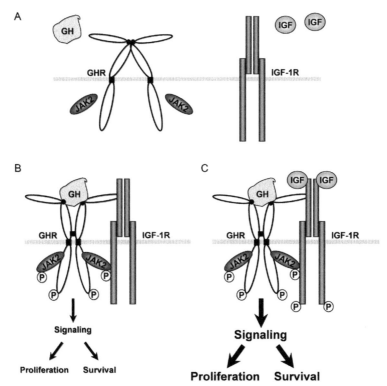

Figure 10.7 *Model for physical and functional interaction of GH and IGF-1 signaling elements in beta cells.* Under resting conditions, GHR (most likely predimerized) and IGF-1R are present on the cell plasma membrane (A). GH engagement causes GHR conformational changes, allowing achievement of an active signaling status (with enhanced GHR–JAK2 interaction). This, in turn, allows the association of GHR–JAK2 with IGF-1R (B). Further IGF-1 binding to the IGF-1R within such a complex facilitates, stabilizes, and enhances the GHR signaling (augmenting STAT5 activation) and simultaneously activates IGF-1R (yielding synergistic Akt activation), which in part contributes to enhanced beta-cell proliferation and enhanced survival under the condition of GH and IGF-1 cotreatment (C). For details, see Ref. 140. *Reprinted with permission from Molecular Endocrinology.*

survival, differentiation, and insulin secretion will allow development of therapeutic strategies for diabetes by preservation and expansion of beta-cell mass and improvement of beta-cell function. During the past several decades, the number of studies analyzing the molecular mechanisms including growth factor/hormone signaling pathways that govern islet beta-cell growth and survival *in vitro* and *in vivo* has increased exponentially.

In this review chapter, several aspects of GH, PRL, and IGF-1 signaling and their effects on beta-cell mass and function are discussed. In particular, we focus on the collaborative roles of GH/GHR and IGF-1/IGF-1R signaling in regulation of beta-cell proliferation and antiapoptosis. Increasing evidence implies that a cross talk between GH/GHR and IGF-1/IGF-1R systems does exist in beta cells. Specifically, we demonstrate that GH induces formation of GHR–JAK2–IGF-1R protein complex. GH and IGF-1 synergize in activating STAT5 and Akt pathways. Concomitantly, more numbers of beta cells are in proliferating stages and fewer cells undergo apoptosis when exposed to GH and IGF-1 in combination (cotreatment) versus single hormonal regimens.[140] Although it is difficult to assign a specific pathway to beta-cell growth and/or survival at this point, our findings in beta cells are consistent with the general view of STATs and Akt as potent mediators of mitogenic[157,158] and antiapoptotic[106,159] signals, respectively. Future studies will further elucidate the precise roles of GH–IGF-1 cross talk in regulation of beta-cell mass and function. The discovery of the GHR–JAK2–IGF-1R protein complex formation upon GH engagement to GHR, together with recent data from osteoblasts,[148,155] raises the possibility that IGF-1R may be a proximal cofactor in GH/GHR signaling. A specific role of the complex formation in beta-cell function and diabetes remains to be explored.

GH and IGF-1 signaling play critical roles in many physiological and pathophysiological conditions including diabetes. In concert with recent observations that GH increases beta-cell proliferation in transplanted human and fetal rat islets,[160] our novel findings of synergy in GH and IGF-1 signaling and resultant augmentation in beta-cell proliferation and survival[140] have important biological and therapeutic implications.

ACKNOWLEDGMENTS

This work was supported by grants from American Heart Association (10BGIA4050019), Science Foundation Arizona (CAA0259-08), St. Joseph's Foundation Startup Package (to Y. H.), and National Institutes of Health (R01GM085237, to Y. C.). We apologize to those whose work has not been covered and cited due to space constraints.

REFERENCES

1. DeFronzo RA. Pathogenesis of type 2 diabetes: metabolic and molecular implications for identifying diabetes genes. *Diabetes Rev.* 1997;5:177–269.
2. DeFronzo RA. Pathogenesis of type 2 diabetes mellitus. *Med Clin North Am.* 2004;88:787–835.
3. Prentki M, Nolan C. Islet beta cell failure in type 2 diabetes. *J Clin Invest.* 2006;116:1802–1812.

4. Lingohr MK, Buettner R, Rhodes CJ. Pancreatic cell growth and survival—a role in obesity-linked type 2 diabetes? *Trends Mol Med*. 2002;8:375–384.
5. Zimmet P, Alberti K, Shaw J. Global and societal implications of the diabetes epidemic. *Nature*. 2001;414:782–787.
6. Yoon K, Lee J, Kim J, et al. Epidemic obesity and type 2 diabetes in Asia. *Lancet*. 2006;368:1681–1688.
7. Kahn S, Hull R, Utzschneider K. Mechanisms linking obesity to insulin resistance and type 2 diabetes. *Nature*. 2006;444:840–846.
8. Mokdad A, Bowman B, Ford E, Vinicor F, Marks J, Koplan J. The continuing epidemics of obesity and diabetes in the United States. *JAMA*. 2001;286:1195–1200.
9. Nielsen JH, Galsgaard ED, Moldrup A, et al. Regulation of cell mass by hormones and growth factors. *Diabetes*. 2001;50:S25–S29.
10. Nielsen JH, Serup P. Molecular basis for islet development, growth, and regeneration. *Curr Opin Endocrinol Diabetes*. 1998;5:97–107.
11. Nielsen JH, Svensson C, Galsgaard ED, Møldrup A, Billestrup N. Beta cell proliferation and growth factors. *J Mol Med*. 1999;77:62–66.
12. Dickson LM, Rhodes CJ. Pancreatic beta-cell growth and survival in the onset of type 2 diabetes: a role for protein kinase B in the Akt? *Am J Physiol Endocrinol Metab*. 2004;287: E192–E198.
13. Kahn SE. Clinical review 135: the importance of beta-cell failure in the development and progression of type 2 diabetes. *J Clin Endocrinol Metab*. 2001;86:4047–4058.
14. Rhodes C. Type 2 diabetes—a matter of cell life and death? *Science*. 2005;307:380–384.
15. Karaca M, Magnan C, Kargar C. Functional pancreatic beta-cell mass: involvement in type 2 diabetes and therapeutic intervention. *Diabetes Metab*. 2009;35:77–84.
16. Dor Y, Brown J, Martinez OI, Melton DA. Adult pancreatic beta-cells are formed by self-duplication rather than stem-cell differentiation. *Nature*. 2004;429:41–46.
17. Teta M, Rankin MM, Long SY, Stein GM, Kushner JA. Growth and regeneration of adult beta cells does not involve specialized progenitors. *Dev Cell*. 2007;12:817–826.
18. Bonner-Weir S. Islet growth and development in the adult. *J Mol Endocrinol*. 2000;24:297–302.
19. Newgard CB, McGarry JD. Metabolic coupling factors in pancreatic beta-cell signal transduction. *Annu Rev Biochem*. 1995;64:689–719.
20. Prentki M. New insights into pancreatic beta-cell metabolic signaling in insulin secretion. *Eur J Endocrinol*. 1996;134:272–286.
21. Jackerott M, Møldrup A, Thams P, et al. STAT5 activity in pancreatic beta-cells influences the severity of diabetes in animal models of type 1 and 2 diabetes. *Diabetes*. 2006;55:2705–2712.
22. Vasavada RC, Gonzalez-Pertusa JA, Fujinaka Y, Fiaschi-Taesch N, Cozar-Castellano I, Garcia-Ocaña A. Growth factors and beta cell replication. *Int J Biochem Cell Biol*. 2006;38:931–950.
23. Scaglia L, Smith FE, Bonner-Weir S. Apoptosis contributes to the involution of beta cell mass in the post partum rat pancreas. *Endocrinology*. 1995;136:5461–5468.
24. Tura A, Grassi A, Winhofer Y, et al. Progression to type 2 diabetes in women with former gestational diabetes: time trajectories of metabolic parameters. *PLoS One*. 2012;7:e50419.
25. Leung DW, Spencer SA, Cachianes G, et al. Growth hormone receptor and serum binding protein: purification, cloning and expression. *Nature*. 1987;330:537–543.
26. Bazan JF. Structural design and molecular evolution of a cytokine receptor superfamily. *Proc Natl Acad Sci U S A*. 1990;87:6934–6938.
27. Argetsinger LS, Campbell GS, Yang X, et al. Identification of JAK2 as a growth hormone receptor-associated tyrosine kinase. *Cell*. 1993;74:237–244.
28. Ihle JN. The Janus protein tyrosine kinase family and its role in cytokine signaling. *Adv Immunol*. 1995;60:1–35.

29. Frank SJ, Messina JL. Growth hormone receptor. In: Oppenheim JJ, Feldman M, eds. *Cytokine Reference On-Line*. Harcourt, London: Academic Press; 2002:1–21.
30. Carter-Su C, Schwartz J, Smit LS. Molecular mechanism of growth hormone action. *Annu Rev Physiol*. 1996;58:187–207.
31. Waxman DJ, Frank SJ. Growth hormone action: signaling via a JAK/STAT-coupled receptor. In: Conn PM, Means A, eds. *Principles of Molecular Regulation*. Totowa, NJ: Humana Press; 2000:55–83.
32. Smit LS, Meyer DJ, Billestrup N, Norstedt G, Schwartz J, Carter-Su C. The role of the growth hormone (GH) receptor and JAK1 and JAK2 kinases in the activation of Stats 1, 3, and 5 by GH. *Mol Endocrinol*. 1996;10:519–533.
33. Udy GB, Towers RP, Snell RG, et al. Requirement of STAT5b for sexual dimorphism of body growth rates and liver gene expression. *Proc Natl Acad Sci U S A*. 1997;94:7239–7244.
34. Woelfle J, Billiard J, Rotwein P. Acute control of insulin-like growth factor-I gene transcription by growth hormone through Stat5b. *J Biol Chem*. 2003;278:22696–22702.
35. Davey HW, Xie T, McLachlan MJ, Wilkins RJ, Waxman DJ, Grattan DR. STAT5b is required for GH-induced liver IGF-I gene expression. *Endocrinology*. 2001;142:3836–3841.
36. Davey HW, McLachlan MJ, Wilkins RJ, Hilton DJ, Adams TE. STAT5b mediates the GH-induced expression of SOCS-2 and SOCS-3 mRNA in the liver. *Mol Cell Endocrinol*. 1999;158:111–116.
37. Bergad PL, Shih HM, Towle HC, Schwarzenberg SJ, Berry SA. Growth hormone induction of hepatic serine protease inhibitor 2.1 transcription is mediated by a Stat5-related factor binding synergistically to two gamma-activated sites. *J Biol Chem*. 1995;270:24903–24910.
38. Huang Y, Kim SO, Yang N, Jiang J, Frank SJ. Physical and functional interaction of growth hormone and insulin-like growth factor-1 signaling elements. *Mol Endocrinol*. 2004;18:1471–1485.
39. Huang Y, Kim SO, Jiang J, Frank SJ. Growth hormone-induced phosphorylation of epidermal growth factor (EGF) receptor in 3T3-F442A cells. Modulation of EGF-induced trafficking and signaling. *J Biol Chem*. 2003;278:18902–18913.
40. Huang Y, Chang Y, Wang X, Jiang J, Frank SJ. Growth hormone alters epidermal growth factor receptor binding affinity via activation of ERKs in 3T3-F442A cells. *Endocrinology*. 2004;145:3297–3306.
41. Goffin V, Shiverick K, Kelly P, Martial J. Sequence-function relationships within the expanding family of prolactin, growth hormone, placental lactogen, and related proteins in mammals. *Endocr Rev*. 1996;17:385–410.
42. Ben-Jonathan N, Mershon JL, Allen DL, Steinmetz RW. Extrapituitary prolactin: distribution, regulation, functions, and clinical aspects. *Endocr Rev*. 1996;17:639–669.
43. Freeman ME, Kanyicska B, Lerant A, Nagy G. Prolactin: structure, function, and regulation of secretion. *Physiol Rev*. 2000;80:1523–1631.
44. Rui H, Kirken RA, Farrar WL. Activation of receptor-associated tyrosine kinase JAK2 by prolactin. *J Biol Chem*. 1994;269:5364–5368.
45. Bole-Feysot C, Goffin V, Edery M, Binart N, Kelly PA. Prolactin (PRL) and its receptor: actions, signal transduction pathways and phenotypes observed in PRL receptor knockout mice. *Endocr Rev*. 1998;19:225–268.
46. Clevenger CV, Furth PA, Hankinson SE, Schuler LA. The role of prolactin in mammary carcinoma. *Endocr Rev*. 2003;24:1–27.
47. Lee RC, Walters JA, Reyland ME, Anderson SM. Constitutive activation of the prolactin receptor results in the induction of growth factor-independent proliferation and constitutive activation of signaling molecules. *J Biol Chem*. 1999;274:10024–10034.
48. Aksamitiene E, Achanta S, Kolch W, Kholodenko BN, Hoek JB, Kiyatkin A. Prolactin-stimulated activation of ERK1/2 mitogen-activated protein kinases is controlled by PI3-kinase/Rac/PAK signaling pathway in breast cancer cells. *Cell Signal*. 2011;23:1794–1805.

49. Huang Y, Li X, Jiang J, Frank SJ. Prolactin modulates phosphorylation, signaling and trafficking of epidermal growth factor receptor in human T47D breast cancer cells. *Oncogene.* 2006;25:7565–7576.
50. Hughes JP, Friesen HG. The nature and regulation of the receptors for pituitary growth hormone. *Annu Rev Physiol.* 1985;47:469–482.
51. Cunningham BC, Bass S, Fuh G, Wells JA. Zinc mediation of the binding of human growth hormone to the human prolactin receptor. *Science.* 1990;250:1709–1712.
52. Somers W, Ultsch M, De Vos AM, Kossiakoff AA. The X-ray structure of a growth hormone–prolactin receptor complex. *Nature.* 1994;372:478–481.
53. Nielsen JH, Moldrup A, Billestrup N, Petersen ED, Allevato G, Stahl M. The role of growth hormone and prolactin in beta cell growth and regeneration. *Adv Exp Med Biol.* 1992;321:9–17.
54. Nielsen JH. Effects of growth hormone, prolactin, and placental lactogen on insulin content and release, and deoxyribonucleic acid synthesis in cultured pancreatic islets. *Endocrinology.* 1982;110:600–606.
55. Sun AM, Lin BJ, Haist RE. Studies on the effects of growth hormone and thyroxine on proinsulin synthesis and insulin formation in the isolated islets of Langerhans of the rat. *Can J Physiol Pharmacol.* 1972;50:1147–1151.
56. Whittaker PG, Taylor KW. Direct effects of rat growth hormone in rat islets of Langerhans in tissues culture. *Diabetologia.* 1980;18:323–328.
57. Pierluissi J, Pierluissi R, Ashcroft SJ. Effects of growth hormone on insulin release in the rat. *Diabetologia.* 1980;19:391–396.
58. Tesone M, Oliveira-Filho RM, Charreau EH. Prolactin binding in rat Langerhans islets. *J Recept Res.* 1980;1:355–372.
59. Nielsen JH, Linde S, Welinder BS, Billestrup N, Madsen OD. Growth hormone is a growth factor for the differentiated pancreatic beta-cell. *Mol Endocrinol.* 1989;3:165–173.
60. Swenne I, Hill DJ, Strain AJ, Milner RD. Growth hormone regulation of somatomedin C/insulin-like growth factor I production and DNA replication in fetal rat islets in tissue culture. *Diabetes.* 1987;36:288–294.
61. Brelje TC, Parsons JA, Sorenson RL. Regulation of islet beta-cell proliferation by prolactin in rat islets. *Diabetes.* 1994;43:263–273.
62. Brelje TC, Sorenson RL. Role of prolactin versus growth hormone on islet B-cell proliferation in vitro: implications for pregnancy. *Endocrinology.* 1991;128:45–57.
63. Rabinovitch A, Quigley C, Rechler MM. Growth hormone stimulates islet B-cell replication in neonatal rat pancreatic monolayer cultures. *Diabetes.* 1983;32:307–312.
64. Møldrup A, Petersen ED, Nielsen JH. Effects of sex and pregnancy hormones on growth hormone and prolactin receptor gene expression in insulin-producing cells. *Endocrinology.* 1993;133:1165–1172.
65. Sorenson RL, Brelje TC. Adaptation of islets of Langerhans to pregnancy: beta-cell growth, enhanced insulin secretion and the role of lactogenic hormones. *Horm Metab Res.* 1997;29:301–307.
66. Møldrup A, Billestrup N, Nielsen JH. Rat insulinoma cells express both a 115-kDa growth hormone receptor and a 95-kDa prolactin receptor structurally related to the hepatic receptors. *J Biol Chem.* 1990;265:8686–8690.
67. Asfari M, De W, Postel-Vinay M, Czernichow P. Expression and regulation of growth hormone (GH) and prolactin (PRL) receptors in a rat insulin producing cell line (INS-1). *Mol Cell Endocrinol.* 1995;107:209–214.
68. Swenne I, Hill DJ, Strain AJ, Milner RD. Effects of human placental lactogen and growth hormone on the production of insulin and somatomedin C/insulin-like growth factor I by human fetal pancreas in tissue culture. *J Endocrinol.* 1987;113:297–303.

69. Hansen LH, Wang X, Kopchick JJ, et al. Identification of tyrosine residues in the intracellular domain of the growth hormone receptor required for transcriptional signaling and Stat5 activation. *J Biol Chem.* 1996;271:12669–12673.
70. Cousin SP, Hugl SR, Myers MG, White MF, Reifel-Miller A, Rhodes CJ. Stimulation of pancreatic beta-cell proliferation by growth hormone is glucose-dependent: signal transduction via Janus kinase 2 (JAK2)/signal transducer and activator of transcription 5 (STAT5) with no crosstalk to insulin receptor substrate-mediate mitogenic signalling. *Biochem J.* 1999;344:649–658.
71. Billestrup N, Nielsen J. The stimulatory effect of growth hormone, prolactin, and placental lactogen on beta-cell proliferation is not mediated by insulin-like growth factor-I. *Endocrinology.* 1991;129:883–888.
72. Sekine N, Wollheim CB, Fujita T. GH signaling in pancreatic beta-cells. *Endocr J.* 1998;45:S33–S40.
73. Galsgaard ED, Friedrichsen BN, Nielsen JH, Møldrup A. Expression of dominant-negative STAT5 inhibits growth hormone- and prolactin-induced proliferation of insulin-producing cells. *Diabetes.* 2001;50:S40–S41.
74. Galsgaard ED, Gouilleux F, Groner B, Serup P, Nielsen JH, Billestrup N. Identification of a growth hormone-responsive STAT5-binding element in the rat insulin 1 gene. *Mol Endocrinol.* 1996;10:652–660.
75. Friedrichsen BN, Galsgaard ED, Nielsen JH, Møldrup A. Growth hormone- and prolactin-induced proliferation of insulinoma cells, INS-1, depends on activation of STAT5 (signal transducer and activator of transcription 5). *Mol Endocrinol.* 2001;15:136–148.
76. Friedrichsen BN, Richter HE, Hansen JA, et al. Signal transducer and activator of transcription 5 activation is sufficient to drive transcriptional induction of cyclin D2 gene and proliferation of rat pancreatic beta-cells. *Mol Endocrinol.* 2003;17:945–958.
77. Stout LE, Svensson AM, Sorenson RL. Prolactin regulation of islet-derived INS-1 cells: characteristics and immunocytochemical analysis of STAT5 translocation. *Endocrinology.* 1997;138:1592–1603.
78. Hügl SR, Merger M. Prolactin stimulates proliferation of the glucose-dependent beta-cell line INS-1 via different IRS-proteins. *JOP.* 2007;8:739–752.
79. Iwahashi H, Hanafusa T, Eguchi Y, et al. Cytokine induced apoptotic cell-death in a mouse pancreatic beta-cell line: inhibition by bcl-2. *Diabetologia.* 1996;39:530–536.
80. Rabinovitch A, Suarez-Pinzon W, Strynadka K, et al. Transfection of human pancreatic islets with anti-apoptotic gene (bcl-2) proteins beta-cells from cytokine-induced destruction. *Diabetes.* 1999;48:1223–1229.
81. Karlsen AE, Galsgarrd ED, Nerup J, Nielsen JH. Effect of growth hormone and prolactin on cytokine induced toxicity in the INS-1 beta cell line (abstract). *Eur J Endocrinol.* 2000;142:10.
82. Terra LF, Garay-Malpartida MH, Wailemann RA, Sogayar MC, Labriola L. Recombinant human prolactin promotes human beta cell survival via inhibition of extrinsic and intrinsic apoptosis pathways. *Diabetologia.* 2011;54:1388–1397.
83. Fujinaka Y, Takane K, Yamashita H, Vasavada RC. Lactogens promote beta cell survival through JAK2/STAT5 activation and Bcl-XL upregulation. *J Biol Chem.* 2007;282:30707–30717.
84. Naka T, Narazaki M, Hirata M, et al. Structure and function of a new STAT-induced STAT inhibitor. *Nature.* 1997;387:924–929.
85. Starr R, Willson TA, Viney EM, et al. A family of cytokine-inducible inhibitors of signalling. *Nature.* 1997;387:917–921.
86. Endo TA, Masuhara M, Yokouchi M, et al. A new protein containing an SH2 domain that inhibits JAK kinases. *Nature.* 1997;387:921–924.

87. Uyttendaele I, Lemmens I, Verhee A, et al. Mammalian protein-protein interaction trap (MAPPIT) analysis of STAT5, CIS, and SOCS2 interactions with the growth hormone receptor. *Mol Endocrinol.* 2007;21:2821–2831.
88. Adams TE, Hansen JA, Starr R, Nicola NA, Hilton DJ, Billestrup N. Growth hormone preferentially induces the rapid, transient expression of SOCS-3, a novel inhibitor of cytokine receptor signaling. *J Biol Chem.* 1998;273:1285–1287.
89. Hansen JA, Lindberg K, Hilton DJ, Nielsen JH, Billestrup N. Mechanism of inhibition of growth hormone receptor signaling by suppressor of cytokine signaling proteins. *Mol Endocrinol.* 1999;13:1832–1843.
90. Metcalf D, Greenhalgh CJ, Viney E, et al. Gigantism in mice lacking suppressor of cytokine signalling-2. *Nature.* 2000;405:1069–1073.
91. Kawazoe Y, Naka T, Fujimoto M, et al. Signal transducer and activator of transcription (STAT)-induced STAT inhibitor 1 (SSI-1)/suppressor of cytokine signaling 1 (SOCS1) inhibits insulin signal transduction pathway through modulating insulin receptor substrate 1 (IRS-1) phosphorylation. *J Exp Med.* 2001;193:263–269.
92. Bruun C, Heding PE, Rønn SG, et al. Suppressor of cytokine signalling-3 inhibits tumor necrosis factor-alpha induced apoptosis and signalling in beta cells. *Mol Cell Endocrinol.* 2009;311:32–38.
93. Liu JL, Coschigano KT, Robertson K, et al. Disruption of growth hormone receptor gene causes diminished pancreatic islet size and increased insulin sensitivity in mice. *Am J Physiol Endocrinol Metab.* 2004;287:405–413.
94. Freemark M, Avril I, Fleenor D, et al. Targeted deletion of the PRL receptor: effects on islet development, insulin production, and glucose tolerance. *Endocrinology.* 2002;143:1378–1385.
95. Lee JY, Gavrilova O, Davani B, Na R, Robinson GW, Hennighausen L. The transcription factors Stat5a/b are not required for islet development but modulate pancreatic beta-cell physiology upon aging. *Biochim Biophys Acta.* 2007;1773:1455–1461.
96. Wu Y, Liu C, Sun H, et al. Growth hormone receptor regulates β cell hyperplasia and glucose-stimulated insulin secretion in obese mice. *J Clin Invest.* 2011;121:2422–2426.
97. Yakar S, Liu JL, Stannard B, et al. Normal growth and development in the absence of hepatic insulin-like growth factor I. *Proc Natl Acad Sci U S A.* 1999;96:7324–7329.
98. Sjögren K, Liu JL, Blad K, et al. Liver-derived insulin-like growth factor I (IGF-I) is the principal source of IGF-I in blood but is not required for postnatal body growth in mice. *Proc Natl Acad Sci U S A.* 1999;96:7088–7092.
99. Salmon WD, Daughaday WH. A hormonally controlled serum factor which stimulates sulfate incorporation by cartilage *in vitro. J Lab Clin Med.* 1957;49:825–836.
100. Daughaday WH. Growth hormone axis overview—somatomedin hypothesis. *Pediatr Nephrol.* 2000;14:537–540.
101. Wang J, Zhou J, Powell-Braxton L, Bondy C. Effects of Igf1 gene deletion on postnatal growth patterns. *Endocrinology.* 1999;140:3391–3394.
102. Liu JL, LeRoith D. Insulin-like growth factor 1 is essential for postnatal growth in response to growth hormone. *Endocrinology.* 1999;140:5178–5184.
103. Favelyukis S, Till JH, Hubbard SR, Miller WT. Structure and autoregulation of the insulin-like growth factor 1 receptor kinase. *Nat Struct Biol.* 2001;8:1058–1063.
104. Laviola L, Natalicchio A, Giorgino F. The IGF-I signaling pathway. *Curr Pharm Des.* 2007;13:663–669.
105. Krishna M, Narang H. The complexity of mitogen-activated protein kinases (MAPKs) made simple. *Cell Mol Life Sci.* 2008;65:3525–3544.
106. Manning BD, Cantley LC. AKT/PKB signaling: navigating downstream. *Cell.* 2007;129:1261–1274.

107. Zong CS, Chan J, Levy DE, Horvath C, Sadowski HB, Wang LH. Mechanism of STAT3 activation by insulin-like growth factor I receptor. *J Biol Chem.* 2000;275:15099–15105.
108. Peraldi P, Filloux C, Emanuelli B, Hilton DJ, Van Obberghen E. Insulin induces suppressor of cytokine signaling-3 tyrosine phosphorylation through janus-activated kinase. *J Biol Chem.* 2001;276:24614–24620.
109. Velloso LA, Carvalho CR, Rojas FA, Folli F, Saad MJ. Insulin signalling in heart involves insulin receptor substrates-1 and -2, activation of phosphatidylinositol 3-kinase and the JAK 2-growth related pathway. *Cardiovasc Res.* 1998;40:96–102.
110. Giorgetti-Peraldi S, Peyrade F, Baron V, Van Obberghen E. Involvement of Janus kinases in the insulin signaling pathway. *Eur J Biochem.* 1995;234:656–660.
111. Kulkarni R. Receptors for insulin and insulin-like growth factor-1 and insulin receptor substrate-1 mediate pathways that regulate islet function. *Biochem Soc Trans.* 2002;30:317–322.
112. Withers DJ, Gutierrez JS, Towery HH, et al. Disruption of IRS-2 causes type 2 diabetes in mice. *Nature.* 1998;391:900–904.
113. Withers DJ, Burks DJ, Towery HH, Altamuro SL, Flint CL, White MF. IRS-2 coordinates IGF-1 receptor-mediated β-cell development and peripheral insulin signaling. *Nat Genet.* 1999;23:32–40.
114. Rhodes CJ, White MF. Molecular insights into insulin action and secretion. *Eur J Clin Invest.* 2002;32:3–13.
115. Smith F, Rosen K, Villa-Komarof L, Weir GC, Bonner-Weir S. Enhanced IGF-1 expression in regenerating rat pancreas is localized to capillaries and proliferating duct cells. *Diabetes.* 1990;39(Suppl 1):66A.
116. Maake C, Reinecke M. Immunohistochemical localization of insulin-like growth factor 1 and 2 in the endocrine pancreas of rat, dog, and man, and their coexistence with classical islet hormones. *Cell Tissue Res.* 1993;273:249–259.
117. Jevdjovic T, Maake C, Eppler E, Zoidis E, Reinecke M, Zapf J. Effects of insulin-like growth factor-I treatment on the endocrine pancreas of hypophysectomized rats: comparison with growth hormone replacement. *Eur J Endocrinol.* 2004;151:223–231.
118. Jevdjovic T, Maake C, Zwimpfer C, et al. The effect of hypophysectomy on pancreatic islet hormone and insulin-like growth factor I content and mRNA expression in rat. *Histochem Cell Biol.* 2005;123:179–188.
119. Kahn SE. The relative contributions of insulin resistance and beta-cell dysfunction to the pathophysiology of Type 2 diabetes. *Diabetologia.* 2003;46:3–19.
120. Weir GC, Laybutt DR, Kaneto H, Bonner-Weir S, Sharma A. Beta-cell adaptation and decompensation during the progression of diabetes. *Diabetes.* 2001;50:S154–S159.
121. Burks DJ, White MF. Beta-cell mass and function in type 2 diabetes: IRS proteins and beta-cell function. *Diabetes.* 2001;50:S140–S145.
122. Araki E, Lipes M, Patti M, et al. Alternative pathway of insulin signaling in mice with targeted disruption of the IRS-1 gene. *Nature.* 1994;372:186–190.
123. Tamemoto H, Kadowaki T, Tobe K, et al. Insulin resistance and growth retardation in mice lacking insulin receptor substrate-1. *Nature.* 1994;372:182–186.
124. Kulkarni R, Holzenberger M, Shih D, et al. beta-cell-specific deletion of the Igf1 receptor leads to hyperinsulinemia and glucose intolerance but does not alter beta-cell mass. *Nat Genet.* 2002;31:111–115.
125. Xuan S, Kitamura T, Nakae J, et al. Defective insulin secretion in pancreatic beta cells lacking type 1 IGF receptor. *J Clin Invest.* 2002;110:1011–1019.
126. Ueki K, Okada T, Hu J, et al. Total insulin and IGF-I resistance in pancreatic beta cells causes overt diabetes. *Nat Genet.* 2006;38:583–588.
127. Hügl SR, White MF, Rhodes CJ. Insulin-like growth factor I (IGF-I)-stimulated pancreatic beta-cell growth is glucose-dependent. *J Biol Chem.* 1998;273:17771–17779.

128. Rhodes CJ. Introduction: the molecular cell biology of insulin production. *Semin Cell Dev Biol.* 2000;11:223–225.

129. Cousin S, Hugl S, Wrede C, Kjio H, Myers M, Rhodes C. Free fatty acid-induced inhibition of glucose and insulin-like growth factor 1-induced deoxyribonucleic acid synthesis in the pancreatic beta-cell line INS-1. *Endocrinology.* 2001;142:229–240.

130. Lingohr M, Dickson L, McCuaig J, Hugl S, Twardzik D, Rhodes C. Activation of IRS-2 mediated signal transduction by IGF-1, but not by TGF or EGF, augments pancreatic cell proliferation. *Diabetes.* 2002;51:966–976.

131. Briaud I, Dickson LM, Lingohr MK, McCuaig JF, Lawrence JC, Rhodes CJ. Insulin receptor substrate-2 proteasomal degradation mediated by a mammalian target of rapamycin (mTOR)-induced negative feedback down-regulates protein kinase B-mediated signaling pathway in beta-cells. *J Biol Chem.* 2005;280:2282–2293.

132. Lingohr MK, Dickson LM, Wrede CE, et al. Decreasing IRS-2 expression in pancreatic cells (INS-1) promotes apoptosis, which can be compensated for by introduction of IRS-4 expression. *Mol Cell Endocrinol.* 2003;209:17–31.

133. Tuttle RL, Gill NS, Pugh W, et al. Regulation of pancreatic beta-cell growth and survival by the serine/threonine protein kinase Akt1/PKBalpha. *Nat Med.* 2001;7:1133–1137.

134. Wrede CE, Dickson LM, Lingohr MK, Briaud I, McCuaig JF, Rhodes CJ. Protein kinase B/Akt prevents fatty acid induced apoptosis in pancreatic beta-cells (INS-1). *J Biol Chem.* 2002;277:49676–49684.

135. da Silva Xavier G, Qian Q, Cullen P, Rutter G. Distinct roles for insulin and insulin-like growth factor-1 receptors in pancreatic beta-cell glucose sensing revealed by RNA silencing. *Biochem J.* 2004;377:149–158.

136. Sekine N, Ullrich S, Regazzi R, Pralong W, Wollheim C. Postreceptor signaling of growth hormone and prolactin and their effects in the differentiated insulin-secreting cell line, INS-1. *Endocrinology.* 1996;137:1841–1850.

137. Harrison M, Dunger A, Berg S, et al. Growth factor protection against cytokine-induced apoptosis in neonatal rat islets of Langerhans: role of Fas. *FEBS Lett.* 1998;435:207–210.

138. Yakar S, Rosen CJ, Beamer WG, et al. Circulating levels of IGF-1 directly regulate bone growth and density. *J Clin Invest.* 2002;110:771–781.

139. Lupu F, Terwilliger JD, Lee K, Segre GV, Efstratiadis A. Roles of growth hormone and insulin-like growth factor 1 in mouse postnatal growth. *Dev Biol.* 2001;229:141–162.

140. Ma F, Wei Z, Shi C, et al. Signaling cross talk between growth hormone (GH) and insulin-like growth factor-I (IGF-I) in pancreatic islet β-Cells. *Mol Endocrinol.* 2011;25:2119–2133.

141. de Vos AM, Ultsch M, Kossiakoff AA. Human growth hormone and extracellular domain of its receptor: crystal structure of the complex. *Science.* 1992;255:306–312.

142. Brown RJ, Adams JJ, Pelekanos RA, et al. Model for growth hormone receptor activation based on subunit rotation within a receptor dimer. *Nat Struct Mol Biol.* 2005;12:814–821.

143. van den Eijnden MJ, Lahaye LL, Strous GJ. Disulfide bonds determine growth hormone receptor folding, dimerisation and ligand binding. *J Cell Sci.* 2006;119:3078–3086.

144. Frank SJ, Fuchs SY. Modulation of growth hormone receptor abundance and function: roles for the ubiquitin–proteasome system. *Biochim Biophys Acta.* 2008;1782:785–794.

145. Kopchick JJ, Parkinson C, Stevens EC, Trainer PJ. Growth hormone receptor antagonists: discovery, development, and use in patients with acromegaly. *Endocr Rev.* 2002;23:623–646.

146. Gual P, Baron V, Lequoy V, Van Obberghen E. Interaction of Janus kinases JAK-1 and JAK-2 with the insulin receptor and the insulin-like growth factor-1 receptor. *Endocrinology.* 1998;139:884–893.

147. Zhang Y, Jiang J, Kopchick JJ, Frank SJ. Disulfide linkage of growth hormone (GH) receptors (GHR) reflects GH-induced GHR dimerization. Association of JAK2 with the GHR is enhanced by receptor dimerization. *J Biol Chem*. 1999;274:33072–33084.

148. Gan Y, Zhang Y, Digirolamo DJ, et al. Deletion of IGF-I receptor (IGF-IR) in primary osteoblasts reduces GH-induced STAT5 signaling. *Mol Endocrinol*. 2010;24:644–656.

149. Edmondson SR, Russo VC, McFarlane AC, Wraight CJ, Werther GA. Interactions between growth hormone, insulin-like growth factor I, and basic fibroblast growth factor in melanocyte growth. *J Clin Endocrinol Metab*. 1999;84:1638–1644.

150. Ashcom G, Gurland G, Schwartz J. Growth hormone synergizes with serum growth factors in inducing c-fos transcription in 3T3-F442A cells. *Endocrinology*. 1992;131:1915–1921.

151. Ling Z, Hannaert JC, Pipeleers D. Effect of nutrients, hormones and serum on survival of rat islet beta cells in culture. *Diabetologia*. 1994;37:15–21.

152. McClenaghan NH. Physiological regulation of the pancreatic beta-cell: functional insights for understanding and therapy of diabetes. *Exp Physiol*. 2007;92:481–496.

153. Hohmeier H, Newgard C. Cell lines derived from pancreatic islets. *Mol Cell Endocrinol*. 2004;228:121–128.

154. Sotiropoulos A, Ohanna M, Kedzia C, et al. Growth hormone promotes skeletal muscle cell fusion independent of insulin-like growth factor 1 up-regulation. *Proc Natl Acad Sci U S A*. 2006;103:7315–7320.

155. DiGirolamo DJ, Mukherjee A, Fulzele K, et al. Mode of growth hormone action in osteoblasts. *J Biol Chem*. 2007;282:31666–31674.

156. LeRoith D, Bondy C, Yakar S, Liu JL, Butler A. The somatomedin hypothesis: 2001. *Endocr Rev*. 2001;22:53–74.

157. Ihle JN. The Stat family in cytokine signaling. *Curr Opin Cell Biol*. 2001;13:211–217.

158. Levy DE, Darnell JEJ. Stats: transcriptional control and biological impact. *Nat Rev Mol Cell Biol*. 2002;3:651–662.

159. Sale EM, Sale GJ. Protein kinase B: signalling roles and therapeutic targeting. *Cell Mol Life Sci*. 2008;65:113–127.

160. Höglund E, Mattsson G, Tyrberg B, Andersson A, Carlsson C. Growth hormone increases beta-cell proliferation in transplanted human and fetal rat islets. *JOP*. 2009;10:242–248.

CHAPTER ELEVEN

The Impact of Dietary Methionine Restriction on Biomarkers of Metabolic Health

Manda L. Orgeron, Kirsten P. Stone, Desiree Wanders, Cory C. Cortez, Nancy T. Van, Thomas W. Gettys

Laboratory of Nutrient Sensing and Adipocyte Signaling, Pennington Biomedical Research Center, Baton Rouge, Louisiana, USA

Contents

Abstract

Calorie restriction without malnutrition, commonly referred to as dietary restriction (DR), results in a well-documented extension of life span. DR also produces significant, long-lasting improvements in biomarkers of metabolic health that begin to accrue soon after its introduction. The improvements are attributable in part to the effects of DR on energy balance, which limit fat accumulation through reduction in energy intake. Accumulation of excess body fat occurs when energy intake chronically exceeds the energy costs for growth and maintenance of existing tissue. The resulting obesity promotes the development of insulin resistance, disordered lipid metabolism, and increased expression of inflammatory markers in peripheral tissues. The link between the life-extending effects of DR and adiposity is the subject of an ongoing debate, but it is clear that decreased fat accumulation improves insulin sensitivity and produces beneficial effects on overall metabolic health. Over the last 20 years, dietary methionine restriction (MR) has emerged as a promising DR mimetic because it produces a comparable extension in life span, but surprisingly, does not require food restriction. Dietary MR also reduces adiposity but does so through a paradoxical increase in both energy intake and expenditure. The increase in energy expenditure fully compensates for increased energy intake and effectively limits fat deposition. Perhaps more importantly, the diet increases metabolic flexibility and overall insulin sensitivity and improves lipid metabolism while decreasing systemic inflammation. In this chapter, we describe recent advances in our understanding of the mechanisms and effects of dietary MR and discuss the remaining obstacles to implementing MR as a treatment for metabolic disease.

ABBREVIATIONS

ACC-1 acetyl-CoA carboxylase 1
ANCOVA analysis of covariance
BAT brown adipose tissue
DR dietary restriction
EAAs essential amino acids
EE energy expenditure
FASN fatty acid synthase
FGF-21 fibroblast growth factor 21
GCN2 general control nonderepressible 2
IR insulin resistance
IWAT inguinal white adipose tissue
MR methionine restriction
MsrA and MsrB methionine sulfoxide reductase A and B
RER respiratory exchange ratio
SCD-1 stearoyl-CoA desaturase 1
SNS sympathetic nervous system
TCA tricarboxylic acid
TF transcription factor
UCP1 uncoupling protein 1
WAT white adipose tissue

1. RATIONALE FOR STUDY OF DIETARY METHIONINE RESTRICTION

The essential amino acids (EAAs) (e.g., methionine, lysine, leucine, isoleucine, tryptophan, valine, threonine, phenylalanine, and histidine) cannot be synthesized endogenously, so 10–20 mg/kg body weight of each must be obtained in the diet each day from consumed protein. Moreover, dietary protein sources must contain the full array of EAAs because proteins deficient in one or more EAAs quickly produce an aversive feeding response that results in a significant decrease in consumption of the diet.

Semisynthetic diets absent a single EAA have been used to explore the sensing and signaling mechanisms that mediate the behavioral and physiological responses to EAA deprivation.[1–6] An important implication of this work is that dietary amino acids are functioning much like receptor ligands in the sense that specific concentration ranges engage signaling systems linked to molecular responses that have biochemical and physiological consequences. An important distinction is that the absence of an EAA (e.g., EAA deprivation) is the signal that initiates the full signaling response, suggesting that limitation of an EAA must attain some threshold of restriction for triggering the response. This assumption has been supported by substantial empirical evidence, but the somewhat surprising finding is that dietary restriction (DR) of EAAs within narrowly defined ranges has proven highly beneficial to metabolic status and overall health. In particular, the beneficial responses that result from restricting normal intakes of dietary methionine within a defined range are the subject of this chapter.

2. ORIGINS OF DIETARY METHIONINE RESTRICTION AS AN EXPERIMENTAL APPROACH TO INCREASE LONGEVITY

The initial reports of the health benefits of dietary methionine restriction (MR) were from the Orentreich group.[7,8] They found that removing cysteine and reducing dietary methionine from control levels of 0.86% (i.e., 8.6 g/kg diet) to restricted levels of 0.17% (1.7 g/kg diet) increased longevity by 30–40%. A significant difference between this and other models of DR is that no food restriction was involved and the rats were provided their diets ad libitum. In an important follow-up to their initial work, the authors showed that dietary MR increased longevity in a variety of rat strains with differing pathological profiles.[9] These findings showed that dietary MR decreased mortality from all causes of death and support the view that this dietary approach affects the overall rate of aging. In two of their studies,[7,9] Orentreich et al. pair-fed a second control group the control diet to the level of intake of the MR group. They found that life span was not extended in the pair-fed group, showing that restriction of methionine, and not overall DR, was responsible for the increase in longevity in the MR group.

Dietary MR also increases longevity in mice[10,11] and flies[12] so it seems likely that MR, like DR, will prove to be efficacious across multiple species. Grandison et al.[12] used Drosophila as a model organism to identify the nutrients being limited during DR that mediated the increase in longevity and decrease in fecundity. In nature, Drosophila eats yeast so implementation of DR is accomplished by diluting the yeasts and allowing unlimited consumption of the diluted diet. Using this approach, enhanced longevity and decreased fecundity appear to be mediated primarily by specific nutrients from the yeasts and independent of caloric intake.[13] Using flies as their model organism, Grandison et al.[12] investigated which nutrients were responsible for increased longevity and decreased fecundity by systematically adding back nutrients to the restricted diet. They found that adding back carbohydrate or fat had no effect, but adding back all 10 EAAs reduced longevity and increased fecundity to the same extent as full feeding. Interestingly, adding back methionine alone fully restored fecundity but did not decrease longevity. Thus, the limitation of methionine by DR was solely responsible for reduced fecundity, while limitation of methionine and additional EAAs was required for increased longevity.[12] Further study of the comparative mechanisms engaged by DR in flies and MR in rodents will be required to understand how limiting methionine functions to affect longevity and health span in the two models.

Additional work pointing to a role for methionine metabolism in longevity comes from approaches in model organisms, where manipulation of genes involved in the reduction of oxidized methionine residues affected life span.[14,15] Oxidation of free methionine or methionine in proteins occurs naturally and produces a mixture of the two diastereomers, methionine-R-sulfoxide and methionine-S-sulfoxide. Methionine sulfoxide accumulates in tissue proteins over time and progressively compromises their function. The oxidation of methionine is normally reversed by methionine sulfoxide reductases A and B (MsrA and MsrB), which reduce the two diastereoisomers back to methionine. In yeast, overexpression of either MsrA or MsrB increased replicative life span of yeast, while deletion of either isoform reduced yeast life span.[14] To test the hypothesis that these manipulations were affecting longevity by increasing the supply of methionine, Koc et al.[14] manipulated media concentrations of methionine and found that increasing methionine decreased life span while limiting methionine increased the replicative life span of yeast. In complementary studies with Drosophila, Ruan et al.[15] showed that transgenic overexpression of MsrA markedly extends fly life span. In addition, targeted disruption of MsrA in mice significantly reduced their life span, increased their susceptibility to oxidative stress, and accentuated accumulation of oxidized proteins in their tissues.[16] Collectively, these studies make the case that genetic manipulation of the genes involved in repair of oxidized methionine affects longevity in multiple species. Although the underlying mechanisms are far from clear, limiting dietary methionine mimics the effects achieved by genetic enhancement of methionine repair, while increased dietary methionine accentuates the accumulation of oxidized methionine. As presented in subsequent sections, restriction of dietary methionine produces a number of other short- and long-term metabolic responses that may be equally important as mediators of its effects on longevity. Evaluation of their relative significance is the subject of growing interest in the field.

3. ANTI-INFLAMMATORY RESPONSES TO DIETARY METHIONINE RESTRICTION AND LONGEVITY

The original studies of dietary MR in rats showed that in addition to increasing life span, the diet produced a lifelong reduction in body weight and accumulation of adipose tissue.[7,8] Calorie restriction to 40% of ad libitum intake (DR) also produces a robust extension of mean and maximal life span across species,[17–19] along with the expected reduction in fat accumulation. In addition to comparable effects on longevity and adiposity, DR and MR share other beneficial outcomes including increased insulin sensitivity and a comprehensive improvement in biomarkers of metabolic health.[8,10,20–24] Previous studies have established that the reduction in adiposity produced by DR is associated with reduced expression of proinflammatory markers in peripheral tissues.[25] Thus, an emerging hypothesis is that DR delays death in part by reducing comorbidities associated with chronic inflammatory states such as obesity, diabetes, and cardiovascular disease.[26] The impact of dietary MR on inflammation has been largely unexplored, and a significant unanswered question is whether decreased adiposity, irrespective of mechanism, produces comparable reductions in systemic inflammation. Using transcriptional profiling of peripheral tissues after long-term DR and MR, we recently examined the systems biology of 59 networks annotated to the inflammatory process.[27] Despite comparable reductions in adiposity with both diets, the anti-inflammatory responses to MR were far more extensive than DR and targeted different inflammatory processes in both liver and white adipose tissue (WAT). In particular, the primary pathways affected by MR in inguinal white adipose tissue (IWAT) involved phagocyte and macrophage migration, and the majority of genes within these pathways were downregulated by MR.[27] The primary pathways affected by MR in liver involved accumulation, activation, and morphology of leukocytes and macrophages, and like IWAT, the majority of affected genes were downregulated.[27] Another important observation is that the transcriptional changes appear temporally unrelated to improvements in metabolic biomarkers and occur well after the diet reduces fat deposition, suggesting that the anti-inflammatory effects of MR are not responsible for improved insulin sensitivity.[21,27] Collectively, these findings suggest the interesting possibility that the delays in age-associated inflammation by MR are secondary to the metabolic effects of the diet, rather than a direct result of reduced adiposity.

4. EFFECTS OF DIETARY MR ON ENERGY INTAKE AND ADIPOSITY

The initial reports that dietary MR increased longevity in rats reported that the diet decreased accumulation of body weight by ~40% over 2 years.[7,8] Paradoxically, the weight-adjusted food consumption of the MR group was ~90% higher than controls during the first 3 months and 62% higher after 2 years.[7] Dietary MR produces a similar increase in energy intake in mice, whether expressed per mouse (~20% increase) or on a weight-adjusted basis (~60% increase).[28,29] To evaluate the impact of dietary MR on energy balance and test for metabolic effects of the diet, the Orentreich group used a pair-feeding approach by feeding a third group of rats the control diet to the amounts consumed by the MR group.[7,22] The pair-fed group gained 75 g more than the MR group over the first 3 months[7] and 115 g more than the MR group after 63 weeks.[22] These findings provide prima facie evidence that dietary MR increases the energy costs of maintaining body weight and are highly suggestive that the mechanism involves increased energy expenditure (EE).

The effects of dietary MR on body composition were initially assessed by comparing fat pad weights to control rats after different times on the diet. For example, dietary MR for 18 months reduced dissectible visceral fat mass from 72 g in the control group to 20 g in the MR group. Expressed as a percent of body weight, the MR diet reduced visceral fat pad weights from 13.7% to 7.4% based on this method of assessing body composition.[22] More accurate and comprehensive assessments of body composition using both DEXA- and NMR-based methods have shown that in control rats, adiposity increases from ~16% to near 30% over a 2-year period after weaning.[21] In rats consuming the MR diet, the increase in adiposity is limited to a 4% increase from 16% to 20% over the same period.[21] This 4% increase in the MR group is achieved in the first 3 months after introduction of MR, so from 3 to 20 months, adiposity is essentially clamped at 20% in the MR group. In contrast, adiposity increases from 16% to 22% during the first 3 months on the control diet and from 22% to 26% from 3 to 9 months.[21] When dietary MR is initiated after physical maturity (e.g., 6 months of age), their initial adiposity (23%) remains unchanged over the following 6 months, while adiposity in the control group increases from 23% to 27% over the same period.[21] Body weight is also unchanged in the 6 months following introduction of dietary MR, while body weight in the control group increases from 385 to 480 g.[21] These studies show that introduction of MR in either a juvenile or adult context effectively limits fat deposition and restricts expansion of adipose tissue mass.

Postweaning dietary MR in mice produces similar but strain-specific changes in body weight and adiposity. In FVB mice, dietary MR effectively limited increases in body weight or adiposity during a 10-week study, whereas the adiposity of control mice increased from ~14% to ~25% over the same period.[28] Increases in adiposity were similarly limited by MR over a 10-week study in C57BL/6J mice and the increase in body weight was limited to 3 g (e.g., from ~18 to ~21 g).[29] In mice on the control diet, body weight increased from 18 to 28 g over the same period, while adiposity increased from 12% to 21%.[29] As well documented in rats, mice also respond to dietary MR by increasing both absolute and weight-adjusted food consumption,[28,29] supporting the view that the diet limits fat deposition despite producing an increase in energy intake. Collectively, these findings make a compelling case that dietary MR, regardless of when it is initiated, impacts nutrient partitioning between fat and protein deposition through effects on both energy intake and EE.

5. EFFECTS OF DIETARY MR ON ENERGY EXPENDITURE

Group differences in adiposity not attributable to differences in energy intake infer group differences in EE, and expanded access to small-animal indirect calorimetry has made it the method of choice for measuring EE in rodents. The computational challenges inherent in application of this method are substantial in cases where the genotype or diet results in significant group differences in body size and/or composition. One of the immutable laws of calorimetry, beginning with the eighteenth-century work of Lavoisier and Laplace and extending through the subsequent work of Rubner, Brody, and Kleiber,[30,31] is that EE is proportional to some function of body size. Thus, EE must be scaled accordingly to determine whether group differences remain after correcting for size. This is precisely the problem encountered after long-term MR when group differences in body weight and composition are vastly different. It is well accepted that such differences are not inconsequential since fat and lean tissue make unique mass-specific contributions to overall EE. Scaling to body weight assumes all tissues have the same rate of metabolism, while scaling to lean mass assumes that adipose tissue is metabolically inert. The respective assumptions are demonstrably incorrect, but the error introduced by scaling to body weight is far more significant and increases in proportion to the difference in mass of the groups being compared.[32-36] It is clear that scaling to body weight after long-term MR is particularly inappropriate.[22] Although not ideal, scaling EE to lean body mass introduces far less bias into group comparisons. Application of this approach in rats after 3 months of MR showed that nighttime EE (kJ/h/kg lean body mass) was 70% higher in the MR group than controls while daytime EE was only 25% higher.[21] After 9 months on the diet, both night- and daytime EE were ∼90% higher in the MR group, whereas the difference decreased to 25% after 20 months on the diet.[21] When MR was introduced after physical maturity, the diet increased EE by 21% relative to controls and produced a matching 20% higher energy intake.[21] The comparable increases in energy intake and EE are consistent with the stable body weight and composition that was observed in the MR group over the 6-month period of study.[21]

Measurements of EE in mice after various intervals of dietary MR present the same challenges of ratio-based scaling of EE with groups differing in size and composition. For example, at the end of a 10-week study, FVB mice on the control diet are 19% heavier and have 80% more fat in their carcasses than mice in the MR group.[28] However, despite these differences, lean body mass is comparable between mice on the two diets. EE scaled to lean mass was 31% higher in the MR group than controls by the end of the study.[28] Viewed collectively, the observations provide compelling evidence that dietary MR limits growth and fat deposition by decreasing metabolic efficiency through a mechanism that increases EE per unit lean mass.

An alternative to ratio-based normalization of EE is provided by analysis of covariance (ANCOVA), which uses least squares analysis to assess the impact of variation in fixed (e.g., genotype) and continuous (body composition, intake, and activity) variables in relation to variation in EE between animals. ANCOVA was initially recommended for analysis of rodent indirect calorimetry data in 2006,[33] and a series of recent papers have used the approach to assess the relative contributions of fat-free mass, fat mass, activity, and energy intake to total EE.[34,35,37] Application of ANCOVA to analyze the effects of dietary MR on EE, incorporating body composition, activity, and energy intake into the model, showed that 10 weeks of dietary MR increased EE in C57BL/6J mice by ~35%.[29] Although it may be somewhat surprising that ANCOVA and ratio-based scaling of EE provided similar assessments of the effects of MR on EE, two additional points should be considered. First, application of ANCOVA to indirect calorimetry data in both humans and rodents has shown that lean body mass accounts for ~70–75% of the variation in total daily EE.[38,39] Second, variation in voluntary activity accounts for ~20% of the variation in EE in humans[39,40] and presumably rodents. Therefore, if dietary MR produced changes in voluntary activity, accounting for this source of variation within ANCOVA would diminish the variation in EE attributable to MR. However, analyses of the responses of mice[28,29] and rats[21] to MR provide no evidence that the diet alters voluntary activity. This indicates that dietary MR is affecting energy balance through changes in activity-independent components of EE.

Lastly, several recent publications have addressed additional design considerations when applying indirect calorimetry to test for treatment differences in EE.[34,37,41] One recommendation is to test for group differences in EE in animals before the emergence of treatment-induced changes in size and/or body composition. In recent studies in the authors' laboratory, we have taken this approach and evaluated the acute effects of dietary MR on EE after first obtaining baseline measures of EE in mice on the control diet before introduction of dietary MR. We have found that on day 6 after initiation of MR, nighttime EE (kJ/h/mouse) is 20% higher than controls and the increase in EE occurs in conjunction with an increase in energy intake. Then, by day 10, both night- and daytime EEs are ~30% higher in mice on the MR diet compared to controls (authors' unpublished data). The observed increases in EE occurred well before diet-induced changes in body weight or composition, further supporting the conclusion that the MR-induced increases in EE are not an analytic artifact of ratio-based scaling of indirect calorimetry data. Lastly, continuous monitoring of voluntary activity before and after initiation of MR shows that activity did not differ between the dietary groups at any time during the study. Collectively, analysis of EE using indirect calorimetry after both short- and long-term dietary MR provides convincing support for the conclusion that the diet produces a rapidly developing and long-lasting increase in EE that uncouples peripheral fuel oxidation and increases the energy costs of maintenance and growth. A significant ongoing challenge is to determine the molecular mechanism(s) of the uncoupling and identify the tissue sites where it is occurring.

6. EFFECTS OF DIETARY MR ON FUEL SELECTION AND METABOLIC FLEXIBILITY

The respiratory exchange ratio (RER) provides a real-time index of substrate utilization during the phases of the metabolic cycle. RER is calculated from the molar ratios of O_2 consumed and CO_2 produced during the oxidation of glucose (1.00), lipid (0.70), and protein (0.80).[42–44] RERs typically approach 1 during the switch to glucose utilization in the fed state and towards 0.7 during the switch to fat utilization during fasting. Metabolic flexibility measures how effectively substrate switching occurs during the transitions between fasted and fed states. The concept arose from the recognition by Kelley and Mandarino[45] that the daily shifts from fat to carbohydrate utilization were impaired by insulin resistance (IR). The impairment is recognized as a diminution in the normal increase in RER that occurs upon refeeding and has been attributed to compromised insulin–dependent glucose uptake in peripheral tissues.[46,47] The increase in glucose utilization associated with improved insulin sensitivity is reflected by a concomitant increase of RER in the fed state. Studies in rats show that dietary MR expands the dynamic range of RER excursions during the transitions between fed and fasted states,[21] supporting the conclusion that metabolic flexibility is significantly enhanced by dietary MR.

An increase in metabolic flexibility is indicative of an increase in overall insulin sensitivity and is often accompanied by a reduction in both fasting and postprandial insulin levels. Malloy et al.[22] reported that 80 weeks of dietary MR reduced basal insulin levels in rats by sevenfold compared to controls. This difference occurs because insulin does not increase with age in the MR group as it does in controls, where insulin levels increase from ~1 ng/ml after weaning to over 7 ng/ml after 80 weeks.[22] The authors also evaluated the ability of rats to clear a glucose challenge after 23 and 72 weeks on the respective diets and found modest improvements in glucose clearance in the MR group only after 72 weeks.[22] However, fasting insulin was lower in the MR group at both 23 and 72 weeks, and the increase in insulin required to clear the glucose challenge was far less than the increase in the control group at both time points.[22]

In mice, the reduction in basal insulin produced by dietary MR can be detected within 7 days of introduction of the diet, and after 8 weeks, the reduction is ~fourfold.[29] As observed in rats,[21] the dynamic range of day to night excursions in RER is also enhanced in mice after 8 weeks of dietary

MR.[28] Collectively, these observations make a compelling case that overall insulin sensitivity is enhanced by dietary MR, but provide no insight into how the diet has changed insulin sensitivity in individual tissues. In recent studies conducted in collaboration with the Mouse Metabolic Phenotyping Center at Vanderbilt University, hyperinsulinemic–euglycemic clamps conducted after 8 weeks of dietary MR provided evidence that the diet significantly enhanced insulin-dependent glucose uptake in all peripheral tissues. The glucose infusion rate required to maintain euglycemia during the clamp was threefold higher in the MR group compared to controls (authors' unpublished results). This increase in glucose utilization is consistent with the three- to fourfold decrease in basal insulin observed after 8 weeks of dietary MR. More importantly, the clamp data show that MR enhanced both the ability of insulin to suppress hepatic glucose production and insulin-dependent uptake of 2-deoxyglucose in muscle and adipose tissue (authors' unpublished data). An important ongoing challenge is to determine the specific mechanism(s) engaged by MR that produce this uniform improvement in insulin responsiveness among all tissues.

One attractive hypothesis is that endocrine changes produced by the diet work through a common mechanism in all tissues to enhance insulin sensitivity. For example, both the adipocyte hormone (adiponectin) and the liver hormone (fibroblast growth factor 21, FGF-21) improve insulin sensitivity and both are significantly increased by dietary MR.[21,22,24,29,48] The actions of both hormones are complex and include functions additional to regulation of insulin sensitivity. The role of each hormone in mediating the physiological responses to dietary MR is currently unknown but is being evaluated using loss-of-function approaches with adiponectin- and FGF-21-null mice. Clearly, the ability of dietary MR to limit expansion of adipose tissue also plays a contributing role and it seems likely that the endocrine changes produced by MR are not entirely secondary to reduced adiposity. In contrast, emerging evidence supports a role for FGF-21 in remodeling of adipose tissue, increases in adiponectin, regulation of EE, and enhancement of insulin action.[49–53] Therefore, the hypothesis that dietary MR produces several of its metabolic effects through modulation of circulating FGF-21 is intriguing and will no doubt be evaluated in the near future.

Dietary MR has been studied in preclinical settings that involve introduction of the diet soon after weaning and after physical maturity, but the translational potential of dietary MR will most certainly involve evaluations of its efficacy in an adult context and likely with obese subjects that manifest markers of metabolic disease. The evidence from preclinical experiments in rats involving introduction of dietary MR after physical maturity shows that the diet is fully effective in reducing basal insulin.[21] The efficacy of dietary MR has also been evaluated in obesity-prone Osborne–Mendel rats and C57BL/6J mice, both of which are sensitive to the obesogenic effects of high-fat diets. In both reports, methionine was limited in 60 kcal% high-fat diets and provided for ~4 months after weaning.[21,48] Dietary MR reduced basal insulin by ~threefold in each case, and in mice, the authors reported that glucose tolerance and the ability of insulin to lower blood glucose were both enhanced by the diet.[48] Thus, in both models of obesity, dietary MR is able to ameliorate the development of obesity and the associated development of IR.

The important remaining question is whether dietary MR can ameliorate and/or reverse existing obesity and IR. This is the most relevant translational context for dietary MR. Male C57BL/6J mice are one of the best studied preclinical models of diet-induced metabolic disease, developing IR and dysregulated lipid metabolism when fed a diet high in saturated fat.[54] The progression and severity of the IR depends on the amount of dietary fat and how long the diet is consumed. For example, very high-fat diet formulations (e.g., 55–60 kcal% fat) produce rapid deterioration in whole-body insulin sensitivity that becomes progressively worse and involves both hepatic and peripheral IR.[55–58] In a recent collaborative study with the Mouse Metabolic Phenotyping Center at Vanderbilt University, we tested the efficacy of short-term (e.g., 8 weeks) dietary MR to reverse IR established in C57BL/6J mice after chronic consumption (e.g., 16 weeks) of a very high-fat diet (e.g., 58 kcal%). Hyperinsulinemic–euglycemic clamps conducted after remediating the mice for 8 subsequent weeks of MR revealed that the MR diet improved overall insulin sensitivity by 50% and reduced body weight by ~18% (authors' unpublished data). The improvement in overall insulin sensitivity was primarily the result of increased suppression of hepatic glucose production by insulin. Additional experiments with this model will be required to assess the ability of MR to enhance insulin sensitivity in other tissues. It will also be interesting to determine if longer-term consumption of dietary MR will fully reverse diet-induced obesity and IR. The collective view is that dietary MR is highly effective at enhancing insulin sensitivity in multiple biological contexts but more importantly is able to reverse preexisting IR associated with diet-induced obesity.

7. EFFECTS OF DIETARY MR ON HEPATIC LIPID METABOLISM: TRANSCRIPTIONAL MECHANISMS

Dysregulation of lipid metabolism and ectopic lipid accumulation in peripheral tissues are hallmarks of metabolic syndrome, while reductions in circulating and tissue lipid levels are biomarkers of improved metabolic health. The effects of dietary MR on lipid metabolism were initially described by Malloy et al.[22] who reported that MR attenuated the age-associated increase in circulating cholesterol and triglycerides in rats. The authors examined the responses in rats that had consumed the MR diet for different times after weaning. They found that serum triglycerides did not increase between week 16 and week 81 in rats on the MR diet, while in rats on the control diet, levels increased by ~70% over the same period. Serum cholesterol also increased by 65% in the controls between 16 and 105 weeks, while the increase was limited to 16% in the MR group.[22] In studies of shorter duration, MR produced a threefold reduction in serum triglycerides that was evident 1 month after introduction of the diet.[59] These findings indicate that dietary MR reduces serum lipids soon after its introduction, but they also show that it prevents the subsequent age-dependent increase that normally occurs in conjunction with increasing adiposity.[22,59]

Changes in serum lipids are often indicative of changes in hepatic lipogenesis and/or release, coupled with changes in the capacity of adipose tissue to release and/or store triglyceride. In addition to reducing serum lipids, dietary MR reduces hepatic lipids by two- to threefold in both mice[29] and rats,[60] suggesting the liver as an important target of MR and potentially responsible for the reduction in serum lipids. Excess fat accumulation in the liver typically results from some combination of increased delivery, increased synthesis, and decreased oxidation or decreased export.[61] Therefore, the dysfunctional steps that lead to increased liver fat provide a useful framework for evaluating the mechanisms through which dietary MR produces the opposite effect and reduces both plasma and hepatic triglycerides.

Studies to date have primarily addressed the transcriptional effects of MR on genes involved in lipogenesis or fatty acid oxidation[60,62] and have been coupled with ex vivo and in vivo assays to evaluate the biological significance of observed changes of genes within particular pathways. For example, the rate-limiting enzymes for de novo lipogenesis (acetyl-CoA carboxylase 1, ACC-1; fatty acid synthase, FASN) and triglyceride synthesis (stearoyl-CoA desaturase 1, SCD-1) were identified as transcriptional targets of

MR in the liver.[62] Thus, the observed decreases in FASN and ACC-1 mRNAs were predicted to reduce the capacity for hepatic *de novo* lipogenesis, while the reduction in SCD-1 mRNA should reduce formation of the preferred monounsaturated fatty acids (C16:1 and C18:1) used in the first committed step of TG synthesis. Therefore, the observed reductions in hepatic triglyceride by dietary MR are consistent with the predicted reductions in hepatic lipogenic capacity, but they do not establish a cause and effect relationship or that *in vivo* rates of hepatic lipogenesis are decreased by MR. However, recent studies in our laboratory assessed the physiological significance of the transcriptional changes in hepatic lipogenic genes[60] by measuring ^{2}H-enrichment in palmitate 12 h after injecting control and MR mice with ^{2}H$_2$O (authors' unpublished data). As expected, the calculated rates of *de novo* lipogenesis in livers from MR mice were three- to fourfold lower than control mice. The effects of dietary MR on hepatic fatty acid oxidation have also been measured in rats where, after 9 months on the diet, the capacity to oxidize ^{14}C-palmitate was increased ∼40% compared to controls.[60] Lastly, the effects of MR on hepatic export of triglyceride have not been measured, but based on the significant reduction in serum lipids, it seems likely that the combination of increased oxidative capacity and reduced lipogenic function are the most dominant effects of the diet on hepatic lipid metabolism.

More comprehensive analyses of the transcriptional responses to dietary MR have used high-content microarray-based approaches to identify biological processes affected. For example, Perrone *et al.*[24] used gene set enrichment analysis of their microarray data to identify biological process that were differentially affected in peripheral tissues by 3 months of MR. A discussion of the extensive transcriptional effects of MR is beyond the scope of this chapter, but if attention is restricted to hepatic lipid metabolism, it is somewhat surprising that their gene set enrichment algorithm failed to identify lipid or fatty acid synthesis as pathways that were downregulated by MR. In contrast, both hepatic fatty acid oxidation and cholesterol/isoprenoid metabolism were identified as downregulated pathways.[24] These conclusions are somewhat at odds with their previous data showing significant downregulation of hepatic lipogenic genes, upregulation of oxidative genes, and increased hepatic citrate synthase activity.[59,62] To assess the functional significance of observed changes in gene expression, Perrone *et al.*[24] coupled unbiased metabolomics analysis of tissues and serum with their transcriptional analysis to assess the impact of dietary MR on the metabolome. This comprehensive analysis should provide a valuable ongoing resource for probing how transcriptional changes of genes in specific tissues affect the levels of associated metabolites in the serum.

Another recent high-content evaluation of the transcriptional responses to chronic MR identified lipid metabolism as among the top biological processes affected by the diet in liver and adipose tissue.[60] Recent developments in bioinformatics software not only include better annotation of the systems biology of transcriptional changes but also provide new algorithms that seek to identify transcriptional mechanisms from the observed changes in gene expression. This analytic strategy seeks to identify transcription factors (TFs) and nuclear receptors mediating dietary responses based on expected causal effects of TFs/nuclear receptors on known target genes relative to observed changes of those genes within the dataset. By examining direction of change in target gene expression, the algorithm predicts whether specific TFs/nuclear receptors are activated or inhibited. In the liver, the coordinated downregulation of over 30 genes involved in lipid metabolism led to the prediction that SREBF1, SFEBF2, and MLXIPL were inhibited by MR.[60] Moreover, the fact that the majority of repressed genes receive transcriptional input from two or more of these TFs further corroborates the interconnectedness of the regulatory network. The diet-induced decrease in expression of hepatic SREBP-1c is consistent with the repressive effect of MR on lipogenic gene expression. SREBP-1c and ChRE-BP are also subject to regulation by cholesterol and glucose, and both are decreased by dietary MR.[23] However, it remains unclear whether decreases in cholesterol and glucose are the cause or product of transcriptional responses to MR. In either case, hepatic SREBP-1c appears to be a key target of the mechanism through which MR reduces hepatic *de novo* lipogenesis, triglyceride synthesis, and lipid content.[60] An important remaining challenge is to identify additional TFs/nuclear receptors recruited by dietary MR and explore the signaling mechanisms involved in their recruitment as mediators of the transcriptional responses to the diet.

8. TRANSCRIPTIONAL EFFECTS OF DIETARY MR ON ADIPOSE TISSUE

Dietary MR produces significant changes in adipose tissue mass, cell morphology, mitochondrial content, and endocrine function.[21,23,28,29,60,62] An evaluation of the transcriptional responses of peripheral tissues to MR (e.g., liver, IWAT, skeletal muscle, and brown adipose tissue (BAT)) shows that >75% of all differentially expressed genes were found in the liver and

IWAT.[60] Lipid metabolism was the top molecular and cellular process affected by MR in both tissues, but interestingly, the diet produced opposite effects on genes associated with lipid synthesis in the two tissues.[60] For example, the downregulation of lipogenic genes in the liver is mirrored by a reciprocal upregulation of key lipogenic genes in WAT depots.[60] QRT-PCR and Western blots of the rate-limiting enzymes for *de novo* lipogenesis (FASN and ACC-1) and triglyceride synthesis (SCD-1) show that MR increased their mRNA and protein expression in all WAT depots.[60] These findings indicate that dietary MR has transformed WAT into a potentially important site of lipid synthesis to compensate for the loss of lipogenic capacity in the liver.[60] This conclusion is supported by two lines of evidence. First, RERs consistently exceed 1 at night in the MR group, indicative of high rates of glucose utilization coupled to interconversion to fat by *de novo* lipogenesis.[63,64] Given that lipogenic capacity of the liver is severely compromised by MR, it seems likely that WAT is the site where glucose interconversion to lipid is occurring. The second line of evidence comes from measuring ^{2}H-enrichment in palmitate in mice acutely injected with $2H_2O$. As predicted, the calculated rates of *de novo* lipogenesis in retroperitoneal WAT, IWAT, epididymal WAT, and BAT from MR mice were three- to fourfold higher than control mice (authors' unpublished observations), establishing that both WAT and BAT are important sites of *de novo* lipogenesis after MR. Considered together, these findings indicate that dietary MR has produced a fundamental change in the respective functions of liver and adipose tissue, particularly in the fed state where the roles of adipose tissue and the liver in *de novo* lipogenesis have been reversed.

The transcriptional responses to MR in WAT also included genes involved in lipid oxidation, tricarboxylic acid (TCA) cycle, respiratory chain function, and adaptive thermogenesis.[21,24,28,60] To evaluate the impact of these changes, the *ex vivo* capacity of freshly isolated tissues to oxidize fatty acids was measured using ^{14}C-palmitate, coupled with measures of citrate synthase activity as a surrogate of TCA cycle flux. MR produced a fivefold increase in both palmitate oxidation and citrate synthase activity in IWAT, along with a doubling of mitochondrial numbers.[60] The transcriptional remodeling of WAT is similar to the "browning" of WAT that occurs in WAT during cold exposure and involves many of the same changes seen with MR, including reduction in cell size, formation of multilocular adipocytes, and increased uncoupling protein 1 (UCP1) expression.[21,28] It is well established that cold exposure, acting via norepinephrine, elicits a simultaneous increase in glucose uptake, lipogenesis, and β-oxidation in BAT.[65,66] Moreover, after chronic cold exposure, the increased number of brown adipocytes in WAT enhances glucose uptake and lipogenic function within these depots. Recent work has emphasized the importance of BAT to triglyceride clearance while documenting the regulatory role of sympathetic nervous system (SNS) input in the process.[67] Thus, the simultaneous increase in lipogenic and oxidative gene expression in WAT may be reflective of extensive remodeling of WAT depots produced by dietary MR through effects on SNS activity. This conclusion is supported by our previous work showing that dietary MR increased UCP1 mRNA 3- to 10-fold among WAT depots,[21,28] and recent observations that MR doubled mitochondrial density among WAT depots and increased oxidative capacity of IWAT by fivefold.[60] Many of these responses characteristic of the "browning" response in WAT are also produced or enhanced by FGF-21.[53] Given the significant upregulation of hepatic FGF-21 expression and release by MR,[24] it will be interesting to assess the relative roles of the SNS and FGF-21 as mediators of the remodeling of WAT and increase in EE produced by the diet.

Dietary MR increases EE by increasing uncoupled respiration in peripheral tissues, but a full accounting of sites involved and tissue-specific mechanisms is incomplete. The evidence is compelling that induction of UCP1 in BAT and WAT by MR increases uncoupled respiration in both tissues. The increase accounts for ~50% of the overall diet-induced increase in EE,[21,28] leaving ~50% of the MR-induced increase in total EE being mediated through UCP1-independent mechanisms. Recent studies with cold-adapted UCP1-null mice show that high rates of uncoupled respiration do not require the presence of UCP1.[68] While activation of thermogenic respiration during cold exposure increases β-oxidation of fatty acids, it also produces a coordinated increase in glucose utilization coupled with a paradoxical increase in de novo lipogenesis.[66,69,70] An interesting feature of the interconversion of glucose to lipid prior to oxidation is that it produces more heat and captures less of the potential energy normally obtained from direct oxidation of glucose.[66,70] Thus, by increasing heat loss and decreasing net ATP generation from glucose during interconversion to lipid, de novo lipogenesis represents a metabolically inefficient substrate cycle capable of making a significant UCP1-independent contribution to nonshivering thermogenesis, particularly during periods of high glucose utilization.[70] The interconversion of glucose to lipid prior to oxidation produces an RER of 1.0 when rates of de novo lipogenesis and lipid oxidation are equal.[43,44] Thus, the impact of this inefficient conversion of glucose to lipid is minimal when rates of fatty acid oxidation are low (e.g., fed state at ambient temperature). However, rates of fatty acid oxidation increase more than 12-fold in UCP1-null mice during cold exposure,[68] and nighttime RQs reached or exceeded 1.0, guaranteeing that rates of glucose conversion to lipid were proportionately increased to match or exceed the increase in lipid oxidation. Stated another way, the increased flux of glucose through this metabolically inefficient pathway is capable of significantly increasing heat production and EE through a mechanism that does not require but can be enhanced by UCP1. In our studies with MR, the diet produces a consistent enhancement of *de novo* lipogenesis that is temporally matched with nighttime increases in EE. We hypothesize that enhanced substrate cycling of glucose through this mechanism at night is an important component of the uncoupled respiration and metabolic inefficiency produced by dietary MR. However, EE in animals on the MR diet is also higher than controls during the day when fat becomes the primary metabolic fuel and substrate cycling through a glucogenic/lipogenic mechanism is precluded. And since rodents primarily sleep during the day, group differences in resting EE suggest that dietary MR is also impacting overall EE by reducing the metabolic efficiency of fat oxidation. It is unclear whether this uncoupling of fat oxidation is restricted to specific tissues or whether MR is inducing metabolic inefficiency in multiple tissues. These questions are being explored using loss-of-function approaches with UCP1-null mice and with mice lacking β-adrenergic receptors. The respective mouse lines will allow us to determine the proportions of the increased EE that are dependent on UCP1 and the extent to which increased SNS activation is required for transcriptional remodeling of adipose tissue. Given the role of the SNS in regulating endocrine function of adipose tissue, it also seems likely that MR may be using the SNS as a motor arm to produce components of its energy balance phenotype through reduction of leptin expression and release.[21,22,28] For example, the reductions in plasma leptin produced by MR are disproportionate to the reductions in adipose tissue mass produced by the diet,[21,22,28] producing a strong orexigenic signal that is perhaps responsible in part for the hyperphagic response to dietary MR. An attractive model to explore this hypothesis is the *ob/ob* mouse, which lacks the ability to express functional leptin because of a mutation within the gene.[71] However, the *ob* mutation results in secondary changes in β-adrenergic signaling in adipose tissue that compromise the ability of both BAT and WAT to fully respond to SNS input.[72,73] We hypothesize that the MR-dependent increase in SNS input to adipose tissue is an important component of the mechanism through which the diet functions to affect energy balance.

Therefore, it will be interesting to determine experimentally the extent to which the metabolic derangements of the *ob/ob* mouse compromise its ability to respond to dietary MR.

When considered in a physiological context, the transcriptional and morphological remodeling of WAT by dietary MR produces fundamental changes in the way WAT functions in both fed and fasted states. In the fed state, WAT becomes an important site of glucose uptake and utilization for interconversion to lipid, while in the fasted state, the increase in oxidative capacity expands its involvement in fatty acid oxidation. The net effect is to limit overall expansion of adipose tissue mass and ectopic lipid accumulation, but the changes also modify the respective roles of adipose tissue and the liver, particularly in terms of lipogenesis. When viewed collectively, the tissue-specific transcriptional responses to dietary MR make a compelling case that the diet has effectively remodeled the integration of lipid metabolism between the liver and adipose tissue in a manner that is beneficial to the overall metabolic profile of the animal.

9. SENSING OF DIETARY MR

An important unanswered question is how restriction of dietary methionine is detected and how sensing of the restriction is translated into highly integrated transcriptional responses in the liver and WAT. Restricting availability of EAAs effectively limits charging of tRNA with its cognate amino acid and activates the highly conserved and ubiquitously expressed protein kinase, GCN2 (general control nonderepressible 2), which limits ribosomal translation of most mRNAs.[74–76] Transcriptional effects of EAA deprivation on lipogenic genes were initially identified in human HepG2 cells, where media lacking single EAAs decreases transcriptional initiation and expression of FASN.[77] These studies suggest the interesting possibility that MR functions through GCN2 to decrease expression of lipogenic genes in the liver. However, MR increased lipogenic gene expression in WAT and muscle, arguing against a role for GCN2 and suggesting involvement of additional sensing systems in these tissues. Alternatively, WAT and muscle may be responding to endocrine signals originated in different tissues or to neural signals resulting from central sensing of reducing circulating methionine. These questions are being explored in GCN2-null mice at present and findings to date suggest that many of the responses to dietary MR are intact in the absence of GCN2. Although preliminary, these findings provide compelling evidence that sensing and signaling systems in addition to GCN2 are involved in detecting and mediating the physiological, biochemical, and transcriptional responses to dietary MR.

10. PERSPECTIVES AND FUTURE DIRECTIONS

In our work on dietary MR to date, we have identified a range of dietary methionine concentrations that produces profound improvements in biomarkers of metabolic health. The important next steps are to refine our understanding of the degree of MR linked to each component and extend this work to dietary restriction of other EAAs. A significant body of work has been devoted to the study of dietary leucine deprivation. We recently published a detailed accounting of the similarities and differences between the responses produced by the two diets.[29] One critical difference is that dietary MR produces hyperphagia while dietary leucine deprivation causes significant food aversion and a rapidly developing negative energy balance that cannot be sustained beyond a few weeks. It should be noted that restriction of dietary methionine to levels much lower than the 0.17% methionine provided in the MR diet formulation produces food aversion and the same detrimental effects as leucine deprivation. Therefore, a comparison of the physiological responses to dietary MR and leucine deprivation suggests that the EAAs are in a sense functioning as ligands with responses determined by the degree of the restriction. It follows from this that an important future objective will be to examine systematically the physiological responses to incremental restriction of methionine, leucine, and other EAAs. In addition to identifying a range of restrictions that are most beneficial, this approach could also provide important mechanistic insights regarding EAA-specific sensing mechanisms. For example, in preliminary studies to date, we have found that dietary leucine restriction reproduces some but not all of the physiological responses produced by dietary MR. These models promise to provide a fruitful approach to dissect and identify the unique mechanisms engaged by dietary restriction of methionine to improve metabolic health.

Another experimental strategy being taken to better understand the components of the complex responses to dietary MR is to study their spatial and temporal organization. For example, it is difficult to distinguish the direct effects of MR in a particular tissue from responses that are modulated by detection of the restriction in another anatomical site that may then provide secondary signaling or endocrine input to the initial site. Thus, in addition to its spatial organization, the individual component of the response to

MR is also temporally organized, developing in a reproducible progression after introduction of the diet. In a recent review,[29] we proposed four potential sites of sensing dietary EAA restriction: (1) direct sensing of luminal or absorbed EAAs in the gut, (2) sensing of EAAs in the portal circulation and/or liver, (3) direct sensing of reduced EAAs by tissues, and (4) sensing of EAAs in various regions of the brain. There are enormous gaps in our understanding of how these and as yet unknown sensing components function together to mediate the integrated physiological responses to changes in dietary EAA content. It will be particularly important to identify the central amino acid-sensing systems and map how they are organized to provide integrated regulation of the components of energy balance and communication to peripheral tissues. A better understanding of how dietary MR enhances tissue-specific and overall insulin sensitivity is also a central focus. The overall metabolic phenotype produced by EAA deprivation versus restriction is the product of a series of responses that are anatomically and temporally organized and, in many cases, interdependent. Therefore, a significant ongoing challenge within the field will be to develop experimental approaches that distinguish between the direct, tissue-specific responses to MR and the responses perceived in one anatomical site and modulated in another.

Lastly, the translational potential of the concepts developed in preclinical studies of dietary MR was recently evaluated in a human cohort meeting the criteria for metabolic syndrome.[78] Dietary MR was accomplished using the semisynthetic medical food, Hominex-2® (Abbott Nutrition, Columbus, OH) in a short-term study (16 weeks) to evaluate the metabolic consequences of limiting dietary methionine from 35 mg/kg BW/d to 2 mg/kg BW/d. The experimental diet (Hominex-2®) is a commercial food designed to provide nutritional support for patients with pyridoxine-unresponsive homocystinuria or hypermethioninemia. It is comprised in part of elemental amino acids, and their associated low palatability resulted in high withdrawal rates and raised questions about compliance and achieving the desired degree of MR. Another poststudy concern stems from the fact that although Hominex-2® is methionine-deficient, it contains methionine-sparing cystine.[79] This could be significant because the rodent MR diet lacks cystine and a recent study reported that adding it back to the rodent MR formulation reversed many of the beneficial metabolic effects of MR.[59] Thus, the cystine in Hominex-2® may have limited the full efficacy of the MR achieved with this approach. Notwithstanding these experimental limitations, we found that dietary MR increased fat oxidation and reduced hepatic lipid content in subjects with metabolic disease.[78] Development of methods to produce highly palatable, methionine-depleted proteins represents a better approach because it will solve both the cystine and palatability problems. This represents an area of intense interest in our laboratory and is likely to provide the best strategy for testing the translational potential of dietary MR in the clinic.

ACKNOWLEDGMENTS

Grants or fellowships supporting the writing of the chapter: This work was supported in part by ADA 1-12-BS-58 (T. W. G) and NIH RO1DK-096311 (T. W. G.). This work also made use of the Genomics and Cell Biology & Bioimaging core facilities supported by NIH P20-GM103528 (T. W. G.) and NIH P30 DK072476. D. W. is supported by NIH Institutional Training Grant, 5T32DK064584-10 and ADA Mentor-Based Postdoctoral Fellowship 7-13-MI-05.

REFERENCES

1. Anthony TG, McDaniel BJ, Byerley RL, et al. Preservation of liver protein synthesis during dietary leucine deprivation occurs at the expense of skeletal muscle mass in mice deleted for eIF2 kinase GCN2. *J Biol Chem.* 2004;279:36553–36561.
2. Guo F, Cavener DR. The GCN2 eIF2alpha kinase regulates fatty-acid homeostasis in the liver during deprivation of an essential amino acid. *Cell Metab.* 2007;5:103–114.
3. Xiao F, Huang Z, Li H, et al. Leucine deprivation increases hepatic insulin sensitivity via GCN2/mTOR/S6K1 and AMPK pathways. *Diabetes.* 2011;60:746–756.
4. Xia T, Cheng Y, Zhang Q, et al. S6K1 in the central nervous system regulates energy expenditure via MC4R/CRH pathways in response to deprivation of an essential amino acid. *Diabetes.* 2012;61:2461–2471.
5. Cheng Y, Meng Q, Wang C, et al. Leucine deprivation decreases fat mass by stimulation of lipolysis in white adipose tissue and upregulation of uncoupling protein 1 (UCP1) in brown adipose tissue. *Diabetes.* 2010;59:17–25.
6. Cheng Y, Zhang Q, Meng Q, et al. Leucine deprivation stimulates fat loss via increasing CRH expression in the hypothalamus and activating sympathetic nervous system. *Mol Endocrinol.* 2011;25:1624–1635.
7. Orentreich N, Matias JR, DeFelice A, Zimmerman JA. Low methionine ingestion by rats extends life span. *J Nutr.* 1993;123:269–274.
8. Richie Jr JP, Leutzinger Y, Parthasarathy S, Malloy V, Orentreich N, Zimmerman JA. Methionine restriction increases blood glutathione and longevity in F344 rats. *FASEB J.* 1994;8:1302–1307.
9. Zimmerman JA, Malloy V, Krajcik R, Orentreich N. Nutritional control of aging. *Exp Gerontol.* 2003;38:47–52.
10. Miller RA, Buehner G, Chang Y, Harper JM, Sigler R, Smith-Wheelock M. Methionine-deficient diet extends mouse lifespan, slows immune and lens aging, alters glucose, T4, IGF-I and insulin levels, and increases hepatocyte MIF levels and stress resistance. *Aging Cell.* 2005;4:119–125.
11. Sun L, Sadighi Akha AA, Miller RA, Harper JM. Life-span extension in mice by pre-weaning food restriction and by methionine restriction in middle age. *J Gerontol A Biol Sci Med Sci.* 2009;64:711–722.
12. Grandison RC, Piper MD, Partridge L. Amino-acid imbalance explains extension of lifespan by dietary restriction in Drosophila. *Nature.* 2009;462:1061–1064.

13. Mair W, Piper MD, Partridge L. Calories do not explain extension of life span by dietary restriction in Drosophila. *PLoS Biol.* 2005;3:e223.

14. Koc A, Gasch AP, Rutherford JC, Kim HY, Gladyshev VN. Methionine sulfoxide reductase regulation of yeast lifespan reveals reactive oxygen species-dependent and -independent components of aging. *Proc Natl Acad Sci U S A.* 2004;101:7999–8004.

15. Ruan H, Tang XD, Chen ML, et al. High-quality life extension by the enzyme peptide methionine sulfoxide reductase. *Proc Natl Acad Sci U S A.* 2002;99:2748–2753.

16. Moskovitz J, Bar-Noy S, Williams WM, Requena J, Berlett BS, Stadtman ER. Methionine sulfoxide reductase (MsrA) is a regulator of antioxidant defense and lifespan in mammals. *Proc Natl Acad Sci U S A.* 2001;98:12920–12925.

17. Deyl Z, Juricova M, Stuchlikova E. The effect of nutritional regimes upon collagen concentration and survival of rats. *Adv Exp Med Biol.* 1975;53:359–369.

18. McCay CM, Crowell MF, Maynard LA. The effect of retarded growth upon the length of life span and upon the ultimate body size. 1935. *Nutrition.* 1989;5:155–171.

19. Kemnitz JW. Calorie restriction and aging in nonhuman primates. *ILAR J.* 2011;52:66–77.

20. Barzilai N, Gabriely I. The role of fat depletion in the biological benefits of caloric restriction. *J Nutr.* 2001;131:903S–906S.

21. Hasek BE, Stewart LK, Henagan TM, et al. Dietary methionine restriction enhances metabolic flexibility and increases uncoupled respiration in both fed and fasted states. *Am J Physiol Regul Integr Comp Physiol.* 2010;299:R728–R739.

22. Malloy VL, Krajcik RA, Bailey SJ, Hristopoulos G, Plummer JD, Orentreich N. Methionine restriction decreases visceral fat mass and preserves insulin action in aging male Fischer 344 rats independent of energy restriction. *Aging Cell.* 2006;5:305–314.

23. Perrone CE, Mattocks DA, Hristopoulos G, Plummer JD, Krajcik RA, Orentreich N. Methionine restriction effects on 11β-HSD1 activity and lipogenic/lipolytic balance in F344 rat adipose tissue. *J Lipid Res.* 2008;49:12–23.

24. Perrone CE, Mattocks DA, Plummer JD, et al. Genomic and metabolic responses to methionine-restricted and methionine-restricted, cysteine-supplemented diets in Fischer 344 rat inguinal adipose tissue, liver and quadriceps muscle. *J Nutrigenet Nutrigenomics.* 2012;5:132–157.

25. Park S, Park NY, Valacchi G, Lim Y. Calorie restriction with a high-fat diet effectively attenuated inflammatory response and oxidative stress-related markers in obese tissues of the high diet fed rats. *Mediators Inflamm.* 2012;2012:984643.

26. Omodei D, Fontana L. Calorie restriction and prevention of age-associated chronic disease. *FEBS Lett.* 2011;585:1537–1542.

27. Wanders D, Ghosh S, Stone K, Van NT, Gettys TW. Transcriptional impact of dietary methionine restriction on systemic inflammation: relevance to biomarkers of metabolic disease during aging. *Biofactors.* 2013. http://dx.doi.org/10.1002/biof.1111.

28. Plaisance EP, Henagan TM, Echlin H, et al. Role of β-adrenergic receptors in the hyperphagic and hypermetabolic responses to dietary methionine restriction. *Am J Physiol Regul Integr Comp Physiol.* 2010;299:R740–R750.

29. Anthony TG, Morrison CD, Gettys TW. Remodeling of lipid metabolism by dietary restriction of essential amino acids. *Diabetes.* 2013;62:2635–2644.

30. Kleiber M. *The Fire of Life. An Introduction to Animal Energetics.* New York, NY: John Wiley & Sons, Inc.; 1961

31. Brody S. *Bioenergetics and Growth.* New York, NY: Reinhold; 1945.

32. Butler AA, Kozak LP. A recurring problem with the analysis of energy expenditure in genetic models expressing lean and obese phenotypes. *Diabetes.* 2010;59:323–329.

33. Arch JR, Hislop D, Wang SJ, Speakman JR. Some mathematical and technical issues in the measurement and interpretation of open-circuit indirect calorimetry in small animals. *Int J Obes (Lond).* 2006;30:1322–1331.

34. Tschop MH, Speakman JR, Arch JR, et al. A guide to analysis of mouse energy metabolism. *Nat Methods*. 2011;9:57–63.
35. Kaiyala KJ, Schwartz MW. Toward a more complete (and less controversial) understanding of energy expenditure and its role in obesity pathogenesis. *Diabetes*. 2011;60:17–23.
36. Kaiyala KJ, Morton GJ, Leroux BG, Ogimoto K, Wisse B, Schwartz MW. Identification of body fat mass as a major determinant of metabolic rate in mice. *Diabetes*. 2010;59:1657–1666.
37. Even PC, Nadkarni NA. Indirect calorimetry in laboratory mice and rats: principles, practical considerations, interpretation and perspectives. *Am J Physiol Regul Integr Comp Physiol*. 2012;303:R459–R476.
38. Bogardus C, Lillioja S, Ravussin E, et al. Familial dependence of the resting metabolic rate. *N Engl J Med*. 1986;315:96–100.
39. Ravussin E, Lillioja S, Anderson TE, Christin L, Bogardus C. Determinants of 24-hour energy expenditure in man. Methods and results using a respiratory chamber. *J Clin Invest*. 1986;78:1568–1578.
40. Redman LM, Heilbronn LK, Martin CK, et al. Metabolic and behavioral compensations in response to caloric restriction: implications for the maintenance of weight loss. *PLoS One*. 2009;4:e4377.
41. Munzberg H, Henagan TM, Gettys TW. Animal models of obesity: challenges to translating technological advances to mechanistic insights. In: Bray GA, Bouchard C, eds. *Handbook of Obesity*. 3rd ed. London: Informa Books, Inc.; 2013
42. Ferrannini E. The theoretical bases of indirect calorimetry: a review. *Metabolism*. 1988;37:287–301.
43. Elia M, Livesey G. Theory and validity of indirect calorimetry during net lipid synthesis. *Am J Clin Nutr*. 1988;47:591–607.
44. Simonson DC, DeFronzo RA. Indirect calorimetry: methodological and interpretative problems. *Am J Physiol*. 1990;258:E399–E412.
45. Kelley DE, Mandarino LJ. Fuel selection in human skeletal muscle in insulin resistance: a reexamination. *Diabetes*. 2000;49:677–683.
46. Galgani JE, Heilbronn LK, Azuma K, et al. Metabolic flexibility in response to glucose is not impaired in people with type 2 diabetes after controlling for glucose disposal rate. *Diabetes*. 2008;57:841–845.
47. Galgani JE, Moro C, Ravussin E. Metabolic flexibility and insulin resistance. *Am J Physiol Endocrinol Metab*. 2008;295:E1009–E1017.
48. Ables GP, Perrone CE, Orentreich D, Orentreich N. Methionine-restricted C57BL/6J mice are resistant to diet-induced obesity and insulin resistance but have low bone density. *PLoS One*. 2012;7:e51357.
49. Holland WL, Adams AC, Brozinick JT, et al. An FGF21-adiponectin-ceramide axis controls energy expenditure and insulin action in mice. *Cell Metab*. 2013;17:790–797.
50. Lin Z, Tian H, Lam KS, et al. Adiponectin mediates the metabolic effects of FGF21 on glucose homeostasis and insulin sensitivity in mice. *Cell Metab*. 2013;17:779–789.
51. Veniant MM, Hale C, Helmering J, et al. FGF21 promotes metabolic homeostasis via white adipose and leptin in mice. *PLoS One*. 2012;7:e40164.
52. Berglund ED, Li CY, Bina HA, et al. Fibroblast growth factor 21 controls glycemia via regulation of hepatic glucose flux and insulin sensitivity. *Endocrinology*. 2009;150:4084–4093.
53. Fisher FM, Kleiner S, Douris N, et al. FGF21 regulates PGC-1alpha and browning of white adipose tissues in adaptive thermogenesis. *Genes Dev*. 2012;26:271–281.
54. Surwit RS, Kuhn CM, Cochrane C, McCubbin JA, Feinglos MN. Diet-induced type II diabetes in C57BL/6J mice. *Diabetes*. 1988;37:1163–1167.

55. Park SY, Cho YR, Kim HJ, et al. Unraveling the temporal pattern of diet-induced insulin resistance in individual organs and cardiac dysfunction in C57BL/6 mice. *Diabetes*. 2005;54:3530–3540.

56. Storlien LH, James DE, Burleigh KM, Chisholm DJ, Kraegen EW. Fat feeding causes widespread in vivo insulin resistance, decreased energy expenditure, and obesity in rats. *Am J Physiol*. 1986;251:E576–E583.

57. Neschen S, Morino K, Hammond LE, et al. Prevention of hepatic steatosis and hepatic insulin resistance in mitochondrial acyl-CoA:glycerol-sn-3-phosphate acyltransferase 1 knockout mice. *Cell Metab*. 2005;2:55–65.

58. Samuel VT, Liu ZX, Qu X, et al. Mechanism of hepatic insulin resistance in nonalcoholic fatty liver disease. *J Biol Chem*. 2004;279:32345–32353.

59. Elshorbagy AK, Valdivia-Garcia M, Mattocks DA, et al. Cysteine supplementation reverses methionine restriction effects on rat adiposity: significance of stearoyl-coenzyme A desaturase. *J Lipid Res*. 2011;52:104–112.

60. Hasek BE, Boudreau A, Shin J, et al. Remodeling the integration of lipid metabolism between liver and adipose tissue by dietary methionine restriction in rats. *Diabetes*. 2013;62:3362–3372.

61. Postic C, Girard J. Contribution of de novo fatty acid synthesis to hepatic steatosis and insulin resistance: lessons from genetically engineered mice. *J Clin Invest*. 2008;118:829–838.

62. Perrone CE, Mattocks DA, Jarvis-Morar M, Plummer JD, Orentreich N. Methionine restriction effects on mitochondrial biogenesis and aerobic capacity in white adipose tissue, liver, and skeletal muscle of F344 rats. *Metabolism*. 2009;59:1000–1011.

63. Jequier E, Felber JP. Indirect calorimetry. *Baillieres Clin Endocrinol Metab*. 1987;1:911–935.

64. Livesey G, Elia M. Estimation of energy expenditure, net carbohydrate utilization, and net fat oxidation and synthesis by indirect calorimetry: evaluation of errors with special reference to the detailed composition of fuels. *Am J Clin Nutr*. 1988;47:608–628.

65. Inokuma K, Ogura-Okamatsu Y, Toda C, Kimura K, Yamashita H, Saito M. Uncoupling protein 1 is necessary for norepinephrine-induced glucose utilization in brown adipose tissue. *Diabetes*. 2005;54:1385–1391.

66. Yu XX, Lewin DA, Forrest W, Adams SH. Cold elicits the simultaneous induction of fatty acid synthesis and beta-oxidation in murine brown adipose tissue: prediction from differential gene expression and confirmation in vivo. *FASEB J*. 2002;16:155–168.

67. Bartelt A, Bruns OT, Reimer R, et al. Brown adipose tissue activity controls triglyceride clearance. *Nat Med*. 2011;17:200–205.

68. Ukropec J, Anunciado RP, Ravussin Y, Hulver MW, Kozak LP. UCP1-independent thermogenesis in white adipose tissue of cold-acclimated Ucp1 −/− mice. *J Biol Chem*. 2006;281:31894–31908.

69. Trayhurn P. Fuel selection in brown adipose tissue. *Proc Nutr Soc*. 1995;54:39–47.

70. Masoro EJ. Role of lipogenesis in nonshivering thermogenesis. *Fed Proc*. 1963;22:868–873.

71. Zhang Y, Proenca R, Maffei M, Barone M, Leopold L, Friedman JM. Positional cloning of the mouse obese gene and its human homologue. *Nature*. 1994;372:425–432.

72. Gettys TW, Ramkumar V, Uhing RJ, Seger L, Taylor IL. Alterations in mRNA levels, expression, and function of GTP-binding regulatory proteins in adipocytes from obese mice (C57BL/6J-ob/ob). *J Biol Chem*. 1991;266:15949–15955.

73. Collins S, Daniel KW, Rohlfs EM, Ramkumar V, Taylor IL, Gettys TW. Impaired expression and functional activity of the β-3 and β-1 adrenergic receptor in adipose tissue of congenitally obese (C57BLJ6-ob/ob) mice. *Mol Endocrinol*. 1994;8:518–527.

74. Deval C, Chaveroux C, Maurin AC, et al. Amino acid limitation regulates the expression of genes involved in several specific biological processes through GCN2-dependent and GCN2-independent pathways. *FEBS J.* 2009;276:707–718.

75. Palii SS, Kays CE, Deval C, Bruhat A, Fafournoux P, Kilberg MS. Specificity of amino acid regulated gene expression: analysis of genes subjected to either complete or single amino acid deprivation. *Amino Acids.* 2009;37:79–88.

76. Shan J, Ord D, Ord T, Kilberg MS. Elevated ATF4 expression, in the absence of other signals, is sufficient for transcriptional induction via CCAAT enhancer-binding protein-activating transcription factor response elements. *J Biol Chem.* 2009;284:21241–21248.

77. Dudek SM, Semenkovich CF. Essential amino acids regulate fatty acid synthase expression through an uncharged transfer RNA-dependent mechanism. *J Biol Chem.* 1995;270:29323–29329.

78. Plaisance EP, Greenway FL, Boudreau A, et al. Dietary methionine restriction increases fat oxidation in obese adults with metabolic syndrome. *J Clin Endocrinol Metab.* 2011;96: E836–E840.

79. Brosnan JT, Brosnan ME. The sulfur-containing amino acids: an overview. *J Nutr.* 2006;136:1636S–1640S.

CHAPTER TWELVE

Carbohydrate Metabolism and Pathogenesis of Diabetes Mellitus in Dogs and Cats

Margarethe Hoenig
College of Veterinary Medicine, University of Illinois, Urbana, Illinois, USA

Contents

Abstract

Diabetes mellitus (DM) is a common disease in dogs and cats and its prevalence is increasing in both species, probably due to an increase in obesity, although only in cats has obesity been clearly identified as a major risk factor for diabetes. While the classification of diabetes in dogs and cats has been modeled after that of humans, many aspects are different. Autoimmune destruction of beta cells, a feature of type 1 DM in people, is common in dogs; however, in contrast to what is seen in people, the disease occurs in older dogs. Diabetes also occurs in older cats but islet pathology in those species is characterized by the presence of amyloid, the hallmark of type 2 DM. Despite being overweight or obese, most naive diabetic cats, contrary to type 2 diabetic humans, present with low insulin concentrations. The physiology of carbohydrate metabolism and pathogenesis of diabetes, including histopathologic findings, in dogs and cats are discussed in this chapter.

1. INTRODUCTION

Diabetes mellitus (DM) is a common endocrine disease in dogs and cats and occurring primarily in older animals. It presents clinically in both

Progress in Molecular Biology and Translational Science, Volume 121
ISSN 1877-1173
http://dx.doi.org/10.1016/B978-0-12-800101-1.00012-0
377

species with many similarities; however, there are also striking differences in the physiology and pathophysiology of carbohydrate metabolism and species-specific differences in the pathogenesis and presentation of the disease.

2. THE DOG

2.1. Physiology of carbohydrate metabolism

2.1.1 Tasting, digestion, and absorption of glucose in dogs

Dogs are carnivores; however, they have adapted to diets also including vegetables and grains. They, like cats, do not contain salivary amylase but amylase, originating from the exocrine pancreas, is found in large amounts in chyme, and they are able to digest starch well.[1] Dogs, like humans, are attracted to sweets and have functional sweet receptors. This is different from cats, which lack functional sweet receptors and are not attracted to compounds that taste sweet. Sweet receptors are heterodimers of G protein-coupled receptors. Their structure in dogs is similar to that of humans.[2] Dogs are attracted to most sugars, including sucrose, glucose, fructose, and lactose but not maltose, although a preference for fructose and sucrose was not seen by all investigators.[3] Glucose absorption from the intestinal tract is similar to that seen in humans, that is, ~70% of labeled glucose in a meal is detected in the blood stream. After feeding a mixed meal, ~3× more glucose were taken up from the systemic circulation by nonsplanchnic tissues (presumably the skeletal muscle) than by splanchnic tissues.[4,5] The intestinal sodium-dependent glucose transporter-1 (SGLT1) amino acid sequence was found to be most closely related to that of cats with 92–97% similarity and 84–93% identity; however, dogs expressed significantly more SGLT1 than cats and the properties of intestinal glucose transport were also different between the two species. The V_{max} was about twofold higher in the dog compared to the cat. In addition, most disaccharidases were shown to have about a two- to threefold higher activity in the dog than cat.[6]

2.1.2 Cellular uptake and metabolism of glucose in dogs

Intestinal glucose transport appears to be as rapid in dogs as that in people or cats. When dogs were given an oral glucose solution (1 g/kg in 50 ml of water), glucose concentrations increased by ~2.5 mmol/L,[7] which is similar to the response of people to approximately the same amount of glucose.[8] Maximal glucose concentrations were reached at 30 min and declined to baseline values at 90 min. Glucose transporters of various tissues in healthy or naïve diabetic dogs have not been compared, to the authors knowledge. Glucose, once inside the cells, is trapped through phosphorylation by several hexokinases. Glucokinase is a high K_m hexokinase of the brain, liver, pancreas (beta cells), kidney, and intestine, with narrow substrate specificity. It has been purified in dogs and was found to have similar kinetic properties as seen in other species, including humans.[9] The kinetic features of glucokinase allow for rapid phosphorylation of high amounts of glucose, and glucokinase is not inhibited by its product, glucose-6-phosphate. This is different to what is seen with other hexokinases, which are low K_m enzymes with a wider substrate specificity. Glucokinase has been named the glucosensor of beta and neuronal cells because it is the rate-limiting enzyme for insulin and neurotransmitter release.[10,11] In the liver, glucokinase activity is required for glycogen synthesis.[12] Like in other mammals, excluding the cat, the dog's glucokinase activity is adaptive, that is, it changes with feeding or starvation.[13] Insulin is required for maximal activity of glucokinase.[14] This has also been shown in a recent study where diabetes in dogs was cured with the injection of adeno-associated viral vectors of serotype 1 encoding for glucokinase and insulin, whereas gene transfer for insulin or glucokinase alone failed to achieve complete correction of the diabetes.[15]

2.2. Pathogenesis of DM in dogs

2.2.1 Prevalence, breed susceptibility, and risk factors

DM is one of the most common endocrine diseases in dogs. The prevalence of canine DM in dogs presented to veterinary teaching hospitals in the United States increased from < 0.02% to 0.064% between 1970 and 1999.[16] An Italian study saw an even higher prevalence (1.33%),[17] whereas the number of reported diabetes cases was less when insurance cohorts were examined in the United Kingdom (0.032%).[18] Investigators of a Swedish insurance cohort concluded that the estimated cumulative proportion of dogs that would develop DM before the age of 12 years in their country was 1.2%.[19] It is difficult to compare studies from European countries with those of the United States because the majority of dogs in the United States are neutered, and neutering is less common in dogs in European countries, exposing them to the greater influence of sex hormones, known to influence the development of diabetes.

There is also some disparity when comparing breed predilection for DM between European countries and the United States. At a major US veterinary teaching hospital, Samoyeds, miniature schnauzers, miniature poodles, pugs, and toy poodles were found to be at high risk for developing DM. Dog breeds found to be at low risk for developing DM were German shepherd, golden retriever, and American pit bull terrier.[20] In an earlier study from the United States, poodles were also found to be at increased risk and German shepherds, cocker spaniels, boxers, and collies were found at decreased risk. Intact females and neutered females were at higher risk compared to intact males. But in males, neutering also increased the risk for diabetes.[21] In a study examining canine diabetes trends in the United States over a 30-year time span, Samoyed, Australian terrier, miniature and standard schnauzer, and fox terrier were at highest risk for diabetes.[16] In a large Swedish study examining 180,000 insured dogs, Australian terriers, Samoyeds, Swedish elkhounds, and Swedish lapphunds were found to have the highest incidence. It was also interesting that 72% of those diabetic dogs were female. Swedish elkhounds, beagles, Norwegian elkhounds, and border collies with diabetes were almost exclusively female. Breed, female sex, and previous hyperadrenocorticism were found to be the biggest risk factors for the development of DM in that study.[19] Interestingly, the US study found the Norwegian elkhound to be a breed at low risk to develop diabetes,[16] possibly indicating some geographic influence on the susceptibility for the disease. Samoyeds seem to be universally seen as the breed at high risk for diabetes and topped the UK Canine Diabetes Register and Archive list.[18] The breed susceptibility is listed in Table 12.1.

Table 12.1 Dogs at high and low risk to develop diabetes mellitus (alphabetical order)[16,20,21]

High risk	Low risk
Australian terrier	American pit bull terrier
Fox terrier	Boxer
Miniature poodles	Cocker spaniel
Miniature schnauzers	Collie
Pugs	German shepherd
Samoyeds	Golden retriever
Standard schnauzer	
Swedish elkhound	
Swedish lapphund	
Toy poodles	

2.2.2 Genetic and environmental influences

It is thought that the majority of dogs have a form of diabetes similar to type 1 DM in people, that is, involving a genetic component, which is influenced by environmental input. The possibility of a genetic basis for canine diabetes has been suggested by the occurrence of the disease in dog families. Breed prevalence of diabetes also indicates that genetic factors might play a role in the susceptibility of dogs to develop diabetes and suggests that differences in genetic risk factors are breed-specific. A small number of research groups have examined the genetic basis of canine diabetes and their findings have been summarized in a recent review article.[22] In one study of 460 diabetic dogs and 1047 controls, the DLA haplotypes DLA-DRB1*009-DQA1*001-DQB1*008 (DRB1*009), DLA-DRB1*015-DQA1*006-DQB1*.23 (DRB1*015), and DLA-DRB1*002-DQA1*009-DQB1*001 (DRB1*002) seemed to confer the risk of diabetes, whereas the DLA-DQA1*004-DQB1*013 haplotype seemed to confer protection against DM and was significantly reduced in frequency in diabetic dogs. Clearly, diabetic Samoyeds frequently expressed either DLA-DRB1*009/DRB1*015 (15/42 dogs) or just DLA-DRB1*-015 (16/42 dogs). The latter was also common in diabetic cairn and Tibetan terriers. In general, the frequency of the DRB1*009/DQA1*001 haplotype was higher in those breeds with higher diabetes risk and that of DQA1*004/DQB1*13 was lower. Conversely, the DRB1*009/DQA1*001 haplotype was absent in golden retrievers, boxers, and German shepherds, which have a low predisposition for diabetes. In other breeds, which are at moderate risk to develop DM, those high–risk DLA haplotypes were uncommon. It has been cautioned, however, by Catchpole and coworkers that MHC gene associations are not specific for diabetes,[23] because similar DLA genotypes have been found in association with susceptibility to hypothyroidism[24] and hypoadrenocorticism.[25]

Apart from the MHC region, type 1 DM in people is also characterized by polymorphisms of the region flanking the 5′-end of the human insulin gene on chromosome 11 and several investigators have shown linkage of the insulin variable number tandem repeat (VNTR)/insulin gene with type 1 DM in humans.[26] The insulin VNTR is not found in the canine insulin promoter but was found at the 5′-end of intron 2; however, it does not appear to play a significant role in the susceptibility of diabetes in high–risk breeds.[22] In a study of 483 diabetic dogs and 869 controls,[27] alleles of a number of individual SNPs were identified, which showed clear association with canine diabetes, but more studies are needed in larger cohorts to confirm the association between a marker and disease susceptibility.

It has been known for almost four decades that type 1 DM in people is an immune-mediated disease that leads to the selective destruction of beta cells.[28] B lymphocytes and CD4 and CD8 T lymphocytes are an integral part of this pathological process, which is called insulitis. Today, several beta cell proteins are known to be target antigens.[29,30] In a recent landmark study evaluating dogs with recent onset type 1 DM (mean age 11.7 years), it was found that CD8[+] cytotoxic T cells were the most abundant population during insulitis, observed during the early stage of beta cell destruction. Macrophages (CD68[+]) were also present during both early and later insulitis, and CD20[+] cells were recruited to islets as beta cell death progressed. Once beta cell destruction was complete and the islets became insulin-deficient, the number of all immune cells declined rapidly, and the authors concluded that immune cells are only present when viable beta cells are present.[31] Because glucagon staining was normal in those islets, they suggested that immune cell recruitment might start within the beta cell, and once these cells are lost, the stimulus declines. Antibodies originating from the autoimmune destruction of beta cell proteins can be applied in the diagnosis and in the prediction of this disease in people. At the time of clinical diagnosis, ~80–90% of type 1 diabetic patients are positive for antibodies to B cell antigens.[32]

Because of the similarity between human and canine type 1 DM, investigators have examined dogs for the presence of beta cell antibodies. In the first such study,[33] purified beta cells from a radiation-induced transplantable rat insulinoma were used to detect beta cell antibodies in serum from untreated diabetic dogs. Serum from dogs in which anti-beta cell antibodies were induced by injecting a purified beta cell suspension subcutaneously was used as positive control. Approximately 50% of the diabetic dogs showed a strongly positive reaction. There was no correlation between sex and the occurrence of antibodies in the diabetic dogs. There was also no correlation to the age of the dogs. The identity of the antigen was not investigated in that study. Later, 30 dogs were found to be positive for antibodies against the 65 kDa isoform of canine glutamic acid decarboxylase and two of them also had antibodies against the insulinoma-associated antigen-2, both frequently found in human type 1 diabetics.[34] In a recent study, 5/40 of newly diagnosed dogs had antibodies against insulin.[35] Despite the similarity in pathogenesis, islet pathological changes differ from those of humans (see succeeding text).

Environmental factors, for example, viruses, dietary factors, or toxins, have been implicated in the pathogenesis of type 1 DM in people as both triggers and potentiators of beta cell destruction.[36] While one can assume that similar factors might also play a role in the etiopathogenesis of canine diabetes, this has not been investigated.

2.2.3 Classification of canine DM

The human diabetes classification is used in dogs; however, there are some features that are different from what is seen in people and others that have not been studied in detail, making a comparison difficult. A classification into insulin-dependent and insulin-independent diabetes has been proposed[37]; however, there are no reports that glucose control in diabetic dogs can be achieved with oral antihyperglycemic agents, and all dogs require insulin to survive. Glucose-induced insulin responses in diabetic dogs have been examined in a small number of studies and the information has been used to assign the dogs to different categories. Insulin measurements in dogs were first reported in the 1970s, when several investigators started utilizing radioimmunoassay.[38] The most extensive study of glucose clearance and insulin-secretory responses is likely the one by Mattheeuws and coworkers from 1984,[39] who already had shown earlier[40] that diabetic dogs belong to one of three types based on their response to intravenous glucose tolerance testing (IVGTT) (Table 12.1). They compared 71 diabetic dogs with 20 control dogs, defining diabetes as fasting blood glucose concentrations above 6.7 mmol/L (121 mg/dL) (Table 12.2). Type 1 diabetic dogs had severe clinical signs of diabetes and insulin could not be detected in the basal state or in response to the IVGTT. All 32 dogs were glucosuric and half also ketonuric. The average age of those dogs was 8 years and the range was large (3–14 years). All type 1 dogs had lost weight and none of them was obese at the time of presentation. Type 2 diabetics differed from type 1 dogs only regarding their basal insulin concentrations, which were similar to those seen

Table 12.2 Different types of diabetes mellitus in dogs[39]

Type of DM	Weight	Fasting glucose	Fasting insulin	Glucose intolerance	Insulin response to glucose
Type 1	Weight loss	High	ND	Yes	Not detectable
Type 2	Nonobese	High	Normal	Yes	Not detectable
Type 2	Obese	High	High	Yes	Not detectable
Type 3	Nonobese	Mild increase	Normal	Yes	Decreased
Type 3	Obese	Mild increase	High	Yes	High

in the healthy control dogs in nonobese type 2 dogs, but sixfold higher in the obese type 2 dogs. However, no insulin response was seen when IVGTT was performed, regardless of body condition. Glucosuria was seen in all 15 dogs and ketonuria in half of the dogs, similar to the finding in type 1 dogs. The average age was higher than in type 1 dogs (9 years), and the lowest age was 7 years. Finally, type 3 dogs were mildly hyperglycemic (8 mmol/L) and glucose-intolerant during the IVGTT but did not show clinical signs associated with diabetes. They were neither glucosuric nor ketonuric. The basal insulin values of those dogs were not statistically different from control dogs if their body condition was normal. However, during an IVGTT, their insulin response was only about 50% of that seen in control dogs. If the dogs were obese, their basal and stimulated insulin concentrations were higher than those of control dogs in type 3 group. All abnormal values of the latter group normalized when they lost weight. These dogs were younger than the other two groups with an average age of ~6 years, ranging from 3 to 10 years. Most of the dogs in these early studies were intact and it is quite certain that not only obesity but also reproductive hormones influenced the response pattern, especially in type 2 and 3 groups.

From these studies, it can be concluded that the majority of diabetic dogs cannot respond to an intravenous glucose challenge with insulin secretion, indicating severe beta cell failure. This is also supported by other studies showing that untreated and treated diabetic dogs cannot respond to an intravenous challenge with glucagon, and their insulin and C-peptide concentrations remained low, despite an increase in blood glucose.[37,41]

In a recent publication by Catchpole and coworkers,[23] the following classification of diabetes in dogs was proposed:

a. *Insulin deficiency diabetes* includes beta cell loss due to autoimmune processes, beta cell hypoplasia or abiotrophy, exocrine pancreatic disease (such as pancreatitis), or idiopathic causes.

b. *Insulin-resistant diabetes* results from the insulin antagonistic effect of either endogenous or exogenous hormones, which oppose the effect of insulin and/or decrease insulin secretion such as sex steroids, glucocorticoids, and growth hormone, among others. The authors point out that dogs in the insulin-resistant group can progress to having insulin deficiency diabetes.

This classification also has limits because insulin-secretion profiles in dogs with concomitant endocrinopathies are largely lacking. However, many times, dogs with diabetes and concomitant diseases leading to insulin resistance are ketonuric upon first presentation, which would suggest that these dogs are also insulin-deficient.

It has been suggested that canine diabetes is similar to the latent autoimmune diabetes of adult (LADA)[42]; however, there is little to support such notion. Although some, but not all, LADA patients develop diabetes at a later age and diabetes is also seen mostly in older dogs, LADA patients also present initially with clinical diabetes, which is not insulin-requiring.[43,44] Clinical diabetes in dogs, however, is always insulin-requiring. In addition, insulitis has been found in LADA patients, which is not seen frequently in diabetic dogs.

In people, the oral glucose tolerance test (OGTT) is the mainstay of glucose tolerance testing. It appears that the variations in blood glucose and insulin concentrations in the dog after OGTT are too large to allow for a meaningful interpretation of results.[7,45] In one study, using a 4 g/kg glucose oral stimulus, individual glucose and insulin profiles showed extreme variance and no apparent pattern.[45] This might have been due to large variations in gastric emptying time, possibly due to intestinal hypertonicity evoked by the high glucose load.

2.2.4 Diseases leading to insulin resistance and beta cell dysfunction

The relationship between insulin sensitivity and insulin secretion has been described as hyperbolic,[46] that is, individuals with lower insulin sensitivity compensate with higher insulin secretion to maintain normal glucose tolerance. It is not known which events finally lead to the point where insulin becomes inadequate to overcome the insulin resistance and to maintain glucose control. However, it is known from longitudinal studies in people that insulin action and secretion and endogenous glucose output are all abnormal when diabetes occurs.[47]

2.2.4.1 Endocrine diseases and endocrine drugs

Diabetes in dogs occurs in connection with hypersecretion of insulin-counterregulatory hormones, especially hypersomatotropism and hypercortisolism. The exogenous administration of glucocorticoids and progestins can also lead to diabetes in dogs. Obesity in dogs has in the past not been suggested as a risk factor for diabetes in dogs but epidemiological studies of large cohorts are needed to revisit this issue in view of recent developments.

2.2.4.1.1 Hypersomatotropism Acromegaly, the clinical syndrome caused by growth hormone excess, is most often caused in dogs by endogenous or exogenous progestins and is most prevalent in areas where female dogs are not spayed at an early age. This progestin-induced growth hormone originates from the foci of hyperplastic ductular epithelium in the mammary tissue[48] and is biochemically identical to the growth hormone, which is produced by the pituitary gland.[49] Unlike the pituitary-derived growth hormone, however, the progestin-induced growth hormone is not released in a pulsatile manner, does not respond to stimulation with growth hormone-releasing hormone, and is not inhibited by somatostatin.[50] While an increase in progesterone in intact bitches during diestrus is a normal physiological event, some older dogs develop diabetes because of the anti-insulin action of the growth hormone.[51] This could suggest that beta cell mass is marginal in these dogs and insulin secretion is sufficient only when the dogs are not resistant (i.e., not in diestrus) but that secretion cannot be increased in response to this increased demand during the luteal phase. Some of these dogs revert to normal glucose control, especially if they are spayed early, that is, at a time when beta cells have not been damaged irreversibly by hyperglycemia.

2.2.4.1.2 Hypercortisolism

Hypercortisolism in dogs can be of pituitary or adrenal origin. It can also be caused by the exogenous administration of glucocorticoids. In dogs with a combination of spontaneously occurring hypercortisolism and diabetes, it is often impossible to identify which disease occurred first. Both conditions are seen in middle-aged to older dogs and have similar clinical signs. In a study by Hess and coworkers,[52] 38% of dogs diagnosed within the first months of insulin treatment and 9% diagnosed before diabetes was diagnosed also have hypercortisolism based on results from low-dose dexamethasone suppression testing. The remainder were diagnosed to have hypercortisolism at least 1 month after the diagnosis of diabetes had been made. Because hypercortisolism is a slowly progressive disease and clinical signs might not be detected early in the disease process, these data suggest that high cortisol concentrations likely play a role in the pathogenesis of diabetes. Cortisol's effects on intermediary metabolism antagonize those of insulin, leading to hyperinsulinemia and insulin resistance. It also increases endogenous glucose production through the stimulation of gluconeogenesis, lipolysis, and proteolysis. If present in the long term, this could lead to beta exhaustion and overt beta cell failure. One cannot predict which diabetic dogs will revert to normal glucose control once hypercortisolism has been treated adequately, because it depends on the residual function of the beta cells. In a large study of 60 dogs with untreated hypercortisolism,[53] it was shown that dogs with hypercortisolism indeed have different beta cell function. Eight of the dogs had normal glucose and insulin concentrations, 24 dogs had normal glucose but high insulin concentrations suggestive of compensated insulin resistance, 23 dogs were hyperglycemic and hyperinsulinemic suggestive of uncompensated insulin resistance, and finally, five dogs were ketoacidotic and had low levels of insulin. Those five dogs remained diabetic after hypercortisolism had been corrected, indicating irreversible beta cell damage. In people with hypercortisolism, insulin resistance often persists for years after successful treatment because of central obesity.[54] This relationship has not been investigated in dogs.

2.2.4.1.3 Hyperthyroidism Contrary to cats, where hyperthyroidism is the most common endocrine disease, hypersecretion of the thyroid hormone is only seen in ~10% of dogs with thyroid tumors.[55] There are no reports linking canine hyperthyroidism to DM.

2.2.4.1.4 Hypothyroidism Hypothyroidism has been seen concomitantly with diabetes.[52,56,57] DM was the most common concurrent disease and was seen in 10% of hypothyroid dogs by Dixon and coworkers.[57] In the study by Hess and coworkers,[52] 4% of the diabetic dogs also had hypothyroidism. One of the dogs had lymphocytic thyroiditis, suggesting a polyglandular autoimmune process.[24] It has recently been shown that hypothyroidism leads to marked insulin resistance.[58] In this study, hypothyroidism was induced by intravenous administration of radioiodine in eight dogs, which resulted in low insulin sensitivity, high baseline growth hormone and high insulin-like growth factor-1 (IGF-1) concentrations, and abdominal obesity. Despite the insulin resistance, glucose control was maintained because insulin output increased sufficiently in the basal and stimulated state to overcome the defect. Because of the concomitant increase in weight and obesity, it is difficult to say if the effects were due to hypothyroidism or obesity or both. Hypothyroidism was not found to be a risk factor for diabetes in a large Swedish study.[19]

2.2.4.1.5 Pheochromocytoma Pheochromocytoma is a catecholamine-secreting neoplasm of the adrenal or extra-adrenal chromaffin tissue. Approximately one-third of human patients with pheochromocytoma have diabetes. The diabetes may be related to catecholamine-induced insulin resistance and/or suppression of insulin secretion.[59,60] However, in dogs, it appears that the incidence of pheochromocytoma-associated diabetes is low. Feldman and Nelson examined 98 dogs with pheochromocytoma and found that 3 of them also had diabetes.[61] In another study of 50 dogs, none of them had an increase in blood glucose and clinical signs of diabetes.[60]

2.2.4.1.6 Obesity Obesity is the most common nutritional disorder in dogs and also one of the fastest growing health problems. Environmental factors associated with canine obesity included the socioeconomic status of the owner, owner age, frequency of treats, and amount of exercise the dog receives.[62] From 2007 to 2012, a 37% increase in obesity was reported in the US dog population by Banfield Pet Hospitals, a large chain of veterinary hospitals.[63] However, while obesity is a major risk factor for type 2 DM in people and increases the diabetes risk over two- to fourfold in cats, it has rarely been identified as a risk factor in dogs.[64] More epidemiological studies of a large number of diabetic dogs are clearly needed to look at this issue, especially in view of the fact that Banfield Pet Hospitals also saw an increase of 32% in diabetes in dogs over the same time frame, which strongly suggests obesity as a risk factor in dogs. Obese dogs show changes in glucose and insulin that are correlated with the degree of obesity. Dogs with a degree of obesity less than 40% had insulin and glucose concentrations similar to normal body condition dogs.[65] For dogs with a degree of obesity of 40–70%, hyperinsulinemia was evident but not glucose intolerance. The latter was only seen when the degree of obesity exceeded 70%. It has been stated that, contrary to obese cats and humans, obese dogs are protected from the progression to diabetes because they have higher insulin-secretion rates during the early phase of a glucose tolerance test and increase insulin secretion despite insulin resistance.[66] However, increasing insulin output in response to insulin resistance is an appropriate response to the hyperbolic relationship between resistance and secretion and is seen not only in obese dogs but also in cats.[46,67–69] In obese people, a lowering of first-phase secretion is also not seen when glucose tolerance is normal.[46,70]

One characteristic, which is strikingly different between dogs and cats (and humans), is the absence of islet amyloid in obese and diabetic dogs. Islet amyloid is found in a large percentage of diabetic cats and humans.[71,72] It also has been shown to be increased in obese cats[73] (and is thought to contribute to the pathogenesis of diabetes). The amino acid sequence of islet amyloid polypeptide (IAPP) in dogs is very similar to that of cats and humans and also contains the fibrillogenic sequence Gly-Ala-Ile-Leu-Ser, which is present in human and feline IAPP.[74] However, while hypersecretion of insulin in cats leads to hypersecretion of IAPP and eventually deposition of islet amyloid in and around beta cells and their destruction,[72] there is no report of islet amyloid in obese or diabetic dogs, although hypersecretion of insulin in dogs with insulinomas causes islet amyloidosis.[75] It is not known why obese or diabetic dogs do not develop islet amyloid and seem to be able to process IAPP normally, but it was already recognized over 20 years ago that a species-specific IAPP structural motif alone was not adequate for the conversion of IAPP to amyloid fibrils *in vivo*.[76]

It is unclear what role adipokines play in the pathogenesis of diabetes in dogs. In a comprehensive analysis of cytokines and adipokines in diabetic dogs, it was found that adiponectin, leptin, and resistin concentrations were not different between healthy and diabetic dogs.[77] In type 1 diabetic people, adiponectin levels have been not only shown to increase in most studies[78,79] but also shown to decrease[80] or stay the same.[81] An increase in adiponectin levels has been attributed not only to lower renal clearance and microvascular problems but also to long-term diabetes treatment, where high levels were found irrespective of glucose control.[79]

2.2.4.2 Diseases of the exocrine pancreas

Acute pancreatitis was diagnosed in 28 (13%) diabetic dogs on the basis of clinical signs and appropriate ultrasonographic (27/28) and/or histological findings (5/28).[52] In a postmortem analysis of diabetic dogs,[82] the number of dogs with histological evidence of acute pancreatitis was much higher (28%). This number increased to 33% when chronic pancreatitis cases were added. Severe acute or chronic inflammation of the pancreas can lead to various degrees of islet destruction and transient or permanent diabetes. There is a strong correlation between pancreatitis and hyperlipidemia in miniature schnauzers in the United States, a breed that is also at high risk to develop diabetes.[52]

2.2.5 Pathological changes

Despite evidence that diabetes in the dog is in part due to an autoimmune destruction of beta cells, lymphocytic infiltration of islets is detected infrequently in diabetic dogs. It is possible that lymphocytic infiltration may occur early in the disease progression and is no longer seen by the time animals are presented with clinical diabetes; however, lymphocytic infiltration was also not found in young diabetic dogs.[83,84] These dogs had no islets or scant shrunken islets, suggesting abnormal development. In a study of 18 diabetic dogs,[82] the pancreas was not visible in two dogs and had been replaced by fibrous tissue. In three other dogs, there was extensive interstitial sclerosis associated with widespread loss of acinar cells suggestive of relapsing pancreatitis. Periductal and perivascular fibroses were observed in five dogs, including two dogs that were <6 months old. Active inflammation was only observed in two dogs. In all dogs, a marked reduction of beta cells was observed, whereas the number of alpha and delta cells appeared normal in >70% and 85%, respectively. Only two dogs had a reduced number of islets. This is different from findings by Gepts and Toussant[85] who examined 30 diabetic dogs and found that both islets and beta cell numbers were markedly reduced. Beta cells were often degranulated and showed hydropic changes. In dogs with long-standing disease, islet numbers were very low and beta cells were not detectable anymore.

2.2.6 Summary
Diabetes in the dog is a multifaceted disease. In virtually all diabetic dogs, exogenous insulin is required to maintain glucose control, suggesting absolute insulin deficiency. Many breeds show a genetic predisposition, and environmental factors may play a modulating role, although those factors have not been identified. There is evidence that diabetes in the dog is in part due to an autoimmune destruction of beta cells. While obesity is prevalent in dogs and is increasing at a rapid rate, it is not currently known definitively to be a risk factor for the development of diabetes. However, the fact that an almost parallel increase in canine obesity and diabetes was seen by Banfield Pet Hospitals, a large chain of veterinary hospitals, strongly suggests a correlation. It will be important to obtain more information about the physiological and pathological changes of islets from long-term obese and naive diabetic dogs with a history or physical examination finding of obesity to learn more about the pathogenesis of obesity-related diabetes. It is interesting that obese or diabetic dogs do not develop islet amyloid, which is thought to contribute to the pathogenesis of diabetes in humans and cats. Dogs have the amino acid sequence within IAPP, which is necessary for fibril formation, but islet amyloid has only been seen in canine insulinoma cases.

3. THE CAT
3.1. Physiology of glucose control
3.1.1 Tasting, digestion, and absorption of glucose in cats
The biochemistry and physiology of carbohydrate metabolism in cats have similarity to that in some mammalian species, but not to others. Cats are obligatory carnivores and as such, their natural diet consists primarily of

fat and protein and the amount of carbohydrates is small. Contrary to dogs, cats lack functional sweet receptors.[2,86] They neither are attracted to nor show avoidance of the taste of sweet carbohydrates and high-intensity sweeteners (such as saccharin and cyclamate) and avoid stimuli that taste either bitter or very sour to humans.[1] Sweet taste receptors consist of T1R2 and T1R3 proteins. In cats, T1R3 is an expressed and likely functional receptor, whereas T1R2 is an unexpressed pseudogene.[86]

Cats, like dogs, also lack salivary amylase, which is the enzyme involved in the digestion of starch. However, the impact of this difference on glucoregulation is not known because amylase is also found in feline pancreas and chyme.[87] Compared to dogs, the activities of intestinal disaccharidases (sucrose and maltase) are lower in cats, whereas lactase activity is lower in some parts of the feline small intestine and much higher than in dogs in others.[6] Despite these differences, cats are capable of digesting cooked starch and various carbohydrates with an apparent digestibility of > 94% in one study[4] and 89–100% in another.[88] In the intestine, the sugars glucose and galactose, products of disaccharidase digestion, are absorbed into enterocytes by SGLT1 against an electrochemical gradient, whereas fructose is absorbed by facilitated diffusion by glucose transporter 5 (GLUT5). All are transported from enterocytes into the blood by GLUT2, a high K_m low-affinity glucose transporter with characteristics similar to the GLUT2 in the liver, pancreas, and kidney. Many species can upregulate the capacity of the intestine to absorb glucose in response to high concentrations of dietary carbohydrate.[6,89–92] It was shown in one study[89] that cats are unable to upregulate intestinal sugar absorption. However, the number of animals used was small (two and three per group on a high-protein and a high-carbohydrate diet, respectively) and the cats were only 3 months old. Batchelor and coworkers showed that the sweet receptor subunit T1R2 was also not expressed in the feline intestine, limiting the capacity to upregulate the transport of sugars with increased intake. They suggested that the level of SGLT1 of cats was sufficient for absorbing the carbohydrate content of their natural diet, that is, a low-carbohydrate diet,[6] but that high-carbohydrate diets might be unsuitable for cats because the V_{max} of cats is about 50% that of dogs. However, they also pointed to the possibility that the feline intestine, similar to other species, might have the capacity to absorb glucose via a constitutively active pathway, which is not sodium-dependent. This pathway, however, has not been examined in cats. It is not known which concentration of sugars would actually be present in the small intestine after carbohydrate-containing diets and it would be difficult to predict if V_{max} would be exceeded with commercially available dry or canned diets.

3.1.2 Cellular uptake and metabolism of glucose in cats

Cats do not have glucokinase activity; however, the activity of other hexokinases, which have a higher affinity for glucose than glucokinase, and that of other enzymes within the glycolytic pathway has been shown to be upregulated.[93–95] A lack of glucokinase activity in beta cells would phenotypically be similar to maturity onset diabetes of the young type 2 in people, which is characterized by mild to severe hyperglycemia, depending on the mutation of the GK gene.[96] This, however, is not seen in healthy cats. In fact, cats have fasting glucose concentrations that are not different from those of humans or other mammals, and they are able to respond very rapidly to an intravenous or oral glucose bolus with an insulin-secretion pattern, which is similar to that of other mammals.[97,98] It has been suggested that glucose clearance in cats is lower than that of humans and dogs after an oral or intravenous glucose challenge, in part due to the lack of glucokinase.[99] However, when the response to a low glucose bolus is compared, there does not seem to be any difference between humans, dogs, and cats.[98,100,101] When examining a high dose of glucose (1 g/kg body weight) in dogs and cats, results are not as uniform. In dogs, a return to baseline was seen at ~60 min in some studies,[7,66] whereas other investigators still reported high glucose concentrations with that glucose dose at that time.[102] Glucose concentrations in cats after the 1 g/kg dose have been shown to reach baseline levels at ~90 min,[98,103] which is similar to results from O'Brien and coworkers.[102] A direct comparison of IVGTT results between species is difficult, in part, because of technical aspects in the administration of the test. Alterations are likely not a reflection of decreased glucokinase activity but rather indicate a reduction of glucose uptake by insulin-dependent and insulin-independent means, including elimination through the kidney. Based on the response pattern to a glucose stimulus, it can be concluded that cats compensate well for the lack of beta cell glucokinase activity.

They also seem to compensate well for the lack of liver glucokinase deficiency. Hyperglycemia would also be the clinical sign for animals with isolated hepatic glucokinase deficiency and is due to impaired hepatic glycogen synthesis.[104] However, liver glycogen content in cats is similar to that of humans and dogs.[105,106]

It has been stated that cats are unable to adjust their metabolism and are always gluconeogenic. This is based primarily on the results from an older study by Rogers and coworkers[107] in which three cats fed a low- or high-protein diet lacked the ability to adapt the levels of enzymes regulating amino acid catabolism, gluconeogenesis, and ureagenesis. Kettelhut and coworkers confirmed that the gluconeogenic capacity of cats on a high-protein diet was already high in the fed state and no further increase was seen during fasting. However, they also showed that cats were indeed metabolically flexible because, similar to omnivorous animals, when cats were fasted after feeding a high-carbohydrate diet, they increased gluconeogenesis and reacted, therefore, appropriately to the different macronutrients.[105] A further indication that cats can adjust their metabolic fluxes was the observation that cats on a high-carbohydrate diet have higher glycogen deposits and lower phosphoenolpyruvate kinase activity than those on a high-protein diet.[105] Results from other studies also support the notion that cats can adapt to variations in macronutrients in the diet. In several studies, it was shown that cats adapt to increased protein by increasing amino acid oxidation and the activation of related enzymes[108,109]; other investigators have shown that cats adapt to varying dietary fat concentrations.[110] In our laboratory, we also have documented that cats show metabolic flexibility and increase glucose oxidation, glycogenesis, and lipogenesis when insulin concentrations are high. In lean male cats, the respiratory exchange ratio increased to >1 during a euglycemic hyperinsulinemic clamp indicating that these cats can replete their glycogen and lipid stores in response to insulin.[111] In a recent study from our laboratory,[106] we showed that the magnitude of postprandial gluconeogenesis and glycogenolysis in cats is not different from that seen in people. Six hours after food intake, glycogenolysis in cats contributed about 45% to total glucose production and glycogenolysis about 55%; in people, after intake of a 1000 kcal meal, almost identical values were seen at approximately the same postprandial period,[112] demonstrating that the cat is not the only species in which gluconeogenesis plays an important role, even in the postprandial state.

3.2. Pathogenesis of DM in cats

3.2.1 Prevalence, breed disposition, and risk factors

Diabetes is a relatively common endocrinopathy in cats. In a recent study in the United Kingdom by McCann and coworkers, the prevalence within an insured cat population was 1 in 230.[113] Male cats were significantly more likely to be diabetic than females, with the exception of the Burmese breed where male and female cats were at equal risk. Neutered cats were more likely than intact cats to develop diabetes. There was also a significant increase in the frequency of diabetes with increasing body weight and in cats that had received glucocorticoids. Burmese cats had a significantly greater frequency of occurrence (1 in 57) of diabetes than Persian cats or domestic shorthair cats, not only in the United Kingdom but also in Australia and New Zealand.[114] No breed predilection has been documented in a study of 333 cats in the United States.[115] In that study, over 50% of the cats were >10 years old, and age was identified as the single most important risk factor. Neutered and sexually intact males were at higher risk than neutered or intact females, as were obese cats. It was found that a body weight >6.8 kg contributed a 2.2-fold increase in risk.[115] An even higher risk factor associated with obesity (3.9) was seen in a study by Scarlett and Donaghue.[116] More recently,[117] the prevalence of diabetes was reported to be ~1.2% and significant risk factors included male gender, increasing age for both gender, increasing weight for males but not females, and mixed breed for females. Diet has also been suggested as risk factor for diabetes in cats,[118] especially diets high in carbohydrates, but there is little scientific evidence to support this notion. In a large study of 96 diabetic and 192 matched control cats, no diet effect was found but the investigators saw a significant correlation between DM and low physical activity and indoor confinement.[119] Inactivity was also associated with greater risk for diabetes in the study by McCann and coworkers[113] and the consumption of wet (low-carbohydrate content) and dry (high-carbohydrate content) diets, but not a combination of the two.

3.2.2 Genetic and environmental influences

A genetic basis has been suggested in Burmese cats, which has a much higher incidence of diabetes in the United Kingdom, New Zealand, and Australia than in other countries, but the gene or genes, which might be involved, have not been determined.[120] Genome-wide association studies have revealed a variety of type 2 DM susceptibility genes in people[121] but similar studies have not been performed in cats. Environmental influences in people range from exposure to endocrine-disrupting pollutants to shortened sleep duration to physical inactivity to excess caloric intake.[122] Physical inactivity and excess caloric intake leading to obesity play a role in the risk to develop diabetes in cats as already mentioned; other influences (e.g., exposure to chemicals and changes in light/dark cycles) have not been studied in cats.

3.2.3 Classification of feline diabetes

It is thought that diabetic cats have primarily type 2 DM, based on the fact that most diabetic cats have islet amyloid, which has been called the hallmark of type 2 DM.[123,124] Type 1 DM due to autoimmune destruction does not appear to exist in cats because antibodies to beta cell proteins have yet to be documented.[125] Lymphocytic infiltration of beta cells, which is seen in human type 1 diabetic patients,[126] is extremely rare in cats.[127] The majority of diabetic cats have low or undetectable insulin concentrations at the time of diagnosis. Kirk and coworkers showed that 23 of 30 diabetic cats showed no detectable insulin secretion when stimulated with an intravenous glucagon bolus.[128] Similarly, O'Brien and coworkers found low normal fasting insulin levels in a small series of diabetic cats and no response to a glucose challenge.[102] In a recent study of 21 naive diabetic cats, nine cats had insulin concentrations below the normal reference interval. In two of those cats, the concentrations were undetectable (<2 μU/mL). None of the cats had insulin values above 12 μU/ml (reference interval 4–15 μU/ml).[129] These findings suggest that fasting hyperinsulinemia is not a feature of overt diabetes in cats. This is contrary to what is seen in people during the early phase of type 2 DM, in which fasting insulin concentrations are usually higher than those seen in healthy individuals, although still inappropriately low in relation to the high blood glucose concentration.[130,131] During the course of the disease in people, insulin secretion decreases and beta cell function might eventually be lost completely.[132] The fact that most naive diabetic cats have low or undetectable insulin concentrations suggests that beta cell function deteriorates more rapidly in cats than in people. This is also suggested by recent findings showing that there was a clear separation of fructosamine concentrations between obese and naive diabetic cats (as discussed in the succeeding text). The insulin deficiency, however, does not appear to be permanent in many cats, and the islets seem to be able to recover to a degree that allows them to maintain euglycemia, because it has been shown that cats with transient diabetes responded with insulin secretion when they were no longer diabetic but showed no insulin response to glucagon when diabetic.[133] The remission rate of diabetic cats has been reported to vary between 25 and >80%, supporting the ability of beta cells to recover.[134–136]

3.2.4 Diseases leading to insulin resistance and beta cell dysfunction

Insulin resistance and a progressive decline in beta cell function are characteristics of type 2 DM. It has been reported that cats are very sensitive to the diabetogenic effect of some hormones. Already as early as 1942, Lukens and Dohan documented the development of diabetes in cats after injection with pituitary extract.[137] Many spontaneous diseases have since been documented to cause diabetes. Although this has not been studied in detail, they seem to primarily cause insulin resistance and hyperinsulinemia in the early phase of the disease, which is then followed by beta cell dysfunction and a rise in glucose concentrations, if the primary disease is not eliminated. Administration of growth hormone, progestogens, and glucocorticoids may also lead to diabetes in some cats,[72,138,139] but the effects are variable, perhaps because of variability in the underlying beta cell mass and function.

3.2.4.1 Endocrine diseases and endocrine drugs

3.2.4.1.1 Hypersomatotropism Hypersomatotropism is a well-recognized syndrome in cats and most afflicted cats present with diabetes. In a recent study of diabetic cats in the United Kingdom, one of three diabetic cats was found to have high IGF-1 concentrations strongly suggestive of acromegaly.[140] Contrary to the dog, where hypersomatotropism is usually caused by growth hormone secreted from the mammary tissue,[50] the predominant cause of feline hypersomatotropism is a functional somatotropic adenoma in the pars distalis of the anterior pituitary gland resulting in excessive growth hormone and IGF-1 concentrations and insulin resistance. Indeed, a pituitary tumor was found in 94% of the cats with high IGF-1 levels.[140,141] The early signs of hypersomatotropism may not be recognized initially by the owner and, therefore, the insulin resistance may be present for a long time before the cat becomes overtly diabetic. Once diabetic, blood glucose concentrations are usually difficult to regulate with insulin injections and very high insulin doses may have to be administered to overcome the resistance. Diabetes may resolve in some cats, when treatment is initiated early enough and beta cell dysfunction not yet permanent.

3.2.4.1.2 Hypercortisolism Hypercortisolism in cats is very rare; fewer than 100 cases of spontaneously occurring hypercortisolism have been documented since the first case report in 1975.[142] It can be caused by hyperfunction of the pituitary or adrenal gland. Most of those cases (~80–90%) presented with hyperglycemia. Diabetes can also be caused iatrogenically through the administration of glucocorticoids. We showed that the administration of a combination of dexamethasone and growth hormone leads to deterioration of beta cell function and finally overt diabetes in partially pancreatectomized cats. It was interesting to follow these cats during the progression to the diabetic state. It became obvious that baseline glucose concentrations remained normal for a long time and only increased when insulin output had dropped about fourfold from that seen in normal cats. At that time, the insulin-secretion profile was erratic.[72] In a study where prednisolone was given at an anti-inflammatory dose (2 mg/kg body weight/day) for 8 days or megestrol acetate (at a dose of 5 mg/cat), both drugs caused a significant but similar decrease in glucose clearance and three cats treated with prednisolone and one cat treated with megestrol acetate developed fasting hyperglycemia.[139] Similar negative effects on glucose control by these drugs have been seen by other investigators.[138]

3.2.4.1.3 Hyperthyroidism
Hyperthyroidism is now the most common endocrine disease of cats in the United States and one of the most common diseases in older cats. It is not known how frequently diabetes occurs in hyperthyroid cats, but anecdotally, it is well known that they can occur concomitantly. There are, however, no epidemiological or pathophysiological data linking hyperthyroidism with diabetes, nor is an increase in fasting blood glucose a common feature of the hyperthyroid cat. In a study comparing healthy with hyperthyroid cats, glucose clearance after a high glucose load (1 g/kg) was decreased in hyperthyroid cats, and insulin secretion was increased during an intravenous glucose tolerance test.[143] This pattern is characteristic for peripheral insulin resistance, which causes decreased uptake of glucose into the muscle and fat tissue. However, under normal circumstances, even mild hyperglycemia is rarely seen in hyperthyroid cats. It is known that hyperthyroidism by itself can cause increased hepatic glucose production and a dramatic increase in Krebs cycle flux,[144,145] and one might therefore expect excess glucose production in thyrotoxicosis. However, we have shown in cats that pyruvate cycling flux, a futile cycle, is also stimulated by the thyroid hormone, thereby negating an effect on gluconeogenesis.[146] It is conceivable that in hyperthyroid cats, gluconeogenesis is kept low and fasting blood glucose is kept in the normal range through enhancement of this futile cycle but this has not been examined.

3.2.4.1.4 Obesity
Obesity is a risk factor for diabetes in cats. It has been shown by Banfield Pet Hospitals that, in their practices, cat obesity has increased 90% since 2007 and diabetes 16%.[63] The reason for the rapid increase in obesity is likely a combination of inactivity[113,119] and unlimited access to food, which owners even provide to cats that are already overweight or obese.[129] The overall prevalence of feline obesity is now thought to be about 40–50%,[147,148] although a recent study from the United Kingdom found a much lower prevalence (10%).[149] It is interesting that the prevalence of obesity in people and cats is about equal, yet about 8% of humans developed type 2 DM, whereas the prevalence of diabetes in cats is only about 1% or less. The reasons for the discrepancy are not clear but likely

involve the longer life span of humans, but there could be other factors, such as longer maintenance of hepatic insulin sensitivity in cats than people and also perhaps a difference in islet cell resilience.

Obesity leads to peripheral insulin resistance in cats, which describes the loss of insulin action primarily in the muscle and adipose tissue. Although insulin resistance is a major factor in the pathogenesis of diabetes, most insulin-resistant people do not become diabetic nor do most insulin-resistant cats. It appears that in most obese cats, the beta cells respond adequately to the increased demand, even in the long term, and increase insulin secretion to overcome the peripheral insulin resistance and maintain euglycemia.[67] The loss of insulin sensitivity is, in part, due to a partitioning of fatty acids away from the adipose tissue to the muscle resulting in higher amounts of intra- and extramyocellular fat.[150,151] Obese cats also have higher liver fat than lean cats (6.8% vs. 1.3%, respectively),[152] which in people has been shown to directly measure hepatic insulin sensitivity and predict type 2 DM, independent of other cardiovascular risk factors.[153] It is unlikely that liver fat plays the same role in cats because obese cats maintain hepatic insulin sensitivity.[106,146] Insulin-sensitive glucose transporters 4, GLUT4, are decreased in both the muscle and fat tissue in obese cats, whereas GLUT1, which is not dependent on insulin, is not altered.[154] As already mentioned, despite the insulin resistance in the peripheral tissues, hepatic insulin sensitivity is maintained in obese cats, in both the fasted and postprandial state.[106,146] This leads to lower endogenous glucose production in obese cats in both conditions compared to lean cats. The lower endogenous glucose production is partially accomplished through higher activity of pyruvate cycling, a futile pathway that acts as a controlling mechanism to modulate endogenous glucose production by limiting gluconeogenesis.[106,146]

It has been calculated that every kilogram increase in body weight leads to a 15–30% increase in insulin resistance.[155,156] However, likely because of their ability to regulate endogenous glucose production, most obese cats have normal fasting and postprandial glucose concentrations. This was shown in a study where normal body condition and obese cats were observed for 7 days with a continuous glucose-monitoring system during their daily routine (i.e., feeding and exercise, among others). No differences were detected in daily variations in glucose concentrations between lean and obese cats, confirming that obese cats compensate well for the peripheral insulin resistance (Fig. 12.1).[157] In a recent study, we were able to show that there was a clear distinction between overweight/obese cats and naive diabetic cats regarding blood glucose and fructosamine concentrations.

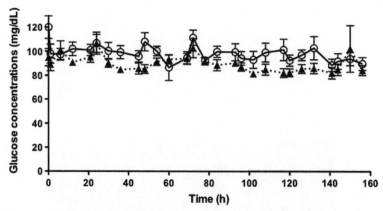

Figure 12.1 Comparison of glucose concentrations of six lean and seven obese cats during a 156-h period.[157] The cats were fed every 24 h starting at time 3 h. Arrows indicate time of feeding. The diet composition was carbohydrate 6.4, fat 4.14, and protein 9.76 g/100 kcal ME. *Reprinted with permission from the Am. Vet. Med. Assoc.*

Fructosamine concentrations in healthy normal body condition cats and obese cats were all in the low normal range (181–282 µmol/L and 164–280 µmol/L, respectively). In naive diabetic cats, they ranged from 341 to 806 µmol/L. This implies that the switch to diabetes occurs quickly in obese cats. If the progression would be slow, a more graded scale of fructosamine concentrations would have been observed between obese and naive diabetic cats.[129] The finding that naive diabetic cats have low insulin secretion[102,129] despite the fact that 2/3 of them were overweight or obese[129] also indicates that the switch to diabetes likely occurs quickly. It involves beta cell failure leading to an increase in hepatic glucose production, in addition to the already present loss of glucose uptake into insulin-sensitive tissues due to peripheral insulin resistance. It is not clear what role IAPP and islet amyloid play in this switch. IAPP secretion has been shown to be increased in obese cats[158] and it has also been shown that islet amyloid is increased (Fig. 12.2).[73] Because early islet amyloid formation takes place intracellularly,[75] one might speculate that in the beta cells of cats and humans, it interferes with intracellular signaling events that are important in the orderly secretion process of insulin. It is thought that increased IAPP production is, in part, involved in the pathogenesis of diabetes because it can lead to misfolding of the IAPP molecule, consequent formation of toxic IAPP oligomers, and eventually the formation of amyloid fibrils, causing a loss of beta cells[158]; however, it is also known that many obese cats and

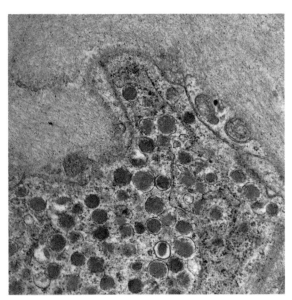

Figure 12.2 Scanning electron micrograph of a pancreatic beta cell and surrounding amyloid from an obese cat.

humans adaptively increase insulin and IAPP secretion and do not develop diabetes.

Inflammation has been suggested as an important pathogenic factor in the development of obesity-related type 2 DM in people.[159] While it has been implied to also be a factor in the pathogenesis in cats,[160] there are currently no data to support this. We have recently shown that inflammatory cytokines (IL-1, IL-6, and TNF-α) and catalase, superoxide dismutase, glutathione peroxidase, and urinary isoprostane concentrations did not change with the development of obesity. Inflammatory cells are also not a feature of adipose tissue in obese cats. This lack of an inflammatory response to obesity in cats might contribute to their lack to develop cardiovascular problems and atherosclerosis in obesity and diabetes, despite alterations in lipid particle size and number usually seen in obese people with cardiovascular problems.[156,161]

The concentrations of adipokines adiponectin and leptin change with changes in body condition and contribute to alterations in metabolism in obese cats. Leptin is positively and adiponectin is negatively correlated with insulin resistance in cats similar to other species.[156,162,163] Leptin acts in the hypothalamus to signal satiety and to increase basal metabolic rate. Adiponectin is the most abundant gene product of the adipose tissue. In both humans and cats, adiponectin mRNA expression and serum levels decrease in obesity and rise with weight loss.[163] Adiponectin is a marker of the activity of the peroxisome proliferator-activated receptor gamma, a nuclear transcription factor that is a key regulator of glucose metabolism, lipid metabolism, and adipogenesis.[164]

Because we have found that most owners, even those of overweight, obese, or diabetic cats, feed cats ad libitum,[129] owner education about controlling caloric intake is the most important aspect of lowering the risk for diabetes in cats because weight loss reverses obesity-induced insulin resistance.[163]

3.2.4.2 Diseases of the exocrine pancreas

It is difficult to diagnose pancreatitis in cats and many times, the diagnosis is made on postmortem examination. Most of the studies have examined diabetic cats that had been diabetic in the long term and data on the number of naive diabetics presented with pancreatitis, either acute or chronic, are lacking. As such, it is impossible to know if pancreatitis might have contributed to the development of diabetes in some cases. In a study evaluating 37 diabetic cats, which underwent necropsy, chronic pancreatitis was diagnosed in 17 and acute to subacute pancreatitis in two cats. Another seven cats had exocrine pancreatic adenocarcinoma and one cat had an adenoma.[165] In a study examining 42 cases that presented with ketoacidosis, pancreatitis was confirmed in seven and a tentative diagnosis was made in nine.[166] In a third study, DM was identified in 5 of 33 cats with chronic pancreatitis but only 1 of 30 cats with acute necrotizing pancreatitis.[167] It can be seen that the results are quite variable and an answer to the question about the role of pancreatitis in the pathogenesis of diabetes or vice versa is not obvious.

3.2.5 Pathology

Islet amyloid is the hallmark of type 2 DM and has been the focus of many studies in cats and humans.[71,123] As mentioned in the text earlier, islet amyloid contains the amyloidogenic peptide IAPP, a hormone cosecreted with insulin. Amyloid forms under conditions that lead to misfolding of the hormone into fibrillar structures. Their precursors can be toxic, and amyloid itself may present a physical barrier, contributing to both beta cell dysfunction and loss of beta cell mass.[168] The contribution of amyloid in beta cell dysfunction has become more controversial in recent years and the focus has shifted to the cytotoxic IAPP oligomers, which are formed when IAPP secretion is high.[71,123] It is clear that IAPP and islet amyloid are important factors in the pathogenesis of diabetes in cats; however, it is also clear that they are not the sole contributors. Approximately 50% of nondiabetic aged cats also have islet amyloid but have normal glucose concentrations.[169,170]

In an early study of five diabetic cats by Gepts and Toussaint,[85] the histology of the pancreas was completely normal in one cat; four of the cats were found to have degranulation and hydropic changes; in two cases, the authors found large hyaline deposits (presumably amyloid) and in one cat, they saw lymphocytic infiltration. In a study of six diabetic cats by Nakayama,[171] amyloid was found in most islets of two cats and in some islets of another cat. Vacuolation was seen in four cats and occurred mostly in insulin-secreting cells, whereas alpha cells remained intact. Granules seen in the vacuolated cells were suggestive of glycogen accumulation, which would indicate the presence of high insulin concentrations. One cat showed signs of chronic pancreatitis and lymphocytic infiltration. O'Brien and coworkers examined 16 cats, of which seven were diabetic and nine had impaired glucose tolerance tests using a high dose of glucose.[102] Amyloid was marked in three diabetic cats and slight and moderate in one and two cats, respectively. There was no correlation between the extent of amyloid and glucose clearance or insulin concentrations. One cat had a pancreatic adenocarcinoma, which led to the complete destruction of the pancreas. Interestingly, all diabetic cats had significantly higher mean serum glucagon concentrations than normoglycemic control cats or cats with impaired glucose tolerance indicating that amyloid deposition only affected beta cell function. In a later study of six diabetic cats, they found that insular amyloidosis was associated with a significant decrease in cell volume fractions of both alpha and beta cells but suggested that the reduction in beta cell mass did not appear sufficient to be the sole cause of diabetes. They postulated that amyloid formation is preceded by a defect in beta cell function, which is unrelated to amyloid deposition.[172]

3.2.6 Summary

Diabetes in cats is a fascinating disease complex. It is similar to type 2 DM in humans because the majority of diabetic cats, similar to humans, have islet amyloid. However, dissimilar to human type 2 DM is the fact that most naive diabetic cats already have very low insulin concentrations at the time of diagnosis, despite the fact that most are overweight or obese. Obesity, together with age and gender, is one of the most important risk factors in cats and is increasing at a very rapid rate. Because most owners, even those of overweight, obese, or diabetic cats, feed cats ad libitum, owner education

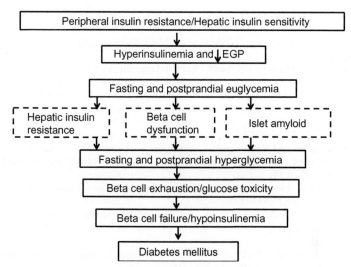

Figure 12.3 Proposed sequence of events in the progression from obesity to diabetes in cats. Abbreviation: EGP, endogenous glucose output.

about controlling caloric intake is the most important aspect of lowering the risk for diabetes in cats because weight loss reverses obesity-induced insulin resistance. Figure 12.3 summarizes the proposed sequence of events in the progression from obesity to DM.

Initially, obesity causes insulin resistance of the muscle and adipose tissue in cats to which the islets respond appropriately by increasing insulin secretion. Hepatic insulin sensitivity is maintained in obese cats. The hyperinsulinemia causes the liver to lower glucose production and allows cats to maintain euglycemia in the fasted and postprandial state despite the peripheral insulin resistance. Many obese cats remain in this perfectly compensated state and never progress to diabetes. Some cats, however, progress, probably through a combination of factors. Beta cells become dysfunctional and insulin secretion decreases. Either because of the decreased insulin or through a separate pathway leading to hepatic insulin resistance, endogenous glucose output increases. Lastly, misfolding of IAPP and formation of toxic oligomers and amyloid fibrils contribute to beta cell dysfunction. The end result is diabetes characterized by hyperglycemia and hypoinsulinemia.

REFERENCES

1. Murray SM, Flickinger EA, Patil AR, Merchen NR, Brent Jr JL, Fahey Jr GC. *In vitro* fermentation characteristics of native and processed cereal grains and potato starch using ileal chyme from dogs. *J Anim Sci.* 2001;79:435–444.

2. Li X, Li W, Wang H, et al. Cats lack a sweet taste receptor. *J Nutr.* 2006;136:1932S–1934S.
3. Glaser D. Specialization and phyletic trends of sweetness reception in animals. *Pure Appl Chem.* 2002;74:1153–1158.
4. Moore MC, Pagliassotti MJ, Swift LL, et al. Disposition of a mixed meal by the conscious dog. *Am J Physiol Endocrinol Metab.* 1994;266:E666–E675.
5. Pagliassotti MJ, Holste LC, Moore MC, Neal DW, Cherrington AD. Comparison of the time courses of insulin and the portal signal on hepatic glucose and glycogen metabolism in the conscious dog. *J Clin Invest.* 1996;97:81–91.
6. Batchelor DJ, Al-Rammahi M, Moran AW, et al. Sodium/glucose cotransporter-1, sweet receptor, and disaccharidase expression in the intestine of the domestic dog and cat: two species of different dietary habit. *Am J Physiol Regul Integr Comp Physiol.* 2011;300:R67–R75.
7. Church DB. A comparison of intravenous and oral glucose tolerance tests in the dog. *Res Vet Sci.* 1980;29:353–359.
8. Anderwald C, Gastaldelli A, Tura A, et al. Mechanism and effects of glucose absorption during an oral glucose tolerance test among females and males. *J Clin Endocrinol Metab.* 2011;96:515–524.
9. Maccioni R, Babul J. Purification and characterization of dog liver glucokinase. *Arch Biol Med Exp (Santiago).* 1980;13:271–286.
10. Sweet IR, Matschinsky FM. Mathematical model of beta-cell glucose metabolism and insulin release. I. Glucokinase as glucosensor hypothesis. *Am J Physiol Endocrinol Metab.* 1995;268:E775–E788.
11. Levin BE, Routh VH, Kang L, Sanders NM, Dunn-Meynell AA. Neuronal glucosensing: what do we know after 50 years? *Diabetes.* 2004;53:2521–2528.
12. de la Iglesia N, Veiga-da-Cunha M, Van Schaftingen E, Guinovart JJ, Ferrer JC. Glucokinase regulatory protein is essential for the proper subcellular localisation of liver glucokinase. *FEBS Lett.* 1999;456:332–338.
13. Hornichter R, Brown J, Snow H. Effects of starvation and diabetes on glucokinase activity in dog liver. *Clin Res.* 1967;15:109.
14. Niemeyer H, Ureta T, Clark-Turri L. Adaptive character of liver glucokinase. *Mol Cell Biochem.* 1975;6:109–126.
15. Callejas D, Mann CJ, Ayuso E, et al. Treatment of diabetes and long-term survival following insulin and glucokinase gene therapy. *Diabetes.* 2013;62:1718–1729.
16. Guptill L, Glickman L, Glickman N. Time trends and risk factors for diabetes mellitus in dogs: analysis of veterinary medical data base records (1970–1999). *Vet J.* 2003;165:240–247.
17. Fracassi F, Pietra M, Boari A, Aste G, Giunti M, Famigli-Bergamini P. Breed distribution of canine diabetes mellitus in Italy. *Vet Res Commun.* 2004;28(suppl 1):339–342.
18. Davison LJ, Herrtage ME, Catchpole B. Study of 253 dogs in the United Kingdom with diabetes mellitus. *Vet Rec.* 2005;156:467–471.
19. Fall T, Hamlin HH, Hedhammar A, Kampe O, Egenvall A. Diabetes mellitus in a population of 180,000 insured dogs: incidence, survival, and breed distribution. *J Vet Intern Med.* 2007;21:1209–1216.
20. Hess RS, Kass PH, Ward CR. Breed distribution of dogs with diabetes mellitus admitted to a tertiary care facility. *J Am Vet Med Assoc.* 2000;216:1414–1417.
21. Marmor M, Willeberg P, Glickman LT, Priester WA, Cypess RH, Hurvitz AI. Epizootiologic patterns of diabetes mellitus in dogs. *Am J Vet Res.* 1982;43:465–470.
22. Catchpole B, Adams JP, Holder AL, Short AD, Ollier WE, Kennedy LJ. Genetics of canine diabetes mellitus: are the diabetes susceptibility genes identified in humans involved in breed susceptibility to diabetes mellitus in dogs? *Vet J.* 2013;195:139–147.

23. Catchpole B, Kennedy LJ, Davison LJ, Ollier WE. Canine diabetes mellitus: from phenotype to genotype. *J Small Anim Pract.* 2008;49:4–10.

24. Kennedy LJ, Quarmby S, Happ GM, et al. Association of canine hypothyroidism with a common major histocompatibility complex DLA class II allele. *Tissue Antigens.* 2006;68:82–86.

25. Hughes AM, Jokinen P, Bannasch DL, Lohi H, Oberbauer AM. Association of a dog leukocyte antigen class II haplotype with hypoadrenocorticism in Nova Scotia Duck Tolling Retrievers. *Tissue Antigens.* 2010;75:684–690.

26. Owerbach D, Gabbay KH. Localization of a type I diabetes susceptibility locus to the variable tandem repeat region flanking the insulin gene. *Diabetes.* 1993;42:1708–1714.

27. Short AD, Catchpole B, Kennedy LJ, et al. Analysis of candidate susceptibility genes in canine diabetes. *J Hered.* 2007;98:518–525.

28. Bottazzo GF, Florin-Christensen A, Doniach D. Islet-cell antibodies in diabetes mellitus with autoimmune polyendocrine deficiencies. *Lancet.* 1974;2:1279–1283.

29. Roep BO, Peakman M. Antigen targets of type 1 diabetes autoimmunity. *Cold Spring Harb Perspect Med.* 2012;2:a007781.

30. Roep BO, Peakman M. Diabetogenic T lymphocytes in human Type 1 diabetes. *Curr Opin Immunol.* 2011;23:746–753.

31. Willcox A, Richardson SJ, Bone AJ, Foulis AK, Morgan NG. Analysis of islet inflammation in human type 1 diabetes. *Clin Exp Immunol.* 2009;155:173–181.

32. Batstra MR, Aanstoot HJ, Herbrink P. Prediction and diagnosis of type 1 diabetes using beta-cell autoantibodies. *Clin Lab.* 2001;47:497–507.

33. Hoenig M, Dawe DL. A qualitative assay for beta cell antibodies. Preliminary results in dogs with diabetes mellitus. *Vet Immunol Immunopathol.* 1992;32:195–203.

34. Davison LJ, Weenink SM, Christie MR, Herrtage ME, Catchpole B. Autoantibodies to GAD65 and IA-2 in canine diabetes mellitus. *Vet Immunol Immunopathol.* 2008;126:83–90.

35. Davison LJ, Walding B, Herrtage ME, Catchpole B. Anti-insulin antibodies in diabetic dogs before and after treatment with different insulin preparations. *J Vet Intern Med.* 2008;22:1317–1325.

36. Knip M, Veijola R, Virtanen SM, Hyoty H, Vaarala O, Akerblom HK. Environmental triggers and determinants of type 1 diabetes. *Diabetes.* 2005;54(Suppl 2):S125–S136.

37. Montgomery T, Nelson R, Feldman E, Robertson K, Polonsky K. Basal and glucagon-stimulated plasma C-peptide concentrations in healthy dogs, dogs with diabetes mellitus, and dogs with hyperadrenocorticism. *J Vet Intern Med.* 1996;10:116–122.

38. Manns JG, Martin CL. Plasma insulin, glucagon, and nonesterified fatty acids in dogs with diabetes mellitus. *Am J Vet Res.* 1972;33:981–985.

39. Mattheeuws D, Rottiers R, Kaneko JJ, Vermeulen A. Diabetes mellitus in dogs: relationship of obesity to glucose tolerance and insulin response. *Am J Vet Res.* 1984;45:98–103.

40. Kaneko JJ, Mattheeuws D, Rottiers RP, Vermeulen A. Glucose tolerance and insulin response in diabetes mellitus of dogs. *J Small Anim Pract.* 1978;19:85–94.

41. Fall T, Holm B, Karlsson A, Ahlgren KM, Kampe O, von Euler H. Glucagon stimulation test for estimating endogenous insulin secretion in dogs. *Vet Rec.* 2008;163:266–270.

42. Catchpole B, Ristic JM, Fleeman LM, Davison LJ. Canine diabetes mellitus: can old dogs teach us new tricks? *Diabetologia.* 2005;48:1948–1956.

43. Pozzilli P, Di Mario U. Autoimmune diabetes not requiring insulin at diagnosis (latent autoimmune diabetes of the adult): definition, characterization, and potential prevention. *Diabetes Care.* 2001;24:1460–1467.

44. Redondo MJ. LADA: time for a new definition. *Diabetes.* 2013;62:339–340.

45. Irvine AJ, Butterwick R, Watson T, Millward DJ, Morgan LM. Determination of insulin sensitivity in the dog: an assessment of three methods. *J Nutr.* 2002;132:1706S–1708S.

46. Kahn SE, Prigeon RL, McCulloch DK, et al. Quantification of the relationship between insulin sensitivity and beta-cell function in human subjects. Evidence for a hyperbolic function. *Diabetes.* 1993;42:1663–1672.

47. Weyer C, Bogardus C, Mott DM, Pratley RE. The natural history of insulin secretory dysfunction and insulin resistance in the pathogenesis of type 2 diabetes mellitus. *J Clin Invest.* 1999;104:787–794.

48. Mol JA, van Garderen E, Selman PJ, Wolfswinkel J, Rijinberk A, Rutteman GR. Growth hormone mRNA in mammary gland tumors of dogs and cats. *J Clin Invest.* 1995;95:2028–2034.

49. Timmermans-Sprang EP, Rao NA, Mol JA. Transactivation of a growth hormone (GH) promoter-luciferase construct in canine mammary cells. *Domest Anim Endocrinol.* 2008;34:403–410.

50. Selman PJ, Mol JA, Rutteman GR, van Garderen E, Rijnberk A. Progestin-induced growth hormone excess in the dog originates in the mammary gland. *Endocrinology.* 1994;134:287–292.

51. Eigenmann JE, Eigenmann RY, Rijnberk A, van der Gaag I, Zapf J, Froesch ER. Progesterone-controlled growth hormone overproduction and naturally occurring canine diabetes and acromegaly. *Acta Endocrinol (Copenh).* 1983;104:167–176.

52. Hess RS, Saunders HM, Van Winkle TJ, Ward CR. Concurrent disorders in dogs with diabetes mellitus: 221 cases (1993–1998). *J Am Vet Med Assoc.* 2000;217:1166–1173.

53. Peterson ME, Altszuler N, Nichols CE. Decreased insulin sensitivity and glucose tolerance in spontaneous canine hyperadrenocorticism. *Res Vet Sci.* 1984;36:177–182.

54. Colao A, Pivonello R, Spiezia S, et al. Persistence of increased cardiovascular risk in patients with Cushing's disease after five years of successful cure. *J Clin Endocrinol Metab.* 1999;84:2664–2672.

55. Leav I, Schiller AL, Rijnberk A, Legg MA, der Kinderen PJ. Adenomas and carcinomas of the canine and feline thyroid. *Am J Pathol.* 1976;83:61–122.

56. Ford SL, Nelson RW, Feldman EC, Niwa D. Insulin resistance in three dogs with hypothyroidism and diabetes mellitus. *J Am Vet Med Assoc.* 1993;202:1478–1480.

57. Dixon RM, Reid SW, Mooney CT. Epidemiological, clinical, haematological and biochemical characteristics of canine hypothyroidism. *Vet Rec.* 1999;145:481–487.

58. Hofer-Inteeworn N, Panciera DL, Monroe WE, et al. Effect of hypothyroidism on insulin sensitivity and glucose tolerance in dogs. *Am J Vet Res.* 2012;73:529–538.

59. La Batide-Alanore A, Chatellier G, Plouin PF. Diabetes as a marker of pheochromocytoma in hypertensive patients. *J Hypertens.* 2003;21:1703–1707.

60. Gilson SD, Withrow SJ, Wheeler SL, Twedt DC. Pheochromocytoma in 50 dogs. *J Vet Intern Med.* 1994;8:228–232.

61. Feldman E, Nelson R. Pheochromocytoma and multiple endocrine neoplasia. In: Feldman E, Nelson RW, eds. *Canine and Feline Endocrinology and Reproduction.* 3rd ed. Philadelphia, PA: Saunders; 2004.

62. Courcier EA, Thomson RM, Mellor DJ, Yam PS. An epidemiological study of environmental factors associated with canine obesity. *J Small Anim Pract.* 2010;51:362–367.

63. Howard B. Banfield sees bump in fat pets. *DVM Newsmagazine.* 2012;.

64. Klinkenberg H, Sallander MH, Hedhammar A. Feeding, exercise, and weight identified as risk factors in canine diabetes mellitus. *J Nutr.* 2006;136:1985S–1987S.

65. Mattheeuws D, Rottiers R, Baeyens D, Vermeulen A. Glucose tolerance and insulin response in obese dogs. *J Am Anim Hosp Assoc.* 1982;20:287–293.

66. Verkest KR, Fleeman LM, Rand JS, Morton JM. Evaluation of beta-cell sensitivity to glucose and first-phase insulin secretion in obese dogs. *Am J Vet Res.* 2011;72:357–366.

67. Kley S, Caffall Z, Tittle E, Ferguson DC, Hoenig M. Development of a feline proinsulin immunoradiometric assay and a feline proinsulin enzyme-linked immunosorbent assay (ELISA): a novel application to examine beta cell function in cats. *Domest Anim Endocrinol*. 2008;34:311–318.
68. Coradini M, Rand JS, Morton JM, Rawlings JM. Effects of two commercially available feline diets on glucose and insulin concentrations, insulin sensitivity and energetic efficiency of weight gain. *Br J Nutr*. 2011;106(suppl 1):S64–S77.
69. Cohn LA, Dodam JR, McCaw DL, Tate DJ. Effects of chromium supplementation on glucose tolerance in obese and nonobese cats. *Am J Vet Res*. 1999;60:1360–1363.
70. Taniguchi A, Nakai Y, Doi K, et al. Insulin sensitivity, insulin secretion, and glucose effectiveness in obese subjects: a minimal model analysis. *Metabolism*. 1995;44:1397–1400.
71. Westermark P, Andersson A, Westermark GT. Islet amyloid polypeptide, islet amyloid, and diabetes mellitus. *Physiol Rev*. 2011;91:795–826.
72. Hoenig M, Hall G, Ferguson D, et al. A feline model of experimentally induced islet amyloidosis. *Am J Pathol*. 2000;157:2143–2150.
73. Gal A, Hoenig M, Wallig MA, Singh K. Histopathology of pancreata from life-long dietary induced obese cats and lean controls. *Vet Pathol*. 2010;47:40S (A150).
74. Jordan K, Murtaugh MP, O'Brien TD, Westermark P, Betsholtz C, Johnson KH. Canine IAPP cDNA sequence provides important clues regarding diabetogenesis and amyloidogenesis in type 2 diabetes. *Biochem Biophys Res Commun*. 1990;169:502–508.
75. Paulsson JF, Andersson A, Westermark P, Westermark GT. Intracellular amyloid-like deposits contain unprocessed pro-islet amyloid polypeptide (proIAPP) in beta cells of transgenic mice overexpressing the gene for human IAPP and transplanted human islets. *Diabetologia*. 2006;49:1237–1246.
76. Johnson KH, O'Brien TD, Westermark P. Newly identified pancreatic protein islet amyloid polypeptide. What is its relationship to diabetes? *Diabetes*. 1991;40:310–314.
77. O'Neill S, Drobatz K, Satyaraj E, Hess R. Evaluation of cytokines and hormones in dogs before and after treatment of diabetic ketoacidosis and in uncomplicated diabetes mellitus. *Vet Immunol Immunopathol*. 2012;148:276–283.
78. Maahs DM, Ogden LG, Snell-Bergeon JK, et al. Determinants of serum adiponectin in persons with and without type 1 diabetes. *Am J Epidemiol*. 2007;166:731–740.
79. Lindstrom T, Frystyk J, Hedman CA, Flyvbjerg A, Arnqvist HJ. Elevated circulating adiponectin in type 1 diabetes is associated with long diabetes duration. *Clin Endocrinol (Oxf)*. 2006;65:776–782.
80. Stefan N, Bunt JC, Salbe AD, Funahashi T, Matsuzawa Y, Tataranni PA. Plasma adiponectin concentrations in children: relationships with obesity and insulinemia. *J Clin Endocrinol Metab*. 2002;87:4652–4656.
81. Pereira RI, Snell-Bergeon JK, Erickson C, et al. Adiponectin dysregulation and insulin resistance in type 1 diabetes. *J Clin Endocrinol Metab*. 2012;97:E642–E647.
82. Alejandro R, Feldman EC, Shienvold FL, Mintz DH. Advances in canine diabetes mellitus research: etiopathology and results of islet transplantation. *J Am Vet Med Assoc*. 1988;193:1050–1055.
83. Atkins CE, Hill JR, Johnson RK. Diabetes mellitus in the juvenile dog: a report of four cases. *J Am Vet Med Assoc*. 1979;175:362–368.
84. Atkins CE, Chin HP. Insulin kinetics in juvenile canine diabetics after glucose loading. *Am J Vet Res*. 1983;44:596–600.
85. Gepts W, Toussaint D. Spontaneous diabetes in dogs and cats. A pathological study. *Diabetologia*. 1967;3:249–265.
86. Li X, Li W, Wang H, et al. Pseudogenization of a sweet-receptor gene accounts for cats' indifference toward sugar. *PLoS Genet*. 2005;1:27–35.
87. Kienzle E. Carbohydrate metabolism in the cat. Activity of amylase in the gastrointestinal tract of the cat. *J Anim Phys An Nutr*. 1993;69:92.

88. de-Oliveira LD, Carciofi AC, Oliveira MC, et al. Effects of six carbohydrate sources on diet digestibility and postprandial glucose and insulin responses in cats. *J Anim Sci.* 2008;86:2237–2246.

89. Buddington RK, Chen JW, Diamond JM. Dietary regulation of intestinal brush-border sugar and amino acid transport in carnivores. *Am J Physiol.* 1991;261:R793–R801.

90. Dyer J, Daly K, Salmon KS, et al. Intestinal glucose sensing and regulation of intestinal glucose absorption. *Biochem Soc Trans.* 2007;35:1191–1194.

91. Shirazi-Beechey SP, Hirayama BA, Wang Y, Scott D, Smith MW, Wright EM. Ontogenic development of lamb intestinal sodium-glucose co-transporter is regulated by diet. *J Physiol.* 1991;437:699–708.

92. Wood IS, Dyer J, Hofmann RR, Shirazi-Beechey SP. Expression of the Na^+/glucose co-transporter (SGLT1) in the intestine of domestic and wild ruminants. *Pflugers Arch.* 2000;441:155–162.

93. Ballard F. Glucose utilization in mammalian liver. *Comp Biochem Physiol.* 1965;14:437–443.

94. Arai T, Kawaue T, Abe M, et al. Comparison of glucokinase activities in the peripheral leukocytes between dogs and cats. *Comp Biochem Physiol C Pharmacol Toxicol Endocrinol.* 1998;120:53–56.

95. Tanaka A, Inoue A, Takeguchi A, Washizu T, Bonkobara M, Arai T. Comparison of expression of glucokinase gene and activities of enzymes related to glucose metabolism in livers between dog and cat. *Vet Res Commun.* 2005;29:477–485.

96. Hussain K. Mutations in pancreatic ß-cell glucokinase as a cause of hyperinsulinaemic hypoglycaemia and neonatal diabetes mellitus. *Rev Endocr Metab Disord.* 2010;11:179–183.

97. Hoenig M, Jordan ET, Ferguson DC, de Vries F. Oral glucose leads to a differential response in glucose, insulin, and GLP-1 in lean versus obese cats. *Domest Anim Endocrinol.* 2010;38:95–102.

98. Hoenig M, Alexander S, Holson J, Ferguson DC. Influence of glucose dosage on interpretation of intravenous glucose tolerance tests in lean and obese cats. *J Vet Intern Med.* 2002;16:529–532.

99. Farrow H, Rand JS, Morton JM, Sunvold G. Postprandial glycaemia in cats fed a moderate carbohydrate meal persists for a median of 12 hours—female cats have higher peak glucose concentrations. *J Feline Med Surg.* 2012;14:706–715.

100. Bergman RN, Phillips LS, Cobelli C. Physiologic evaluation of factors controlling glucose tolerance in man: measurement of insulin sensitivity and beta-cell glucose sensitivity from the response to intravenous glucose. *J Clin Invest.* 1981;68:1456–1467.

101. Rottiers R, Mattheeuws D, Kaneko JJ, Vermeulen A. Glucose uptake and insulin secretory responses to intravenous glucose loads in the dog. *Am J Vet Res.* 1981;42:155–158.

102. O'Brien TD, Hayden DW, Johnson KH, Stevens JB. High dose intravenous glucose tolerance test and serum insulin and glucagon levels in diabetic and non-diabetic cats: relationships to insular amyloidosis. *Vet Pathol.* 1985;22:250–261.

103. Backus RC, Cave NJ, Ganjam VK, Turner JB, Biourge VC. Age and body weight effects on glucose and insulin tolerance in colony cats maintained since weaning on high dietary carbohydrate. *J Anim Physiol Anim Nutr (Berl).* 2010;94:e318–e328.

104. Velho G, Petersen KF, Perseghin G, et al. Impaired hepatic glycogen synthesis in glucokinase-deficient (MODY-2) subjects. *J Clin Invest.* 1996;98:1755–1761.

105. Kettelhut IC, Foss MC, Migliorini RH. Glucose homeostasis in a carnivorous animal (cat) and in rats fed a high-protein diet. *Am J Physiol Regul Integr Comp Physiol.* 1980;239:R437–R444.

106. Hoenig M. Effect of macronutrients, age, and obesity on 6 and 24-hour post-prandial glucose metabolism in cats. *Am J Physiol Regul Integr Comp Physiol.* 2011;301: R1798–R1807.

107. Rogers QR, Morris JG, Freedland RA. Lack of hepatic enzymatic adaptation to low and high levels of dietary protein in the adult cat. *Enzyme*. 1977;22:348–356.
108. Russell K, Lobley GE, Millward DJ. Whole-body protein turnover of a carnivore, *Felis silvestris catus*. *Br J Nutr*. 2003;89:29–37.
109. Green AS, Ramsey JJ, Villaverde C, Asami DK, Wei A, Fascetti AJ. Cats are able to adapt protein oxidation to protein intake provided their requirement for dietary protein is met. *J Nutr*. 2008;138:1053–1060.
110. Lester T, Czarnecki-Maulden G, Lewis D. Cats increase fatty acid oxidation when iso-calorically fed meat-based diets with increasing fat content. *Am J Physiol Regul Integr Comp Physiol*. 1999;277:R878–R886.
111. Hoenig M, Thomaseth K, Waldron M, Ferguson DC. Fatty acid turnover, substrate oxidation, and heat production in lean and obese cats during the euglycemic hyper-insulinemic clamp. *Domest Anim Endocrinol*. 2007;32:329–338.
112. Petersen KF, Price T, Cline GW, Rothman DL, Shulman GI. Contribution of net hepatic glycogenolysis to glucose production during the early postprandial period. *Am J Physiol Endocrinol Metab*. 1996;270:E186–E191.
113. McCann TM, Simpson KE, Shaw DJ, Butt JA, Gunn-Moore DA. Feline diabetes mellitus in the UK: the prevalence within an insured cat population and a questionnaire-based putative risk factor analysis. *J Feline Med Surg*. 2007;9:289–299.
114. Rand JS, Bobbermien LM, Hendrikz JK, Copland M. Over representation of Burmese cats with diabetes mellitus. *Aust Vet J*. 1997;75:402–405.
115. Panciera DL, Thomas CB, Eicker SW, Atkins CE. Epizootiologic patterns of diabetes mellitus in cats: 333 cases (1980–1986). *J Am Vet Med Assoc*. 1990;197:1504–1508.
116. Scarlett JM, Donoghue S. Associations between body condition and disease in cats. *J Am Vet Med Assoc*. 1998;212:1725–1731.
117. Prahl A, Guptill L, Glickman NW, Tetrick M, Glickman LT. Time trends and risk factors for diabetes mellitus in cats presented to veterinary teaching hospitals. *J Feline Med Surg*. 2007;9:351–358.
118. Rand JS, Fleeman LM, Farrow HA, Appleton DJ, Lederer R. Canine and feline diabetes mellitus: nature or nurture? *J Nutr*. 2004;134:2072S–2080S.
119. Slingerland LI, Fazilova VV, Plantinga EA, Kooistra HS, Beynen AC. Indoor confine-ment and physical inactivity rather than the proportion of dry food are risk factors in the development of feline type 2 diabetes mellitus. *Vet J*. 2009;179:247–253.
120. Lederer R, Rand JS, Jonsson NN, Hughes IP, Morton JM. Frequency of feline diabetes mellitus and breed predisposition in domestic cats in Australia. *Vet J*. 2009;179: 254–258.
121. Bonnefond A, Froguel P, Vaxillaire M. The emerging genetics of type 2 diabetes. *Trends Mol Med*. 2010;16:407–416.
122. Ershow AG. Environmental influences on development of type 2 diabetes and obesity: challenges in personalizing prevention and management. *J Diabetes Sci Technol*. 2009;3:727–734.
123. Hull RL, Westermark GT, Westermark P, Kahn SE. Islet amyloid: a critical entity in the pathogenesis of type 2 diabetes. *J Clin Endocrinol Metab*. 2004;89:3629–3643.
124. Jurgens CA, Toukatly MN, Fligner CL, et al. Beta-cell loss and beta-cell apoptosis in human type 2 diabetes are related to islet amyloid deposition. *Am J Pathol*. 2011;178:2632–2640.
125. Hoenig M, Reusch C, Peterson ME. Beta cell and insulin antibodies in treated and untreated diabetic cats. *Vet Immunol Immunopathol*. 2000;77:93–102.
126. van Belle TL, Coppieters KT, von Herrath MG. Type 1 diabetes: etiology, immunol-ogy, and therapeutic strategies. *Physiol Rev*. 2011;91:79–118.
127. Hall DG, Kelley LC, Gray ML, Glaus TM. Lymphocytic inflammation of pancreatic islets in a diabetic cat. *J Vet Diagn Invest*. 1997;9:98–100.

128. Kirk C, Feldman EC, Nelson RW. Diagnosis of naturally acquired type-I and type-II diabetes mellitus in cats. *Am J Vet Res.* 1993;54:463–467.

129. Hoenig M, Traas A, Schaeffer D. Evaluation of routine hematology profiles, fructosamine, thyroxine, insulin, and proinsulin concentrations in lean, overweight, obese, and diabetic cats. *J Am Vet Med Assoc.* 2013;243:1302–1309.

130. Hotta K, Funahashi T, Arita Y, et al. Plasma concentrations of a novel, adipose-specific protein, adiponectin, in type 2 diabetic patients. *Arterioscler Thromb Vasc Biol.* 2000;20:1595–1599.

131. Cerasi E. b-cell dysfunction vs insulin resistance in type 2 diabetes: the eternal "chicken and egg" question. *Medicographia.* 2011;33:35–41.

132. Saad MF, Knowler WC, Pettitt DJ, Nelson RG, Mott DM, Bennett PH. Sequential changes in serum insulin concentration during development of non-insulin-dependent diabetes. *Lancet.* 1989;1:1356–1359.

133. Nelson RW, Griffey SM, Feldman EC, Ford SL. Transient clinical diabetes mellitus in cats: 10 cases (1989–1991). *J Vet Intern Med.* 1999;13:28–35.

134. Kraus MS, Calvert CA, Jacobs GJ, Brown J. Feline diabetes mellitus: a retrospective mortality study of 55 cats (1982–1994). *J Am Anim Hosp Assoc.* 1997;33:107–111.

135. Roomp K, Rand J. Intensive blood glucose control is safe and effective in diabetic cats using home monitoring and treatment with glargine. *J Feline Med Surg.* 2009;11:668–682.

136. Zini E, Hafner M, Osto M, et al. Predictors of clinical remission in cats with diabetes mellitus. *J Vet Intern Med.* 2010;24:1314–1321.

137. Lukens F, Dohan F. Pituitary-diabetes in the cat. Recovery following insulin or dietary therapy. *Endocrinology.* 1942;30:175–201.

138. Moise NS, Reimers TJ. Insulin therapy in cats with diabetes mellitus. *J Am Vet Med Assoc.* 1983;182:158–164.

139. Middleton DJ, Watson AD. Glucose intolerance in cats given short-term therapies of prednisolone and megestrol acetate. *Am J Vet Res.* 1985;46:2623–2625.

140. Niessen SJ, Petrie G, Gaudiano F, et al. Feline acromegaly: an underdiagnosed endocrinopathy? *J Vet Intern Med.* 2007;21:899–905.

141. Berg RI, Nelson RW, Feldman EC, Kass PH, Pollard R, Refsal KR. Serum insulin-like growth factor-I concentration in cats with diabetes mellitus and acromegaly. *J Vet Intern Med.* 2007;21:892–898.

142. Fox J, Beatty J. A case report of complicated diabetes mellitus in a cat. *J Am Anim Hosp Assoc.* 1975;11:129–134.

143. Hoenig M, Ferguson DC. Impairment of glucose tolerance in hyperthyroid cats. *J Endocrinol.* 1989;121:249–251.

144. Karlander SG, Khan A, Wajngot A, Torring O, Vranic M, Efendic S. Glucose turnover in hyperthyroid patients with normal glucose tolerance. *J Clin Endocrinol Metab.* 1989;68:780–786.

145. Klieverik LP, Sauerwein HP, Ackermans MT, Boelen A, Kalsbeek A, Fliers E. Effects of thyrotoxicosis and selective hepatic autonomic denervation on hepatic glucose metabolism in rats. *Am J Physiol Endocrinol Metab.* 2008;294:E513–E520.

146. Kley S, Hoenig M, Glushka J, et al. The impact of obesity, sex, and diet on hepatic glucose production in cats. *Am J Physiol Regul Integr Comp Physiol.* 2009;296: R936–R943.

147. Courcier EA, O'Higgins R, Mellor DJ, Yam PS. Prevalence and risk factors for feline obesity in a first opinion practice in Glasgow, Scotland. *J Feline Med Surg.* 2010;12:746–753.

148. Russell K, Sabin R, Holt S, Bradley R, Harper EJ. Influence of feeding regimen on body condition in the cat. *J Small Anim Pract.* 2000;41:12–17.

149. Courcier EA, Mellor DJ, Pendlebury E, Evans C, Yam PS. An investigation into the epidemiology of feline obesity in Great Britain: results of a cross-sectional study of 47 companion animal practises. *Vet Rec.* 2012;171:560.

150. Hoenig M, McGoldrick JB, deBeer M, Demacker PN, Ferguson DC. Activity and tissue-specific expression of lipases and tumor-necrosis factor alpha in lean and obese cats. *Domest Anim Endocrinol.* 2006;30:333–344.

151. Wilkins C, Long Jr RC, Waldron M, Ferguson DC, Hoenig M. Assessment of the influence of fatty acids on indices of insulin sensitivity and myocellular lipid content by use of magnetic resonance spectroscopy in cats. *Am J Vet Res.* 2004;65:1090–1099.

152. Clark M, Larsen R, Lu W, Hoenig M. Investigation of 1H MRS for quantification of hepatic triglyceride in lean and obese cats. *Res Vet Sci.* 2013;95:678–680.

153. Yki-Jarvinen H. Fat in the liver and insulin resistance. *Ann Med.* 2005;37:347–356.

154. Brennan CL, Hoenig M, Ferguson DC. GLUT4 but not GLUT1 expression decreases early in the development of feline obesity. *Domest Anim Endocrinol.* 2004;26:291–301.

155. Hoenig M, Thomaseth K, Brandao J, Waldron M, Ferguson DC. Assessment and mathematical modeling of glucose turnover and insulin sensitivity in lean and obese cats. *Domest Anim Endocrinol.* 2006;31:373–389.

156. Hoenig M, Pach N, Thomaseth K, Le A, Schaeffer D, Ferguson DC. Cats differ from other species in their cytokine and antioxidant enzyme response when developing obesity. *Obesity (Silver Spring).* 2013;21:E407–E414.

157. Hoenig M, Pach N, Thomaseth K, Devries F, Ferguson DC. Evaluation of long-term glucose homeostasis in lean and obese cats by use of continuous glucose monitoring. *Am J Vet Res.* 2012;73:1100–1106.

158. Henson MS, Hegstad-Davies RL, Wang Q, et al. Evaluation of plasma islet amyloid polypeptide and serum glucose and insulin concentrations in nondiabetic cats classified by body condition score and in cats with naturally occurring diabetes mellitus. *Am J Vet Res.* 2011;72:1052–1058.

159. Sjoholm A, Nystrom T. Inflammation and the etiology of type 2 diabetes. *Diabetes Metab Res Rev.* 2006;22:4–10.

160. Laflamme DP. Companion Animals Symposium: obesity in dogs and cats: what is wrong with being fat? *J Anim Sci.* 2012;90:1653–1662.

161. Jordan E, Kley S, Le NA, Waldron M, Hoenig M. Dyslipidemia in obese cats. *Domest Anim Endocrinol.* 2008;35:290–299.

162. Harle P, Straub RH. Leptin is a link between adipose tissue and inflammation. *Ann N Y Acad Sci.* 2006;1069:454–462.

163. Hoenig M, Thomaseth K, Waldron M, Ferguson DC. Insulin sensitivity, fat distribution, and adipocytokine response to different diets in lean and obese cats before and after weight loss. *Am J Physiol Regul Integr Comp Physiol.* 2007;292:R227–R234.

164. Yki-Jarvinen H. Thiazolidinediones. *N Engl J Med.* 2004;351:1106–1118.

165. Goossens MM, Nelson RW, Feldman EC, Griffey SM. Response to insulin treatment and survival in 104 cats with diabetes mellitus (1985–1995). *J Vet Intern Med.* 1998;12:1–6.

166. Bruskiewicz KA, Nelson RW, Feldman EC, Griffey SM. Diabetic ketosis and ketoacidosis in cats: 42 cases (1980–1995). *J Am Vet Med Assoc.* 1997;211:188–192.

167. Ferreri JA, Hardam E, Kimmel SE, et al. Clinical differentiation of acute necrotizing from chronic nonsuppurative pancreatitis in cats: 63 cases (1996–2001). *J Am Vet Med Assoc.* 2003;223:469–474.

168. Clark A, Cooper GJ, Lewis CE, et al. Islet amyloid formed from diabetes-associated peptide may be pathogenic in type-2 diabetes. *Lancet.* 1987;2:231–234.

169. Yano BL, Hayden DW, Johnson KH. Feline insular amyloid: association with diabetes mellitus. *Vet Pathol.* 1981;18:621–627.

170. Yano BL, Hayden DW, Johnson KH. Feline insular amyloid: incidence in adult cats with no clinicopathologic evidence of overt diabetes mellitus. *Vet Pathol.* 1981;18:310–315.
171. Nakayama H, Uchida K, Ono K, Goto N. Pathological observation of six cases of feline diabetes mellitus. *Nihon Juigaku Zasshi.* 1990;52:819–822.
172. O'Brien TD, Hayden DW, Johnson KH, Fletcher TF. Immunohistochemical morphometry of pancreatic endocrine cells in diabetic, normoglycaemic glucose-intolerant and normal cats. *J Comp Pathol.* 1986;96:357–369.

CHAPTER THIRTEEN

Impaired Insulin Signaling and Mechanisms of Memory Loss

Jenna Bloemer[1], Subhrajit Bhattacharya[1], Rajesh Amin,
Vishnu Suppiramaniam
Department of Pharmacal Sciences, Harrison School of Pharmacy, Auburn University, Auburn,
Alabama, USA
[1]These authors contributed equally to this work.

Contents

Progress in Molecular Biology and Translational Science, Volume 121
ISSN 1877-1173
http://dx.doi.org/10.1016/B978-0-12-800101-1.00013-2

Abstract

Insulin is secreted from the β-cells of the pancreas and helps maintain glucose homeostasis. Although secreted peripherally, insulin also plays a profound role in cognitive function. Increasing evidence suggests that insulin signaling in the brain is necessary to maintain health of neuronal cells, promote learning and memory, decrease oxidative stress, and ultimately increase neuronal survival. This chapter summarizes the different facets of insulin signaling necessary for learning and memory and additionally explores the association between cognitive impairment and central insulin resistance. The role of impaired insulin signaling in the advancement of cognitive dysfunction is relevant to the current debate of whether the shared pathophysiological mechanisms between diabetes and cognitive impairment implicate a direct relationship. Here, we summarize a vast amount of literature that suggests a strong association between impaired brain insulin signaling and cognitive impairment.

ABBREVIATIONS

AChE acetylcholinesterase
AD Alzheimer's disease
ADAS-cog Alzheimer's Disease Assessment Scale-cognitive subscale
ADDLs amyloid-beta-derived diffusible ligands
AGE advanced glycation end product
AMPAR α-amino-3-hydroxy-5-methyl-4-isoxazolepropionic acid receptor
Aβ amyloid-beta
AβPP amyloid-beta precursor protein
BACE1 beta-secretase 1
BAD Bcl-2-associated death promoter
BBB blood–brain barrier
ChAT choline acetyltransferase
CNS central nervous system
ERK1/2 extracellular signal-regulated kinase 1/2
FoxO forkhead box O
GH growth hormone
GLP-1 glucagon-like peptide-1
GLP-1R glucagon-like peptide-1 receptor
GSK3β glycogen synthase kinase-3-beta
HbA1c glycated hemoglobin

ic-STZ intracerebral STZ
IDE insulin-degrading enzyme
IGF insulin-like growth factor
IL-6 interleukin-6
ILK integrin-linked kinase
IQCODE Informant Questionnaire on Cognitive Decline in the Elderly
IR insulin receptor
IRS-1 insulin receptor substrate-1
IU international unit
LRP-1 lipoprotein receptor-related protein
LTD long-term depression
LTP long-term potentiation
MAPK mitogen-activated protein kinase
MCI mild cognitive impairment
MMSE mini-mental state examination
mTOR mammalian target of rapamycin
NFTs neurofibrillary tangles
NF-κB nuclear factor-κB
NIRKO neuron-specific insulin receptor knockout
NMDAR N-methyl-D-aspartate receptor
PI3K phosphatidylinositide 3-kinase
PKB protein kinase B
PKC protein kinase C
PPAR peroxisome proliferator-activated receptor
RAGE receptor for AGE
ROS reactive oxygen species
S100B S100 calcium-binding protein B
STZ streptozotocin
T1DM type 1 diabetes mellitus
T2DM type 2 diabetes mellitus

1. INTRODUCTION

The brain has historically been classified as an insulin-insensitive organ;[1] however, a number of recent studies indicate that brain insulin signaling is required for normal cognitive function. Human and animal studies suggest that the dysfunction of central insulin signaling causes a wide range of cognitive impairments. Discoveries concerning regulation of brain function by insulin have led researchers to hypothesize that certain types of memory loss may be closely related to diabetes, because these conditions arise from strikingly similar pathologies.[2–5] Furthermore, patients with diabetes have an increased risk of developing mild cognitive impairment (MCI).[6] MCI is characterized as cognitive decline that is more severe than normal age-related memory loss, but not significant enough to greatly interfere with daily activities.[7] MCI can further develop into Alzheimer's disease (AD), the most common form of dementia.[8] Recent discoveries concerning the underlying insulin signaling pathologies present in AD have led some researchers to propose that AD is "type 3 diabetes."[5]

Diabetes mellitus, the well-established insulin signaling pathology, is a complex metabolic and immunologic disorder characterized by insulin resistance or deficiency leading to impaired glucose homeostasis. Type 1 diabetes mellitus (T1DM) develops due to autoimmune processes leading to the destruction of insulin-producing pancreatic beta cells, whereas type 2 diabetes mellitus (T2DM) is characterized by desensitization of insulin receptors (IRs) and relative insulin deficiency. An estimated 67% of the US population is diabetic or prediabetic,[9] implying that well over half of the population displays peripheral insulin signaling deficits. As diabetes affects 25.8 million people in the United States alone and AD is thought to currently affect around 34 million people worldwide,[10] it is imperative that the common pathophysiological links between AD and diabetes be carefully examined with the goal of developing effective treatment strategies to target these chronic and debilitating disease states.

2. DIABETES AND COGNITIVE DEFICITS

2.1. Type 1 diabetes mellitus

In the early 1900s, it was noted that diabetic patients showed impaired performance on neuropsychological tests,[11] and since then, numerous clinical studies have examined the effects of diabetes on cognitive abilities with somewhat conflicting results.[4] There is consensus, however, that cognitive impairment is more prevalent in those with diabetes, and a recent long-term study found that diabetes increases the risk for development of dementia by approximately 75%.[6] In particular, studies have shown that some of the most common cognitive deficits in patients with T1DM include decreased speed in information processing and impaired mental flexibility, which is the ability to apply previously acquired knowledge in novel situations.[4,12] The severity of cognitive decline in T1DM patients is inversely related to age of onset.[13,14]

Interestingly, a large clinical study found that frequency and duration of severe hypoglycemic episodes had no overall effect on cognitive decline over time when comparisons were made within T1DM patient populations.[15] However, a follow-up study showed that hyperglycemia over time leads to cognitive impairment, which was demonstrated by assessment of glycated hemoglobin (HbA1c) levels, the standard clinical parameter to determine long-term blood glucose levels. Participants with higher HbA1c levels demonstrated decreased performance in psychomotor efficacy and motor speed.[15] These studies show that within T1DM patients, the severity of cognitive decline is at least partially dependent on the level of blood glucose control, which is directly related to insulin levels in the body.

2.2. Type 2 diabetes mellitus

A number of epidemiological studies have discovered a positive correlation of T2DM with both vascular and neurodegenerative types of cognitive impairment.[16-20] T2DM patients have an increased occurrence of AD,[21] which may be due to increased incidences of vascular dementia.[22] A study involving 233 diabetic patients screened for cognitive impairment by the mini-mental state examination (MMSE) and Informant Questionnaire on Cognitive Decline in the Elderly (IQCODE) found that cognitive impairment was much higher in these patients compared to the general population.[23] During acute hypoglycemic events, immediate and working memory is impaired,[24,25] and studies have suggested that verbal memory is especially disrupted in T2DM-related cognitive impairment.[26,27] Heightened adiposity is also positively correlated with an increased risk for cognitive dysfunction in these patients.[28] T2DM and cognitive impairment share a complex interrelationship with many possible linking factors, such as microvascular complications,[29,30] hyperglycemia,[4,31] hyperinsulinemia,[32] and even clinical depression,[33] among many others (Fig. 13.1). For example, hyperinsulinemia decreases expression of IRs on the blood–brain barrier (BBB) leading to decreased insulin uptake in the brain and possibly contributing to impaired brain insulin signaling.[34-36] Although peripheral insulin resistance is correlated with cognitive dysfunction, it might not serve as a direct indicator of central insulin resistance, as the two conditions have been found to exist as separate entities.[37]

3. CENTRAL INSULIN REGULATION

3.1. The insulin signaling pathway

A familiarity with the insulin signaling pathway is essential when considering the association between central insulin resistance and cognitive function.

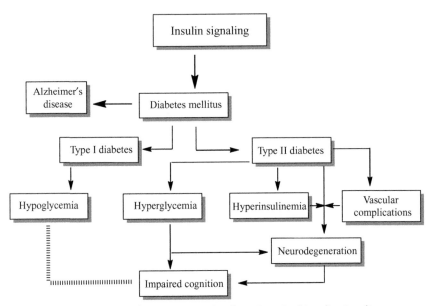

Figure 13.1 Progression of events stemming from impaired insulin signaling.

The presence of IRs in the mammalian brain was first reported in 1978, and IRs were subsequently found to be present throughout the central nervous system (CNS).[38] In postmortem studies, neuronal IR expression is decreased in a progressive manner with the severity of neurodegeneration in AD.[39] The receptors are found at higher concentration on neuronal cells than on glial cells and are located at both the cell body and synapses of neurons. These receptors are responsive to insulin and insulin-like growth factor (IGF).[40] The IR is a transmembrane tyrosine kinase that exists as a tetramer bound covalently by disulfide bonds.[41–43] The extracellular α-subunits contain the insulin-binding sites, while the transmembrane β-subunits serve to anchor the receptor in the cell membrane and contain intracellular phosphorylation sites essential for signal transduction.[43] Insulin signaling is highly evolutionarily conserved, which highlights its significance in cellular processes.[5]

The two major pathways that are activated by insulin signaling are the phosphatidylinositide 3-kinase (PI3K)/Akt pathway and the Ras/mitogen-activated protein kinase (MAPK) pathway.[44,45] Binding of insulin to the α-subunit induces autophosphorylation of the intracellular portion of the β-subunit, specifically at Tyr^{1146}, Tyr^{1150}, and Tyr^{1151},[41] in addition to two other sites that do not seem to be important for activation of the signaling cascade.[46] Autophosphorylation is followed by phosphorylation of insulin receptor substrate-1 (IRS-1) at specific sites.[45,47,48] IRS-1 induces activation of the PI3K pathway, in which PI3K phosphorylates Akt, also known as protein kinase B (PKB),[49] leading to a number of physiological effects such as increased glycogen formation and protein synthesis.[50,51] Activated Akt phosphorylates glycogen synthase kinase-3-beta (GSK3β) at Ser,[52] thus inactivating it. In its active state, GSK3β inhibits glycogen synthase, so the inactivation of GSK3β leads to increased glycogen synthesis.[53–56] Akt additionally induces protein synthesis via downstream activation of mammalian target of rapamycin (mTOR) signaling[49,57] and cell survival through inhibition of several apoptotic agents.[58] Activated IRS-1 also activates the Ras/MAPK pathway that leads to increased cell proliferation and decreased levels of apoptosis[59,60] (see also Fig. 13.2).

3.2. The source of insulin in the brain

The major source of central insulin is a topic of debate. Some evidence indicates that brain insulin levels are solely dependent on peripheral insulin levels, while other studies suggest that insulin is produced in the brain.[61] Circulating insulin is thought to cross the BBB through a saturable carrier-mediated transporter[62–64]; hence, central insulin levels are likely to vary based on production of insulin by the β-cells of the pancreas. There are also a number of correlative studies suggesting that insulin levels in the brain are dependent on local neuronal production.[65–68] Indicators of central insulin production including insulin-like mRNA,[69,70] preproinsulin I and II mRNA,[71–73] and C-peptide are present in mammalian neuronal tissues.[66,74] C-Peptide is critical in the formation of the proinsulin molecule[75] and is not thought to cross the BBB.[67,76] In addition, insulin is synthesized in cultured rat neurons from certain areas of the brain, including the hippocampus, prefrontal cortex, and olfactory bulb,[77] areas where IR density is the greatest.[78] The presence of such molecules in the brain may serve as positive indicators of central insulin synthesis. The evidence concerning central insulin production is inconclusive,[79] and further work is needed to determine whether neurons are able to synthesize insulin.

3.3. Role of IRS-1 in central insulin resistance

The phosphorylation of specific serine (Ser or S) and tyrosine (Tyr or Y) residues on IRS-1 may regulate central insulin sensitivity (Fig. 13.3). A recent study found that insulin resistance in the hippocampus is significantly associated with increased levels of phosphorylated IRS-1 at Ser[616] and Ser[636/639].[37] Similarly, peripheral insulin resistance exhibited in T2DM is thought to be

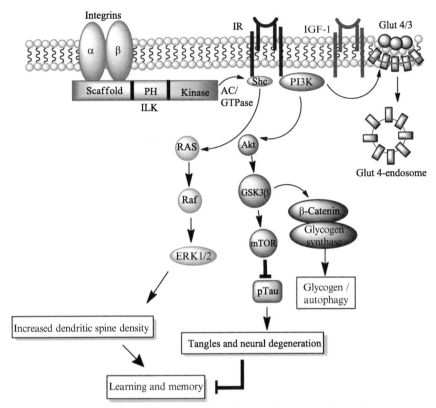

Figure 13.2 Integrin, IR, and IGF-1 pathways. The transduction pathways for insulin signaling involve integrins, scaffolding protein integrin-linked kinase (ILK), and IR/IGF-1. The downstream signaling of Ras/Raf/ERK1/2 leads to increased dendritic spine density. The parallel pathway is mediated by Akt/GSK and mTOR, inhibiting neural degeneration and neurofibrillary tangle formation. An offshoot from this latter pathway progresses through the interaction of catenins and glycogen synthase. Ultimately, these signaling pathways modulate learning and memory. (See color plate.)

due to chronic upregulation of insulin signaling pathways leading to increasingly persistent phosphorylation of specific serine residues on the IR.[82–84] Feedback inhibition via phosphorylation of these residues is induced by the downstream signaling molecules ERK1/2 (extracellular signal–regulated kinase 1/2), GSK3β, mTOR, and PKC (protein kinase C).[85,86] Central levels of IRS-1 pS^{616} and IRS-1 $pS^{636/639}$ were found to increase in correlation with severity of cognitive impairment and levels of deposited oligomeric Aβ plaques, one of the major hallmarks of AD.[37] In studies utilizing postmortem hippocampal tissues from histologically confirmed cases of AD, baseline levels of insulin, IR, and downstream regulatory molecules were similar to the

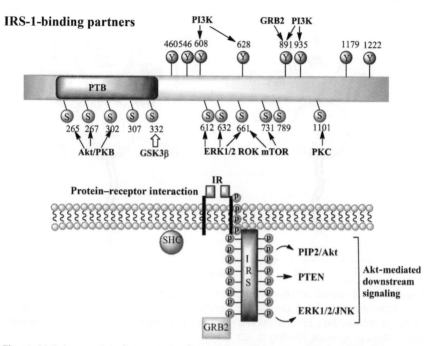

Figure 13.3 Interactions between insulin receptor substrate-1 (IRS-1) and signaling proteins. Upper panel shows important serine (S) and tyrosine (Y) sites on the IRS-1 molecule that interact with multiple kinases. GSK3β and PKC are negative modulators of IRS-1 activity. Several kinases such as Akt/PKB, ERK, ROK, and mTOR have been reported to act as negative or positive modulators in specific situations at certain phosphorylation sites. PI3K acts as a positive modulator of IRS-1.[80] Modulation of IRS-1 activity is thought to play a role in insulin sensitivity, with serine phosphorylation overall leading to increased insulin resistance[81] and tyrosine phosphorylation leading to increased insulin sensitivity. Lower panel shows protein interaction with IRS culminating into downstream signaling pathways. *Adapted by permission from Macmillan Publishers Ltd: Ref. 80.* (See color plate.)

levels found in controls, indicating that there was no significant change in expression of proteins required for the insulin signaling cascade.[37] Instead, the mechanisms for central insulin resistance seem to be based on IR sensitivity and nonquantitative dysfunction of signaling.

4. INSULIN RESISTANCE AND HALLMARKS OF NEURODEGENERATION

4.1. Tau pathology

A major determinant of severity of neurodegeneration in AD and related neurological disorders is the accumulation neurofibrillary tangles (NFTs)

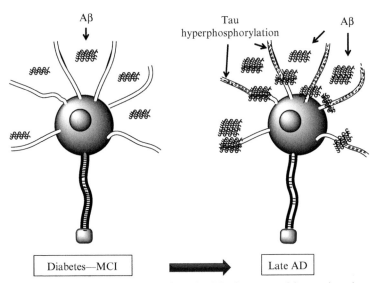

Figure 13.4 Neuronal accumulation of amyloid-β plaques and hyperphosphorylated tau. Hyperphosphorylated tau and Aβ accumulation leads to neurodegeneration. In diabetes-induced MCI, accumulation is minimal but increases with severity as cognitive impairment progresses, leading to AD.

in the brain, comprised mainly of insoluble, hyperphosphorylated tau protein (Fig. 13.4).[87-89] Tau is a microtubule-associated protein, and its major function is to stabilize microtubules in neuronal axons and facilitate axonal transport.[90] Insulin deficiency or resistance leading to dysfunction of insulin signaling impacts a number of downstream proteins that significantly interact with tau expression and phosphorylation.[91] Decreased insulin signaling in the brain results in the reduction of Akt signaling, which causes prolonged activation of GSK3β.[92] The GSK3β overactivation is partially responsible for hyperphosphorylation of tau and resultant misfolding and aggregation leading to NFTs.[93] Studies indicate that tau expression is also mediated by insulin and IGF signaling.[94] Peripheral hyperinsulinemia leads to increased tau phosphorylation *in vivo*[32] possibly due to the downregulation of IR expression that results via homeostatic regulation to compensate for the elevated insulin levels.[95]

4.2. Amyloid-β pathology

The links between insulin signaling and amyloid-β (Aβ) are complex, and the dysfunction of one will greatly affect the other (Fig. 13.5). Protease-mediated

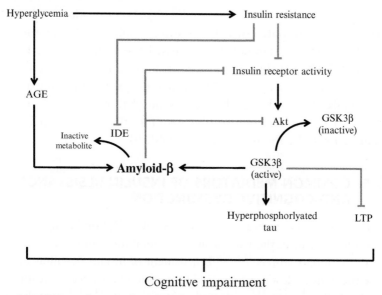

Figure 13.5 Proposed mechanisms for the role of amyloid-β in insulin signaling. Aβ plays a complex role in the shared pathology of insulin resistance and cognitive dysfunction. Hyperglycemia leads to insulin resistance and decreased insulin signaling.[4,31] Activation of Akt is thus decreased,[96] which potentiates the activity of GSK3β, and leads to increased hyperphosphorylated tau protein,[93] decreased LTP,[97] and increased production of Aβ.[98] Increased Aβ is thought to directly inhibit insulin receptors[99] and Akt.[96] Additionally, hyperglycemia increases Aβ production via AGEs.[100] Insulin resistance is thought to decrease IDE, an enzyme that breaks down Aβ, so Aβ plaques accumulate.[101] These processes, specifically Aβ accumulation, hyperphosphorylated tau accumulation, and decreases in LTP contribute to memory impairment.[79] (For color version of this figure, the reader is referred to the online version of this chapter.)

cleavage of Aβ precursor protein (AβPP) forms the Aβ peptide, which self-aggregates to form insoluble extracellular plaques contributing to the pathogenesis of AD.[102] Insulin and Aβ are both metabolized by insulin-degrading enzyme (IDE)[103,104]; therefore, each will act as a competitive inhibitor of the other. IDE knockout mice have a 50% decrease in Aβ degradation leading to accumulation of Aβ plaques in the brain[105] and also exhibit hyperinsulinemia.[106] Likewise, IDE was found to be significantly reduced in postmortem studies of AD patients.[101] In addition, insulin has been shown to decrease intracellular Aβ accumulation via the MAPK pathway by enhancing trafficking of Aβ and AβPP from the Golgi apparatus to the cell membrane.[107] Interestingly, Aβ causes deterioration of neuronal IRs in mature cultured hippocampal cells,[108] and Aβ is thought to act as a competitive inhibitor at IRs by blocking the binding of insulin and thus inhibiting the insulin signaling cascade.[99] Subsequent studies have demonstrated that the infusion of soluble Aβ oligomers into the rat hippocampus significantly decreased the expressions of IRs and IRS-1 and PI3K signaling molecules.[109] Furthermore, intracellular Aβ interferes with the activation of Akt, which propagates prolonged activation of GSK3β,[96] and thus hyperphosphorylation of tau. It has also been shown that IGF-1 reduces the harmful effects of intracellular Aβ on neuronal cells by the inactivation of GSK3β.[110–112]

5. COMMON MEDIATORS OF INSULIN RESISTANCE AND COGNITIVE DYSFUNCTION

Imaging studies indicate that patients with MCI or AD have reduced glucose metabolism in the parietal and temporal lobes of the brain, particularly in the hippocampus.[113–115] Impairment of insulin signaling and thus glucose metabolism leads to increased formation of advanced glycation end products (AGEs),[100,116–118] increased oxidative stress,[119–121] mitochondrial dysfunction,[122] increased neuronal apoptosis,[123] and impairments in the BBB.[124–127] These are all known contributors to cognitive dysfunction associated with neurodegeneration. Therefore, it is not surprising that some researchers classify AD as a metabolic disorder.[123]

5.1. Advanced glycation end products

AGEs are formed by irreversible nonenzymatic reactions between reducing sugars, such as glucose, and amino groups in proteins, lipids, and nucleic acids.[100] The interaction between AGE and the receptor for AGE (RAGE) is thought to ultimately lead to oxidative stress.[128] Hyperglycemia exacerbates AGE formation, since the major source of endogenous AGE is the result of autoxidation of circulating glucose.[116] In the STZ-diabetic rodent model, peripheral AGE levels are significantly increased.[129] Furthermore, research indicates that glycation of Aβ and tau by AGE contributes to the pathology of AD by increasing plaque and NFTs.[118] Formation of AGEs also induces production of reactive oxygen species (ROS), which contribute to oxidative stress.[130,131]

5.2. Mitochondrial dysfunction

Mitochondrial dysfunction is indicated in the pathophysiology of diabetes and AD,[115,132,133] possibly due to damage caused by oxidative stress.[134] Altered insulin signaling was found to reduce the ability of mitochondria to store Ca^{2+} in diabetic rats.[135] The Ca^{2+} storage potential of mitochondria is critical in maintaining the calcium homeostasis, which is essential for normal neuronal function.[136,137] In addition, the "calcium hypothesis" proposes that dysfunction of calcium regulation plays a central role in the development of neurodegenerative diseases.[138–142] One hypothesis for the link between insulin resistance, oxidative stress, and cognitive dysfunction is that hyperglycemia due to peripheral insulin resistance induces high levels of mitochondrial oxidative phosphorylation. This in turn leads to overproduction of ROS in the mitochondria and thus structurally damages the mitochondria leading to neurodegenerative cognitive impairment.[122,133]

5.3. Damage to the BBB

The BBB becomes damaged after prolonged periods of hyperglycemia, and this has been linked to cognitive impairment.[124,126] A brain imaging study found that diabetic patients exhibited impaired BBB compared to controls.[125] Patients with diabetes were also found to have increased levels of antibodies against two CNS proteins, S100B (S100 calcium-binding protein B) and NSE,[143] which indicates that the BBB integrity may be compromised in these patients.[127] Hyperglycemia contributes to the formation of ROS as mentioned earlier, which causes generation of inflammatory mediators such as interleukin-6 (IL-6) and C-reactive protein.[144,145] These inflammatory cytokines may cause damage in the cerebral vesicles and contribute to the loss of structure of the BBB.[146] Likewise, imaging studies have found that BBB dysfunction is present in MCI[147] and that the severity of the BBB impairment positively correlates with progression of cognitive dysfunction.[148] It is also important to note that a few studies have not found a significant difference in BBB structure between cognitively impaired patients and controls.[149,150] In addition, the low-density lipoprotein receptor–related protein (LRP-1) transporter located on the BBB is responsible for Aβ clearance from the brain,[151] and damage to this transporter is linked to AD.[152,153] Therefore, it is hypothesized that damage to the BBB may cause decreased clearance of Aβ from the brain, contributing to cognitive dysfunction.[154–158]

5.4. Formation of toxic lipids

Another metabolic link between insulin dysfunction and cognitive impairment is irregular hepatic metabolism of lipids. Because the insulin pathway is involved in the synthesis of lipids, peripheral insulin resistance leads to abnormal lipid metabolism.[159,160] Downregulation of the insulin cascade stimulates lipolysis, which increases the production of toxic lipids,[161,162] and further propagates insulin resistance.[163,164] A class of toxic lipids that has recently been investigated as a major contributor to cognitive dysfunction is the ceramides. Ceramides cause damage to neuronal cells by inducing activation of proinflammatory cytokines,[163] and they can also directly interfere with the Akt pathway.[165,166] Ceramide levels are increased in pathological states, such as obesity,[167] and in diet-induced obesity models, circulating ceramide levels are greatly increased.[163,168] Systemic administration of ceramides in rodents caused neurodegeneration along with deficits in spatial memory.[169] It is hypothesized that in obesity and diabetes, the generation of these toxic lipids due to increased lipolysis leads to increased peripheral levels of the substances, which can easily cross the BBB and induce neurodegeneration similar to that seen in AD.[123,170]

5.5. Neuronal apoptosis

The antiapoptotic effects of insulin signaling are essential for the health and maintenance of neurons.[123,171,172] The PI3K/Akt (Fig. 13.6) and the MAPK signaling pathways are both involved in antiapoptosis.[123] Forkhead box O (FoxO) is a transcription factor that regulates gluconeogenesis and is

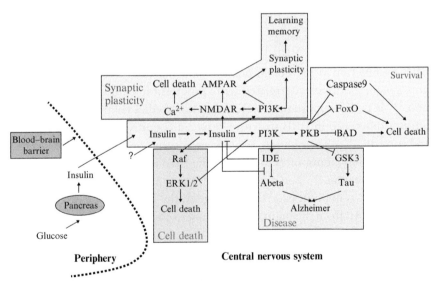

Figure 13.6 Summary of pathways affected by insulin in the CNS. Central insulin signaling plays an essential role in learning and memory, cell survival, and inhibition of Aβ and tau accumulation.[79] *Reprinted from Ref. 79, with permission from Elsevier. (For color version of this figure, the reader is referred to the online version of this chapter.)*

also responsible for transcription of a number of apoptotic agents.[173] Normal insulin signaling activates Akt that then phosphorylates and thus inactivates FoxO.[174] Upon decreased insulin signaling, FoxO remains continually active,[175] leading to accelerated cell death.[79] Akt also inactivates Bcl-2-associated death promoter (BAD),[176] another proapoptotic agent. In addition, insulin signaling has been found to activate nuclear factor-κB (NF-κB) via Raf-1 kinase.[177,178] The actions of NF-κB vary greatly through the body, and it has even been shown to increase apoptosis via certain signaling pathways while inhibiting the process in others.[179] However, NF-κB activation and subsequent signaling in the insulin pathway have been found to inhibit apoptosis.[172] NF-κB exhibits neuroprotective effects,[180] and it is thought that membrane peroxidation due to oxidative stress,[181] as occurs in neurodegenerative disorders, suppresses NF-κB and leads to increased cell death.[180] In addition, neuronal NF-κB has been found to protect against the neurodegenerative effects of Aβ in neuronal cultures.[182]

6. INSULIN RESISTANCE AND HIPPOCAMPAL DYSFUNCTION

6.1. Regulation of cholinergic neurons

Memory loss is related to the dysfunction of cholinergic homeostasis in the brain; therefore, determining the effects of insulin signaling and insulin resistance on cholinergic neurons is critical in evaluating a possible relationship between diabetes and cognitive dysfunction. The hippocampus is one of the major areas associated with learning and memory, and a decrease in cholinergic synaptic terminals in the hippocampus is correlated with memory loss in many neurodegenerative diseases.[183–185] In the CA1 area of the rat hippocampus, choline acetyltransferase (ChAT), the rate-limiting enzyme in ACh formation, is colocalized with essential insulin signaling proteins such as IR subunits, IRS-1, Akt, and GSK3β,[186] which indicates that insulin signaling plays a key role in cholinergic neurons. High levels of neuronal Aβ deposits lead to the disruption of cholinergic neurons via dysfunction of insulin signaling pathways.[187] In postmortem studies of histologically proven cases of late-onset AD, ChAT activity is significantly decreased, especially in the hippocampus.[188,189]

The significance of cholinergic signaling in the pathophysiology of AD is evident as the typical drugs used in its treatment are acetylcholinesterase inhibitors, which inhibit the breakdown of ACh in the synaptic cleft. A common rodent model of central insulin resistance is intracerebral streptozotocin (ic-STZ), and these rodents have decreased expression of ChAT mRNA and increased expression of AChE mRNA, both of which equate to reduced cholinergic neurotransmission.[5,190] These findings provide evidence that streptozotocin (STZ), a DNA alkylator, is targeting specific pathways, since a nonspecific DNA alkylator would be expected to decrease expression of mRNA indiscriminately. Furthermore, in ic-STZ rats treated with peroxisome proliferator-activated receptor (PPAR) agonists, common agents in the treatment of diabetes that increase IR sensitivity, ChAT expression is increased and performance in spatial memory tasks such as the Morris water maze is greatly improved.[5] These results consistently implicate that insulin pathways control the expression of ChAT and that the disruption of these pathways can lead to cognitive impairment.

6.2. Synaptic plasticity

Insulin regulates learning and memory through interactions with the glutamatergic system. The glutamatergic system plays a well-elucidated role in cognition, and it interacts with the cholinergic system to regulate a number of cognitive processes.[191–194] Communications between neurons can be strengthened or reduced in an activity-dependent manner, a phenomenon termed synaptic plasticity.[195] A precise cellular definition of memory remains elusive due to the complexity of neuronal networks and biochemical processes involved. Though the definition of memory is inherently abstract, synaptic plasticity is a fundamental representation of memory formation and storage.[196] Insulin signaling is considered essential for hippocampal synaptic plasticity mechanisms including long-term potentiation (LTP) and long-term depression (LTD).[197–201] LTP is a sustained enhancement of synaptic strength, and in contrast, LTD is a form of activity-dependent mitigated synaptic strength. Evidence suggests that the activation of cholinergic receptors enhances LTP via increased expression and function of the glutamate receptor subtypes, N-methyl-D-aspartate receptors (NMDARs) and α-amino-3-hydroxy-5-methyl-4-isoxazolepropionic acid receptors (AMPARs).[202,203] A recent study found that disruption of the medial septum, the major cholinergic input to the hippocampus, greatly reduces hippocampal LTP by modulating glutamatergic neurotransmission.[199,204] In addition, ERK1/2, a downstream molecule in the insulin signaling pathway, is important for the enhancement of LTP.[205]

The ic-STZ insulin resistance model shows postsynaptic dysfunction of glutamatergic neurotransmission leading to decreased LTP resulting in cognitive deficits.[197–199] A key connection between synaptic plasticity and insulin signaling pathways is GSK3β. The inhibition of GSK3β is essential for the induction of hippocampal LTP.[97,206,207] In the ic–STZ model, there is decreased expression and sensitivity of IRs,[190] leading to prolonged activation of GSK3β. Two major inactivators of GSK3β are integrin-linked kinase (ILK) and Akt, which phosphorylate GSK3β at Ser^9.[97] ILK expression is decreased in the hippocampus in ic–STZ rodents, while Akt levels remained unchanged, suggesting that ILK may be one of the major kinases involved in the inactivation of GSK3β.[190] Decreased levels of ILK in hippocampal neurons correlated well with impaired LTP, possibly due to alterations in AMPAR and NMDAR expression and function.[199] Inhibition of LTP as demonstrated in ic–STZ rats may be a key component of cognitive dysfunction induced by central insulin resistance.

7. ROLE OF IGF IN COGNITIVE FUNCTION

7.1. IGF-1, insulin, and cognitive impairment

In addition to insulin, IGF-1 has been implicated as an important regulator of glucose homeostasis and cognitive function. IGF-1 activates the insulin signaling pathway by binding to IRs. IGF-1 is also a high-affinity ligand for the IGF-1 receptor, which induces Akt/PI3K and Ras–MAPK signaling cascades.[208,209] Studies utilizing animal models found that elimination of hepatic IGF-1 production leads to hyperinsulinemia and abnormal glucose metabolism.[210,211] IGF-1 plays a significant role in cognitive processes. A meta-analysis including 1795 healthy elderly subjects found that circulating IGF-1 levels have significant positive correlation to cognitive function as measured by the MMSE.[212] IGF-1 levels are also associated with increases in speed of information processing[213] and working memory.[214] While circulating IGF-1 levels are increased in early stages of AD,[39,215] a decrease is observed in later stages of the disease.[216,217] This indicates that AD is related to IGF-1 resistance, since an initial decrease in receptor sensitivity would lead to an overproduction of IGF-1 as a feedback mechanism.[215]

7.2. Neuroprotective effects of IGF-1

IGF-1 is produced in both the liver[218] and the CNS,[219,220] and circulating IGF-1 has important vasoprotective properties.[221] Since cerebrovascular alterations have a well-established role in cognitive decline, decreased levels of IGF-1 present in the aging population may contribute to vascular complications leading to cognitive dysfunction.[222] Additionally, IGF-1 is important in the proper formation and storage of memories. In liver-specific IGF-1 knockout mice, total circulating IGF levels are around 60% of normal values, similar to the age-related decline of IGF in humans.[223,224] In these rodents, spatial memory is impaired, but the deficit in memory can be fully reversed by IGF-1 administration.[225] The animals also display deficits in hippocampal synaptic plasticity, possibly due to decreased NMDA receptor expression or impairment of the Akt pathway,[224] which can be recovered by systemic IGF-1 administration.[223] IGF-1 also prevents Aβ-derived diffusible ligands (also known as ADDLs) from binding to neuronal synapses, which prevents the buildup of senile plaques associated with neurodegenerative disorders like AD.[226]

7.3. Therapeutic implications of IGF-1

IGF-1 has a complex interrelationship with cognitive dysfunction, much like insulin. The pathways of the two hormones overlap to a great extent, and they even bind the same receptors albeit with differing affinities. Investigations into the possible therapeutic use of IGF-1 in the treatment of cognitive dysfunction are promising, but many barriers remain. In fact, some studies have shown that IGF-1 signaling promotes production of Aβ leading to buildup of senile plaques,[227,228] whereas other studies have shown that IGF-1 increases Aβ clearance.[229] There is also evidence correlating high IGF-1 levels with risk for certain types of cancer.[230–232] Thus, studies relating to manipulation of IGF-1 pathways as therapeutic targets for neurodegenerative diseases must proceed cautiously. In one clinical trial, a potent IGF-1-releasing drug was tested in 416 patients with mild to moderate AD. The substance MK-677 (ibutamoren mesylate) is known to induce growth hormone (GH) secretion, which in turn induces the release of IGF-1. At the end of 12 months, patients receiving MK-677 had a 72.9% increase in circulating IGF-1 levels. However, based on scores from a wide range of neuropsychological tests, no differences in cognitive functions between the control and treatment groups were observed.[233]

8. ANIMAL MODELS OF INSULIN SIGNALING DEFICITS
8.1. NIRKO model

The majority of information relating to the association between cognitive dysfunction and altered brain insulin signaling is derived from studies involving the use of animal models of insulin resistance. Investigations into genetically altered rodent models of insulin resistance or deficiency have presented somewhat conflicting results about the effect of such alterations in behavioral aspects of cognition. Neuron-specific insulin receptor knock-out (NIRKO) mice highlight the importance of central IRs.[234] Unexpectedly, these mice did not show any significant difference in the formation of spatial memories in Morris water maze experiments.[235] However, hyper-phosphorylation of tau protein is considerably increased in this model, possibly due to impairment of the PI3K pathway.[92] Normal inhibition of neuronal apoptosis is also decreased in NIRKO mice due to disruption of the same pathway.[236]

8.2. db/db model

Another model for T2DM is the *db/db* model, which is characterized by defects in the leptin receptor leading to increased plasma insulin and glucose levels, insulin resistance, and obesity.[237] The leptin receptors are found in various areas, including adipose tissue, hippocampus, and hypothalamus.[238,239] The loss of hippocampal leptin receptors leads to impaired insulin signaling and memory deficits.[238] No significant differences were found from results of the Y-maze test between *db/db* rodents and the control group. On the other hand, hippocampal LTP is negatively affected in these rodents.[236] In addition, evidence suggests that central leptin signaling plays a preventive role in Aβ and hyperphosphorylated tau accumulation.[240,241]

8.3. STZ-diabetic model

Researchers have gained novel insights into the common pathophysiology of memory loss and diabetes by generating animal models of diabetes using STZ, an alkylator, which is partially selective for and highly toxic to the insulin-producing β-cells of the pancreas.[242] Systemic administration of high doses of STZ in rats produces an effective model of T1DM by destroying the insulin-producing cells, similar to the damage done by auto-antibodies in T1DM. In STZ-diabetic rats, memory impairments are determined by the Morris water maze task,[190,243,244] the standard behavioral test for assessment of spatial memory. The deficits in spatial memory can be prevented by insulin administration, indicating that the decline in spatial memory is related to central insulin deficiency.[243,245] One study found that after STZ administration, working and reference memory also declined.[190]

These behavioral findings are consistent with the molecular evidence that diabetes and memory impairment arise from insulin deficiency or resistance.

8.4. Intracerebral STZ model

Intracerebral STZ administration causes dysfunction of insulin signaling by interfering with cerebral glucose metabolism. This leads to an altered neurochemical state that mimics sporadic AD.[246–248] The ic-STZ model does not induce peripheral diabetes as the STZ is concentrated centrally and does not reach circulation in significant amounts.[190] This model has overwhelmingly proven to demonstrate long-lasting dysfunction of learning and memory and decreases in LTP.[199,246,249] ic-STZ induces decreased secretion of insulin, decreased expression of IRs, increased hyperphosphorylated tau, and increased AβPP, contributing to neurodegeneration.[250,251] Additionally, ic-STZ rodents express lower neuronal levels of ChAT and higher levels of AChE, both contributing to decreased cholinergic signaling by reducing ACh levels.[190,251] ic-STZ also impairs signaling cascades and leads to decreased trafficking of hippocampal glutamate receptors necessary for memory formation.[199] A recent study found that ic-STZ causes lesions in the corpus callosum.[252] Atrophy of the corpus callosum has been reported in neurodegenerative memory disorders, and the severity of atrophy positively correlates with the degree of cognitive impairment.[253,254] These models of central insulin resistance are essential in examination of the relationship between cognitive impairment and deficits in insulin signaling, augmenting the search for novel therapeutic approaches (Table 13.1).

9. ROLE OF ANTIDIABETES DRUGS IN THE TREATMENT OF COGNITIVE DYSFUNCTION

9.1. Insulin

There is considerable interest in the use of antidiabetes drugs in treating and preventing neurodegeneration. Clinical trials testing the effects of such drugs on cognition have greatly expanded over the last decade. A number of clinical trials have assessed the effects of intranasal insulin on cognitive function.[258–260] Intranasal administration allows insulin to reach the brain without a significant increase in plasma insulin levels.[258,261] The majority of these clinical trials demonstrate that intranasal insulin improves memory in adults with cognitive impairment.[259,260,262] A recent double-blind, placebo-controlled clinical trial found that in patients with MCI or mild to moderate AD, a dose of 20 IU intranasal insulin improved delayed

Table 13.1 Cognitive effects of various antidiabetes agents

Therapeutic agent	Purpose of use	Mechanism of action	Effects on memory/cognition	Other side effects
Biguanides (metformin)	Oral hypoglycemic	Lowers blood glucose levels	Monotherapy leads to increased memory loss but, in conjunction with insulin, improves memory[255]	Lactic acidosis, inhibition of vitamin B_{12} absorption, renal/hepatic impairment
Sulfonylureas (glibenclamide)	Oral hypoglycemic	Stimulates β-cells in pancreas increasing insulin release	No direct evidence found and has effect on K-ATP channels in modulating inhibitory avoidance task–related memory storage[256]	No remarkable side effects; usually safe to use
Thiazolidinediones	Oral hypoglycemic	Binds to peroxisome proliferator-activated receptor-gamma (PPAR–γ)/retinoid X receptor (RXR) complex	Reverses memory decline in AD patients and hippocampal glucocorticoid receptor downregulation[257]	Hepatotoxicity in some cases

memory. Interestingly, a 40 IU dose had no improvement in delayed recall, but there was an increase in caregiver assessment of cognitive ability.[259] There seems to be a correlation between memory improvement with intranasal insulin and lack of the APOE-ε4 allele. The presence of APOE-ε4 allele increases the risk of developing late-onset sporadic AD.[263] In subjects with the APOE-ε4 allele, no improvement in cognitive function was observed after intranasal insulin administration, but patients without the allele showed significant improvement in verbal memory.[262] Intranasal insulin appears to be a promising candidate for treatment of cognitive impairment, but long-term clinical trials are needed.

9.2. PPAR agonists

PPAR-γ agonists have been explored in detail for the treatment of cognitive dysfunction, and recent studies have shown that PPAR-β/δ (also known as PPAR-δ) agonists may also protect against neurodegeneration by reducing neuronal inflammation.[264–266] PPAR-γ agonists improve glucose homeostasis by increasing expression of glucose transporters, enhancing insulin sensitivity, and decreasing tumor necrosis factor-alpha.[267,268] Animal models show significant evidence for the efficacy of PPAR-γ agonists for the improvement of cognitive function.[269–274] Nevertheless, in human studies, the evidence supporting similar positive benefits is variable. A few studies have reported significant cognitive benefits upon administration of the PPAR-γ agonist pioglitazone in AD patients. In one study, after 6 months, patients receiving pioglitazone scored higher on MMSE and Alzheimer's Disease Assessment Scale-cognitive subscale (ADAS-cog) than the previous baseline, while the scores decreased in the control group.[275] Another study presented similar results in patients receiving the PPAR-γ agonist rosiglitazone. AD patients who received the drug and lacked the APOE-ε4 allele scored higher on ADAS-cog than the control group.[257] However, a large Phase III clinical trial containing 693 subjects found no significant difference in AD patients treated with rosiglitazone compared to the patients receiving a placebo.[276] Alternatively, there is growing interest in the possible use of PPAR-β/δ agonists in the treatment of cognitive disorders. A recent study found that PPAR-β/δ knockout mice demonstrated impairment in memory related to object recognition. These rodents also presented higher levels of hyperphosphorylated tau and BACE1, a protease of AβPP that leads to increased formation of Aβ.[277] Therefore, clinical trials evaluating the efficacy of PPAR-δ agonists in the treatment of cognitive disorders may be explored in the near future.

9.3. GLP-1R agonists

Recently, there has been interest in testing the efficacy of glucagon-like peptide-1 receptor (GLP-1R) agonists, traditionally used to treat T2DM, in patients with cognitive dysfunction. GLP-1R agonists control insulin homeostasis by enhancing insulin release during periods of hyperglycemia in diabetic patients.[278] Interestingly, GLP-1R agonists do not have an effect on blood glucose and insulin levels in nondiabetic patients.[279] Additionally, GLP-1R is present in neuronal progenitor cells, and GLP-1 functions as a neuroprotective hormone.[280] Behavioral, biochemical, and electrophysiological data from animal studies support the potential benefits of GLP-1 for cognitive deficits. In behavioral tests, recognition memory and spatial memory were improved in a mouse model of AD after chronic administration of liraglutide, a GLP-1R agonist.[281] In wild-type rats, there was no significant difference in memory after GLP-1R agonist administration, suggesting that this drug may only show benefit in cognitive impairments associated with pathological conditions.[282] GLP-1R agonists significantly enhance hippocampal LTP in memory-impaired rodents.[283] A recent study demonstrated that Aβ-induced neurodegeneration could potentially be reversed by the administration of the GLP-1R agonist Val^8-GLP-1(7–36).[284] Overall, there is substantial potential for the treatment of cognitive disorders by drugs that have traditionally been used to treat diabetes. Numerous clinical trials evaluating the efficacy of these medications are currently underway.

10. SUMMARY

Over the last two decades, research relating to insulin signaling in the brain and its role in cognition has vastly expanded. However, the exploration into shared pathology involving cognitive dysfunction and central insulin signaling remains elusive. Although there is strong correlation between cognitive impairment and central insulin sensitivity, more research is warranted to confirm whether AD can justifiably be called "type 3 diabetes." Some of the most compelling evidence for this generalization arises from the fact that impaired insulin signaling leads to Aβ accumulation, hyperphosphorylation of tau, and deficits in synaptic plasticity. In addition, compelling evidence presented by recent studies for the use of antidiabetes drugs in the treatment of cognitive impairment further supports the notion that AD is diabetes of the brain. Future research on brain insulin signaling should proceed with the ultimate goal of developing novel treatments to alleviate cognitive dysfunction.

ACKNOWLEDGMENTS

The authors would like to thank Mr. Wooseok Lee, Ms. Manal Ali Buabeid, Mr. Stephen Krauss, and Mr. Babatunde Fariyike for their contributions to this chapter.

REFERENCES

1. Goodner CJ, Berrie MA. The failure of rat hypothalamic tissues to take up labeled insulin in vivo or to respond to insulin in vitro. *Endocrinology*. 1977;101:605–612.
2. Tang J, Pei Y, Zhou G. When aging-onset diabetes is coming across with Alzheimer disease: comparable pathogenesis and therapy. *Exp Gerontol*. 2013;48:744–750.
3. Li L, Holscher C. Common pathological processes in Alzheimer disease and type 2 diabetes: a review. *Brain Res Rev*. 2007;56:384–402.
4. Kodl CT, Seaquist ER. Cognitive dysfunction and diabetes mellitus. *Endocr Rev*. 2008;29:494–511.
5. de la Monte SM, Wands JR. Alzheimer's disease is type 3 diabetes-evidence reviewed. *J Diabetes Sci Technol*. 2008;2:1101–1113.
6. Ohara T, Doi Y, Ninomiya T, et al. Glucose tolerance status and risk of dementia in the community: the Hisayama study. *Neurology*. 2011;77:1126–1134.
7. Feldman HH, Jacova C. Mild cognitive impairment. *Am J Geriatr Psychiatry*. 2005;13:645–655.
8. Gandy S. The role of cerebral amyloid beta accumulation in common forms of Alzheimer disease. *J Clin Invest*. 2005;115:1121–1129.
9. Cowie CC, Rust KF, Ford ES, et al. Full accounting of diabetes and pre-diabetes in the U.S. population in 1988-1994 and 2005-2006. *Diabetes Care*. 2009;32:287–294.
10. Barnes DE, Yaffe K. The projected effect of risk factor reduction on Alzheimer's disease prevalence. *Lancet Neurol*. 2011;10:819–828.
11. Miles WR, Root HF. Psychologic tests applied to diabetic patients. *Arch Intern Med*. 1922;30:767–777.
12. Brands AM, Biessels GJ, de Haan EH, Kappelle LJ, Kessels RP. The effects of type 1 diabetes on cognitive performance: a meta-analysis. *Diabetes Care*. 2005;28:726–735.
13. Ryan CM. Memory and metabolic control in children. *Diabetes Care*. 1999;22:1239–1241.
14. Schoenle EJ, Schoenle D, Molinari L, Largo RH. Impaired intellectual development in children with Type I diabetes: association with HbA(1c), age at diagnosis and sex. *Diabetologia*. 2002;45:108–114.
15. Jacobson AM, Musen G, Ryan CM, et al. Long-term effect of diabetes and its treatment on cognitive function. *N Engl J Med*. 2007;356:1842–1852.
16. Yaffe K, Kanaya A, Lindquist K, et al. The metabolic syndrome, inflammation, and risk of cognitive decline. *JAMA*. 2004;292:2237–2242.
17. Luchsinger JA, Tang MX, Stern Y, Shea S, Mayeux R. Diabetes mellitus and risk of Alzheimer's disease and dementia with stroke in a multiethnic cohort. *Am J Epidemiol*. 2001;154:635–641.
18. MacKnight C, Rockwood K, Awalt E, McDowell I. Diabetes mellitus and the risk of dementia, Alzheimer's disease and vascular cognitive impairment in the Canadian Study of Health and Aging. *Dement Geriatr Cogn Disord*. 2002;14:77–83.
19. Luchsinger JA, Reitz C, Patel B, Tang MX, Manly JJ, Mayeux R. Relation of diabetes to mild cognitive impairment. *Arch Neurol*. 2007;64:570–575.
20. Schrijvers EM, Witteman JC, Sijbrands EJ, Hofman A, Koudstaal PJ, Breteler MM. Insulin metabolism and the risk of Alzheimer disease: the Rotterdam Study. *Neurology*. 2010;75:1982–1987.

21. Cukierman T, Gerstein HC, Williamson JD. Cognitive decline and dementia in diabetes—systematic overview of prospective observational studies. *Diabetologia.* 2005;48:2460–2469.

22. Curb JD, Rodriguez BL, Abbott RD, et al. Longitudinal association of vascular and Alzheimer's dementias, diabetes, and glucose tolerance. *Neurology.* 1999;52:971–975.

23. Bruce DG, Casey GP, Grange V, et al. Cognitive impairment, physical disability and depressive symptoms in older diabetic patients: the Fremantle Cognition in Diabetes Study. *Diabetes Res Clin Pract.* 2003;61:59–67.

24. Austin EJ, Deary IJ. Effects of repeated hypoglycemia on cognitive function: a psychometrically validated reanalysis of the Diabetes Control and Complications Trial data. *Diabetes Care.* 1999;22:1273–1277.

25. Sommerfield AJ, Deary IJ, McAulay V, Frier BM. Short-term, delayed, and working memory are impaired during hypoglycemia in individuals with type 1 diabetes. *Diabetes Care.* 2003;26:390–396.

26. Stewart R, Liolitsa D. Type 2 diabetes mellitus, cognitive impairment and dementia. *Diabet Med.* 1999;16:93–112.

27. Strachan MW, Deary IJ, Ewing FM, Frier BM. Is type II diabetes associated with an increased risk of cognitive dysfunction? A critical review of published studies. *Diabetes Care.* 1997;20:438–445.

28. Abbatecola AM, Lattanzio F, Spazzafumo L, et al. Adiposity predicts cognitive decline in older persons with diabetes: a 2-year follow-up. *PLoS One.* 2010;5:e10333.

29. Craft S, Watson GS. Insulin and neurodegenerative disease: shared and specific mechanisms. *Lancet Neurol.* 2004;3:169–178.

30. Boden-Albala B, Cammack S, Chong J, et al. Diabetes, fasting glucose levels, and risk of ischemic stroke and vascular events: findings from the Northern Manhattan Study (NOMAS). *Diabetes Care.* 2008;31:1132–1137.

31. Cox DJ, Kovatchev BP, Gonder-Frederick LA, et al. Relationships between hyperglycemia and cognitive performance among adults with type 1 and type 2 diabetes. *Diabetes Care.* 2005;28:71–77.

32. Freude S, Plum L, Schnitker J, et al. Peripheral hyperinsulinemia promotes tau phosphorylation in vivo. *Diabetes.* 2005;54:3343–3348.

33. Anderson RJ, Freedland KE, Clouse RE, Lustman PJ. The prevalence of comorbid depression in adults with diabetes: a meta-analysis. *Diabetes Care.* 2001;24:1069–1078.

34. Baker LD, Cross DJ, Minoshima S, Belongia D, Watson GS, Craft S. Insulin resistance and Alzheimer-like reductions in regional cerebral glucose metabolism for cognitively normal adults with prediabetes or early type 2 diabetes. *Arch Neurol.* 2011;68:51–57.

35. Craft S. Insulin resistance syndrome and Alzheimer's disease: age- and obesity-related effects on memory, amyloid, and inflammation. *Neurobiol Aging.* 2005;26(suppl 1):65–69.

36. Neumann KF, Rojo L, Navarrete LP, Farias G, Reyes P, Maccioni RB. Insulin resistance and Alzheimer's disease: molecular links & clinical implications. *Curr Alzheimer Res.* 2008;5:438–447.

37. Talbot K, Wang HY, Kazi H, et al. Demonstrated brain insulin resistance in Alzheimer's disease patients is associated with IGF-1 resistance, IRS-1 dysregulation, and cognitive decline. *J Clin Invest.* 2012;122:1316–1338.

38. Havrankova J, Roth J, Brownstein M. Insulin receptors are widely distributed in the central nervous system of the rat. *Nature.* 1978;272:827–829.

39. Rivera EJ, Goldin A, Fulmer N, Tavares R, Wands JR, de la Monte SM. Insulin and insulin-like growth factor expression and function deteriorate with progression of Alzheimer's disease: link to brain reductions in acetylcholine. *J Alzheimers Dis.* 2005;8:247–268.

40. Marks JL, Maddison J, Eastman CJ. Subcellular localization of rat brain insulin binding sites. *J Neurochem.* 1988;50:774–781.

438 Jenna Bloemer et al.

41. Kahn CR, White MF. The insulin receptor and the molecular mechanism of insulin action. *J Clin Invest.* 1988;82:1151–1156.
42. Jacobs S, Cuatrecasas P. Insulin receptor: structure and function. *Endocr Rev.* 1981;2:251–263.
43. Lee J, Pilch PF. The insulin receptor: structure, function, and signaling. *Am J Physiol.* 1994;266:C319–C334.
44. Ullrich A, Bell JR, Chen EY, et al. Human insulin receptor and its relationship to the tyrosine kinase family of oncogenes. *Nature.* 1985;313:756–761.
45. Sun XJ, Rothenberg P, Kahn CR, et al. Structure of the insulin receptor substrate IRS-1 defines a unique signal transduction protein. *Nature.* 1991;352:73–77.
46. Goren HJ, White MF, Kahn CR. Separate domains of the insulin receptor contain sites of autophosphorylation and tyrosine kinase activity. *Biochemistry.* 1987;26:2374–2382.
47. Myers Jr MG, Sun XJ, White MF. The IRS-1 signaling system. *Trends Biochem Sci.* 1994;19:289–293.
48. O'Hare T, Pilch PF. Intrinsic kinase activity of the insulin receptor. *Int J Biochem.* 1990;22:315–324.
49. Okamoto M, Hayashi T, Kono S, et al. Specific activity of phosphatidylinositol 3-kinase is increased by insulin stimulation. *Biochem J.* 1993;290(Pt 2):327–333.
50. Brazil DP, Hemmings BA. Ten years of protein kinase B signalling: a hard Akt to follow. *Trends Biochem Sci.* 2001;26:657–664.
51. Pearson LL, Castle BE, Kehry MR. CD40-mediated signaling in monocytic cells: up-regulation of tumor necrosis factor receptor-associated factor mRNAs and activation of mitogen-activated protein kinase signaling pathways. *Int Immunol.* 2001;13:273–283.
52. Sutherland C, Leighton IA, Cohen P. Inactivation of glycogen synthase kinase-3 beta by phosphorylation: new kinase connections in insulin and growth-factor signalling. *Biochem J.* 1993;296(Pt 1):15–19.
53. Siddle K. Signalling by insulin and IGF receptors: supporting acts and new players. *J Mol Endocrinol.* 2011;47:R1–R10.
54. Balaraman Y, Limaye AR, Levey AI, Srinivasan S. Glycogen synthase kinase 3beta and Alzheimer's disease: pathophysiological and therapeutic significance. *Cell Mol Life Sci.* 2006;63:1226–1235.
55. Lee J, Kim MS. The role of GSK3 in glucose homeostasis and the development of insulin resistance. *Diabetes Res Clin Pract.* 2007;77(suppl 1):S49–S57.
56. Clodfelder-Miller B, De Sarno P, Zmijewska AA, Song L, Jope RS. Physiological and pathological changes in glucose regulate brain Akt and glycogen synthase kinase-3. *J Biol Chem.* 2005;280:39723–39731.
57. Backer JM, Myers Jr MG, Sun XJ, et al. Association of IRS-1 with the insulin receptor and the phosphatidylinositol 3'-kinase. Formation of binary and ternary signaling complexes in intact cells. *J Biol Chem.* 1993;268:8204–8212.
58. Zhuang Z, Zhao X, Wu Y, et al. The anti-apoptotic effect of PI3K-Akt signaling pathway after subarachnoid hemorrhage in rats. *Ann Clin Lab Sci.* 2011;41:364–372.
59. Zhang XF, Settleman J, Kyriakis JM, et al. Normal and oncogenic p21ras proteins bind to the amino-terminal regulatory domain of c-Raf-1. *Nature.* 1993;364:308–313.
60. Lange-Carter CA, Pleiman CM, Gardner AM, Blumer KJ, Johnson GL. A divergence in the MAP kinase regulatory network defined by MEK kinase and Raf. *Science.* 1993;260:315–319.
61. Ghasemi R, Haeri A, Dargahi L, Mohamed Z, Ahmadiani A. Insulin in the brain: sources, localization and functions. *Mol Neurobiol.* 2013;47:145–171.
62. Banks WA. The source of cerebral insulin. *Eur J Pharmacol.* 2004;490:5–12.
63. Woods SC, Seeley RJ, Baskin DG, Schwartz MW. Insulin and the blood-brain barrier. *Curr Pharm Des.* 2003;9:795–800.

64. Pardridge WM, Eisenberg J, Yang J. Human blood-brain barrier insulin receptor. *J Neurochem.* 1985;44:1771–1778.
65. Dorn A, Bernstein HG, Rinne A, Ziegler M, Hahn HJ, Ansorge S. Insulin- and glucagonlike peptides in the brain. *Anat Rec.* 1983;207:69–77.
66. Dorn A, Rinne A, Bernstein HG, Hahn HJ, Ziegler M. Insulin and C-peptide in human brain neurons (insulin/C-peptide/brain peptides/immunohistochemistry/radioimmunoassay). *J Hirnforsch.* 1983;24:495–499.
67. Frolich L, Blum-Degen D, Bernstein HG, et al. Brain insulin and insulin receptors in aging and sporadic Alzheimer's disease. *J Neural Transm.* 1998;105:423–438.
68. Birch NP, Christie DL, Renwick AG. Proinsulin-like material in mouse foetal brain cell cultures. *FEBS Lett.* 1984;168:299–302.
69. Schechter R, Whitmire J, Wheet GS, et al. Immunohistochemical and in situ hybridization study of an insulin-like substance in fetal neuron cell cultures. *Brain Res.* 1994;636:9–27.
70. Devaskar SU, Giddings SJ, Rajakumar PA, Carnaghi LR, Menon RK, Zahm DS. Insulin gene expression and insulin synthesis in mammalian neuronal cells. *J Biol Chem.* 1994;269:8445–8454.
71. Young 3rd WS. Periventricular hypothalamic cells in the rat brain contain insulin mRNA. *Neuropeptides.* 1986;8:93–97.
72. Schechter R, Beju D, Gaffney T, Schaefer F, Whetsell L. Preproinsulin I and II mRNAs and insulin electron microscopic immunoreaction are present within the rat fetal nervous system. *Brain Res.* 1996;736:16–27.
73. Singh BS, Rajakumar PA, Eves EM, Rosner MR, Wainer BH, Devaskar SU. Insulin gene expression in immortalized rat hippocampal and pheochromocytoma-12 cell lines. *Regul Pept.* 1997;69:7–14.
74. Dorn A, Rinne A, Hahn HJ, Bernstein HG, Ziegler M. C-peptide immunoreactive neurons in human brain. *Acta Histochem.* 1982;70:326–330.
75. Wahren J, Ekberg K, Johansson J, et al. Role of C-peptide in human physiology. *Am J Physiol Endocrinol Metab.* 2000;278:E759–E768.
76. Duarte AI, Moreira PI, Oliveira CR. Insulin in central nervous system: more than just a peripheral hormone. *J Aging Res.* 2012;2012:384017.
77. Hoyer S. Memory function and brain glucose metabolism. *Pharmacopsychiatry.* 2003;36(suppl 1):S62–S67.
78. Hopkins DF, Williams G. Insulin receptors are widely distributed in human brain and bind human and porcine insulin with equal affinity. *Diabet Med.* 1997;14:1044–1050.
79. van der Heide LP, Ramakers GM, Smidt MP. Insulin signaling in the central nervous system: learning to survive. *Prog Neurobiol.* 2006;79:205–221.
80. Taniguchi CM, Emanuelli B, Kahn CR. Critical nodes in signalling pathways: insights into insulin action. *Nat Rev Mol Cell Biol.* 2006;7:85–96.
81. Aguirre V, Uchida T, Yenush L, Davis R, White MF. The c-Jun NH(2)-terminal kinase promotes insulin resistance during association with insulin receptor substrate-1 and phosphorylation of Ser(307). *J Biol Chem.* 2000;275:9047–9054.
82. Frojdo S, Vidal H, Pirola L. Alterations of insulin signaling in type 2 diabetes: a review of the current evidence from humans. *Biochim Biophys Acta.* 2009;1792:83–92.
83. Sun XJ, Liu F. Phosphorylation of IRS proteins Yin-Yang regulation of insulin signaling. *Vitam Horm.* 2009;80:351–387.
84. Boura-Halfon S, Zick Y. Phosphorylation of IRS proteins, insulin action, and insulin resistance. *Am J Physiol Endocrinol Metab.* 2009;296:E581–E591.
85. Gual P, Le Marchand-Brustel Y, Tanti JF. Positive and negative regulation of insulin signaling through IRS-1 phosphorylation. *Biochimie.* 2005;87:99–109.
86. Boura-Halfon S, Zick Y. Serine kinases of insulin receptor substrate proteins. *Vitam Horm.* 2009;80:313–349.

87. Stoothoff WH, Johnson GV. Tau phosphorylation: physiological and pathological consequences. *Biochim Biophys Acta.* 2005;1739:280–297.

88. Noble W, Hanger DP, Miller CC, Lovestone S. The importance of tau phosphorylation for neurodegenerative diseases. *Front Neurol.* 2013;4:83.

89. Mi K, Johnson GV. The role of tau phosphorylation in the pathogenesis of Alzheimer's disease. *Curr Alzheimer Res.* 2006;3:449–463.

90. Buee L, Bussiere T, Buee-Scherrer V, Delacourte A, Hof PR. Tau protein isoforms, phosphorylation and role in neurodegenerative disorders. *Brain Res Brain Res Rev.* 2000;33:95–130.

91. Deng Y, Li B, Liu Y, Iqbal K, Grundke-Iqbal I, Gong CX. Dysregulation of insulin signaling, glucose transporters, O-GlcNAcylation, and phosphorylation of tau and neurofilaments in the brain: implication for Alzheimer's disease. *Am J Pathol.* 2009;175:2089–2098.

92. Jolivalt CG, Lee CA, Beiswenger KK, et al. Defective insulin signaling pathway and increased glycogen synthase kinase-3 activity in the brain of diabetic mice: parallels with Alzheimer's disease and correction by insulin. *J Neurosci Res.* 2008;86:3265–3274.

93. Bhat R, Xue Y, Berg S, et al. Structural insights and biological effects of glycogen synthase kinase 3-specific inhibitor AR-A014418. *J Biol Chem.* 2003;278:45937–45945.

94. Schubert M, Brazil DP, Burks DJ, et al. Insulin receptor substrate-2 deficiency impairs brain growth and promotes tau phosphorylation. *J Neurosci.* 2003;23:7084–7092.

95. Moreira PI, Duarte AI, Santos MS, Rego AC, Oliveira CR. An integrative view of the role of oxidative stress, mitochondria and insulin in Alzheimer's disease. *J Alzheimers Dis.* 2009;16:741–761.

96. Drewes G. MARKing tau for tangles and toxicity. *Trends Biochem Sci.* 2004;29:548–555.

97. Zhu LQ, Wang SH, Liu D, et al. Activation of glycogen synthase kinase-3 inhibits long-term potentiation with synapse-associated impairments. *J Neurosci.* 2007;27:12211–12220.

98. DaRocha-Souto B, Coma M, Perez-Nievas BG, et al. Activation of glycogen synthase kinase-3 beta mediates beta-amyloid induced neuritic damage in Alzheimer's disease. *Neurobiol Dis.* 2012;45:425–437.

99. Xie L, Helmerhorst E, Taddei K, Plewright B, Van Bronswijk W, Martins R. Alzheimer's beta-amyloid peptides compete for insulin binding to the insulin receptor. *J Neurosci.* 2002;22:RC221.

100. Singh R, Barden A, Mori T, Beilin L. Advanced glycation end-products: a review. *Diabetologia.* 2001;44:129–146.

101. Cole GM, Frautschy SA. The role of insulin and neurotrophic factor signaling in brain aging and Alzheimer's disease. *Exp Gerontol.* 2007;42:10–21.

102. Stine Jr WB, Dahlgren KN, Krafft GA, LaDu MJ. In vitro characterization of conditions for amyloid-beta peptide oligomerization and fibrillogenesis. *J Biol Chem.* 2003;278:11612–11622.

103. Qiu WQ, Folstein MF. Insulin, insulin-degrading enzyme and amyloid-beta peptide in Alzheimer's disease: review and hypothesis. *Neurobiol Aging.* 2006;27:190–198.

104. Chesneau V, Perlman RK, Li W, Keller GA, Rosner MR. Insulin-degrading enzyme does not require peroxisomal localization for insulin degradation. *Endocrinology.* 1997;138:3444–3451.

105. Bernstein HG, Ansorge S, Riederer P, Reiser M, Frolich L, Bogerts B. Insulin-degrading enzyme in the Alzheimer's disease brain: prominent localization in neurons and senile plaques. *Neurosci Lett.* 1999;263:161–164.

106. Farris W, Mansourian S, Chang Y, et al. Insulin-degrading enzyme regulates the levels of insulin, amyloid beta-protein, and the beta-amyloid precursor protein intracellular domain in vivo. *Proc Natl Acad Sci USA.* 2003;100:4162–4167.

107. Gasparini L, Gouras GK, Wang R, et al. Stimulation of beta-amyloid precursor protein trafficking by insulin reduces intraneuronal beta-amyloid and requires mitogen-activated protein kinase signaling. *J Neurosci.* 2001;21:2561–2570.

108. Zhao WQ, De Felice FG, Fernandez S, et al. Amyloid beta oligomers induce impairment of neuronal insulin receptors. *FASEB J.* 2008;22:246–260.

109. Han X, Ma Y, Liu X, et al. Changes in insulin-signaling transduction pathway underlie learning/memory deficits in an Alzheimer's disease rat model. *J Neural Transm.* 2012;119:1407–1416.

110. Zheng WH, Kar S, Dore S, Quirion R. Insulin-like growth factor-1 (IGF-1): a neuroprotective trophic factor acting via the Akt kinase pathway. *J Neural Transm Suppl.* 2000;261–272.

111. Dore S, Bastianetto S, Kar S, Quirion R. Protective and rescuing abilities of IGF-I and some putative free radical scavengers against beta-amyloid-inducing toxicity in neurons. *Ann NY Acad Sci.* 1999;890:356–364.

112. Dore S, Kar S, Quirion R. Insulin-like growth factor I protects and rescues hippocampal neurons against beta-amyloid- and human amylin-induced toxicity. *Proc Natl Acad Sci USA.* 1997;94:4772–4777.

113. Garrido GE, Furuie SS, Buchpiguel CA, et al. Relation between medial temporal atrophy and functional brain activity during memory processing in Alzheimer's disease: a combined MRI and SPECT study. *J Neurol Neurosurg Psychiatry.* 2002;73:508–516.

114. Small GW, Ercoli LM, Silverman DH, et al. Cerebral metabolic and cognitive decline in persons at genetic risk for Alzheimer's disease. *Proc Natl Acad Sci USA.* 2000;97:6037–6042.

115. Sims-Robinson C, Kim B, Rosko A, Feldman EL. How does diabetes accelerate Alzheimer disease pathology? *Nat Rev Neurol.* 2010;6:551–559.

116. Hunt JV, Wolff SP. Oxidative glycation and free radical production: a causal mechanism of diabetic complications. *Free Radic Res Commun.* 1991;12–13(Pt 1):115–123.

117. Brownlee M. Advanced protein glycosylation in diabetes and aging. *Annu Rev Med.* 1995;46:223–234.

118. Biessels GJ, van der Heide LP, Kamal A, Bleys RL, Gispen WH. Ageing and diabetes: implications for brain function. *Eur J Pharmacol.* 2002;441:1–14.

119. Singh DK, Winocour P, Farrington K. Oxidative stress in early diabetic nephropathy: fueling the fire. *Nat Rev Endocrinol.* 2011;7:176–184.

120. Vincent AM, Russell JW, Low P, Feldman EL. Oxidative stress in the pathogenesis of diabetic neuropathy. *Endocr Rev.* 2004;25:612–628.

121. Russell JW, Berent-Spillson A, Vincent AM, Freimann CL, Sullivan KA, Feldman EL. Oxidative injury and neuropathy in diabetes and impaired glucose tolerance. *Neurobiol Dis.* 2008;30:420–429.

122. Moreira PI, Santos MS, Seica R, Oliveira CR. Brain mitochondrial dysfunction as a link between Alzheimer's disease and diabetes. *J Neurol Sci.* 2007;257:206–214.

123. de la Monte SM. Brain insulin resistance and deficiency as therapeutic targets in Alzheimer's disease. *Curr Alzheimer Res.* 2012;9:35–66.

124. Dietrich WD, Alonso O, Busto R. Moderate hyperglycemia worsens acute blood-brain barrier injury after forebrain ischemia in rats. *Stroke.* 1993;24:111–116.

125. Starr JM, Wardlaw J, Ferguson K, MacLullich A, Deary IJ, Marshall I. Increased blood-brain barrier permeability in type II diabetes demonstrated by gadolinium magnetic resonance imaging. *J Neurol Neurosurg Psychiatry.* 2003;74:70–76.

126. Kamada H, Yu F, Nito C, Chan PH. Influence of hyperglycemia on oxidative stress and matrix metalloproteinase-9 activation after focal cerebral ischemia/reperfusion in rats: relation to blood-brain barrier dysfunction. *Stroke.* 2007;38:1044–1049.

127. Serlin Y, Levy J, Shalev H. Vascular pathology and blood-brain barrier disruption in cognitive and psychiatric complications of type 2 diabetes mellitus. *Cardiovasc Psychiatry Neurol.* 2011;2011:609202.

128. Anderson MM, Requena JR, Crowley JR, Thorpe SR, Heinecke JW. The myeloperoxidase system of human phagocytes generates Nepsilon-(carboxymethyl) lysine on proteins: a mechanism for producing advanced glycation end products at sites of inflammation. *J Clin Invest.* 1999;104:103–113.

129. Civelek S, Gelisgen R, Andican G, et al. Advanced glycation end products and antioxidant status in nondiabetic and streptozotocin induced diabetic rats: effects of copper treatment. *Biometals.* 2010;23:43–49.

130. Scivittaro V, Ganz MB, Weiss MF. AGEs induce oxidative stress and activate protein kinase C-beta(II) in neonatal mesangial cells. *Am J Physiol Renal Physiol.* 2000;278: F676–F683.

131. Ceriello A. Hyperglycaemia: the bridge between non-enzymatic glycation and oxidative stress in the pathogenesis of diabetic complications. *Diabetes Nutr Metab.* 1999;12:42–46.

132. Lustbader JW, Cirilli M, Lin C, et al. ABAD directly links Abeta to mitochondrial toxicity in Alzheimer's disease. *Science.* 2004;304:448–452.

133. Nishikawa T, Edelstein D, Du XL, et al. Normalizing mitochondrial superoxide production blocks three pathways of hyperglycaemic damage. *Nature.* 2000;404:787–790.

134. Lenaz G. Role of mitochondria in oxidative stress and ageing. *Biochim Biophys Acta.* 1998;1366:53–67.

135. Moreira PI, Santos MS, Moreno AM, Seica R, Oliveira CR. Increased vulnerability of brain mitochondria in diabetic (Goto-Kakizaki) rats with aging and amyloid-beta exposure. *Diabetes.* 2003;52:1449–1456.

136. Rizzuto R, Bernardi P, Pozzan T. Mitochondria as all-round players of the calcium game. *J Physiol.* 2000;529(Pt 1):37–47.

137. Biessels GJ, ter Laak MP, Hamers FP, Gispen WH. Neuronal Ca^{2+} disregulation in diabetes mellitus. *Eur J Pharmacol.* 2002;447:201–209.

138. Kostyuk E, Voitenko N, Kruglikov I, et al. Diabetes-induced changes in calcium homeostasis and the effects of calcium channel blockers in rat and mice nociceptive neurons. *Diabetologia.* 2001;44:1302–1309.

139. Khachaturian ZS. Calcium hypothesis of Alzheimer's disease and brain aging. *Ann NY Acad Sci.* 1994;747:1–11.

140. Biessels G, Gispen WH. The calcium hypothesis of brain aging and neurodegenerative disorders: significance in diabetic neuropathy. *Life Sci.* 1996;59:379–387.

141. Berridge MJ. Calcium hypothesis of Alzheimer's disease. *Pflugers Arch.* 2010;459: 441–449.

142. Thibault O, Gant JC, Landfield PW. Expansion of the calcium hypothesis of brain aging and Alzheimer's disease: minding the store. *Aging Cell.* 2007;6:307–317.

143. Hovsepyan MR, Haas MJ, Boyajyan AS, et al. Astrocytic and neuronal biochemical markers in the sera of subjects with diabetes mellitus. *Neurosci Lett.* 2004;369:224–227.

144. Hotamisligil GS. Inflammatory pathways and insulin action. *Int J Obes Relat Metab Disord.* 2003;27(suppl 3):S53–S55.

145. Hak AE, Pols HA, Stehouwer CD, et al. Markers of inflammation and cellular adhesion molecules in relation to insulin resistance in nondiabetic elderly: the Rotterdam study. *J Clin Endocrinol Metab.* 2001;86:4398–4405.

146. Huber JD. Diabetes, cognitive function, and the blood-brain barrier. *Curr Pharm Des.* 2008;14:1594–1600.

147. Wang H, Golob EJ, Su MY. Vascular volume and blood-brain barrier permeability measured by dynamic contrast enhanced MRI in hippocampus and cerebellum of patients with MCI and normal controls. *J Magn Reson Imaging.* 2006;24:695–700.

148. Bowman GL, Kaye JA, Moore M, Waichunas D, Carlson NE, Quinn JF. Blood-brain barrier impairment in Alzheimer disease: stability and functional significance. *Neurology.* 2007;68:1809–1814.

149. Dai J, Vrensen GF, Schlingemann RO. Blood-brain barrier integrity is unaltered in human brain cortex with diabetes mellitus. *Brain Res.* 2002;954:311–316.

150. Horani MH, Mooradian AD. Effect of diabetes on the blood brain barrier. *Curr Pharm Des.* 2003;9:833–840.

151. Shibata M, Yamada S, Kumar SR, et al. Clearance of Alzheimer's amyloid-ss(1-40) peptide from brain by LDL receptor-related protein-1 at the blood-brain barrier. *J Clin Invest.* 2000;106:1489–1499.

152. Iwata N, Higuchi M, Saido TC. Metabolism of amyloid-beta peptide and Alzheimer's disease. *Pharmacol Ther.* 2005;108:129–148.

153. Deane R, Wu Z, Sagare A, et al. LRP/amyloid beta-peptide interaction mediates differential brain efflux of Abeta isoforms. *Neuron.* 2004;43:333–344.

154. Alonzo NC, Hyman BT, Rebeck GW, Greenberg SM. Progression of cerebral amyloid angiopathy: accumulation of amyloid-beta40 in affected vessels. *J Neuropathol Exp Neurol.* 1998;57:353–359.

155. Zlokovic BV. Clearing amyloid through the blood-brain barrier. *J Neurochem.* 2004;89:807–811.

156. Bell RD, Zlokovic BV. Neurovascular mechanisms and blood-brain barrier disorder in Alzheimer's disease. *Acta Neuropathol.* 2009;118:103–113.

157. Bowman GL, Quinn JF. Alzheimer's disease and the blood-brain barrier: past, present and future. *Aging Health.* 2008;4:47–55.

158. Sharma HS, Castellani RJ, Smith MA, Sharma A. The blood-brain barrier in Alzheimer's disease: novel therapeutic targets and nanodrug delivery. *Int Rev Neurobiol.* 2012;102:47–90.

159. Leonard BL, Watson RN, Loomes KM, Phillips AR, Cooper GJ. Insulin resistance in the Zucker diabetic fatty rat: a metabolic characterisation of obese and lean phenotypes. *Acta Diabetol.* 2005;42:162–170.

160. Capeau J. Insulin resistance and steatosis in humans. *Diabetes Metab.* 2008;34:649–657.

161. Langeveld M, Aerts JM. Glycosphingolipids and insulin resistance. *Prog Lipid Res.* 2009;48:196–205.

162. Kraegen EW, Cooney GJ. Free fatty acids and skeletal muscle insulin resistance. *Curr Opin Lipidol.* 2008;19:235–241.

163. Summers SA. Ceramides in insulin resistance and lipotoxicity. *Prog Lipid Res.* 2006;45:42–72.

164. Zierath JR. The path to insulin resistance: paved with ceramides? *Cell Metab.* 2007;5:161–163.

165. Arboleda G, Huang TJ, Waters C, Verkhratsky A, Fernyhough P, Gibson RM. Insulin-like growth factor-1-dependent maintenance of neuronal metabolism through the phosphatidylinositol 3-kinase-Akt pathway is inhibited by C2-ceramide in CAD cells. *Eur J Neurosci.* 2007;25:3030–3038.

166. Bourbon NA, Sandirasegarane L, Kester M. Ceramide-induced inhibition of Akt is mediated through protein kinase Czeta: implications for growth arrest. *J Biol Chem.* 2002;277:3286–3292.

167. Holland WL, Brozinick JT, Wang LP, et al. Inhibition of ceramide synthesis ameliorates glucocorticoid-, saturated-fat-, and obesity-induced insulin resistance. *Cell Metab.* 2007;5:167–179.

168. Consitt LA, Bell JA, Houmard JA. Intramuscular lipid metabolism, insulin action, and obesity. *IUBMB Life.* 2009;61:47–55.

169. de la Monte SM, Tong M, Nguyen V, Setshedi M, Longato L, Wands JR. Ceramide-mediated insulin resistance and impairment of cognitive-motor functions. *J Alzheimers Dis.* 2010;21:967–984.

170. Tong M, de la Monte SM. Mechanisms of ceramide-mediated neurodegeneration. *J Alzheimers Dis.* 2009;16:705–714.
171. Boehm JE, Chaika OV, Lewis RE. Rac-dependent anti-apoptotic signaling by the insulin receptor cytoplasmic domain. *J Biol Chem.* 1999;274:28632–28636.
172. Bertrand F, Atfi A, Cadoret A, et al. A role for nuclear factor kappaB in the antiapoptotic function of insulin. *J Biol Chem.* 1998;273:2931–2938.
173. Zhang X, Tang N, Hadden TJ, Rishi AK. Akt, FoxO and regulation of apoptosis. *Biochim Biophys Acta.* 2011;1813:1978–1986.
174. Yan L, Lavin VA, Moser LR, Cui Q, Kanies C, Yang E. PP2A regulates the pro-apoptotic activity of FOXO1. *J Biol Chem.* 2008;283:7411–7420.
175. Ni YG, Wang N, Cao DJ, et al. FoxO transcription factors activate Akt and attenuate insulin signaling in heart by inhibiting protein phosphatases. *Proc Natl Acad Sci USA.* 2007;104:20517–20522.
176. Yang E, Zha J, Jockel J, Boise LH, Thompson CB, Korsmeyer SJ. Bad, a heterodimeric partner for Bcl-XL and Bcl-2, displaces Bax and promotes cell death. *Cell.* 1995;80:285–291.
177. Bertrand F, Philippe C, Antoine PJ, et al. Insulin activates nuclear factor kappa B in mammalian cells through a Raf-1-mediated pathway. *J Biol Chem.* 1995;270: 24435–24441.
178. Zhou G, Kuo MT. NF-kappaB-mediated induction of mdr1b expression by insulin in rat hepatoma cells. *J Biol Chem.* 1997;272:15174–15183.
179. Kaltschmidt B, Kaltschmidt C, Hofmann TG, Hehner SP, Droge W, Schmitz ML. The pro- or anti-apoptotic function of NF-kappaB is determined by the nature of the apo-ptotic stimulus. *Eur J Biochem.* 2000;267:3828–3835.
180. Barger SW, Mattson MP. Induction of neuroprotective kappa B-dependent transcrip-tion by secreted forms of the Alzheimer's beta-amyloid precursor. *Brain Res Mol Brain Res.* 1996;40:116–126.
181. Camandola S, Poli G, Mattson MP. The lipid peroxidation product 4-hydroxy-2,3-nonenal inhibits constitutive and inducible activity of nuclear factor kappa B in neurons. *Brain Res Mol Brain Res.* 2000;85:53–60.
182. Barger SW, Horster D, Furukawa K, Goodman Y, Krieglstein J, Mattson MP. Tumor necrosis factors alpha and beta protect neurons against amyloid beta-peptide toxicity: evidence for involvement of a kappa B-binding factor and attenuation of peroxide and Ca2+ accumulation. *Proc Natl Acad Sci USA.* 1995;92:9328–9332.
183. Bartus RT, Dean 3rd RL, Beer B, Lippa AS. The cholinergic hypothesis of geriatric memory dysfunction. *Science.* 1982;217:408–414.
184. Daulatzai MA. Early stages of pathogenesis in memory impairment during normal senescence and Alzheimer's disease. *J Alzheimers Dis.* 2010;20:355–367.
185. Araujo DM, Lapchak PA, Robitaille Y, Gauthier S, Quirion R. Differential alteration of various cholinergic markers in cortical and subcortical regions of human brain in Alzheimer's disease. *J Neurochem.* 1988;50:1914–1923.
186. Wang H, Wang R, Zhao Z, et al. Coexistences of insulin signaling-related proteins and choline acetyltransferase in neurons. *Brain Res.* 2009;1249:237–243.
187. Du YF, Yan P, Guo SG, Qu CQ. Effects of fibrillar Abeta(1-40) on the viability of primary cultures of cholinergic neurons and the expression of insulin signaling-related proteins. *Anat Rec (Hoboken).* 2011;294:287–294.
188. Rossor MN, Garrett NJ, Johnson AL, Mountjoy CQ, Roth M, Iversen LL. A post-mortem study of the cholinergic and GABA systems in senile dementia. *Brain.* 1982;105:313–330.
189. Bird TD, Stranahan S, Sumi SM, Raskind M. Alzheimer's disease: choline acetyltransferase activity in brain tissue from clinical and pathological subgroups. *Ann Neurol.* 1983;14:284–293.

190. Lester-Coll N, Rivera EJ, Soscia SJ, Doiron K, Wands JR, de la Monte SM. Intracerebral streptozotocin model of type 3 diabetes: relevance to sporadic Alzheimer's disease. *J Alzheimers Dis*. 2006;9:13–33.
191. Ovsepian SV. Differential cholinergic modulation of synaptic encoding and gain control mechanisms in rat hippocampus. *Neurosci Res*. 2008;61:92–98.
192. Shinoe T, Matsui M, Taketo MM, Manabe T. Modulation of synaptic plasticity by physiological activation of M1 muscarinic acetylcholine receptors in the mouse hippocampus. *J Neurosci*. 2005;25:11194–11200.
193. Aigner TG. Pharmacology of memory: cholinergic-glutamatergic interactions. *Curr Opin Neurobiol*. 1995;5:155–160.
194. Matsuoka N, Aigner TG. Cholinergic-glutamatergic interactions in visual recognition memory of rhesus monkeys. *Neuroreport*. 1996;7:565–568.
195. Bliss TV, Collingridge GL. A synaptic model of memory: long-term potentiation in the hippocampus. *Nature*. 1993;361:31–39.
196. Jodar L, Kaneto H. Synaptic plasticity: stairway to memory. *Jpn J Pharmacol*. 1995;68:359–387.
197. Kamal A, Biessels GJ, Urban IJ, Gispen WH. Hippocampal synaptic plasticity in streptozotocin-diabetic rats: impairment of long-term potentiation and facilitation of long-term depression. *Neuroscience*. 1999;90:737–745.
198. Gardoni F, Kamal A, Bellone C, et al. Effects of streptozotocin-diabetes on the hippocampal NMDA receptor complex in rats. *J Neurochem*. 2002;80:438–447.
199. Shonesy BC, Thiruchelvam K, Parameshwaran K, et al. Central insulin resistance and synaptic dysfunction in intracerebroventricular-streptozotocin injected rodents. *Neurobiol Aging*. 2012;33(430):e5–e18.
200. Schioth HB, Craft S, Brooks SJ, Frey 2nd WH, Benedict C. Brain insulin signaling and Alzheimer's disease: current evidence and future directions. *Mol Neurobiol*. 2012;46:4–10.
201. Huang CC, Lee CC, Hsu KS. The role of insulin receptor signaling in synaptic plasticity and cognitive function. *Chang Gung Med J*. 2010;33:115–125.
202. Auerbach JM, Segal M. Muscarinic receptors mediating depression and long-term potentiation in rat hippocampus. *J Physiol*. 1996;492(Pt 2):479–493.
203. Auerbach JM, Segal M. A novel cholinergic induction of long-term potentiation in rat hippocampus. *J Neurophysiol*. 1994;72:2034–2040.
204. Kanju PM, Parameshwaran K, Sims-Robinson C, et al. Selective cholinergic depletion in medial septum leads to impaired long term potentiation and glutamatergic synaptic currents in the hippocampus. *PLoS One*. 2012;7:e31073.
205. Parameshwaran K, Buabeid M, Bhattacharya S, et al. Long term alterations in synaptic physiology, expression of β2 nicotinic receptors and ERK1/2 signaling in the hippocampus of rats with prenatal nicotine exposure. *Neurobiol Learn Mem*. 2013;106C:102–111.
206. Hooper C, Markevich V, Plattner F, et al. Glycogen synthase kinase-3 inhibition is integral to long-term potentiation. *Eur J Neurosci*. 2007;25:81–86.
207. Peineau S, Taghibiglou C, Bradley C, et al. LTP inhibits LTD in the hippocampus via regulation of GSK3beta. *Neuron*. 2007;53:703–717.
208. Laviola L, Natalicchio A, Giorgino F. The IGF-I signaling pathway. *Curr Pharm Des*. 2007;13:663–669.
209. Werner H, Weinstein D, Bentov I. Similarities and differences between insulin and IGF-I: structures, receptors, and signalling pathways. *Arch Physiol Biochem*. 2008;114:17–22.
210. Sjogren K, Wallenius K, Liu JL, et al. Liver-derived IGF-I is of importance for normal carbohydrate and lipid metabolism. *Diabetes*. 2001;50:1539–1545.
211. Yakar S, Liu JL, Fernandez AM, et al. Liver-specific IGF-1 gene deletion leads to muscle insulin insensitivity. *Diabetes*. 2001;50:1110–1118.

212. Arwert LI, Deijen JB, Drent ML. The relation between insulin-like growth factor I levels and cognition in healthy elderly: a meta-analysis. *Growth Horm IGF Res.* 2005;15:416–422.

213. Aleman A, Verhaar HJ, De Haan EH, et al. Insulin-like growth factor-I and cognitive function in healthy older men. *J Clin Endocrinol Metab.* 1999;84:471–475.

214. Bellar D, Glickman EL, Juvancic-Heltzel J, Gunstad J. Serum insulin like growth factor-1 is associated with working memory, executive function and selective attention in a sample of healthy, fit older adults. *Neuroscience.* 2011;178:133–137.

215. Vardy ER, Rice PJ, Bowie PC, Holmes JD, Grant PJ, Hooper NM. Increased circulating insulin-like growth factor-1 in late-onset Alzheimer's disease. *J Alzheimers Dis.* 2007;12:285–290.

216. Watanabe T, Miyazaki A, Katagiri T, Yamamoto H, Idei T, Iguchi T. Relationship between serum insulin-like growth factor-1 levels and Alzheimer's disease and vascular dementia. *J Am Geriatr Soc.* 2005;53:1748–1753.

217. Moloney AM, Griffin RJ, Timmons S, O'Connor R, Ravid R, O'Neill C. Defects in IGF-1 receptor, insulin receptor and IRS-1/2 in Alzheimer's disease indicate possible resistance to IGF-1 and insulin signalling. *Neurobiol Aging.* 2010;31:224–243.

218. Sjogren K, Liu JL, Blad K, et al. Liver-derived insulin-like growth factor I (IGF-I) is the principal source of IGF-I in blood but is not required for postnatal body growth in mice. *Proc Natl Acad Sci USA.* 1999;96:7088–7092.

219. Rotwein P, Burgess SK, Milbrandt JD, Krause JE. Differential expression of insulin-like growth factor genes in rat central nervous system. *Proc Natl Acad Sci USA.* 1988;85:265–269.

220. Ayer-le Lievre C, Stahlbom PA, Sara VR. Expression of IGF-I and -II mRNA in the brain and craniofacial region of the rat fetus. *Development.* 1991;111:105–115.

221. Ungvari Z, Csiszar A. The emerging role of IGF-1 deficiency in cardiovascular aging: recent advances. *J Gerontol A Biol Sci Med Sci.* 2012;67:599–610.

222. Sonntag WE, Deak F, Ashpole N, et al. Insulin-like growth factor-1 in CNS and cerebrovascular aging. *Front Aging Neurosci.* 2013;5:27.

223. Trejo JL, Piriz J, Llorens-Martin MV, et al. Central actions of liver-derived insulin-like growth factor I underlying its pro-cognitive effects. *Mol Psychiatry.* 2007;12:1118–1128.

224. Deak F, Sonntag WE. Aging, synaptic dysfunction, and insulin-like growth factor (IGF)-1. *J Gerontol A Biol Sci Med Sci.* 2012;67:611–625.

225. Trejo JL, Llorens-Martin MV, Torres-Aleman I. The effects of exercise on spatial learning and anxiety-like behavior are mediated by an IGF-I-dependent mechanism related to hippocampal neurogenesis. *Mol Cell Neurosci.* 2008;37:402–411.

226. Zhao WQ, Lacor PN, Chen H, et al. Insulin receptor dysfunction impairs cellular clearance of neurotoxic oligomeric a{beta}. *J Biol Chem.* 2009;284:18742–18753.

227. Araki W, Kume H, Oda A, Tamaoka A, Kametani F. IGF-1 promotes beta-amyloid production by a secretase-independent mechanism. *Biochem Biophys Res Commun.* 2009;380:111–114.

228. Freude S, Hettich MM, Schumann C, et al. Neuronal IGF-1 resistance reduces Abeta accumulation and protects against premature death in a model of Alzheimer's disease. *FASEB J.* 2009;23:3315–3324.

229. Freude S, Schilbach K, Schubert M. The role of IGF-1 receptor and insulin receptor signaling for the pathogenesis of Alzheimer's disease: from model organisms to human disease. *Curr Alzheimer Res.* 2009;6:213–223.

230. Barnes BB, Chang-Claude J, Flesch-Janys D, et al. Cancer risk factors associated with insulin-like growth factor (IGF)-I and IGF-binding protein-3 levels in healthy women: effect modification by menopausal status. *Cancer Causes Control.* 2009;20:1985–1996.

231. Douglas JB, Silverman DT, Pollak MN, Tao Y, Soliman AS, Stolzenberg-Solomon-RZ. Serum IGF-I, IGF-II, IGFBP-3, and IGF-I/IGFBP-3 molar ratio and risk of pancreatic cancer in the prostate, lung, colorectal, and ovarian cancer screening trial. *Cancer Epidemiol Biomarkers Prev.* 2010;19:2298–2306.

232. Rinaldi S, Cleveland R, Norat T, et al. Serum levels of IGF-I, IGFBP-3 and colorectal cancer risk: results from the EPIC cohort, plus a meta-analysis of prospective studies. *Int J Cancer.* 2010;126:1702–1715.

233. Sevigny JJ, Ryan JM, van Dyck CH, et al. Growth hormone secretagogue MK-677: no clinical effect on AD progression in a randomized trial. *Neurology.* 2008;71:1702–1708.

234. Bruning JC, Gautam D, Burks DJ, et al. Role of brain insulin receptor in control of body weight and reproduction. *Science.* 2000;289:2122–2125.

235. Schubert M, Gautam D, Surjo D, et al. Role for neuronal insulin resistance in neuro-degenerative diseases. *Proc Natl Acad Sci USA.* 2004;101:3100–3105.

236. Dudek H, Datta SR, Franke TF, et al. Regulation of neuronal survival by the serine-threonine protein kinase Akt. *Science.* 1997;275:661–665.

237. Lee GH, Proenca R, Montez JM, et al. Abnormal splicing of the leptin receptor in diabetic mice. *Nature.* 1996;379:632–635.

238. Beccano-Kelly D, Harvey J. Leptin: a novel therapeutic target in Alzheimer's disease? *Int J Alzheimers Dis.* 2012;2012:594137.

239. Harvey J, Shanley LJ, O'Malley D, Irving AJ. Leptin: a potential cognitive enhancer? *Biochem Soc Trans.* 2005;33:1029–1032.

240. Greco SJ, Sarkar S, Johnston JM, et al. Leptin reduces Alzheimer's disease-related tau phosphorylation in neuronal cells. *Biochem Biophys Res Commun.* 2008;376:536–541.

241. Fewlass DC, Noboa K, Pi-Sunyer FX, Johnston JM, Yan SD, Tezapsidis N. Obesity-related leptin regulates Alzheimer's Abeta. *FASEB J.* 2004;18:1870–1878.

242. Szkudelski T. The mechanism of alloxan and streptozotocin action in B cells of the rat pancreas. *Physiol Res.* 2001;50:537–546.

243. Biessels GJ, Kamal A, Ramakers GM, et al. Place learning and hippocampal synaptic plasticity in streptozotocin-induced diabetic rats. *Diabetes.* 1996;45:1259–1266.

244. Popovic M, Biessels GJ, Isaacson RL, Gispen WH. Learning and memory in streptozotocin-induced diabetic rats in a novel spatial/object discrimination task. *Behav Brain Res.* 2001;122:201–207.

245. Biessels GJ, Kamal A, Urban IJ, Spruijt BM, Erkelens DW, Gispen WH. Water maze learning and hippocampal synaptic plasticity in streptozotocin-diabetic rats: effects of insulin treatment. *Brain Res.* 1998;800:125–135.

246. Lannert H, Hoyer S. Intracerebroventricular administration of streptozotocin causes long-term diminutions in learning and memory abilities and in cerebral energy metabolism in adult rats. *Behav Neurosci.* 1998;112:1199–1208.

247. Hoyer S, Muller D, Plaschke K. Desensitization of brain insulin receptor. Effect on glucose/energy and related metabolism. *J Neural Transm Suppl.* 1994;44:259–268.

248. Plaschke K, Hoyer S. Action of the diabetogenic drug streptozotocin on glycolytic and glycogenolytic metabolism in adult rat brain cortex and hippocampus. *Int J Dev Neurosci.* 1993;11:477–483.

249. Mayer G, Nitsch R, Hoyer S. Effects of changes in peripheral and cerebral glucose metabolism on locomotor activity, learning and memory in adult male rats. *Brain Res.* 1990;532:95–100.

250. Grunblatt E, Salkovic-Petrisic M, Osmanovic J, Riederer P, Hoyer S. Brain insulin system dysfunction in streptozotocin intracerebroventricularly treated rats generates hyperphosphorylated tau protein. *J Neurochem.* 2007;101:757–770.

251. de la Monte SM, Tong M, Lester-Coll N, Plater Jr M, Wands JR. Therapeutic rescue of neurodegeneration in experimental type 3 diabetes: relevance to Alzheimer's disease. *J Alzheimers Dis.* 2006;10:89–109.

252. Kraska A, Santin MD, Dorieux O, et al. In vivo cross-sectional characterization of cerebral alterations induced by intracerebroventricular administration of streptozotocin. PLoS One. 2012;7:e46196.
253. Frederiksen KS, Garde E, Skimminge A, et al. Corpus callosum atrophy in patients with mild Alzheimer's disease. Neurodegener Dis. 2011;8:476–482.
254. Janowsky JS, Kaye JA, Carper RA. Atrophy of the corpus callosum in Alzheimer's disease versus healthy aging. J Am Geriatr Soc. 1996;44:798–803.
255. Chen Y, Zhou K, Wang R, et al. Antidiabetic drug metformin (GlucophageR) increases biogenesis of Alzheimer's amyloid peptides via up-regulating BACE1 transcription. Proc Natl Acad Sci USA. 2009;106:3907–3912.
256. Rashidy-Pour A. ATP-sensitive potassium channels mediate the effects of a peripheral injection of glucose on memory storage in an inhibitory avoidance task. Behav Brain Res. 2001;126:43–48.
257. Risner ME, Saunders AM, Altman JF, et al. Efficacy of rosiglitazone in a genetically defined population with mild-to-moderate Alzheimer's disease. Pharmacogenomics J. 2006;6:246–254.
258. Hanson LR, Frey 2nd WH. Intranasal delivery bypasses the blood-brain barrier to target therapeutic agents to the central nervous system and treat neurodegenerative disease. BMC Neurosci. 2008;9(suppl 3):S5.
259. Craft S, Baker LD, Montine TJ, et al. Intranasal insulin therapy for Alzheimer disease and amnestic mild cognitive impairment: a pilot clinical trial. Arch Neurol. 2012;69:29–38.
260. Reger MA, Watson GS, Green PS, et al. Intranasal insulin improves cognition and modulates beta-amyloid in early AD. Neurology. 2008;70:440–448.
261. Born J, Lange T, Kern W, McGregor GP, Bickel U, Fehm HL. Sniffing neuropeptides: a transnasal approach to the human brain. Nat Neurosci. 2002;5:514–516.
262. Reger MA, Watson GS, Frey 2nd WH, et al. Effects of intranasal insulin on cognition in memory-impaired older adults: modulation by APOE genotype. Neurobiol Aging. 2006;27:451–458.
263. Donix M, Small GW, Bookheimer SY. Family history and APOE-4 genetic risk in Alzheimer's disease. Neuropsychol Rev. 2012;22:298–309.
264. Schnegg CI, Greene-Schloesser D, Kooshki M, Payne VS, Hsu FC, Robbins ME. The PPARdelta agonist GW0742 inhibits neuroinflammation, but does not restore neurogenesis or prevent early delayed hippocampal-dependent cognitive impairment after whole-brain irradiation. Free Radic Biol Med. 2013;61C:1–9.
265. Martin HL, Mounsey RB, Sathe K, et al. A peroxisome proliferator-activated receptor-delta agonist provides neuroprotection in the 1-methyl-4-phenyl-1,2,3,6-tetrahydropyridine model of Parkinson's disease. Neuroscience. 2013;240:191–203.
266. Kalinin S, Richardson JC, Feinstein DL. A PPARdelta agonist reduces amyloid burden and brain inflammation in a transgenic mouse model of Alzheimer's disease. Curr Alzheimer Res. 2009;6:431–437.
267. Smith U. Pioglitazone: mechanism of action. Int J Clin Pract Suppl. 2001;13–18.
268. Bogacka I, Xie H, Bray GA, Smith SR. The effect of pioglitazone on peroxisome proliferator-activated receptor-gamma target genes related to lipid storage in vivo. Diabetes Care. 2004;27:1660–1667.
269. Denner LA, Rodriguez-Rivera J, Haidacher SJ, et al. Cognitive enhancement with rosiglitazone links the hippocampal PPARgamma and ERK MAPK signaling pathways. J Neurosci. 2012;32:16725–16735a.
270. Masciopinto F, Di Pietro N, Corona C, et al. Effects of long-term treatment with pioglitazone on cognition and glucose metabolism of PS1-KI, 3xTg-AD, and wild-type mice. Cell Death Dis. 2012;3:e448.

271. Searcy JL, Phelps JT, Pancani T, et al. Long-term pioglitazone treatment improves learning and attenuates pathological markers in a mouse model of Alzheimer's disease. *J Alzheimers Dis.* 2012;30:943–961.

272. O'Reilly JA, Lynch M. Rosiglitazone improves spatial memory and decreases insoluble Abeta(1-42) in APP/PS1 mice. *J Neuroimmune Pharmacol.* 2012;7:140–144.

273. Nicolakakis N, Hamel E. The nuclear receptor PPARgamma as a therapeutic target for cerebrovascular and brain dysfunction in Alzheimer's disease. *Front Aging Neurosci.* 2010;2:21.

274. Escribano L, Simon AM, Perez-Mediavilla A, Salazar-Colocho P, Del Rio J, Frechilla D. Rosiglitazone reverses memory decline and hippocampal glucocorticoid receptor down-regulation in an Alzheimer's disease mouse model. *Biochem Biophys Res Commun.* 2009;379:406–410.

275. Sato T, Hanyu H, Hirao K, Kanetaka H, Sakurai H, Iwamoto T. Efficacy of PPAR-gamma agonist pioglitazone in mild Alzheimer disease. *Neurobiol Aging.* 2011;32:1626–1633.

276. Gold M, Alderton C, Zvartau-Hind M, et al. Rosiglitazone monotherapy in mild-to-moderate Alzheimer's disease: results from a randomized, double-blind, placebo-controlled phase III study. *Dement Geriatr Cogn Disord.* 2010;30:131–146.

277. Barroso E, Del Valle J, Porquet D, et al. Tau hyperphosphorylation and increased BACE1 and RAGE levels in the cortex of PPARbeta/delta-null mice. *Biochim Biophys Acta.* 2013;1832:1241–1248.

278. Gault VA, Flatt PR, O'Harte FP. Glucose-dependent insulinotropic polypeptide analogues and their therapeutic potential for the treatment of obesity-diabetes. *Biochem Biophys Res Commun.* 2003;308:207–213.

279. Vella A, Shah P, Reed AS, Adkins AS, Basu R, Rizza RA. Lack of effect of exendin-4 and glucagon-like peptide-1-(7,36)-amide on insulin action in non-diabetic humans. *Diabetologia.* 2002;45:1410–1415.

280. Parthsarathy V, Holscher C. Chronic treatment with the GLP1 analogue liraglutide increases cell proliferation and differentiation into neurons in an AD mouse model. *PLoS One.* 2013;8:e58784.

281. McClean PL, Parthsarathy V, Faivre E, Holscher C. The diabetes drug liraglutide prevents degenerative processes in a mouse model of Alzheimer's disease. *J Neurosci.* 2011;31:6587–6594.

282. McGovern SF, Hunter K, Holscher C. Effects of the glucagon-like polypeptide-1 analogue (Val8)GLP-1 on learning, progenitor cell proliferation and neurogenesis in the C57B/16 mouse brain. *Brain Res.* 2012;1473:204–213.

283. McClean PL, Gault VA, Harriott P, Holscher C. Glucagon-like peptide-1 analogues enhance synaptic plasticity in the brain: a link between diabetes and Alzheimer's disease. *Eur J Pharmacol.* 2010;630:158–162.

284. Wang XH, Yang W, Holscher C, et al. Val(8)-GLP-1 remodels synaptic activity and intracellular calcium homeostasis impaired by amyloid beta peptide in rats. *J Neurosci Res.* 2013;91:568–577.

CHAPTER FOURTEEN

The Role of PPARδ Signaling in the Cardiovascular System

Yishu Ding, Kevin D. Yang, Qinglin Yang
Department of Nutrition Sciences, University of Alabama at Birmingham, Birmingham, Alabama, USA

Contents

Abstract

Peroxisome proliferator-activated receptors (PPARα, β/δ, and γ), members of the nuclear receptor transcription factor superfamily, play important roles in the regulation of metabolism, inflammation, and cell differentiation. All three PPAR subtypes are expressed in the cardiovascular system with various expression patterns. Among the three PPAR subtypes, PPARδ is the least studied but has arisen as a potential therapeutic target for cardiovascular and many other diseases. It is known that PPARδ is ubiquitously expressed and abundantly expressed in cardiomyocytes. Accumulated evidence illustrates the role of PPARδ in regulating cardiovascular function and determining pathological progression. In this chapter, we will discuss the current knowledge in the role of PPARδ in the cardiovascular system, the mechanistic insights, and the potential therapeutic utilization for treating cardiovascular disease.

Progress in Molecular Biology and Translational Science, Volume 121
ISSN 1877-1173
http://dx.doi.org/10.1016/B978-0-12-800101-1.00014-4

451

1. INTRODUCTION

Peroxisome proliferator-activated receptors (PPARα, β/δ, and γ) are transcription factors of the ligand-activated nuclear receptor superfamily.[1] They diversely regulate gene expression by forming heterodimers with retinoid X receptor (RXR) and binding to the PPAR-responsive element (PPRE) of target genes.[2] A variety of fatty acids and their derivatives can serve as endogenous ligands that activate PPARs.[1–3] The functions of PPARα and PPARγ have been extensively studied due to the clinical applications of subtype-specific ligands. The physiological and pharmacological roles of PPARδ have emerged during the last decade.[4] Among the three PPAR subtypes, PPARδ is the least studied, partly because no PPARδ ligands have been proven yet for clinical use. Accumulating evidence supports the important roles of PPARδ in regulating metabolism, inflammation, and cell proliferation.[5–7] Cellular metabolic homeostasis, inflammation, and proliferation are key components in the development of cardiovascular disease, such as atherosclerosis, hypertrophic cardiomyopathy, and heart failure. A large body of literature related to the role of PPARs, including PPARδ, in the cardiovascular system has accumulated. Therapies targeting PPARδ are under active investigation in basic pharmacological studies and early clinical trials. The purpose of this chapter is to comprehensively summarize and critically discuss current findings on the role of PPARδ in the cardiovascular system, the mechanistic insights, and the potential therapeutic utilization for treating cardiovascular disease.

2. EXPRESSION OF PPARδ IN THE CARDIOVASCULAR SYSTEM

Initial studies on tissue distribution of PPARs demonstrate the ubiquitous expression pattern of PPARδ.[8–10] Despite this ubiquity, the relative abundance of PPARδ transcript and protein in different tissues and cell types seems to be varied. For example, we found that PPARδ expression is predominant in cardiomyocytes relative to other cell types, at least in the rodent heart.[11] However, little is known about the regulation of PPARδ expression. An early study on the transcriptional regulation of the PPARδ gene revealed that two putative β-catenin/Tcf4-binding sites are located on the promoter and the upregulation of PPARδ is mediated by β-catenin/Tcf4.[12] On the other hand, there are numerous investigations demonstrating that the PPARδ expression may be influenced by other signaling pathways under specific cellular conditions.[13,14] Obviously, further in-depth understanding of the transcriptional regulation of the PPARδ gene will provide new insights into the role of PPARδ in various tissues including the cardiovascular system.

3. THE ROLE OF PPARδ IN REGULATING THE FUNCTION OF ENDOTHELIAL CELLS

Research exploring the impact of PPARδ on endothelial cell function has progressed during the last decade. PPARδ regulates endothelial cell function via several mechanisms. First, PPARδ-selective ligands appear to be able to enhance antioxidative defense of the endothelial cells. Activation of PPARδ elevates the expression of antioxidant genes superoxide dismutase 1 (SOD1), catalase, and thioredoxin. The increased expression of these antioxidant genes thus represses reactive oxygen species (ROS) production in the vascular endothelial cells.[15] Also, PPARδ activation protects endothelial function in diabetic mice.[16,17] The endothelial-protective effect of PPARδ in diabetic mice occurs through phosphatidylinositol-3-kinase (PI3K)/ Akt/eNOS signaling. GW0742, one of the highly potent PPARδ agonists, restores high-glucose-associated endothelial dysfunction repressing superoxide production, NADPH oxidase activity, and mRNA expression of preproendothelin-1, p22(phox), p47(phox), and NOX-1.[16] Treatment with another PPARδ agonist, GW501516, shows similar preventive effects in mouse aorta under high-glucose conditions or in db/db mouse aorta *ex vivo*. PPARδ ligand GW501516 also uncouples eNOS in the cerebral microvessels of hph-1 mice and selectively increases the endothelial expressions of SOD1 and catalase.[18] Heme oxygenase-1 (HO-1) is an inducible, rate-limiting enzyme in the oxidative degradation of heme into biliverdin, releasing free iron (Fe) and carbon monoxide (CO).[19] PPARδ-selective ligand GW501516 induces endothelial HO-1 expression, protecting the vascular endothelium.[20] The antioxidant effect of PPARδ also acts through posttranscriptional pathways. Kim et al. showed that GW501516 dramatically inhibits angiotensin II–induced premature senescence and generation of ROS in human coronary artery endothelial cells (HCAECs) via increasing SIRT1, an NAD^+-dependent class III protein deacetylase that plays important roles in cellular function.[21,22]

PPARδ may also protect the endothelium through its anti-inflammatory effects. Rival et al. demonstrated that the PPARδ agonist, L165041, suppresses TNF-α-induced nuclear translocation of nuclear factor κB (NF-κB), and expression of vascular cell adhesion molecule-1 (VCAM-1) and MCP-1 in EAhy926 endothelial cells.[23] Fan et al. revealed that the PPARδ agonists GW501516 and GW0742 suppress TNF-α-induced expression of VCAM-1 and E-selectin, and the ensuing endothelial-leukocyte adhesion, as well as the interleukin-1β (IL-1β) induced VCAM-1 and E-selectin expression in human umbilical vein endothelial cells (HUVECs).[15] Proinflammatory cytokines can be induced in various cell types by a number of factors, including IL-8. Cytokine induction is a key process of inflammation in response to various stimuli. The influence of PPARδ activators on the production, secretion, and regulation of IL-8 in unstimulated endothelial cells is conveyed by transcriptional, NF-κB-dependent, and posttranscriptional mechanisms that determine mRNA stability.[24] Pretreatment of PPARδ agonists attenuates the levels of IL-6 and IL-8 as a result of increased C-reactive protein (CRP) in HUVECs. This effect of PPARδ may be through the CD32 and NF-κB pathways.[25]

Furthermore, PPARδ may determine endothelial function by regulating angiogenesis. Angiogenesis is the formation of new capillaries from pre-existing blood vessels. Conflicting reports exist regarding the roles of PPARδ in regulating angiogenesis. The early evidence that angiogenesis is regulated by PPARδ is based on an observation that PPARδ ligand GW501516 promotes HUVEC proliferation following the upregulation of vascular endothelial growth factor-α (VEGF-α) and its receptor flt-1.[26] It has also been shown that activation of PPARδ induces endothelial cell proliferation and angiogenesis by upregulating the expression of key vascular endothelial cell growth factors, such as VEGF and adipose differentiation-related protein (ADRP).[27] Most recently, Gaudel et al. found that treatment of GW0742 or PPARδ overexpression promotes angiogenesis in mouse skeletal muscle.[28] Similar effects of PPARδ activation have been reported in endothelial progenitor cells (EPCs). PPARδ activation enhances the proliferation of human EPCs, decreases hypoxia-induced apoptosis, and increases EPCs transendothelial migration and tube formation.[29] PPARδ activates EPCs to induce angiomyogenesis through matrix metalloproteinase (MMP)-9-mediated insulin-like growth factor-1 (IGF-1) paracrine networks.[29] These proangiogenic effects of human EPCs are also through the biosynthesis and release of prostacyclin (PGI$_2$) and subsequent PPARδ activation.[30,31] These effects of PPARδ activation may be due to both genomic and nongenomic activations of the PI3K/Akt pathway.[32] Interestingly, an antiangiogenic effect has also been reported. Meissner et al. showed that treatment with PPARδ agonists inhibits the formation of capillary-like structures and endothelial cell migration, via inhibition of endothelial VEGFR2 protein expression in a time- and concentration-dependent manner. They further elucidated that the inhibitory effects of PPARδ agonists occur through suppression of VEGFR2 promoter activity.[33] Another study demonstrated that PPARδ ligand L-165041 inhibits VEGF-induced angiogenesis, but the antiangiogenic effect may not occur through PPARδ's genomic effects.[34,35] The reasons for the opposite results of PPARδ activation on angiogenesis remain unclear. Given the multiple effects of PPARδ in regulating various metabolic pathways and cell proliferation, it is likely the angiogenic or antiangiogenic effects depend on specific cellular environment and the duration of PPARδ activation. The various PPARδ agonists, doses, and treating duration used in these studies may contribute towards the inconsistent results. Further *in vivo* studies using genetic models in combination with comprehensive ligand treatment protocols are needed to clarify the effects of PPARδ activation on angiogenesis.

PPARδ may protect the endothelial cells through alternative mechanisms, in addition to its roles in antioxidant defense regulation, inflammation, and angiogenesis. PPARδ may exert a direct protective role on the apoptosis pathways. Functional PPRE sequences have been recognized in the promoter region of 14-3-3α, an important antiapoptosis signaling protein.[23] The PPARδ-mediated upregulation of 14-3-3α leads to an increase in the binding of the proapototic protein Bad to 14-3-3α and a reduction in Bad translocation to the mitochondria, thus preventing Bad-triggered apoptosis.[23] A recent study showed that PPARδ protects against ischemia-induced vascular endothelial insults via transcriptional repression at one of the microRNAs, miR-15a, resulting in subsequent increase of the antiapoptotic protein Bcl-2.[36] The cyclooxygenase (COX) pathway in the vascular endothelium plays important roles in thrombosis, atherosclerosis, and vascular inflammation.[37] COX-2-specific inhibitors are reported to be associated with a small but substantial increase in prothrombotic side effects in humans, leading to the withdrawal of these drugs from the market. PPARδ agonists are effective in reducing COX-2 inhibitor-induced cardiovascular side effects.[38] Moreover, PPARδ protects vascular endothelial cells against the deleterious effects of hyperglycemia by downregulating glucose transport.[39]

Taken together, a consensus has been emerging that PPARδ activation largely provides endothelial protection via multiple mechanisms (see Fig. 14.1), although certain aspects of the PPARδ function in angiogenesis need to be further clarified, especially in the *in vivo* context.

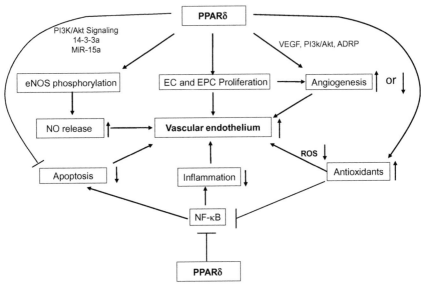

Figure 14.1 The role of PPARδ in regulating the function of endothelial cells. PPARδ regulates endothelial cell function via facilitating endothelial cell (EC) and endothelial progenitor cell (EPC) proliferation, enhancing endothelial nitric oxide synthase (eNOS) phosphorylation and nitric oxide (NO) release, upregulating antioxidant gene expression, inhibiting inflammation and apoptosis, and modulating angiogenesis.

4. THE ROLES OF PPARδ IN THE VASCULAR SMOOTH MUSCLE CELLS

The vascular smooth muscle cells (VSMCs) are the major cell type in the blood vessel walls. The proliferation and migration of VSMCs play a pivotal role in the development of vascular lesions that are commonly found in atherosclerosis and restenosis.[40,41] PPARδ is abundantly expressed in VSMC and is upregulated during vascular lesion formation. VSMC treated with platelet-derived growth factor (PDGF) can induce PPARδ expression.[42,43] However, the effects of the upregulation of PPARδ on the proliferation of VSMC are controversial. Initial study showed that PPARδ overexpression promotes postconfluent cell proliferation in VSMCs by increasing cyclin A and cyclin-dependent kinase 2 (CDK2) and decreasing p57[kip2].[42] Conversely, a few later studies yielded opposite results. Activation of PPARδ by GW501516 significantly inhibits PDGF-induced proliferation in human pulmonary artery smooth muscle cells (HPASMCs) by repressing expression of cyclin D1, cyclin D3, CDK2, and CDK4.[43] GW501516 treatment in these cells also enhances the expression of the cell cycle inhibitory genes G0S2 and p27[kip1].[43] Consistent with this result, PPARδ agonist L-165041 inhibits rat VSMC proliferation via inhibition of PDGF-induced expression of cyclin D1 and CDK4 and thus cell cycle.[44] Furthermore, GW501516 inhibits IL-1β-stimulated proliferation accompanied by cell cycle arrest at the G1 to S phase transition by the induction of p21 and p53, along with decreased expression of CDK4.[45] Similarly, PPARδ ligand (e.g., GW501516) inhibits PDGF- or IL-1β-induced cell migration.[43–45] Differences in species (humans vs. rats), tissues (pulmonary artery vs. aorta), and experiment approaches (ligand treatment vs. PPARδ overexpression) may account for the opposite results. However, at least PPARδ agonists appear to exert largely an inhibiting effect on VSMC proliferation.

VSMC senescence is an independent risk factor contributing to the development of atherosclerosis in human.[46] Aging induces VSMC phenotypic modulation that could have influence on cell senescence and loss of plasticity and reprogramming.[46] Vascular aging is characterized by increased oxidative stress, which is a key factor associated with VSMC senescence.[47,48] The importance of PPARδ signaling in the VSMCs is also related to its role in regulating antioxidant defense. Activation of PPARδ counteracts angiotensin II-induced ROS generation in VSMCs.[49] As in the endothelial cells, PI3K/Akt signaling pathway in VSMC is one of the signaling pathways involved in this process. Ablation of Akt with siRNA further enhances the inhibitory effects of GW501516 in angiotensin II-induced superoxide production. Ligand-activated PPARδ blocks angiotensin II-induced translocation of Rac1 to the cell membrane, inhibiting the activation of NADPH oxidases and consequently ROS generation.[49] Activation of PPARδ by GW501516 substantially attenuates angiotensin II-induced superoxide production in VSMCs following upregulation of genes encoding antioxidant genes, such as glutathione peroxidase 1, thioredoxin 1, manganese SOD2, and HO-1, in VSMCs.[50] Additionally, GW501516 prevents angiotensin II-induced expression of p53 and p21, two key proteins in the senescence pathway, and upregulates expression of PTEN (phosphatase and tensin homolog deleted on chromosome 10), subsequently suppressing the PI3K/Akt pathway.[51] Therefore, these studies demonstrate that PPARδ activation can prevent VSMC senescence via multiple mechanisms.

The preservation of vascular integrity and structure is dependent not only on VSMC proliferation and migration but also on the integrity of the extracellular matrix (ECM). The integrity of the ECM can be a determinant of VSMC apoptosis. PPARδ regulates ECM synthesis and degradation through transforming growth factor-β1 (TGF-β1) and its effector, Smad3, thus augmenting the expression of type I and III collagen, fibronectin, elastin, and tissue inhibitor of metalloproteinases-3 (TIMP-3), but not the expression of TIMP-1, MMP-2, or MMP-9.[52] The MMP family is a class of zinc-dependent endoproteases that degrades structural proteins of ECM.[53] The activation of MMPs is one major pathogenic factor involved in many neurological disorders, such as stroke. A recent study demonstrated that VSMC-specific PPARδ knockout in mice exacerbates cerebrovascular permeability and brain infarction after middle cerebral artery occlusion (MCAO).[54] The reduction of PPARδ is correlated with increased MMP-9 activity in cultured VSMCs after oxygen–glucose deprivation and also in the cerebral cortex of mice following MCAO. PPARδ activation in VSMCs diminishes oxygen–glucose deprivation-induced MMP-9 activity. Since MMP-9 is a direct target of PPARδ-mediated transrepression, PPARδ in VSMCs can prevent ischemic brain injury by inhibiting MMP-9 activation and attenuation of postischemic inflammation.[54] Therefore, it appears that PPARδ-induced upregulation of ECM proteins and suppression of ECM degradation may be a key mechanism underpinning the protective role of PPARδ in the vascular system.

Taken together, current studies suggest that PPARδ plays a crucial role in maintaining VSMC homeostasis by repressing proliferation and migration; enhancing antioxidant, antiapoptosis, and anti-inflammation; and maintaining ECM (Fig. 14.2).

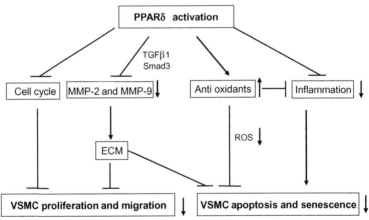

Figure 14.2 The role of PPARδ in regulating the function of vascular smooth muscle cells. PPARδ regulates smooth muscle cells via suppressing VSMCs proliferation, inhibiting VSMC migration with preservation of extracellular matrix (ECM), and inhibiting apoptosis and senescence of VSMCs by upregulating antioxidant genes and suppressing inflammation.

5. THE ROLE OF PPARδ IN REGULATING MACROPHAGE FUNCTION

Lipid homeostasis and inflammation are key determinants in atherogenesis, exemplified by the requirement of lipid-laden, foam cell macrophages for atherosclerotic lesion formation. PPARδ has been implicated in macrophage inflammation and hence serves as a therapeutic target in preventing atherosclerosis. Other than its role in regulating cell proliferation, migration, angiogenesis, and ECM of vascular cells, PPARδ has been supported by many lines of evidence that it further regulates multiple metabolic and inflammatory pathways in macrophages that could determine the progression of atherosclerosis. However, the role of PPARδ activation in regulating lipid metabolism in macrophage seems somewhat paradoxical. It has been shown that PPARδ may mediate the transcriptional response to very-low-density lipoprotein (VLDL) in macrophages, resulting in increased lipid loading in macrophages.[55] However, PPARδ agonists, such as GW501516 and L-165041, increase high-density lipoprotein (HDL) levels in obese rhesus monkeys and in db/db mice.[56] It is plausible that PPARδ activation may exert antiatherogenic effects through an indirect increase of plasma HDL and hence elevated lipid efflux.[57,58] Even though PPARδ activation may increase lipid uptake in macrophage, PPARδ-associated upregulation of systemic HDL levels may override this effect in macrophages and exerts the antiatherogenic effects. Supporting this notion, recent reports showed that PPARδ agonists might repress atherosclerosis by attenuating macrophage lipid accumulation induced by VLDL.[59] PPARδ agonists stimulate a transcriptional program resulting in inhibition of lipoprotein lipase (LPL) activity, activation of fatty acid uptake, and enhanced β-oxidation.[59]

Early studies showed that PPARδ agonists do not inhibit lesion and foam cell formation *in vivo* in LDL receptor (LDLR$^{-/-}$)-deficient mice.[60] However, PPARδ agonists do strongly inhibit the expression of genes associated with the development of atherosclerosis, such as VCAM-1, MCP-1, IFN-γ, and ICAM-1, in macrophages.[60,61] Later studies demonstrated that PPARδ agonists significantly reduce atherosclerosis in apoE$^{-/-}$ mice by raising HDL and exerting anti-inflammatory activity within the vessel wall, subsequently suppressing chemoattractant signaling by downregulation of chemokines.[62] Activation of PPARδ induces gene expression of regulator of G protein signaling (RGS), which is implicated in blocking the signal transduction of chemokine transmigration and macrophage inflammatory responses elicited by atherogenic cytokines.[62] PPARδ may exert an important role in inhibiting leukocyte recruitment and thus lesion progression by repressing the expression of chemokines, attenuating chemokine receptor signaling, and inhibiting MMPs. The anti-inflammatory effect of PPARδ in macrophages is further confirmed as a major mechanism underlying the anti-atherogenic effects of PPARδ agonists.[59,63] Administration of the PPARδ agonist substantially attenuates angiotensin II- or VLDL-accelerated atherosclerosis.[59,63] The treatment increases vascular expression of B-cell lymphoma-6 (Bcl-6), RGS4, and RGS5 and then suppresses inflammatory and atherogenic gene expression. *In vitro* studies demonstrated similar changes in angiotensin II-treated macrophages: PPARδ activation increases both total and free Bcl-6 levels and inhibits angiotensin II activation of mitogen-activated protein kinases (MAPKs), p38, and extracellular signal-regulated kinase 1/2 (ERK1/2).[63] Cells treated with PPARδ agonists are completely resistant to VLDL-induced expression of inflammatory cytokines, mediated by normalization of MAPK (ERK1/2) and Akt/forkhead box protein O1 signaling.[59] Studies on the regulation of macrophages in the adipose and liver have further corroborated the essential role of PPARδ in modulating tissue-resident macrophage activation and hence inflammation.[14,64,65]

A remaining unsolved puzzle is whether liganded and unliganded PPARδ behave differently in controlling macrophage inflammation. Potential different effects of liganded and unliganded PPARδ on lipid metabolism and inflammation have been shown.[66,67] It was suggested that GW0742 switches the association of Bcl-6, a transcription repressor, from PPARδ to VCAM-1 promoter. Reduced endogenous PPARδ expression potentiates the suppressive effect of GW0742, implicating that the unliganded PPARδ may have a proinflammation effect. However, studies in other cells and tissues demonstrate that overexpression of wild-type PPARδ exerts anti-inflammatory effects.[68,69] It is possible that the unliganded PPARδ exerts somewhat different effects in modulating inflammation signaling depending on the endogenous conditions, such as accessibility to endogenous ligands and the availability of PPAR coregulators. It is obvious that more studies using genetic manipulation of PPARδ under *in vivo* condition are required to clarify the mechanisms of action of unliganded PPARδ in macrophage. While further investigation is warranted, the beneficial effects of PPARδ agonists are quite consistent among earlier and recent studies. It is generally agreeable that the combined PPARδ-mediated reduction of lipid accumulation and inflammatory

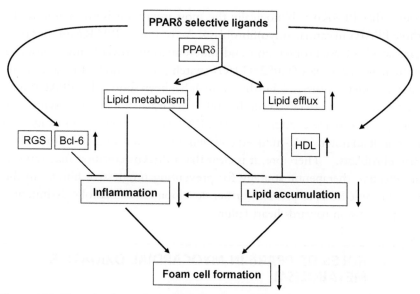

Figure 14.3 The role of PPARδ in regulating the function of macrophages. PPARδ regulates macrophage function via upregulating lipid metabolic genes and facilitating antiinflammation signaling, such as regulators of G protein signaling (RGS) and B-cell lymphoma-6 (Bcl-6), thereby reducing foam cell formation.

cytokine expression (see Fig. 14.3) suggests a novel macrophage-targeted therapeutic option in treating atherosclerosis.[70] Preclinical animal studies support that PPARδ activation exerts antihypertensive effects, restores the vascular structure and function, and reduces the oxidative, proinflammatory, and proatherogenic status in spontaneously hypertensive rats.[71] These results indicate that PPARδ activation antagonizes multiple proinflammatory pathways; hence, PPARδ-selective drugs are good therapeutic candidates for atherosclerosis.

6. THE ROLE OF PPARδ IN FIBROBLAST PROLIFERATION

Heart remodeling in response to pathological stresses includes the development of cardiomyocyte hypertrophy and fibrosis, which aggravates diastolic dysfunction and increases susceptibility to heart failure.[72,73] The proliferation and differentiation of cardiac fibroblasts towards myofibroblasts trigger the process of fibrosis with disproportionate accumulation of ECM components and subsequent diastolic dysfunction.[74,75] Teunissen et al.

found that PPARδ is biologically the most important isoform among the three PPARs in cardiac (myo)fibroblasts. Activation of PPARδ inhibits both cardiac fibroblast proliferation and differentiation towards myofibroblasts with increased levels of G0S2 (G0/G1 switch gene 2) and PTEN, and attenuates collagen synthesis.[76] Furthermore, it has been shown that PPARδ agonists exert inhibitory effects on the ability of cardiac fibroblasts to synthesize collagen in response to angiotensin II. PPARδ is indispensable for its agonists to inhibit angiotensin II-induced collagen type I expression in cultured cardiac fibroblasts.[77] Therefore, it is clear that PPARδ activation may serve as an effective therapeutic strategy for preventing progressive fibrosis in the hypertrophic heart to prevent further development of diastolic dysfunction and progression towards heart failure.

7. ROLES OF PPARδ IN MYOCARDIAL OXIDATIVE METABOLISM

Myocardial fuel preference is highly dependent on the developmental stage and physiology/pathophysiology of the heart. It is determined by transcriptional regulation of a subset of genes encoding key enzymes of cellular substrate utilization. For example, a switch in cardiac substrate preference from glucose to long-chain fatty acids occurs during the fetal to newborn transition when O_2 availability and dietary fat content are substantially elevated.[78,79] In pathological stress-induced cardiac hypertrophy and heart failure, fuel preference may switch from long-chain fatty acids to glucose at least at certain stages of the pathological development (see review[80–83]). It has been proposed that PPARα and PPARδ play crucial roles in the transcriptional regulation of myocardial fuel metabolism.[84–86] PPARδ expression is relatively abundant in cardiomyocytes from early developmental stage to adult.[87,88] Moreover, PPARδ is predominantly expressed in cardiomyocytes relative to other cell types in the myocardium.[88] The roles of PPARα in regulating lipid metabolism in the heart have been well characterized. However, it appears that PPARα is largely dispensable because complete PPARα knockout in mice shows no or only subtle phenotypes at least under basal condition.[84] Eliminating PPARα is not sufficient to prevent the fuel preference switch at the fetal to adult heart stage. Research on the role of PPARδ in regulating myocardial fatty acid metabolism as that of PPARδ has been summarized.[84,89] The first evidence showing the importance of PPARδ in fatty acid oxidation (FAO) of cardiomyocytes has been demonstrated in cultured cardiomyocytes treated with PPARδ-selective ligands[87,88] and with adenovirus-mediated PPARδ overexpression.[88] Both PPARδ-selective ligands and PPARδ overexpression in cardiomyocytes lead to substantial increases in FAO.[88] Since the cardiac expression of PPARδ is not altered in the absence of PPARα,[90] it appears that PPARδ alone is sufficient for maintaining basal myocardial FAO. Both mitochondria-specific (M-CPT I, L-CPT I, UCP2, UCP3, PDK4, 3-ketoacyl-CoA thiolase, MCAD, LCAD, and MCD) and peroxisome-specific (ACO, VLCAD, and 3-ketoacyl-CoA thiolase) FAO genes are regulated by PPARδ in the heart.[11,88,91] PPARδ regulates not only genes encoding key enzymes of FAO but also genes encoding fatty acid uptake proteins, such as long-chain fatty acid synthetase (LACS), fatty acid transport protein 1 (FATP1), fatty acid-binding proteins (FABPs), FAT/CD36, and LPL, in cardiomyocytes.[92] PPARδ-selective ligand treatments and PPARδ overexpression in cultured cardiomyocytes result in elevation of FAO genes and FAO rates in a classic ligand-binding-dependent mechanism.[88] More importantly, our study based on a cardiomyocyte-restricted PPARδ knockout mouse model revealed that PPARδ is required for maintaining constitutive myocardial FAO.[11] In addition to depressed FAO, these mice suffer from cardiac dysfunction, hypertrophy, and heart failure with myocardial lipid accumulation.[11] Therefore, these earlier-mentioned studies suggest that PPARδ could be a pivotal determinant of basal myocardial FAO and it may serve as a "sensor" of intracellular fatty acid content and a constitutive determinant of high-level FAO to fuel the normal adult heart. While PPARα and PPARδ regulate largely a similar set of fatty acid metabolic genes in cardiomyocytes, they are not dependent on each other.[90] Cardiomyocyte-restricted PPARδ knockout mice, but not the conventional PPARα knockout, exhibit myocardial neutral lipid accumulation under baseline condition.[11,93] On the other hand, cardiomyocyte-restricted overexpression of PPARα leads to myocardial lipid accumulation and lipotoxicity, whereas PPARδ overexpression in the heart does not.[91,94] Therefore, it is plausible that differences in the capacities of PPARα and PPARδ to induce genes encoding key enzymes for different aspects of fatty acid metabolism exist. PPARα and PPARδ may induce a mismatched expression of FAO and lipid uptake genes. Greater repression of lipid uptake genes than FAO genes may occur in the PPARα-null heart and greater repression of FAO genes than lipid uptake genes may occur in the PPARδ-null heart. Additionally, PPARδ appears to be an important regulator of glucose metabolism through indirect mechanisms, either via cross-talking with myocyte-enhancing factor 2a (MEF2a)[91] or by induction of PPARγ

coactivator-1α (PGC-1α) expression.[92,94] Therefore, it is evident that
PPARδ is a crucial determinant of myocardial structure/function in the
heart via regulating myocardial substrate metabolism.

Oxidative damages are one of the most prominent pathological developments of cardiac hypertrophy, ischemia/reperfusion, and heart failure.[95,96]
We found that PPARδ regulates the expression of a series of endogenous
antioxidants, such as SOD1 and SOD2.[92,94] PPARδ deficiency in the adult
heart leads to downregulation of antioxidant defense due to the impaired
cardiac expression of endogenous antioxidants.[92] Mitochondrial biogenesis
is the proliferation of preexisting mitochondria and is an essential cellular
response to energy status.[97] We found that mitochondrial biogenesis is
repressed in the heart with PPARδ deficiency and is increased in the heart
with transgenic overexpression of a constitutively active PPARδ.[92,94] This is
consistent with an earlier finding that PPARδ overexpression in the skeletal
muscle induces upregulation of mitochondrial biogenesis.[98] The mechanism
underpinning the effect of PPARδ activation on mitochondrial biogenesis is
likely mediated by one of its target genes, the PGC-1α,[99] which has been
established as a master switch of mitochondrial biogenesis.[100–103] As a result,
mitochondrial function is impaired and mitochondrial volume is
decreased.[92] The decreased mitochondrial function and volume should contribute to the depression of myocardial oxidative metabolism and hence the
accumulation of myocardial lipid in the PPARδ-deficient heart. We further
confirmed that PPARδ activation in cultured cardiomyocytes[104] and in
adult hearts facilitates mitochondrial biogenesis and enhances antioxidant
defense.[94] Others have also reported that PPARδ ligands protect cultured
cardiomyocytes and rats from doxorubicin-induced toxic effects.[105,106] As
a result, the heart is protected from pressure overload-induced cardiac dysfunction and pathological development.[94] Consistently, cardiac-specific
PPARδ overexpression protects the heart from ischemia/reperfusion
injury[91] and PPARδ agonist treatment improves the cardiac function of
right ventricular hypertrophy.[91,107] Although upregulation of glucose utilization in PPARδ overexpressed heart has been proposed as the key mechanism, it is plausible that the role of PPARδ in the regulation of
mitochondrial biogenesis and antioxidant defense may also contribute to
the protection. Since the substantial impairment of mitochondria in the
PPARδ-deficient heart occurs, high-fat diet feeding does not rescue cardiac
phenotype despite dramatically upregulating lipid metabolic gene expression
in these hearts.[108] The overexpression of lipid metabolic gene expression in
the PPARδ-deficient heart in mice subjected to high-fat feeding is via

upregulation of PPARα/PGC-1α signaling,[108] highlighting a potential cross-talk among different transcriptional factors under certain pathophysiological conditions. Nevertheless, the PPARδ expression and its activity in the heart under physiological and pathological conditions remain obscured. Unlike PPARα, cardiac PPARδ expression appears to be relatively stable. However, we did find PPARδ is upregulated in response to high-fat feeding but not to starvation.[108] Therefore, it is now well documented that PPARδ is an essential transcriptional regulator for the structure/function of the heart by regulating myocardial homeostasis not only by regulating fatty acid and glucose metabolism but also by maintaining mitochondrial biogenesis and antioxidant defense.

8. ANTI-INFLAMMATORY EFFECT OF PPARδ IN CARDIOMYOCYTES

As in the research area of vascular biology, a large body of evidence has emerged during the last 10 years illustrating the anti-inflammatory effects of PPARδ and its agonists in cardiomyocytes. Inflammation has been well characterized as one key process during the pathological development of cardiac hypertrophy and heart failure. Similar to the anti-inflammatory effect of PPARδ in the vascular cells, PPARδ activation and overexpression in cultured cardiomyocytes diminish lipopolysaccharide-induced inflammation responses.[68,69] PPARδ activation also abolishes lipid-induced inflammation pathways in mouse heart and human cardiac cells.[109] The anti-inflammatory effect of PPARδ in cardiomyocytes is clearly related to its interfering with NF-κB signaling.[68,69,110,111] PPARδ agonist L-165041 strongly enhances the physical interaction between PPARδ and the p65 subunit of NF-κB.[68] Moreover, adenovirus-mediated overexpression of wild-type PPARδ, but not a mutated PPARδ lacking the ligand-binding domain, represses LPS-induced TNF-α inflammation signaling in cultured cardiomyocytes.[69] Not surprisingly, PPARδ is able to inhibit NF-κB activation in hypertrophic cardiomyocyte[68,112] and therefore contributes to the role of PPARδ in mitigating cardiomyocyte hypertrophic remodeling. In addition, as a new marker for cardiovascular diseases, CRP activates NF-κB-inducing kinase (NIK) and NF-κB pathway, and this protein-induced proinflammation is attenuated by PPARδ agonists in cardiomyocytes and H9C2 cardiomyoblasts.[113] Therefore, the anti-inflammatory effects of PPARδ in cardiomyocytes appear to be through its direct interaction with

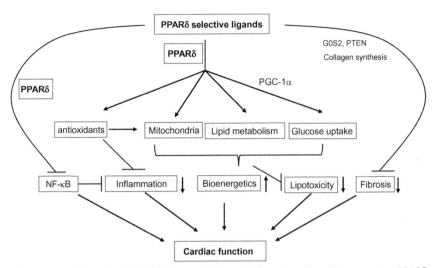

Figure 14.4 The role of PPARδ in regulating the function of cardiomyocytes. PPARδ regulates cardiomyocyte function via activating transcriptional expression of genes encoding proteins of antioxidants, lipid metabolism, glucose utilization, and mitochondrial biogenesis, thereby improving myocardial bioenergetics and repressing lipotoxicity. PPARδ also suppresses inflammation by directly inhibiting NF-κB signaling pathway and repressing cardiac fibrosis by inhibiting cell differentiation and collagen synthesis.

NF-κB signaling and could be facilitated by PPARδ overexpression and PPARδ-selective ligand treatment.

In summary, PPARδ has become an important therapeutic target due to its broad spectrum of protective roles derived from transcriptional regulation of myocardial oxidative metabolism, mitochondrial biogenesis, antioxidant defense, and anti-inflammatory efforts (Fig. 14.4). Other mechanisms related to the myocardial effects of PPARδ activation have also been uncovered. For example, it has been shown that PPARδ activation inhibits angiotensin II–induced cardiomyocyte hypertrophy by suppressing intracellular calcium signaling pathway.[114] Further studies are warranted to explore additional biological actions of PPARδ that may help protect the heart from pathological stress.

9. CONCLUSIONS AND PERSPECTIVES

Taken together, PPARδ activation in general is beneficial in protecting the cardiovascular system from disease conditions. However, the mechanisms underlying the potential detrimental effects of unliganded

PPARδ, especially in the vascular system, remain unclear. The controversy over how PPARδ is involved in the regulation of angiogenesis in the heart needs to be settled with further studies using animal models with cardiovascular-specific PPARδ ablations and agonist treatment. Although considerable progress has been made in the understanding of the roles of PPARδ in the cardiovascular system, many open questions regarding targeting PPARδ as a therapeutic strategy for the treatment of cardiovascular disease linger unanswered. PPARδ expression and activity in the cardiovascular system under pathological conditions have not been well established. A few studies reported alterations of PPARδ protein expression in animal hearts or myocytes under various disease conditions and treatments.[105,106,115] Due to the poor specificity of most commercially available PPARδ antibodies, caution should be heeded when interpreting Western blot results with no information of transcript expression. Additionally, further studies on the mechanistic aspects of the PPARδ action under various conditions of the cardiovascular system will be necessary for the clinical application of the beneficial effects of PPARδ. Further clinical trials with high-affinity PPARδ agonists are required to clarify the remaining issues that prevent the use of this class of compounds in patients. Considering the scope of this chapter, we have not discussed the systemic effects of PPARδ activation. Nonetheless, it is obvious that a better understanding of how PPARδ activation may affect functions of other organs and tissues is essential and obligated in the further therapeutic developments for the treatment of cardiovascular disease.

ACKNOWLEDGMENTS

This work was supported by grants from National Institutes of Health (1R01 HL085499 and 1R01 HL084456) and the American Diabetes Association Basic Science Award (#7-12-BS-208) to Q. Y.

REFERENCES

1. Chawla A, Repa JJ, Evans RM, Mangelsdorf DJ. Nuclear receptors and lipid physiology: opening the x-files. *Science.* 2001;294:1866–1870.
2. Kliewer SA, Xu HE, Lambert MH, Willson TM. Peroxisome proliferator-activated receptors: from genes to physiology. *Recent Prog Horm Res.* 2001;56:239–263.
3. Schoonjans K, Staels B, Auwerx J. Role of the peroxisome proliferator-activated receptor (ppar) in mediating the effects of fibrates and fatty acids on gene expression. *J Lipid Res.* 1996;37:907–925.
4. Fredenrich A, Grimaldi PA. Ppar delta: an uncompletely known nuclear receptor. *Diabetes Metab.* 2005;31:23–27.
5. Barish GD, Narkar VA, Evans RM. Ppar delta: a dagger in the heart of the metabolic syndrome. *J Clin Invest.* 2006;116:590–597.

6. Bishop-Bailey D. Peroxisome proliferator-activated receptor beta/delta goes vascular. *Circ Res.* 2008;102:146–147.

7. Bishop-Bailey D, Bystrom J. Emerging roles of peroxisome proliferator-activated receptor-beta/delta in inflammation. *Pharmacol Ther.* 2009;124:141–150.

8. Lemberger T, Braissant O, Juge-Aubry C, et al. Ppar tissue distribution and interactions with other hormone-signaling pathways. *Ann NY Acad Sci.* 1996;804:231–251.

9. Braissant O, Foufelle F, Scotto C, Dauca M, Wahli W. Differential expression of peroxisome proliferator-activated receptors (ppars): tissue distribution of ppar-alpha, -beta, and -gamma in the adult rat. *Endocrinology.* 1996;137:354–366.

10. Escher P, Braissant O, Basu-Modak S, Michalik L, Wahli W, Desvergne B. Rat ppars: quantitative analysis in adult rat tissues and regulation in fasting and refeeding. *Endocrinology.* 2001;142:4195–4202.

11. Cheng L, Ding G, Qin Q, et al. Cardiomyocyte-restricted peroxisome proliferator-activated receptor-delta deletion perturbs myocardial fatty acid oxidation and leads to cardiomyopathy. *Nat Med.* 2004;10:1245–1250.

12. He TC, Chan TA, Vogelstein B, Kinzler KW. Ppardelta is an apc-regulated target of nonsteroidal anti-inflammatory drugs. *Cell.* 1999;99:335–345.

13. Pedchenko TV, Gonzalez AL, Wang D, DuBois RN, Massion PP. Peroxisome proliferator-activated receptor beta/delta expression and activation in lung cancer. *Am J Respir Cell Mol Biol.* 2008;39:689–696.

14. Kang K, Reilly SM, Karabacak V, et al. Adipocyte-derived th2 cytokines and myeloid ppardelta regulate macrophage polarization and insulin sensitivity. *Cell Metab.* 2008;7:485–495.

15. Fan Y, Wang Y, Tang Z, et al. Suppression of pro-inflammatory adhesion molecules by ppar-delta in human vascular endothelial cells. *Arterioscler Thromb Vasc Biol.* 2008;28:315–321.

16. Quintela AM, Jimenez R, Gomez-Guzman M, et al. Activation of peroxisome proliferator-activated receptor-beta/-delta (pparbeta/delta) prevents endothelial dysfunction in type 1 diabetic rats. *Free Radic Biol Med.* 2012;53:730–741.

17. Tian XY, Wong WT, Wang N, et al. Ppardelta activation protects endothelial function in diabetic mice. *Diabetes.* 2012;61:3285–3293.

18. Santhanam AV, d'Uscio LV, He T, Katusic ZS. Ppardelta agonist gw501516 prevents uncoupling of endothelial nitric oxide synthase in cerebral microvessels of hph-1 mice. *Brain Res.* 2012;1483:89–95.

19. Ryter SW, Alam J, Choi AM. Heme oxygenase-1/carbon monoxide: from basic science to therapeutic applications. *Physiol Rev.* 2006;86:583–650.

20. Ali F, Ali NS, Bauer A, et al. Ppardelta and pgc1alpha act cooperatively to induce haem oxygenase-1 and enhance vascular endothelial cell resistance to stress. *Cardiovasc Res.* 2010;85:701–710.

21. Michan S, Sinclair D. Sirtuins in mammals: insights into their biological function. *Biochem J.* 2007;404:1–13.

22. Olmos Y, Sánchez-Gómez FJ, Wild B, et al. SirT1 regulation of antioxidant genes is dependent on the formation of a FoxO3a/PGC-1α complex. *Antioxid Redox Signal.* 2013;19(13):1507–1521.

23. Rival Y, Beneteau N, Taillandier T, et al. Pparalpha and ppardelta activators inhibit cytokine-induced nuclear translocation of nf-kappab and expression of vcam-1 in eahy926 endothelial cells. *Eur J Pharmacol.* 2002;435:143–151.

24. Meissner M, Hrgovic I, Doll M, et al. Peroxisome proliferator-activated receptor delta activators induce il-8 expression in nonstimulated endothelial cells in a transcriptional and posttranscriptional manner. *J Biol Chem.* 2010;285:33797–33804.

25. Liang YJ, Liu YC, Chen CY, et al. Comparison of ppardelta and ppargamma in inhibiting the pro-inflammatory effects of c-reactive protein in endothelial cells. *Int J Cardiol.* 2010;143:361–367.

26. Stephen RL, Gustafsson MC, Jarvis M, et al. Activation of peroxisome proliferator-activated receptor delta stimulates the proliferation of human breast and prostate cancer cell lines. *Cancer Res.* 2004;64:3162–3170.

27. Piqueras L, Reynolds AR, Hodivala-Dilke KM, et al. Activation of pparbeta/delta induces endothelial cell proliferation and angiogenesis. *Arterioscler Thromb Vasc Biol.* 2007;27:63–69.

28. Gaudel C, Schwartz C, Giordano C, Abumrad NA, Grimaldi PA. Pharmacological activation of pparbeta promotes rapid and calcineurin-dependent fiber remodeling and angiogenesis in mouse skeletal muscle. *Am J Physiol Endocrinol Metab.* 2008;295: E297–E304.

29. Han JK, Kim HL, Jeon KH, et al. Peroxisome proliferator-activated receptor-{delta} activates endothelial progenitor cells to induce angio-myogenesis through matrix metallo-proteinase-9-mediated insulin-like growth factor-1 paracrine networks. *Eur Heart J.* 2011;34:1755–1765.

30. Asahara T, Murohara T, Sullivan A, et al. Isolation of putative progenitor endothelial cells for angiogenesis. *Science.* 1997;275:964–967.

31. He T, Lu T, d'Uscio LV, Lam CF, Lee HC, Katusic ZS. Angiogenic function of prostacyclin biosynthesis in human endothelial progenitor cells. *Circ Res.* 2008;103:80–88.

32. Han JK, Lee HS, Yang HM, et al. Peroxisome proliferator-activated receptor-delta agonist enhances vasculogenesis by regulating endothelial progenitor cells through genomic and nongenomic activations of the phosphatidylinositol 3-kinase/akt pathway. *Circulation.* 2008;118:1021–1033.

33. Meissner M, Hrgovic I, Doll M, Kaufmann R. Ppardelta agonists suppress angiogenesis in a vegfr2-dependent manner. *Arch Dermatol Res.* 2011;303:41–47.

34. Park JH, Lee KS, Lim HJ, Kim H, Kwak HJ, Park HY. The ppardelta ligand l-165041 inhibits vegf-induced angiogenesis, but the antiangiogenic effect is not related to ppardelta. *J Cell Biochem.* 2012;113:1947–1954.

35. Kim MY, Kang ES, Ham SA, et al. The ppardelta-mediated inhibition of angiotensin ii-induced premature senescence in human endothelial cells is sirt1-dependent. *Biochem Pharmacol.* 2012;84:1627–1634.

36. Yin KJ, Deng Z, Hamblin M, et al. Peroxisome proliferator-activated receptor delta regulation of mir-15a in ischemia-induced cerebral vascular endothelial injury. *J Neurosci.* 2010;30:6398–6408.

37. Simmons DL, Botting RM, Hla T. Cyclooxygenase isozymes: the biology of prostaglandin synthesis and inhibition. *Pharmacol Rev.* 2004;56:387–437.

38. Ghosh M, Wang H, Ai Y, et al. Cox-2 suppresses tissue factor expression via endocannabinoid-directed ppardelta activation. *J Exp Med.* 2007;204:2053–2061.

39. Riahi Y, Sin-Malia Y, Cohen G, et al. The natural protective mechanism against hyperglycemia in vascular endothelial cells: roles of the lipid peroxidation product 4-hydroxydodecadienal and peroxisome proliferator-activated receptor delta. *Diabetes.* 2010;59:808–818.

40. Ross R. The pathogenesis of atherosclerosis: a perspective for the 1990s. *Nature.* 1993;362:801–809.

41. Schwartz RS, Murphy JG, Edwards WD, Camrud AR, Vliestra RE, Holmes DR. Restenosis after balloon angioplasty. A practical proliferative model in porcine coronary arteries. *Circulation.* 1990;82:2190–2200.

42. Zhang J, Fu M, Zhu X, et al. Peroxisome proliferator-activated receptor delta is up-regulated during vascular lesion formation and promotes post-confluent cell proliferation in vascular smooth muscle cells. *J Biol Chem.* 2002;277:11505–11512.

43. Liu G, Li X, Li Y, et al. Ppardelta agonist gw501516 inhibits pdgf-stimulated pulmonary arterial smooth muscle cell function related to pathological vascular remodeling. *Biomed Res Int.* 2013;2013:903947.

44. Lim HJ, Lee S, Park JH, et al. Ppar delta agonist l-165041 inhibits rat vascular smooth muscle cell proliferation and migration via inhibition of cell cycle. *Atherosclerosis.* 2009;202:446–454.

45. Kim HJ, Kim MY, Hwang JS, et al. Ppardelta inhibits il-1beta-stimulated proliferation and migration of vascular smooth muscle cells via up-regulation of il-1ra. *Cell Mol Life Sci.* 2010;67:2119–2130.

46. Minamino T, Komuro I. Vascular cell senescence: contribution to atherosclerosis. *Circ Res.* 2007;100:15–26.

47. Matthews C, Gorenne I, Scott S, et al. Vascular smooth muscle cells undergo telomere-based senescence in human atherosclerosis: effects of telomerase and oxidative stress. *Circ Res.* 2006;99:156–164.

48. Ungvari Z, Kaley G, de Cabo R, Sonntag WE, Csiszar A. Mechanisms of vascular aging: new perspectives. *J Gerontol A Biol Sci Med Sci.* 2010;65:1028–1041.

49. Lee H, Ham SA, Kim MY, et al. Activation of ppardelta counteracts angiotensin ii-induced ros generation by inhibiting rac1 translocation in vascular smooth muscle cells. *Free Radic Res.* 2012;46:912–919.

50. Kim HJ, Ham SA, Paek KS, et al. Transcriptional up-regulation of antioxidant genes by ppardelta inhibits angiotensin ii-induced premature senescence in vascular smooth muscle cells. *Biochem Biophys Res Commun.* 2011;406:564–569.

51. Kim HJ, Ham SA, Kim MY, et al. Ppardelta coordinates angiotensin ii-induced senescence in vascular smooth muscle cells through pten-mediated inhibition of superoxide generation. *J Biol Chem.* 2011;286:44585–44593.

52. Kim HJ, Kim MY, Jin H, et al. Peroxisome proliferator-activated receptor delta regulates extracellular matrix and apoptosis of vascular smooth muscle cells through the activation of transforming growth factor-{beta}1/smad3. *Circ Res.* 2009;105:16–24.

53. Nagase H, Woessner Jr JF. Matrix metalloproteinases. *J Biol Chem.* 1999;274:21491–21494.

54. Yin KJ, Deng Z, Hamblin M, Zhang J, Chen YE. Vascular ppardelta protects against stroke-induced brain injury. *Arterioscler Thromb Vasc Biol.* 2011;31:574–581.

55. Chawla A, Lee CH, Barak Y, et al. Ppardelta is a very low-density lipoprotein sensor in macrophages. *Proc Natl Acad Sci U S A.* 2003;100:1268–1273.

56. Leibowitz MD, Fievet C, Hennuyer N, et al. Activation of ppardelta alters lipid metabolism in db/db mice. *FEBS Lett.* 2000;473:333–336.

57. Oliver Jr WR, Shenk JL, Snaith MR, et al. A selective peroxisome proliferator-activated receptor delta agonist promotes reverse cholesterol transport. *Proc Natl Acad Sci U S A.* 2001;98:5306–5311.

58. van der Veen JN, Kruit JK, Havinga R, et al. Reduced cholesterol absorption upon ppardelta activation coincides with decreased intestinal expression of npc1l1. *J Lipid Res.* 2005;46:526–534.

59. Bojic LA, Sawyez CG, Telford DE, Edwards JY, Hegele RA, Huff MW. Activation of peroxisome proliferator-activated receptor delta inhibits human macrophage foam cell formation and the inflammatory response induced by very low-density lipoprotein. *Arterioscler Thromb Vasc Biol.* 2012;32:2919–2928.

60. Li AC, Binder CJ, Gutierrez A, et al. Differential inhibition of macrophage foam-cell formation and atherosclerosis in mice by pparalpha, beta/delta, and gamma. *J Clin Invest.* 2004;114:1564–1576.

61. Graham TL, Mookherjee C, Suckling KE, Palmer CN, Patel L. The ppardelta agonist gw0742x reduces atherosclerosis in ldlr(-/-) mice. *Atherosclerosis.* 2005;181:29–37.

62. Barish GD, Atkins AR, Downes M, et al. Ppardelta regulates multiple proinflammatory pathways to suppress atherosclerosis. *Proc Natl Acad Sci U S A.* 2008;105:4271–4276.

63. Takata Y, Liu J, Yin F, et al. Ppardelta-mediated antiinflammatory mechanisms inhibit angiotensin ii-accelerated atherosclerosis. *Proc Natl Acad Sci U S A.* 2008;105:4277–4282.

64. Desvergne B. Ppardelta/beta: the lobbyist switching macrophage allegiance in favor of metabolism. *Cell Metab.* 2008;7:467–469.

65. Odegaard JI, Ricardo-Gonzalez RR, Red Eagle A, et al. Alternative m2 activation of kupffer cells by ppardelta ameliorates obesity-induced insulin resistance. *Cell Metab.* 2008;7:496–507.

66. Vosper H, Patel L, Graham TL, et al. The peroxisome proliferator-activated receptor delta promotes lipid accumulation in human macrophages. *J Biol Chem.* 2001;276:44258–44265.

67. Lee CH, Chawla A, Urbiztondo N, et al. Transcriptional repression of atherogenic inflammation: modulation by ppardelta. *Science.* 2003;302:453–457.

68. Planavila A, Rodriguez-Calvo R, Jove M, et al. Peroxisome proliferator-activated receptor beta/delta activation inhibits hypertrophy in neonatal rat cardiomyocytes. *Cardiovasc Res.* 2005;65:832–841.

69. Ding G, Cheng L, Qin Q, Frontin S, Yang Q. Ppardelta modulates lipopolysaccharide-induced tnfalpha inflammation signaling in cultured cardiomyocytes. *J Mol Cell Cardiol.* 2006;40:821–828.

70. Chawla A. Control of macrophage activation and function by ppars. *Circ Res.* 2010;106:1559–1569.

71. Zarzuelo MJ, Jimenez R, Galindo P, et al. Antihypertensive effects of peroxisome proliferator-activated receptor-beta activation in spontaneously hypertensive rats. *Hypertension.* 2011;58:733–743.

72. Kostin S, Hein S, Arnon E, Scholz D, Schaper J. The cytoskeleton and related proteins in the human failing heart. *Heart Fail Rev.* 2000;5:271–280.

73. Weber KT. Are myocardial fibrosis and diastolic dysfunction reversible in hypertensive heart disease? *Congest Heart Fail.* 2005;11:322–324, quiz 325.

74. Petrov VV, Fagard RH, Lijnen PJ. Stimulation of collagen production by transforming growth factor-beta1 during differentiation of cardiac fibroblasts to myofibroblasts. *Hypertension.* 2002;39:258–263.

75. Tomasek JJ, Gabbiani G, Hinz B, Chaponnier C, Brown RA. Myofibroblasts and mechano-regulation of connective tissue remodelling. *Nat Rev Mol Cell Biol.* 2002;3:349–363.

76. Teunissen BE, Smeets PJ, Willemsen PH, De Windt LJ, Van der Vusse GJ, Van Bilsen M. Activation of ppardelta inhibits cardiac fibroblast proliferation and the trans-differentiation into myofibroblasts. *Cardiovasc Res.* 2007;75:519–529.

77. Zhang H, Pi R, Li R, et al. Pparbeta/delta activation inhibits angiotensin ii-induced collagen type i expression in rat cardiac fibroblasts. *Arch Biochem Biophys.* 2007;460:25–32.

78. Breuer E, Barta E, Zlatos L, Pappova E. Developmental changes of myocardial metabolism. II. Myocardial metabolism of fatty acids in the early postnatal period in dogs. *Biol Neonat.* 1968;12:54–64.

79. Lopaschuk GD, Collins-Nakai RL, Itoi T. Developmental changes in energy substrate use by the heart. *Cardiovasc Res.* 1992;26:1172–1180.

80. van Bilsen M, van Nieuwenhoven FA, van der Vusse GJ. Metabolic remodelling of the failing heart: beneficial or detrimental? *Cardiovasc Res.* 2009;81:420–428.

81. Lopaschuk GD, Ussher JR, Folmes CD, Jaswal JS, Stanley WC. Myocardial fatty acid metabolism in health and disease. *Physiol Rev.* 2010;90:207–258.

82. Lopaschuk GD, Jaswal JS. Energy metabolic phenotype of the cardiomyocyte during development, differentiation, and postnatal maturation. *J Cardiovasc Pharmacol.* 2010;56:130–140.

83. Lionetti V, Stanley WC, Recchia FA. Modulating fatty acid oxidation in heart failure. *Cardiovasc Res.* 2011;90:202–209.

84. Yang Q, Li Y. Roles of ppars on regulating myocardial energy and lipid homeostasis. *J Mol Med.* 2007;85:697–706.

85. Finck BN. The ppar regulatory system in cardiac physiology and disease. *Cardiovasc Res.* 2007;73:269–277.

86. Madrazo JA, Kelly DP. The ppar trio: regulators of myocardial energy metabolism in health and disease. *J Mol Cell Cardiol.* 2008;44:968–975.

87. Gilde AJ, van der Lee KA, Willemsen PH, et al. Peroxisome proliferator-activated receptor (ppar) alpha and pparbeta/delta, but not ppargamma, modulate the expression of genes involved in cardiac lipid metabolism. *Circ Res.* 2003;92:518–524.

88. Cheng L, Ding G, Qin Q, et al. Peroxisome proliferator-activated receptor delta activates fatty acid oxidation in cultured neonatal and adult cardiomyocytes. *Biochem Biophys Res Commun.* 2004;313:277–286.

89. Yang Q, Cheng LH. Molecular regulation of lipotoxicity in the heart. *Drug Discov Today.* 2005;2:101–107.

90. Liu J, Wang P, He L, et al. Cardiomyocyte-restricted deletion of pparbeta/delta in pparalpha-null mice causes impaired mitochondrial biogenesis and defense, but no further depression of myocardial fatty acid oxidation. *PPAR Res.* 2011;2011:372854.

91. Burkart EM, Sambandam N, Han X, et al. Nuclear receptors pparbeta/delta and pparalpha direct distinct metabolic regulatory programs in the mouse heart. *J Clin Invest.* 2007;117:3930–3939.

92. Wang P, Liu J, Li Y, et al. Peroxisome proliferator-activated receptor delta is an essential transcriptional regulator for mitochondrial protection and biogenesis in adult heart. *Circ Res.* 2010;106:911–919.

93. Watanabe K, Fujii H, Takahashi T, et al. Constitutive regulation of cardiac fatty acid metabolism through peroxisome proliferator-activated receptor alpha associated with age-dependent cardiac toxicity. *J Biol Chem.* 2000;275:22293–22299.

94. Liu J, Wang P, Luo J, et al. Peroxisome proliferator-activated receptor beta/delta activation in adult hearts facilitates mitochondrial function and cardiac performance under pressure-overload condition. *Hypertension.* 2011;57:223–230.

95. Takimoto E, Kass DA. Role of oxidative stress in cardiac hypertrophy and remodeling. *Hypertension.* 2007;49:241–248.

96. Zhang Y, Tocchetti CG, Krieg T, Moens AL. Oxidative and nitrosative stress in the maintenance of myocardial function. *Free Radic Biol Med.* 2012;53:1531–1540.

97. Hock MB, Kralli A. Transcriptional control of mitochondrial biogenesis and function. *Annu Rev Physiol.* 2009;71:177–203.

98. Wang YX, Zhang CL, Yu RT, et al. Regulation of muscle fiber type and running endurance by ppardelta. *PLoS Biol.* 2004;2:e294.

99. Hondares E, Pineda-Torra I, Iglesias R, Staels B, Villarroya F, Giralt M. Ppardelta, but not pparalpha, activates pgc-1alpha gene transcription in muscle. *Biochem Biophys Res Commun.* 2007;354:1021–1027.

100. Wu Z, Puigserver P, Andersson U, et al. Mechanisms controlling mitochondrial biogenesis and respiration through the thermogenic coactivator pgc-1. *Cell.* 1999;98:115–124.

101. Lehman JJ, Barger PM, Kovacs A, Saffitz JE, Medeiros DM, Kelly DP. Peroxisome proliferator-activated receptor gamma coactivator-1 promotes cardiac mitochondrial biogenesis. *J Clin Invest.* 2000;106:847–856.

102. Vega RB, Huss JM, Kelly DP. The coactivator pgc-1 cooperates with peroxisome proliferator-activated receptor alpha in transcriptional control of nuclear genes encoding mitochondrial fatty acid oxidation enzymes. *Mol Cell Biol.* 2000;20:1868–1876.

103. Ventura-Clapier R, Garnier A, Veksler V. Transcriptional control of mitochondrial biogenesis: the central role of pgc-1alpha. *Cardiovasc Res.* 2008;79:208–217.
104. Li Y, Yin R, Liu J, et al. Peroxisome proliferator-activated receptor delta regulates mitofusin 2 expression in the heart. *J Mol Cell Cardiol.* 2009;46:876–882.
105. Altieri P, Spallarossa P, Barisione C, et al. Inhibition of doxorubicin-induced senescence by ppardelta activation agonists in cardiac muscle cells: cooperation between ppardelta and bcl6. *PLoS One.* 2012;7:e46126.
106. Chen ZC, Chen LJ, Cheng JT. Doxorubicin-induced cardiac toxicity is mediated by lowering of peroxisome proliferator-activated receptor delta expression in rats. *PPAR Res.* 2013;2013:456042.
107. Jucker BM, Doe CP, Schnackenberg CG, et al. Ppardelta activation normalizes cardiac substrate metabolism and reduces right ventricular hypertrophy in congestive heart failure. *J Cardiovasc Pharmacol.* 2007;50:25–34.
108. Li Y, Cheng L, Qin Q, et al. High-fat feeding in cardiomyocyte-restricted ppardelta knockout mice leads to cardiac overexpression of lipid metabolic genes but fails to rescue cardiac phenotypes. *J Mol Cell Cardiol.* 2009;47:536–543.
109. Alvarez-Guardia D, Palomer X, Coll T, et al. Pparbeta/delta activation blocks lipid-induced inflammatory pathways in mouse heart and human cardiac cells. *Biochim Biophys Acta.* 1811;2011:59–67.
110. Planavila A, Calvo RR, Vazquez-Carrera M. Peroxisome proliferator-activated receptors and the control of fatty acid oxidation in cardiac hypertrophy. *Mini Rev Med Chem.* 2006;6:357–363.
111. Smeets PJ, Planavila A, van der Vusse GJ, van Bilsen M. Peroxisome proliferator-activated receptors and inflammation: take it to heart. *Acta Physiol (Oxf).* 2007;191:171–188.
112. Smeets PJ, Teunissen BE, Planavila A, et al. Inflammatory pathways are activated during cardiomyocyte hypertrophy and attenuated by peroxisome proliferator-activated receptors pparalpha and ppardelta. *J Biol Chem.* 2008;283:29109–29118.
113. Liang YJ, Chen CY, Juang SJ, et al. Peroxisome proliferator-activated receptor delta agonists attenuated the c-reactive protein-induced pro-inflammation in cardiomyocytes and h9c2 cardiomyoblasts. *Eur J Pharmacol.* 2010;643:84–92.
114. Lee KS, Park JH, Lee S, Lim HJ, Park HY. Ppardelta activation inhibits angiotensin ii induced cardiomyocyte hypertrophy by suppressing intracellular ca2+ signaling pathway. *J Cell Biochem.* 2009;106:823–834.
115. Yu BC, Chang CK, Ou HY, Cheng KC, Cheng JT. Decrease of peroxisome proliferator-activated receptor delta expression in cardiomyopathy of streptozotocin-induced diabetic rats. *Cardiovasc Res.* 2008;80:78–87.

INDEX

Note: Page numbers followed by "*f*" indicate figures and "*t*" indicate tables.

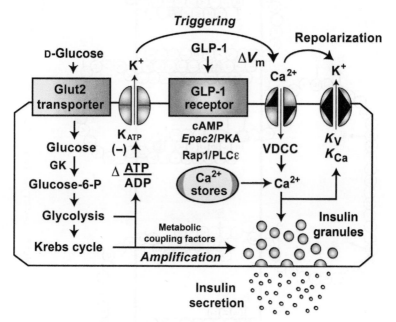

PRASHANT NADKARNI ET AL., FIGURE 2.1

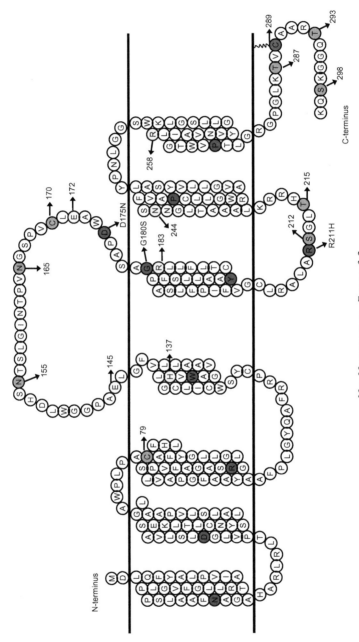

HUI HUANG *ET AL.*, FIGURE 3.2

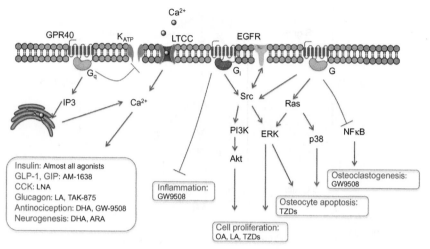

HUI HUANG *ET AL.*, FIGURE 3.3

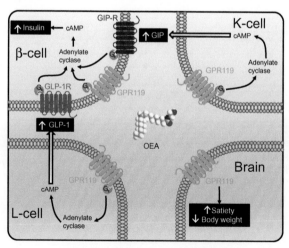

XIU-LEI MO *ET AL.*, FIGURE 4.3

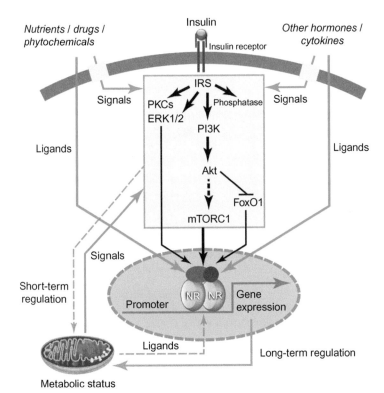

HONG-PING GUAN AND GUOXUN CHEN, FIGURE 6.2

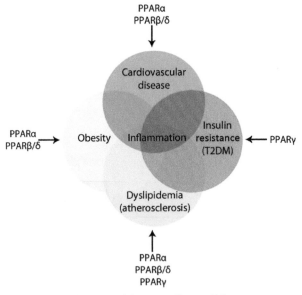

MAHMOUD MANSOUR, FIGURE 7.1

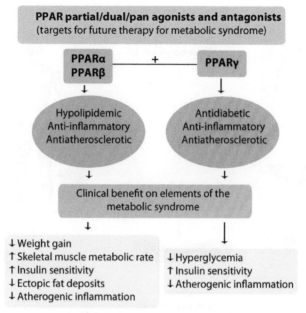

MAHMOUD MANSOUR, FIGURE 7.2

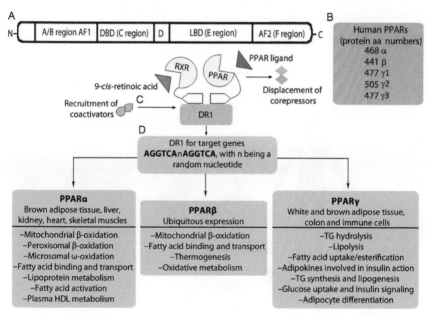

MAHMOUD MANSOUR, FIGURE 7.3

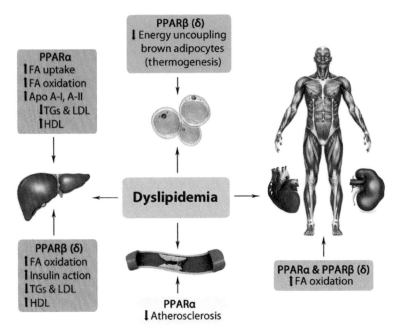

MAHMOUD MANSOUR, FIGURE 7.4

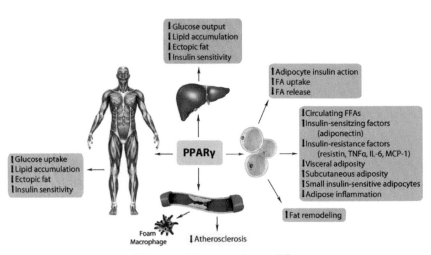

MAHMOUD MANSOUR, FIGURE 7.5

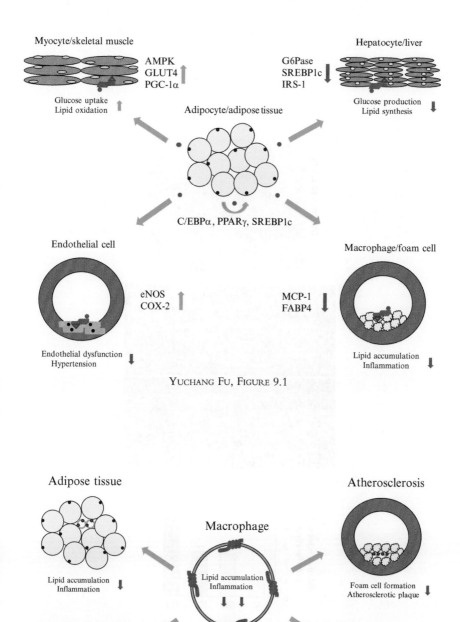

Myocyte/skeletal muscle

Hepatocyte/liver

AMPK
GLUT4 ↑
PGC-1α

G6Pase
SREBP1c ↓
IRS-1

Glucose uptake ↑
Lipid oxidation

Adipocyte/adipose tissue

Glucose production ↓
Lipid synthesis

C/EBPα, PPARγ, SREBP1c

Endothelial cell

Macrophage/foam cell

eNOS ↑
COX-2

MCP-1 ↓
FABP4

Endothelial dysfunction ↓
Hypertension

Lipid accumulation ↓
Inflammation

YUCHANG FU, FIGURE 9.1

Adipose tissue

Atherosclerosis

Macrophage

Lipid accumulation ↓
Inflammation

Lipid accumulation ↓
Inflammation

Foam cell formation ↓
Atherosclerotic plaque

Skeletal muscle

Liver

Lipid oxidation ⇑
Glucose uptake

Lipid synthesis ↓
Glucose production

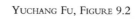

YUCHANG FU, FIGURE 9.2

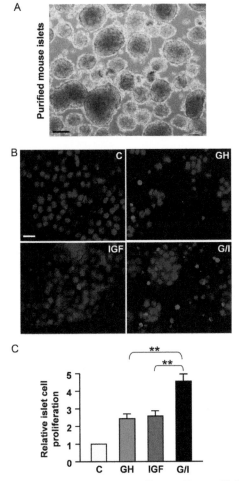

YAO HUANG AND YONGCHANG CHANG, FIGURE 10.6

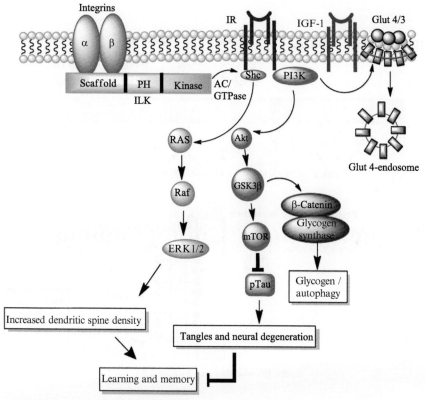

JENNA BLOEMER *ET AL.*, FIGURE 13.2

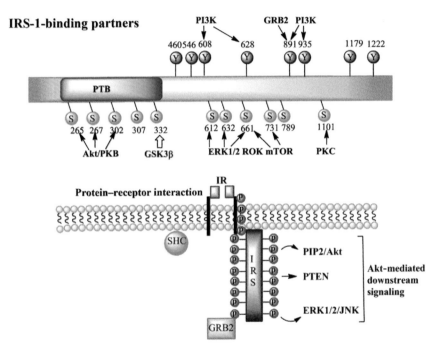

JENNA BLOEMER *ET AL.*, FIGURE 13.3

Printed and bound by CPI Group (UK) Ltd, Croydon, CR0 4YY

08/05/2025

01864964-0001